SEMI-CLASSICAL ANALYSIS FOR NONLINEAR SCHRÖDINGER EQUATIONS

WKB Analysis, Focal Points,
Coherent States
2nd Edition

SEMI-CLASSICAL ANALYSIS FOR NONLINEAR SCHRÖDINGER EQUATIONS

WKB Analysis, Focal Points,
Coherent States
2nd Edition

Rémi Carles

CNRS, France & Université de Rennes 1, France

World Scientific

NEW JERSEY · LONDON · SINGAPORE · BEIJING · SHANGHAI · HONG KONG · TAIPEI · CHENNAI · TOKYO

Published by

World Scientific Publishing Co. Pte. Ltd.
5 Toh Tuck Link, Singapore 596224
USA office: 27 Warren Street, Suite 401-402, Hackensack, NJ 07601
UK office: 57 Shelton Street, Covent Garden, London WC2H 9HE

Library of Congress Cataloging-in-Publication Data
Names: Carles, Rémi, author.
Title: Semi-classical analysis for nonlinear Schrödinger equations : WKB analysis, focal points,
    coherent states / Rémi Carles, CNRS, France & Université de Rennes 1, France.
Description: 2nd edition. | New Jersey : World Scientific, [2021] |
    Includes bibliographical references and index.
Identifiers: LCCN 2020043383 | ISBN 9789811227905 (hardcover) |
    ISBN 9789811227912 (ebook) | ISBN 9789811227929 (ebook other)
Subjects: LCSH: Schrödinger equation. | Nonlinear theories.
Classification: LCC QC174.26.W28 C36 2021 | DDC 530.12/4--dc23
LC record available at https://lccn.loc.gov/2020043383

British Library Cataloguing-in-Publication Data
A catalogue record for this book is available from the British Library.

For any available supplementary material, please visit
https://www.worldscientific.com/worldscibooks/10.1142/12030#t=suppl

Desk Editor: Liu Yumeng

Printed in Singapore

Preface

These pages describe the semi-classical limit for nonlinear Schrödinger equations in the presence of an external potential. The motivation of this study is two-fold. First, it is expected to provide interesting models for physics. For instance, the nonlinear Schrödinger equation is a common model for Bose–Einstein condensation. To describe the physical phenomenon, qualitative properties of the solutions of these equations may be helpful. According to the various régimes considered, different asymptotic behaviors associated to the equations may be interesting. One of them is the semi-classical limit, where the behavior of the wave function as the (rescaled) Planck constant goes to zero is studied. On the other hand, this study also has purely analytical motivations. The semi-classical limit (especially WKB analysis, also called geometrical optics) yields useful informations for problems related to functional analysis, even when there is no Planck constant in the initial problem. For instance, such methods have proven efficient in the construction of parametrices or in studying the propagation of singularities. In these notes, we emphasize some applications of WKB analysis for the study of qualitative properties of super-critical nonlinear Schrödinger equations.

The second edition of this book consists of three parts. The first one is dedicated to the WKB methods and the semi-classical limit before the formation of caustics. The second part treats the semi-classical limit in the presence of caustics, in the special geometric case where the caustic is reduced to a point (or to several isolated points). The third part was not present in the first edition, and addresses the nonlinear propagation of coherent states. The three parts are essentially independent. The first part may be viewed simply as a motivation for the second part, and the third part is essentially complementary. The technical aspects are fairly independent in the three parts. The first part highly relies on techniques coming from the study of hyperbolic partial differential equations, and especially, from quasi-linear ones. The second part is more typical of nonlinear Schrödinger equations in a way, since it involves scattering theory, as well as Strichartz estimates, related to Schrödinger equations. The third part also uses Strichartz estimates, with or without semi-classical parameter: the influence of the external potential is discussed more into details there.

The first two parts correspond to an extended version of a course given in Vienna at the Wolfgang Pauli Institute during a workshop organized by Jean-Claude Saut in May 2007, and in Beijing in October 2007, during a special semester organized by Ping Zhang at the Morningside Center of Mathematics of the Chinese Academy of Sciences. The goal of this course was to give an overview of the current knowledge and techniques in the study of the semi-classical limit for nonlinear Schrödinger equation. These notes are a good opportunity to present several results in a unified way. We have also tried to point out some interesting open questions, when they appear natural in the course of the text. Since the course was addressed to researchers and to graduate students, the text is essentially self-contained, and only for technical details and further developments have we chosen to direct the reader to the original references. Compared to the first edition, the first part was enriched by a new section on multiphase expansions in the case of weakly nonlinear geometric optics, and an application related to this study, concerning instability results for nonlinear Schrödinger equations in negative order Sobolev spaces.

This second edition was the opportunity to fix some identified typos from the first edition, and update some of the results or techniques. The choice of including the third part was not completely obvious, since its connection to the first part is weak, and its connection to the second part essentially absent. However, we believe that it is a nice complement of the first two parts: some comparisons with WKB analysis are made as often as possible. Also, it would have seemed awkward to dedicate a separate volume to the nonlinear propagation of coherent states. Like for the first two parts, this book is the opportunity to rewrite results published in various articles, with a unified presentation, and sometimes simplified proofs.

I wish to thank warmly Thomas Alazard for his careful reading of the initial version of the manuscript, and Clotilde Fermanian for important remarks on the second edition.

General Notations

Functions

By default, the functions that we consider are complex-valued.

The space variable, denoted by x, belongs to $\mathbb{R}^d$. The time variable is denoted by t.

The partial derivatives with respect to the time variable and to the j-th space variable are denoted by ∂_t and ∂_j, respectively.

We denote by Λ the Fourier multiplier $(\mathrm{Id} - \Delta)^{1/2}$, where Δ stands for the Laplacian

$$\Delta = \sum_{j=1}^{d} \partial_j^2 = \sum_{j=1}^{d} \frac{\partial^2}{\partial x_j^2}.$$

Function spaces

We denote by $L^p(\mathbb{R}^d)$, or simply L^p, the usual Lebesgue spaces on $\mathbb{R}^d$. The inner product of $L^2(\mathbb{R}^d)$ is defined as

$$\langle f, g \rangle = \int_{\mathbb{R}^d} f(x)\overline{g}(x)dx.$$

Consider $f = f(t,x)$ a function from $I \times \mathbb{R}^d$ to $\mathbb{C}$, where I is a time interval. If $f \in C(I; L^p(\mathbb{R}^d))$, we write

$$\|f\|_{L^\infty(I;L^p)} = \sup_{t \in I} \|f(t)\|_{L^p(\mathbb{R}^d)}.$$

The Schwartz class of smooth functions $\mathbb{R}^d \to \mathbb{C}$ which decay rapidly as well as all their derivatives is denoted by $\mathcal{S}(\mathbb{R}^d)$.

For $f \in \mathcal{S}(\mathbb{R}^d)$, we define its Fourier transform by

$$\widehat{f}(\xi) = \mathcal{F}f(\xi) = \frac{1}{(2\pi)^{d/2}} \int_{\mathbb{R}^d} e^{-ix \cdot \xi} f(x) dx,$$

so that the inverse Fourier transform is given by

$$\mathcal{F}^{-1}f(x) = \frac{1}{(2\pi)^{d/2}} \int_{\mathbb{R}^d} e^{ix \cdot \xi} f(\xi) d\xi.$$

For $s \in \mathbb{R}$, we define the Sobolev space $H^s(\mathbb{R}^d) = H^s$ as

$$H^s(\mathbb{R}^d) = \left\{ f \in \mathcal{S}'(\mathbb{R}^d) \ ; \ \xi \mapsto \langle \xi \rangle^s \, \widehat{f}(\xi) \in L^2(\mathbb{R}^d) \right\},$$

where we have denoted $\langle \xi \rangle = (1 + |\xi|^2)^{1/2}$. Note that if $s \in \mathbb{N}$, then

$$H^s(\mathbb{R}^d) = \left\{ f \in L^2(\mathbb{R}^d) \ ; \ \partial^\alpha f \in L^2(\mathbb{R}^d), \ \forall \alpha \in \mathbb{N}^d, \ |\alpha| \leqslant s \right\}.$$

For $k \in \mathbb{N}$, denote

$$\Sigma^k = H^k \cap \mathcal{F}(H^k) = \left\{ f \in H^k(\mathbb{R}^d) \ ; \ x \mapsto \langle x \rangle^k f(x) \in L^2(\mathbb{R}^d) \right\}.$$

When $k = 1$, we simply write Σ.
Recall that if $s > d/2$, then $H^s(\mathbb{R}^d)$ is an algebra, and $H^s(\mathbb{R}^d) \subset L^\infty(\mathbb{R}^d)$.

The set $H^\infty(\mathbb{R}^d)$, or simply H^∞, is the intersection of all these spaces:

$$H^\infty = \cap_{s \geqslant 0} H^s(\mathbb{R}^d).$$

This is a Fréchet space, equipped with the distance

$$d(f, g) = \sum_{s \in \mathbb{N}} 2^{-s} \frac{\|f - g\|_{H^s}}{1 + \|f - g\|_{H^s}}.$$

Semi-classical limit

The dependence of functions upon the semi-classical parameter ε is denoted by a superscript. For instance, the wave function is denoted by u^ε.

All the irrelevant constants are denoted by C. In particular, C stands for a constant which is independent of ε, the semi-classical parameter.

Let $(\alpha^h)_{0 < h \leqslant 1}$ and $(\beta^h)_{0 < h \leqslant 1}$ be two families of positive real numbers.

- We write $\alpha^h \ll \beta^h$, or $\alpha^h = o(\beta^h)$, if $\limsup\limits_{h \to 0} \alpha^h/\beta^h = 0$.
- We write $\alpha^h \lesssim \beta^h$, or $\alpha^h = \mathcal{O}(\beta^h)$, if $\limsup\limits_{h \to 0} \alpha^h/\beta^h < \infty$.
- We write $\alpha^h \approx \beta^h$ if $\alpha^h \lesssim \beta^h$ and $\beta^h \lesssim \alpha^h$.

If u^h and v^h are functions, we write $u^h \approx v^h$ if $\|u^h - v^h\| \ll \|v^h\|$, for some norm to be precised (or not, when computations are purely formal).

Contents

PART 1
WKB Analysis

Chapter 1

Preliminary Analysis

We consider nonlinear Schrödinger equations in the presence of a parameter $\varepsilon \in]0,1]$,

$$i\varepsilon\partial_t u^\varepsilon + \frac{\varepsilon^2}{2}\Delta u^\varepsilon = Vu^\varepsilon + f\left(|u^\varepsilon|^2\right)u^\varepsilon, \tag{1.1}$$

where $u^\varepsilon = u^\varepsilon(t,x)$ is complex-valued. Throughout this book, the space variable, denoted by x, lies in the whole Euclidean space $\mathbb{R}^d$, $d \geqslant 1$. Many of the results presented in this first part can easily be adapted to the case of the torus $\mathbb{T}^d$. The external potential $V = V(t,x)$ and the (local) nonlinearity f are supposed to be smooth, *real-valued*, and independent of ε. The aim of these notes is to describe some results about the asymptotic behavior of the solution u^ε as the parameter ε goes to zero. We shall be more precise about the initial data that we consider below. The nonlinearity f is *local* (e.g. power-like nonlinearity): in particular, we choose not to mention results related to nonlocal nonlinearities, such as the Schrödinger–Poisson system

$$i\varepsilon\partial_t u^\varepsilon + \frac{\varepsilon^2}{2}\Delta u^\varepsilon = Vu^\varepsilon + V_p u^\varepsilon \quad ; \quad \Delta V_p = \lambda\left(|u^\varepsilon|^2 - c\right),$$

or the Hartree equation

$$i\varepsilon\partial_t u^\varepsilon + \frac{\varepsilon^2}{2}\Delta u^\varepsilon = Vu^\varepsilon + \lambda\left(\frac{1}{|x|^\gamma}*|u^\varepsilon|^2\right)u^\varepsilon,$$

except in the third part, where the results available for Hartree-type nonlinearities with a smooth kernel are the most spectacular (Chap. 13). We do not consider ε-dependent potentials either, an issue for which the main model we have in mind is that of a lattice periodic potential, whose period is of order ε:

$$i\varepsilon\partial_t u^\varepsilon + \frac{\varepsilon^2}{2}\Delta u^\varepsilon = Vu^\varepsilon + V_\Gamma\left(\frac{x}{\varepsilon}\right)u^\varepsilon + f\left(|u^\varepsilon|^2\right)u^\varepsilon,$$

3

where the potential V_Γ is periodic with respect to some regular lattice $\Gamma \simeq \mathbb{Z}^d$. See for instance [Bensoussan *et al.* (1978); Robert (1998); Teufel (2003)] for an introduction to the asymptotic study in the linear case of the above equation, and [Carles *et al.* (2004)] for an example of asymptotic behavior in a nonlinear régime. Our choice is to focus on (1.1), and to describe as precisely as possible the variety of known phenomena in the limit $\varepsilon \to 0$.

There are several reasons to study the asymptotic behavior of u^ε in the semi-classical limit $\varepsilon \to 0$. Let us mention two. First, (1.1) with $f(|u|^2)u = |u|^4 u$ (quintic nonlinearity) is sometimes used as a model for one-dimensional Bose–Einstein condensation in space dimension $d = 1$ ([Kolomeisky *et al.* (2000)]). When $d = 2$ or 3, a cubic nonlinearity, $f(|u|^2)u = |u|^2 u$, is usually considered. The external potential V can be an harmonic potential (isotropic or anisotropic), or a lattice periodic potential (see e.g. [Dalfovo *et al.* (1999); Pitaevskii and Stringari (2003)]). According to the different physical parameters at stake, the asymptotic behavior of u^ε as $\varepsilon \to 0$ may provide relevant informations to describe u^ε itself. This approach is similar to the theory of geometric optics, developed initially to describe the propagation of electro-magnetic waves, such as light. In that context, the propagation of the wave is also described by partial differential equations, and ε usually corresponds to a wavelength, which is small compared to the other parameters. For Maxwell's equations, ε corresponds to the inverse of the speed of light. We invite the reader to consult [Rauch (2012)] for an overview of this theory, mainly in the context of hyperbolic equations. We shall not develop further on the physical motivations, but rather focus our attention on the mathematical aspects. The term "geometric optics" means that it is expected that the propagation of light is accurately described by rays. For Schrödinger equations, the analogue of this notion is usually called "classical trajectories". These notions are identical, and follow from the notion of bicharacteristic curves. As a consequence, the limit $\varepsilon \to 0$ relates classical and quantum wave equations. In particular, the semi-classical limit $\varepsilon \to 0$ for u^ε is expected to be described by the laws of hydrodynamics. We will come back to this aspect more precisely later.

Another motivation lies in the study the Cauchy problem for nonlinear Schrödinger equations with no small parameter ($V \equiv 0$ and $\varepsilon = 1$ in (1.1), typically). One can prove ill-posedness results for energy-supercritical equations by reducing the problem to semi-classical analysis for (1.1). This aspect is discussed in details in Sec. 5.1 and Sec. 5.2. In Sec. 5.4, we show that multiphase weakly nonlinear geometric optics makes it possible to prove ill-posedness results when negative Sobolev regularity is involved. Note that the application of the theory of geometric optics to functional analysis has a long history. In [Lax (1957)], it was used to construct parametrices. It has also been used to study the propagation of singularities (see e.g. [Taylor (1981)]), or of quasi-singularities [Cheverry (2005)]. In the case of Schrödinger equations, semi-classical analysis has proven useful for instance in control theory [Lebeau (1992)], in the proof of Strichartz estimates [Burq *et al.* (2004)], and in the propagation of singularities for the nonlinear equation [Szeftel (2005)].

We underscore the fact that the WKB analysis for (nonlinear) Schrödinger equations is rather specific to this equation. An important feature is the fact that for gauge invariant nonlinearities, it is possible to describe the solution with one phase and one harmonic only, provided that the initial data are of this form: $u^\varepsilon \approx a e^{i\phi/\varepsilon}$. For several other equations (e.g. Maxwell equations), the analysis is rather different, even on the algebraic level. We invite the reader to consult for instance [Joly *et al.* (1996b); Métivier (2004b); Rauch (2012); Whitham (1999)], and references therein, to have an idea of the important results for equations different from the Schrödinger equation. However, the general framework presented in §1.1 (derivation of the equation, and steps toward a justification) is not specific to the equation: the main specificity of gauge invariant nonlinear Schrödinger equations (as in Eq. (1.1)) is that the equations derived at the formal step look simpler than for other equations, due to the fact that we work with only one phase (and one harmonic).

Before introducing the approach developed in this first part, we present two basic results, which will be used throughout these notes.

Lemma 1.1 (Gronwall lemma and a continuity argument).
(1) *Let $u, a, b \in C([0, T]; \mathbb{R}_+)$ be such that*
$$u(t) \leqslant u(0) + \int_0^t a(\tau) u(\tau) d\tau + \int_0^t b(\tau) d\tau, \quad \forall t \in [0, T].$$
Denote $A(t) = \int_0^t a(\tau) d\tau$. Then
$$u(t) \leqslant u(0) e^{A(t)} + \int_0^t b(s) e^{A(t)-A(s)} ds, \quad \forall t \in [0, T].$$
(2) *Let $u, b \in C([0, T]; \mathbb{R}_+)$ and $f \in C(\mathbb{R}_+; \mathbb{R}_+)$ such that*
$$u(t) \leqslant u(0) + \int_0^t f(u(\tau)) u(\tau) d\tau + \int_0^t b(\tau) d\tau, \quad \forall t \in [0, T].$$

Let $M = \sup\{f(v); \; v \in [0, 2u(0)]\}$. *There exists $\underline{t} \in]0, T]$ such that*

$$u(t) \leqslant u(0)e^{Mt} + \int_0^t b(s)e^{M(t-s)}ds, \quad \forall t \in [0, \underline{t}].$$

Proof. (1) Denote

$$w(t) = u(0) + \int_0^t a(\tau)u(\tau)d\tau + \int_0^t b(\tau)d\tau.$$

By assumption, $w \in C^1([0, T])$ and $w'(t) = a(t)u(t) + b(t) \leqslant a(t)w(t) + b(t)$. Therefore,

$$\left(w(t)e^{-A(t)}\right)' \leqslant b(t)e^{-A(t)},$$

and the first point follows by integrating this inequality, since $u(t) \leqslant w(t)$. (2) Suppose that there exists $t \in]0, T]$ such that $u(t) > 2u(0)$. Since u is continuous, we can define

$$\underline{t} = \min\{\tau \in [0, T]; \; u(\tau) = 2u(0)\} > 0.$$

The assumption implies

$$u(t) \leqslant u(0) + M \int_0^t u(\tau)d\tau + \int_0^t b(\tau)d\tau, \quad \forall t \in [0, \underline{t}].$$

Gronwall lemma then yields

$$u(t) \leqslant u(0)e^{Mt} + \int_0^t b(s)e^{M(t-s)}ds, \quad \forall t \in [0, \underline{t}].$$

The right-hand side is continuous, and is equal to $u(0)$ for $t = 0$. Up to decreasing $\underline{t}$, this right-hand side does not exceed $2u(0)$ for $t \in [0, \underline{t}]$, hence the conclusion of the lemma.

If $u(t) \leqslant 2u(0)$ for all $t \in [0, T]$, then we can trivially take $\underline{t} = T$. $\quad\square$

Lemma 1.2 (Basic energy estimate). *For $\varepsilon > 0$, consider $\mathbf{u}^\varepsilon$ solving*

$$i\varepsilon\partial_t\mathbf{u}^\varepsilon + \frac{\varepsilon^2}{2}\Delta\mathbf{u}^\varepsilon = F^\varepsilon\mathbf{u}^\varepsilon + R^\varepsilon \quad ; \quad \mathbf{u}^\varepsilon_{|t=0} = \mathbf{u}^\varepsilon_0. \tag{1.2}$$

Assume that $F^\varepsilon = F^\varepsilon(t, x)$ is real-valued. Let I be a time interval such that $0 \in I$. Then we have, at least formally:

$$\sup_{t \in I} \|\mathbf{u}^\varepsilon(t)\|_{L^2} \leqslant \|\mathbf{u}^\varepsilon_0\|_{L^2} + \frac{1}{\varepsilon}\int_I \|R^\varepsilon(\tau)\|_{L^2}\, d\tau.$$

Proof. Since the statement is formal, so is the proof. Multiply (1.2) by $\overline{\mathbf{u}}^\varepsilon$, and integrate over $\mathbb{R}^d$:

$$i\varepsilon \int_{\mathbb{R}^d} \overline{\mathbf{u}}^\varepsilon \partial_t \mathbf{u}^\varepsilon dx + \frac{\varepsilon^2}{2} \int_{\mathbb{R}^d} \overline{\mathbf{u}}^\varepsilon \Delta \mathbf{u}^\varepsilon dx = \int_{\mathbb{R}^d} F^\varepsilon |\mathbf{u}^\varepsilon|^2 dx + \int_{\mathbb{R}^d} \overline{\mathbf{u}}^\varepsilon R^\varepsilon dx.$$

Taking the imaginary part, the second term of the left-hand side vanishes, since Δ is self-adjoint. Similarly, since F^ε is real-valued, the first term of the right-hand side disappears, and we have:

$$\varepsilon \frac{d}{dt} \int_{\mathbb{R}^d} |\mathbf{u}^\varepsilon|^2 = \varepsilon \int_{\mathbb{R}^d} \partial_t |\mathbf{u}^\varepsilon|^2 = 2 \operatorname{Im} \int_{\mathbb{R}^d} \overline{\mathbf{u}}^\varepsilon R^\varepsilon.$$

Cauchy–Schwarz inequality yields

$$\varepsilon \frac{d}{dt} \|\mathbf{u}^\varepsilon\|_{L^2}^2 \leqslant 2 \|\mathbf{u}^\varepsilon\|_{L^2} \|R^\varepsilon\|_{L^2}.$$

Let $\delta > 0$. We infer from the above inequality:

$$\varepsilon \frac{d}{dt} \left(\|\mathbf{u}^\varepsilon\|_{L^2}^2 + \delta \right) \leqslant 2 \left(\|\mathbf{u}^\varepsilon\|_{L^2}^2 + \delta \right)^{1/2} \|R^\varepsilon\|_{L^2}.$$

Since $\|\mathbf{u}^\varepsilon\|_{L^2}^2 + \delta \geqslant \delta > 0$, we can simplify:

$$\varepsilon \frac{d}{dt} \left(\|\mathbf{u}^\varepsilon\|_{L^2}^2 + \delta \right)^{1/2} \leqslant \|R^\varepsilon\|_{L^2}.$$

Integration with respect to time yields, for $t \in I$:

$$\varepsilon \left(\|\mathbf{u}^\varepsilon(t)\|_{L^2}^2 + \delta \right)^{1/2} \leqslant \varepsilon \left(\|\mathbf{u}_0^\varepsilon\|_{L^2}^2 + \delta \right)^{1/2} + \int_I \|R^\varepsilon(t)\|_{L^2} dt.$$

The lemma follows by letting $\delta \to 0$. $\qquad\square$

1.1 General presentation

The general approach of WKB expansions (after the three independent papers [Brillouin (1926); Kramers (1926); Wentzel (1926)] — the alphabetical order BKW is usually adopted in French) consists of mainly three steps. The first step, which is described in more details in this section, consists in seeking a function v^ε that solves (1.1) up to a small error term:

$$i\varepsilon \partial_t v^\varepsilon + \frac{\varepsilon^2}{2} \Delta v^\varepsilon = V v^\varepsilon + f \left(|v^\varepsilon|^2 \right) v^\varepsilon + r^\varepsilon,$$

where r^ε should be thought of as a "small" (as $\varepsilon \to 0$) source term. Typically, we require

$$\|r^\varepsilon\|_{L^\infty([-T,T];L^2)} = \mathcal{O} \left(\varepsilon^N \right)$$

for some $T > 0$ independent of ε, and $N > 0$ as large as possible. In this first step, we derive equations that define v^ε, which are hopefully simpler than (1.1). The second step consists in showing that such a v^ε actually exists, that is, in solving the equations derived in the first step. The last step is the study of *stability* (or *consistency*): even if r^ε is small, it is not clear that $u^\varepsilon - v^\varepsilon$ is small too. Typically, we try to prove an error estimate of the form

$$\|u^\varepsilon - v^\varepsilon\|_{L^\infty([-T,T];L^2)} = \mathcal{O}\left(\varepsilon^K\right)$$

for some $K > 0$ (possibly smaller than N). Note also that for the nonlinear problem (1.1), it is not even clear from the beginning that an L^2 solution can be constructed on a time interval independent of $\varepsilon \in]0, 1]$.

The initial data that we consider for WKB analysis are of the form

$$u^\varepsilon(0, x) = \varepsilon^\kappa a_0^\varepsilon(x) e^{i\phi_0(x)/\varepsilon}. \tag{1.3}$$

The phase ϕ_0 is independent of ε and real-valued. The initial amplitude a_0^ε is complex-valued, and may have an asymptotic expansion as $\varepsilon \to 0$,

$$a_0^\varepsilon(x) \underset{\varepsilon \to 0}{\sim} a_0(x) + \varepsilon a_1(x) + \varepsilon^2 a_2(x) + \dots, \tag{1.4}$$

in the sense of formal asymptotic expansions, where the profiles a_j are independent of ε. Note that ε^κ then measures the size of $u^\varepsilon(0, x)$ in $L^\infty(\mathbb{R}^d)$. We shall always consider cases where $\kappa \geqslant 0$. When the nonlinearity is non-trivial, $f \neq 0$, the asymptotic behavior of u^ε as $\varepsilon \to 0$ strongly depends on the value of κ, as is discussed below. An important feature of Schrödinger equations with gauge invariant nonlinearities like in (1.1) is that if the initial data are of the form (1.3), then for small time at least (before caustics), the solution u^ε is expected to keep the same form, at least approximately:

$$u^\varepsilon(t, x) \underset{\varepsilon \to 0}{\sim} \varepsilon^\kappa a^\varepsilon(t, x) e^{i\phi(t,x)/\varepsilon}, \tag{1.5}$$

where a^ε is expected to have an asymptotic expansion as well. This is in sharp contrast with the analogous problems for hyperbolic equations (e.g. Maxwell, wave, Euler): typically, because the solutions of the wave equations are real-valued, the factor $e^{i\phi_0/\varepsilon}$ is replaced, say, by $2\cos(\phi_0/\varepsilon) = e^{i\phi_0/\varepsilon} + e^{-i\phi_0/\varepsilon}$. By nonlinear interaction, other phases are expected to appear, like $e^{ik\phi/\varepsilon}$, $k \in \mathbb{Z}$, for instance. This can be guessed by looking at the first iterates of a Picard's scheme. Moreover, phases different from ϕ might be involved in the description of u^ε, by nonlinear mechanisms too. We will see that unlike for these models, such a phenomenon is ruled out for nonlinear Schrödinger equations, provided that only one phase is considered initially, see (1.3). This is an important geometric feature in this study. On the other hand, studying the asymptotic behavior of u^ε whose initial data are *sums* of initial data as in (1.3) is an interesting open question in general. We present some results in this direction in Sec. 2.6, in the case of very specific (initial) phases.

To describe the expected influence of the parameter κ on the asymptotic behavior of u^ε, assume that the nonlinearity f is homogeneous:

$$f\left(|u^\varepsilon|^2\right)u^\varepsilon = \lambda|u^\varepsilon|^{2\sigma}u^\varepsilon, \quad \lambda \in \mathbb{R}, \ \sigma > 0.$$

The case $\sigma \in \mathbb{N} \setminus \{0\}$ corresponds to a smooth nonlinearity. Even though the parameter κ may be viewed as a measurement of the size of the (initial) wave function, we shall rather consider data of order $\mathcal{O}(1)$, by introducing $\widetilde{u}^\varepsilon = \varepsilon^{-\kappa}u^\varepsilon$. Dropping the tildes, we therefore consider

$$i\varepsilon\partial_t u^\varepsilon + \frac{\varepsilon^2}{2}\Delta u^\varepsilon = Vu^\varepsilon + \lambda\varepsilon^\alpha|u^\varepsilon|^{2\sigma}u^\varepsilon; \quad u^\varepsilon(0,x) = a_0^\varepsilon(x)e^{i\phi_0(x)/\varepsilon}, \quad (1.6)$$

where $\alpha = 2\sigma\kappa \geqslant 0$.

1.2 Formal derivation of the equations

Assuming that the initial data have an asymptotic expansion of the form (1.4), we seek $u^\varepsilon(t,x) \sim a^\varepsilon(t,x)e^{i\phi(t,x)/\varepsilon}$, with

$$a^\varepsilon(t,x) \underset{\varepsilon\to 0}{\sim} a(t,x) + \varepsilon a^{(1)}(t,x) + \varepsilon^2 a^{(2)}(t,x) + \dots$$

We use the convention $a^{(0)} = a$. On a formal level at least, the general idea consists in plugging this asymptotic expansion into (1.6), and then ordering in powers of ε. The lowest powers are the ones we really want to cancel, and if we are left with some extra terms, we want to be able to consider them as small source terms in the limit $\varepsilon \to 0$ (by a perturbative analysis for instance). To summarize, we first find $b^{(0)}, b^{(1)}, \dots$, such that

$$i\varepsilon\partial_t u^\varepsilon + \frac{\varepsilon^2}{2}\Delta u^\varepsilon - Vu^\varepsilon - \lambda\varepsilon^\alpha|u^\varepsilon|^{2\sigma}u^\varepsilon \underset{\varepsilon\to 0}{\sim} \left(b^{(0)} + \varepsilon b^{(1)} + \varepsilon^2 b^{(2)} + \dots\right)e^{i\phi/\varepsilon}.$$

Then we consider the equations $b^{(0)} = 0$, $b^{(1)} = 0$, etc. Note that this makes sense provided that $\alpha \in \mathbb{N}$, for otherwise, non-integer powers of ε appear in the above right-hand side.

Denoting by ∂ a differentiation with respect to the time variable, or any space variable, we compute formally:

$$\partial u^\varepsilon \underset{\varepsilon \to 0}{\sim} \left(i\varepsilon^{-1} \left(a + \varepsilon a^{(1)} + \varepsilon^2 a^{(2)} + \dots \right) \partial\phi \right.$$
$$\left. + \partial a + \varepsilon \partial a^{(1)} + \varepsilon^2 \partial a^{(2)} + \dots \right) e^{i\phi/\varepsilon}.$$

Similarly, for $1 \leqslant j \leqslant d$,

$$\partial_j^2 u^\varepsilon \underset{\varepsilon \to 0}{\sim} \left(-\varepsilon^{-2} \left(a + \varepsilon a^{(1)} + \varepsilon^2 a^{(2)} + \dots \right) (\partial_j\phi)^2 \right.$$
$$+ i\varepsilon^{-1} \left(a + \varepsilon a^{(1)} + \varepsilon^2 a^{(2)} + \dots \right) \partial_j^2\phi$$
$$+ 2i\varepsilon^{-1} \left(\partial_j a + \varepsilon \partial_j a^{(1)} + \varepsilon^2 \partial_j a^{(2)} + \dots \right) \partial_j\phi$$
$$\left. + \partial_j^2 a + \varepsilon \partial_j^2 a^{(1)} + \varepsilon^2 \partial_j^2 a^{(2)} + \dots \right) e^{i\phi/\varepsilon}.$$

Ordering in powers of ε, we infer:

$$i\varepsilon\partial_t u^\varepsilon + \frac{\varepsilon^2}{2}\Delta u^\varepsilon \underset{\varepsilon \to 0}{\sim} \left(-\left(\partial_t\phi + \frac{1}{2}|\nabla\phi|^2 \right) \left(a + \varepsilon a^{(1)} + \varepsilon^2 a^{(2)} + \dots \right) \right.$$
$$+ i\varepsilon \left(\partial_t a + \nabla\phi \cdot \nabla a + \frac{1}{2} a \Delta\phi \right)$$
$$+ i\varepsilon^2 \left(\partial_t a^{(1)} + \nabla\phi \cdot \nabla a^{(1)} + \frac{1}{2} a^{(1)} \Delta\phi - \frac{i}{2}\Delta a \right)$$
$$\vdots$$
$$\left. + i\varepsilon^{j+1} \left(\partial_t a^{(j)} + \nabla\phi \cdot \nabla a^{(j)} + \frac{1}{2} a^{(j)} \Delta\phi - \frac{i}{2}\Delta a^{(j-1)} \right) \right.$$
$$\left. + \dots \right) e^{i\phi/\varepsilon}.$$

For the nonlinear term, we choose to compute only the first two terms:

$$|u^\varepsilon|^{2\sigma} u^\varepsilon \underset{\varepsilon \to 0}{\sim} \left(|a|^{2\sigma} a + \varepsilon \left(|a|^{2\sigma} a^{(1)} + 2\sigma \,\mathrm{Re}\left(\bar{a} a^{(1)} \right) |a|^{2\sigma-2} a \right) + \dots \right) e^{i\phi/\varepsilon}.$$

To simplify the discussion, assume in the following lines that α is an integer, $\alpha \in \mathbb{N}$. Since we want to consider a leading order amplitude a which is not identically zero, it is natural to demand, for the term of order ε^0:

$$\partial_t\phi + \frac{1}{2}|\nabla\phi|^2 + V = \begin{cases} 0 & \text{if } \alpha > 0, \\ -\lambda|a|^{2\sigma} & \text{if } \alpha = 0. \end{cases} \tag{1.7}$$

For the term of order ε^1, we find:

$$\partial_t a + \nabla\phi \cdot \nabla a + \frac{1}{2}a\Delta\phi = \begin{cases} 0 & \text{if } \alpha > 1, \\ -i\lambda|a|^{2\sigma}a & \text{if } \alpha = 1, \\ -2i\lambda\sigma\,\mathrm{Re}\left(\overline{a}a^{(1)}\right)|a|^{2\sigma-2}a & \text{if } \alpha = 0. \end{cases} \quad (1.8)$$

Before giving a rigorous meaning to this approach, we comment on these cases. Intuitively, the larger the α, the smaller the influence of the nonlinearity: for large α, the nonlinearity is not expected to be relevant at leading order as $\varepsilon \to 0$. In terms of the problem (1.1)–(1.3), this means that small initial waves (large κ) evolve linearly at leading order: this corresponds to the general phenomenon that very small nonlinear waves behave linearly at leading order. Here, we see that if $\alpha > 1$, then ϕ and a solve equations which are independent of λ, hence of the nonlinearity. Since at leading order, we expect

$$u^\varepsilon(t,x) \underset{\varepsilon\to 0}{\sim} a(t,x)e^{i\phi(t,x)/\varepsilon},$$

this means that the leading order behavior of u^ε is linear. As a consequence, we also expect

$$u^\varepsilon(t,x) \underset{\varepsilon\to 0}{\sim} u^\varepsilon_{\mathrm{lin}}(t,x),$$

where $u^\varepsilon_{\mathrm{lin}}$ solves the linear problem

$$i\varepsilon\partial_t u^\varepsilon_{\mathrm{lin}} + \frac{\varepsilon^2}{2}\Delta u^\varepsilon_{\mathrm{lin}} = V u^\varepsilon_{\mathrm{lin}} \quad ; \quad u^\varepsilon_{\mathrm{lin}}(0,x) = u^\varepsilon(0,x) = a^\varepsilon_0(x)e^{i\phi_0(x)/\varepsilon}.$$

Decreasing the value of α, the critical threshold corresponds to $\alpha = 1$: the nonlinearity shows up in the equation for a, but not in the equation for ϕ. This régime is referred to as *weakly nonlinear geometric optics*. The term "weakly" means that the phase ϕ is determined independently of the nonlinearity: the equations for a and ϕ are decoupled. We will see that for $\alpha \geqslant 1$, the equation for a can be understood as a transport equation along the classical trajectories (rays of geometric optics) associated to ϕ, which in turn are determined by the initial phase ϕ_0 and the Hamiltonian

$$\tau + \frac{|\xi|^2}{2} + V(t,x).$$

See Sec. 1.3.1 below.

The case $\alpha = 0$ is supercritical, and contains several difficulties. We point out two of those, which show that dealing with the supercritical case requires a different approach. First, the equation for the phase involves the

The projection of the solution (x, ξ) on the physical space, that is $x(t, y)$, is called *classical trajectory*, or *ray*. The Cauchy–Lipschitz Theorem yields:

Lemma 1.3. *Assume that V and ϕ_0 are smooth: $V \in C^\infty(\mathbb{R} \times \mathbb{R}^d; \mathbb{R})$ and $\phi_0 \in C^\infty(\mathbb{R}^d; \mathbb{R})$. Then for all $y \in \mathbb{R}^d$, there exists $T_y > 0$ and a unique solution to (1.11), $(x(t, y), \xi(t, y)) \in C^\infty([-T_y, T_y] \times \mathbb{R}^d; \mathbb{R}^d)^2$.*

The link with (1.10) appears in

Lemma 1.4. *Let ϕ_{eik} be a smooth solution to (1.10). Then necessarily,*

$$\nabla \phi_{\text{eik}}(t, x(t, y)) = \xi(t, y),$$

as long as all the terms remain smooth.

Proof. For ϕ_{eik} a smooth solution to (1.10), introduce the ordinary differential equation

$$\frac{d}{dt}\widetilde{x} = \nabla \phi_{\text{eik}}(t, \widetilde{x}) \quad ; \quad \widetilde{x}\big|_{t=0} = y. \tag{1.12}$$

By the Cauchy–Lipschitz Theorem, (1.12) has a smooth solution $\widetilde{x} \in C^\infty([-\widetilde{T}_y, \widetilde{T}_y])$ for some $\widetilde{T}_y > 0$ possibly very small. Set

$$\widetilde{\xi}(t) := \nabla \phi_{\text{eik}}(t, \widetilde{x}(t)).$$

We compute

$$\frac{d}{dt}\widetilde{\xi} = \nabla \partial_t \phi_{\text{eik}}(t, \widetilde{x}(t)) + \nabla^2 \phi_{\text{eik}}(t, \widetilde{x}(t)) \cdot \frac{d}{dt}\widetilde{x}(t)$$

$$= \nabla \partial_t \phi_{\text{eik}}(t, \widetilde{x}(t)) + \nabla^2 \phi_{\text{eik}}(t, \widetilde{x}(t)) \cdot \nabla \phi_{\text{eik}}(t, \widetilde{x}(t))$$

$$= \nabla \left(\partial_t \phi_{\text{eik}} + \frac{1}{2}|\nabla \phi_{\text{eik}}|^2 \right)(t, \widetilde{x}(t)) = -\nabla V(t, \widetilde{x}(t)).$$

We infer that $(\widetilde{x}, \widetilde{\xi})$ solves (1.11). The lemma then follows from uniqueness for (1.11). □

Note that knowing $\nabla \phi_{\text{eik}}$ suffices to get ϕ_{eik} itself, which is given by

$$\phi_{\text{eik}}(t, x) = \phi_0(x) - \int_0^t \left(\frac{1}{2}|\nabla \phi_{\text{eik}}(\tau, x)|^2 + V(\tau, x) \right) d\tau.$$

The above lemma and the Local Inversion Theorem yield

Lemma 1.5. *Let V and ϕ_0 smooth as in Lemma 1.3. Let $t \in [-T, T]$ and θ_0 an open set of $\mathbb{R}^d$. Denote*

$$\theta_t := \{x(t, y) \in \mathbb{R}^d, y \in \theta_0\} \quad ; \quad \theta := \{(t, x) \in [-T, T] \times \mathbb{R}^d, x \in \theta_t\}.$$

Suppose that for $t \in [-T, T]$, the mapping

$$\theta_0 \ni y \mapsto x(t, y) \in \theta_t$$

is bijective, and denote by $y(t, x)$ its inverse. Assume also that

$$\nabla_x y \in L^\infty_{\text{loc}}(\theta).$$

Then there exists a unique function $\theta \ni (t, x) \mapsto \phi_{\text{eik}}(t, x) \in \mathbb{R}$ that solves (1.10), and satisfies $\nabla^2_x \phi_{\text{eik}} \in L^\infty_{\text{loc}}(\theta)$. Moreover,

$$\nabla \phi_{\text{eik}}(t, x) = \xi(t, y(t, x)). \tag{1.13}$$

Note that the existence time T may depend on the neighborhood θ_0. It actually does in general, as shown by the following example.

Example 1.6. Assume that $V \equiv 0$ and

$$\phi_0(x) = -\frac{1}{(2 + 2\delta)T_c} \left(|x|^2 + 1 \right)^{1+\delta}, \quad T_c > 0, \ \delta \geqslant 0.$$

For $\delta > 0$, integrating (1.11) yields:

$$x(t, y) = y + \int_0^t \xi(s, y)ds = y + \int_0^t \xi(0, y)ds = y - \frac{t}{T_c} \left(|y|^2 + 1 \right)^\delta y$$

$$= y \left(1 - \frac{t}{T_c} \left(|y|^2 + 1 \right)^\delta \right).$$

For $R > 0$, we see that the rays starting from the ball $\{|y| = R\}$ meet at the origin at time

$$T_c(R) = \frac{T_c}{(R^2 + 1)^\delta}.$$

Since R is arbitrary, this shows that several rays can meet arbitrarily fast, thus showing that the above lemma cannot be applied uniformly in space.

Of course, the above issue would not appear if the space variable x belonged to a compact set instead of the whole space $\mathbb{R}^d$. To obtain a local time of existence with is independent of $y \in \mathbb{R}^d$, we have to make an extra assumption, in order to be able to apply a *global* inversion theorem.

Assumption 1.7 (Geometric assumption). *We assume that the potential and the initial phase are smooth, real-valued, and subquadratic:*

- $V \in C^\infty(\mathbb{R} \times \mathbb{R}^d)$, *and* $\partial_x^\alpha V \in C(\mathbb{R}; L^\infty(\mathbb{R}^d))$ *as soon as* $|\alpha| \geqslant 2$.
- $\phi_0 \in C^\infty(\mathbb{R}^d)$, *and* $\partial^\alpha \phi_0 \in L^\infty(\mathbb{R}^d)$ *as soon as* $|\alpha| \geqslant 2$.

We emphasize that we call "subquadratic" functions which are at most quadratic: exactly quadratic functions are included. As a consequence of this assumption on V, if $a_0^\varepsilon \in L^2$, then (1.9) has a unique solution $u^\varepsilon \in C(\mathbb{R}; L^2)$. See e.g. [Reed and Simon (1975)].

The following result can be found in [Schwartz (1969)], or in Appendix A of [Dereziński and Gérard (1997)].

Lemma 1.8. *Suppose that the function* $\mathbb{R}^d \ni y \mapsto x(y) \in \mathbb{R}^d$ *satisfies:*

$$|\det \nabla_y x| \geqslant C_0 > 0 \quad and \quad |\partial_y^\alpha x| \leqslant C, \ |\alpha| = 1, 2.$$

Then x is bijective.

We can then prove

Proposition 1.9. *Under Assumption 1.7, there exists $T > 0$ and a unique solution $\phi_{\mathrm{eik}} \in C^\infty\left([-T, T] \times \mathbb{R}^d\right)$ to (1.10). In addition, this solution is subquadratic: $\partial_x^\alpha \phi_{\mathrm{eik}} \in L^\infty([-T, T] \times \mathbb{R}^d)$ as soon as $|\alpha| \geqslant 2$.*

Proof. We know that we can solve (1.11) locally in time in the neighborhood of any $y \in \mathbb{R}^d$. In order to apply the above global inversion result, differentiate (1.11) with respect to y:

$$\begin{cases} \partial_t \partial_y x(t, y) = \partial_y \xi(t, y) & ; \ \partial_y x(0, y) = \mathrm{Id}, \\ \partial_t \partial_y \xi(t, y) = -\nabla_x^2 V(t, x(t, y)) \partial_y x(t, y) & ; \ \partial_y \xi(0, y) = \nabla^2 \phi_0(y). \end{cases} \quad (1.14)$$

Integrating (1.14) in time, we infer from Assumption 1.7 that for any $T > 0$, there exists C_T such that for $(t, y) \in [-T, T] \times \mathbb{R}^d$:

$$|\partial_y x(t, y)| + |\partial_y \xi(t, y)| \leqslant C_T + C_T \int_0^t \left(|\partial_y x(s, y)| + |\partial_y \xi(s, y)|\right) ds.$$

Gronwall lemma yields:

$$\|\partial_y x(t)\|_{L_y^\infty} + \|\partial_y \xi(t)\|_{L_y^\infty} \leqslant C'(T). \quad (1.15)$$

Similarly,

$$\left\|\partial_y^\alpha x(t)\right\|_{L_y^\infty} + \left\|\partial_y^\alpha \xi(t)\right\|_{L_y^\infty} \leqslant C(\alpha, T), \quad \forall \alpha \in \mathbb{N}^n, \ |\alpha| \geqslant 1. \quad (1.16)$$

Integrating the first line of (1.14) in time, we have:

$$\det \nabla_y x(t, y) = \det \left(\mathrm{Id} + \int_0^t \nabla_y \xi(s, y) \, ds\right).$$

We infer from (1.15) that for $t \in [-T, T]$, provided that $T > 0$ is sufficiently small, we can find $C_0 > 0$ such that:

$$|\det \nabla_y x(t, y)| \geqslant C_0, \quad \forall (t, y) \in [-T, T] \times \mathbb{R}^d. \quad (1.17)$$

Lemma 1.8 shows that we can invert $y \mapsto x(t, y)$ for $t \in [-T, T]$.

To apply Lemma 1.5 with $\theta_0 = \theta = \theta_t = \mathbb{R}^d$, we must check that $\nabla_x y \in L^\infty_{\text{loc}}(\mathbb{R}^d)$. Differentiate the relation

$$x\,(t, y(t, x)) = x$$

with respect to x:

$$\nabla_x y(t, x) \nabla_y x\,(t, y(t, x)) = \text{Id}.$$

Therefore, $\nabla_x y(t, x) = \nabla_y x\,(t, y(t, x))^{-1}$ as matrices, and

$$\nabla_x y(t, x) = \frac{1}{\det \nabla_y x(t, y)} \text{adj}\,(\nabla_y x\,(t, y(t, x)))\,, \tag{1.18}$$

where $\text{adj}\,(\nabla_y x)$ denotes the adjugate of $\nabla_y x$. We infer from (1.15) and (1.17) that $\nabla_x y \in L^\infty(\mathbb{R}^d)$ for $t \in [-T, T]$. Therefore, Lemma 1.5 yields a smooth solution ϕ_{eik} to (1.10); it is local in time and *global in space*: $\phi_{\text{eik}} \in C^\infty([-T, T] \times \mathbb{R}^d)$.

The fact that ϕ_{eik} is subquadratic as stated in Proposition 1.9 then stems from (1.13), (1.16), (1.17) and (1.18). □

Note that Example 1.6 shows that the above result is essentially sharp: if Assumption 1.7 is not satisfied, then the above result fails to be true. Similarly, if we consider $V = V(x) = -x^4$ in space dimension $d = 1$, then, also due to an infinite speed of propagation, the Hamiltonian $-\partial_x^2 - x^4$ is not essentially self-adjoint (see Chap. 13, Sec. 6, Cor. 22 in [Dunford and Schwartz (1963)]). We now give some examples of cases where the phase ϕ_{eik} can be computed explicitly, which also show that in general, the above time T is necessarily finite.

Example 1.10 (Quadratic phase). *Resume Example 1.6, and consider the value $\delta = 0$. In that case, Assumption 1.7 is satisfied, and (1.10) is solved explicitly:*

$$\phi_{\text{eik}}(t, x) = \frac{|x|^2}{2(t - T_c)} - \frac{1}{2T_c}.$$

This shows that we can solve (1.10) globally in space, but only locally in time: as $t \to T_c$, ϕ_{eik} ceases to be smooth. A caustic reduced to a single point (the origin) is formed.

Remark 1.11. More generally, the space-time set where the map $y \mapsto x(t, y)$ ceases to be a diffeomorphism is called *caustic*. The behavior of the solution u^ε to (1.6) with $\lambda = 0$ is given for all time in terms of oscillatory integrals ([Duistermaat (1974); Maslov and Fedoriuk (1981)]). We present results concerning the asymptotic behavior of solutions to (1.6) with $\lambda \neq 0$ in the presence of point caustics in the second part of this book.

Example 1.12 (Harmonic potential). *When $\phi_0 \equiv 0$, and V is independent of time and quadratic, $V = V(x) = \frac{1}{2}\sum_{j=1}^{d}\omega_j^2 x_j^2$, we have:*

$$\phi_{\text{eik}}(t,x) = -\sum_{j=1}^{d}\frac{\omega_j}{2}x_j^2 \tan(\omega_j t).$$

This also shows that we can solve (1.10) *globally in space, but locally in time only. Note that if we replace formally ω_j by $i\omega_j$, then V is turned into $-V$, and the trigonometric functions become hyperbolic functions: we can then solve* (1.10) *globally in space and time.*

Example 1.13 (Plane wave). *If we assume $V \equiv 0$ and $\phi_0(x) = \xi_0 \cdot x$ for some $\xi_0 \in \mathbb{R}^d$, then we find:*

$$\phi_{\text{eik}}(t,x) = \xi_0 \cdot x - \frac{1}{2}|\xi_0|^2 t.$$

Also in this case, we can solve (1.10) *globally in space and time.*

1.3.2   The transport equations

To cancel the ε^1 term, the second step consists in solving (1.8):

$$\partial_t a + \nabla\phi_{\text{eik}} \cdot \nabla a + \frac{1}{2}a\Delta\phi_{\text{eik}} = 0 \quad ; \quad a(0,x) = a_0(x), \tag{1.19}$$

where a_0 is given as the first term in the asymptotic expansion of the initial amplitude (1.4). The equation is a transport equation (see e.g. [Evans (1998)]), since the characteristics for the operator $\partial_t + \nabla\phi_{\text{eik}} \cdot \nabla$ do not meet for $t \in [-T, T]$, by construction. As a matter of fact, this equation can be solved rather explicitly, in terms of the geometric tools that we have used in the previous paragraph.

Introduce the *Jacobi's determinant*

$$J_t(y) = \det\nabla_y x(t,y),$$

where $x(t,y)$ is given by the Hamiltonian flow (1.11). Note that $J_0(y) = 1$ for all $y \in \mathbb{R}^d$. By construction, for $t \in [-T, T]$, the function $y \mapsto J_t(y)$ is uniformly bounded from above and from below:

$$\exists C > 0, \ \frac{1}{C} \leqslant J_t(y) \leqslant C, \quad \forall(t,y) \in [-T, T] \times \mathbb{R}^d.$$

Define the function A by

$$A(t,y) := a\,(t, x(t,y))\,\sqrt{J_t(y)}.$$

Then since for $t \in [-T, T]$, $y \mapsto x(t, y)$ is a global diffeomorphism on $\mathbb{R}^d$, (1.19) is *equivalent* to the equation

$$\partial_t A(t, y) = 0 \quad ; \quad A(0, y) = a_0(y).$$

We obviously have $A(t, y) = a_0(y)$ for all $t \in [-T, T]$, and back to the function a, this yields

$$a(t, x) = \frac{1}{\sqrt{J_t(y(t, x))}} a_0\left(y(t, x)\right), \tag{1.20}$$

where $y(t, x)$ is the inverse map of $y \mapsto x(t, y)$.

Remark 1.14. The computations of Sec. 1.2 show that the amplitudes are given by

$$\partial_t a^{(j)} + \nabla \phi_{\text{eik}} \cdot \nabla a^{(j)} + \frac{1}{2} a^{(j)} \Delta \phi_{\text{eik}} = \frac{i}{2} \Delta a^{(j-1)} \quad ; \quad a^{(j)}_{|t=0} = a_j,$$

with the convention $a^{(-1)} = 0$ and $a^{(0)} = a$. For $j \geqslant 1$, this equation is the inhomogeneous analogue of (1.19). It can be solved by using the same change of variable as above. This shows that when ϕ_{eik} becomes singular (formation of a caustic), *all* the terms computed by this WKB analysis become singular in general. The WKB hierarchy ceases to be relevant at a caustic.

Proposition 1.15. *Let $s \geqslant 0$ and $a_0 \in H^s(\mathbb{R}^d)$. Then (1.19) has a unique solution $a \in C([-T; T]; H^s)$, where $T > 0$ is given by Proposition 1.9.*

Proof. Existence and uniqueness at the L^2 level stem from the above analysis, (1.20). To prove that an H^s regularity is propagated for $s > 0$, we could also use (1.20). We shall use another approach, which will be more natural in the nonlinear setting. To simplify the presentation, we assume $s \in \mathbb{N}$, and prove *a priori* estimates in H^s. Let $\alpha \in \mathbb{N}^d$, with $|\alpha| \leqslant s$. Applying ∂_x^α to (1.19), we find:

$$\partial_t \partial_x^\alpha a + \nabla \phi_{\text{eik}} \cdot \nabla \partial_x^\alpha a = [\nabla \phi_{\text{eik}} \cdot \nabla, \partial_x^\alpha] a - \frac{1}{2} \partial_x^\alpha \left(a \Delta \phi_{\text{eik}}\right) =: R_\alpha, \tag{1.21}$$

where $[P, Q] = PQ - QP$ denotes the commutator of the operators P and Q. Take the inner product of (1.21) with $\partial_x^\alpha a$, and consider the real part:

$$\frac{1}{2} \frac{d}{dt} \|\partial_x^\alpha a\|_{L^2}^2 + \text{Re} \int_{\mathbb{R}^d} \overline{\partial_x^\alpha a} \nabla \phi_{\text{eik}} \cdot \nabla \partial_x^\alpha a \leqslant \|R_\alpha\|_{L^2} \|a\|_{H^s}.$$

Notice that we have

$$\left| \text{Re} \int_{\mathbb{R}^d} \overline{\partial_x^\alpha a} \nabla \phi_{\text{eik}} \cdot \nabla \partial_x^\alpha a \right| = \frac{1}{2} \left| \int_{\mathbb{R}^d} \nabla \phi_{\text{eik}} \cdot \nabla \left|\partial_x^\alpha a\right|^2 \right|$$

$$= \frac{1}{2} \left| \int_{\mathbb{R}^d} \left|\partial_x^\alpha a\right|^2 \Delta \phi_{\text{eik}} \right| \leqslant C \|a\|_{H^s}^2,$$

since $\Delta\phi_{\text{eik}} \in L^\infty([-T,T] \times \mathbb{R}^d)$ from Proposition 1.9. Summing over α such that $|\alpha| \leqslant s$, we infer:

$$\frac{d}{dt}\|a\|_{H^s}^2 \leqslant C\|a\|_{H^s}^2 + \|R_\alpha\|_{H^s}^2.$$

To apply Gronwall lemma, we need to estimate the last term: we use the fact that the derivatives of order at least two of ϕ_{eik} are bounded, from Proposition 1.9, to have:

$$\|R_\alpha\|_{L^2} \leqslant C\|a\|_{H^s}.$$

We can then conclude:

$$\|a\|_{L^\infty([-T,T];H^s)} \leqslant C\|a_0\|_{H^s},$$

which completes the proof of the proposition. □

Let us examine what can be deduced at this stage, and see which rigorous meaning can be given to the relation $u^\varepsilon \sim ae^{i\phi_{\text{eik}}/\varepsilon}$. Let

$$v_1^\varepsilon(t,x) := a(t,x)e^{i\phi_{\text{eik}}(t,x)/\varepsilon}.$$

Proposition 1.16. *Let $s \geqslant 2$, $a_0 \in H^s(\mathbb{R}^d)$, and Assumption 1.7 be satisfied. Suppose that*

$$\|a_0^\varepsilon - a_0\|_{H^{s-2}} = \mathcal{O}\left(\varepsilon^\beta\right)$$

for some $\beta > 0$. Then there exists $C > 0$ independent of $\varepsilon \in \,]0,1]$ such that

$$\sup_{t\in[-T,T]} \|u^\varepsilon(t) - v_1^\varepsilon(t)\|_{L^2} \leqslant C\varepsilon^{\min(1,\beta)},$$

where T is given by Proposition 1.9. If in addition $s > d/2 + 2$, then

$$\sup_{t\in[-T,T]} \|u^\varepsilon(t) - v_1^\varepsilon(t)\|_{L^\infty} \leqslant C\varepsilon^{\min(1,\beta)}.$$

Proof. Let $w_1^\varepsilon := u^\varepsilon - v_1^\varepsilon$. By construction, is solves

$$i\varepsilon\partial_t w_1^\varepsilon + \frac{\varepsilon^2}{2}\Delta w_1^\varepsilon = Vw_1^\varepsilon - \frac{\varepsilon^2}{2}e^{i\phi_{\text{eik}}/\varepsilon}\Delta a \quad ; \quad w_{1|t=0}^\varepsilon = a_0^\varepsilon - a_0. \quad (1.22)$$

By Lemma 1.2, which can be made rigorous in the present setting (exercise), we have:

$$\sup_{t\in[-T,T]} \|w_1^\varepsilon(t)\|_{L^2} \leqslant \|a_0^\varepsilon - a_0\|_{L^2} + \frac{\varepsilon}{2}\int_{-T}^T \|\Delta a(\tau)\|_{L^2}d\tau \leqslant C\left(\varepsilon^\beta + \varepsilon\right),$$

where we have used the assumption on $a_0^\varepsilon - a_0$ and Proposition 1.15. This yields the first estimate of the proposition.

To prove the second estimate, we want to use the Sobolev embedding $H^s \subset L^\infty$ for $s > d/2$. A first idea could be to differentiate (1.22) with respect to space variables, and use Lemma 1.2. However, this direct approach fails, because the source term

$$\frac{\varepsilon^2}{2} e^{i\phi_{\mathrm{eik}}/\varepsilon} \Delta a$$

is of order $\mathcal{O}(\varepsilon^2)$ in L^2, but of order $\mathcal{O}(\varepsilon^{2-s})$ in H^s, $s \geqslant 0$. This is due to the rapidly oscillatory factor $e^{i\phi_{\mathrm{eik}}/\varepsilon}$. Moreover, under our assumptions, it is not guaranteed that $\nabla\phi_{\mathrm{eik}}\Delta a \in C([-T,T]; L^2)$, since $\nabla\phi_{\mathrm{eik}}$ may grow linearly with respect to the space variable, as shown by Examples 1.10 and 1.12. We therefore adopt a different point of view, relying on the remark:

$$\left| u^\varepsilon - v_1^\varepsilon \right| = \left| u^\varepsilon - a e^{i\phi_{\mathrm{eik}}/\varepsilon} \right| = \left| u^\varepsilon e^{-i\phi_{\mathrm{eik}}/\varepsilon} - a \right|.$$

Set $a^\varepsilon := u^\varepsilon e^{-i\phi_{\mathrm{eik}}/\varepsilon}$. We check that it solves

$$\partial_t a^\varepsilon + \nabla\phi_{\mathrm{eik}} \cdot \nabla a^\varepsilon + \frac{1}{2} a^\varepsilon \Delta\phi_{\mathrm{eik}} = i\frac{\varepsilon}{2}\Delta a^\varepsilon \quad ; \quad a^\varepsilon_{|t=0} = a_0^\varepsilon.$$

Therefore, $r^\varepsilon = a^\varepsilon - a = w_1^\varepsilon e^{-i\phi_{\mathrm{eik}}/\varepsilon}$ solves

$$\partial_t r^\varepsilon + \nabla\phi_{\mathrm{eik}}\cdot\nabla r^\varepsilon + \frac{1}{2} r^\varepsilon \Delta\phi_{\mathrm{eik}} = i\frac{\varepsilon}{2}\Delta r^\varepsilon + i\frac{\varepsilon}{2}\Delta a \quad ; \quad r^\varepsilon_{|t=0} = a_0^\varepsilon - a_0. \quad (1.23)$$

Note that this equation is very similar to the transport equation (1.19), with two differences. First, the presence of the operator $i\varepsilon\Delta$ acting on r^ε on the right-hand side. Second, the source term $i\varepsilon\Delta a$, which makes the equation inhomogeneous.

We know by construction that $r^\varepsilon \in C([-T,T]; L^2)$, and we seek *a priori* estimates in $C([-T,T]; H^k)$. These are established along the same lines as in the proof of Proposition 1.15. We note that since the operator $i\Delta$ is skew-symmetric on H^s, the term $i\varepsilon\Delta r^\varepsilon$ vanishes from the energy estimates in H^s. Then, the source term is of order ε in $C([-T;T]; H^{s-2})$ from Proposition 1.15. We infer:

$$\sup_{t\in[-T,T]} \|r^\varepsilon(t)\|_{H^{s-2}} \lesssim \varepsilon^\beta + \varepsilon.$$

Note that this estimate, along with a standard continuation argument, shows that $a^\varepsilon \in C([-T;T]; H^{s-2})$ for $\varepsilon > 0$ sufficiently small. Since $s - 2 > d/2$, we deduce

$$\sup_{t\in[-T,T]} \|r^\varepsilon(t)\|_{L^\infty} \lesssim \varepsilon^\beta + \varepsilon,$$

which completes the proof of the proposition. $\qquad\square$

Before analyzing the accuracy of higher order approximate solutions, let us examine the candidate v_1^ε in the case of the examples given in Sec. 1.3.1.

Example 1.17 (Quadratic phase). *Resume Example 1.10. In this case, we compute, for $t < T_c$,*

$$a(t,x) = \left(\frac{T_c}{T_c - t}\right)^{n/2} a_0\left(\frac{T_c}{T_c - t}x\right).$$

As $t \to T_c$, not only ϕ_{eik} ceases to be smooth, but also a. This is a general feature of the formation of caustics: all the terms constructed by the usual WKB analysis become singular.

Example 1.18 (Harmonic potential). *Resume Example 1.12. If $|t|$ is sufficiently small so that ϕ_{eik} remains smooth on $[0, t]$, we find:*

$$a(t,x) = \prod_{j=1}^{d}\left(\frac{1}{\cos(\omega_j t)}\right)^{1/2} a_0\left(\frac{x_1}{\cos(\omega_1 t)}, \ldots, \frac{x_n}{\cos(\omega_n t)}\right).$$

Here again, ϕ_{eik} and a become singular simultaneously.

Example 1.19 (Plane wave). *If we assume $V \equiv 0$ and $\phi_0(x) = \xi_0 \cdot x$ for some $\xi_0 \in \mathbb{R}^d$, then we find:*

$$a(t,x) = a_0(x - \xi_0 t).$$

The initial amplitude is simply transported with constant velocity.

We can continue this analysis to arbitrary order:

Proposition 1.20. *Let $k \in \mathbb{N} \setminus \{0\}$ and $s \geqslant 2k + 2$. Let $a_0, a_1, \ldots, a_k$ with $a_j \in H^{s-2j}(\mathbb{R}^d)$, and let Assumption 1.7 be satisfied. Suppose that*

$$\|a_0^\varepsilon - a_0 - \varepsilon a_1 - \ldots - \varepsilon^k a_k\|_{H^{s-2k-2}} = \mathcal{O}\left(\varepsilon^{k+\beta}\right)$$

for some $\beta > 0$. Then we can find $a^{(1)}, \ldots, a^{(k)}$, with

$$a^{(j)} \in C([-T, T]; H^{s-2j}),$$

such that if we set

$$v_{k+1}^\varepsilon = \left(a + \varepsilon a^{(1)} + \ldots + \varepsilon^k a^{(k)}\right) e^{i\phi_{\text{eik}}/\varepsilon},$$

there exists $C > 0$ independent of $\varepsilon \in]0, 1]$ such that

$$\sup_{t \in [-T,T]}\left\|\left(u^\varepsilon(t) - v_{k+1}^\varepsilon(t)\right) e^{-i\phi_{\text{eik}}(t)/\varepsilon}\right\|_{H^{s-2k-2}} \leqslant C\varepsilon^{\min(k+1, k+\beta)},$$

where T is given by Proposition 1.9.

Proof. We simply sketch the proof, since it follows arguments which have been introduced above. First, the computations presented in Sec. 1.2 show that to cancel the term in ε^{j+1}, $1 \leqslant j \leqslant k$, we naturally impose:

$$\partial_t a^{(j)} + \nabla \phi_{\text{eik}} \cdot \nabla a^{(j)} + \frac{1}{2} a^{(j)} \Delta \phi_{\text{eik}} = \frac{i}{2} \Delta a^{(j-1)} \quad ; \quad a^{(j)}_{|t=0} = a_j.$$

This equation is the inhomogeneous analogue of (1.19). Using the same arguments as in the proof of Proposition 1.15, it is easy to see that it has a unique solution $a^{(j)} \in C([-T,T]; L^2)$, whose spatial regularity is that of $a^{(j-1)}$, minus 2. Starting an induction with Proposition 1.15, we construct

$$a^{(j)} \in C([-T,T]; H^{s-2j}).$$

To prove the error estimate, introduce $r_k^\varepsilon = a^\varepsilon - a - \varepsilon a^{(1)} - \ldots - \varepsilon^k a^{(k)}$, where we recall that $a^\varepsilon = u^\varepsilon e^{-i\phi_{\text{eik}}/\varepsilon}$. By construction, the remainder r_k^ε is in $C([-T;T]; L^2)$ since $s \geqslant 2k+2$, and it solves:

$$\begin{cases} \partial_t r_k^\varepsilon + \nabla \phi_{\text{eik}} \cdot \nabla r_k^\varepsilon + \frac{1}{2} r_k^\varepsilon \Delta \phi_{\text{eik}} = i \frac{\varepsilon}{2} \Delta r_k^\varepsilon + i \frac{\varepsilon^{k+1}}{2} \Delta a^{(k)}, \\ r_k^\varepsilon{}_{|t=0} = a_0^\varepsilon - a_0 - \ldots - \varepsilon^k a_k. \end{cases}$$

We can then mimic the end of the proof of Proposition 1.16. $\qquad\square$

To conclude, we see that we can construct an arbitrarily accurate (as $\varepsilon \to 0$) approximation of u^ε on $[-T,T]$, provided that the initial profiles a_j are sufficiently smooth. The goal now is to see how this approach can be adapted to a nonlinear framework.

1.4   Basic results in the nonlinear case

Before presenting a WKB analysis in the case $f \neq 0$ in (1.1), we recall a few important facts about the nonlinear Cauchy problem for (1.1). We shall simply gather classical results, which can be found for instance in [Cazenave and Haraux (1998); Cazenave (2003); Ginibre and Velo (1985a); Kato (1989); Tao (2006)]. Several notions of solutions are available. According to the cases, we will work with the notion of strong solutions (Chaps. 2, 4 and 4.5), of weak solutions (Chaps. 2.6.4 and 4.5) or of mild solutions (especially in the second and third parts of this book).

In this section, one should think that the parameter $\varepsilon > 0$ is *fixed*. The dependence upon ε is discussed in the forthcoming sections.

1.4.1 Formal properties

Since V and f are real-valued, the L^2 norm of u^ε is formally independent of time:

$$\|u^\varepsilon(t)\|_{L^2} = \|u^\varepsilon(0)\|_{L^2}. \tag{1.24}$$

This can be seen from the proof of Lemma 1.2, with $F^\varepsilon = V + f\left(|u^\varepsilon|^2\right)$ and $R^\varepsilon = 0$. This relation yields an *a priori* bound for the L^2 norm of u^ε.

When the potential V is time-independent, $V = V(x)$, (1.1) has a Hamiltonian structure. Introduce

$$F(y) = \int_0^y f(\eta)d\eta.$$

The following energy is formally independent of time:

$$\begin{aligned} E^\varepsilon(u^\varepsilon(t)) &= \frac{1}{2}\|\varepsilon\nabla u^\varepsilon(t)\|_{L^2}^2 + \int_{\mathbb{R}^d} F\left(|u^\varepsilon(t,x)|^2\right)dx \\ &\quad + \int_{\mathbb{R}^d} V(x)|u^\varepsilon(t,x)|^2 dx. \end{aligned} \tag{1.25}$$

We see that if E^ε is finite, and if $V \geqslant 0$ and $F \geqslant 0$, then this yields an *a priori* bound on $\|\varepsilon\nabla u^\varepsilon(t)\|_{L^2}$.

Example 1.21. If $V = V(x) \geqslant$ and $f(y) = \lambda y^\sigma$, then (1.25) becomes

$$E^\varepsilon = \frac{1}{2}\|\varepsilon\nabla u^\varepsilon(t)\|_{L^2}^2 + \frac{\lambda}{\sigma+1}\int_{\mathbb{R}^d}|u^\varepsilon(t,x)|^{2\sigma+2}dx + \int_{\mathbb{R}^d} V(x)|u^\varepsilon(t,x)|^2 dx.$$

If $\lambda \geqslant 0$ (*defocusing nonlinearity*), this yields an *a priori* bound on $\|\varepsilon\nabla u^\varepsilon(t)\|_{L^2}$. On the other hand, if $\lambda < 0$, then $\|\varepsilon\nabla u^\varepsilon(t)\|_{L^2}$ may become unbounded in finite time: this is the *finite time blow-up* phenomenon (see e.g. [Cazenave (2003); Sulem and Sulem (1999)]). Since the L^2 norm of u^ε is conserved, one can replace the assumption $V \geqslant 0$ with $V \geqslant -C$ for some $C > 0$, and leave the above discussion unchanged.

Example 1.22. If V is unbounded from below, the conservation of the energy does not seem to provide interesting informations. For instance, if $V(x) = -|x|^2$, then even in the linear case $f = 0$, the energy is not a positive energy functional (see [Carles (2003a)] though, for the nonlinear Cauchy problem).

1.4.2 Strong solutions

A remarkable fact is that if the external potential V is subquadratic in the sense of Assumption 1.7, then one can define a strongly continuous semigroup for the linear equation (1.9). As we have mentioned already, if no sign assumption is made on V, then Assumption 1.7 is essentially sharp: if $d = 1$ and $V(x) = -x^4$, then $-\partial_x^2 + V$ is not essentially self-adjoint on the set of test functions ([Dunford and Schwartz (1963)]). Under Assumption 1.7, one defines $U^\varepsilon(t, s)$ such that $u^\varepsilon_{\text{lin}}(t, x) = U^\varepsilon(t, s)\varphi^\varepsilon(x)$, where

$$i\varepsilon\partial_t u^\varepsilon_{\text{lin}} + \frac{\varepsilon^2}{2}\Delta u^\varepsilon_{\text{lin}} = V u^\varepsilon_{\text{lin}} \quad ; \quad u^\varepsilon_{\text{lin}}(s, x) = \varphi^\varepsilon(x).$$

Note that $U^\varepsilon(t, t) = \text{Id}$. The existence of $U^\varepsilon(t, s)$ is established in [Fujiwara (1979)], along with the following properties:

- The map $(t, s) \mapsto U^\varepsilon(t, s)$ is strongly continuous.
- $U^\varepsilon(t, s)^* = U^\varepsilon(t, s)^{-1}$.
- $U^\varepsilon(t, \tau)U^\varepsilon(\tau, s) = U^\varepsilon(t, s)$.
- $U^\varepsilon(t, s)$ is unitary on L^2: $\|U^\varepsilon(t, s)\varphi^\varepsilon\|_{L^2} = \|\varphi^\varepsilon\|_{L^2}$.

We construct strong solutions which are (at least) in $H^s(\mathbb{R}^d)$, for $s > d/2$. Recall that H^s is then an algebra, embedded into $L^\infty(\mathbb{R}^d)$. We shall also use the following version of Schauder's lemma:

Lemma 1.23 (Schauder's lemma). *Suppose that $G : \mathbb{C} \to \mathbb{C}$ is a smooth function, such that $G(0) = 0$. Then the map $u \mapsto G(u)$ sends $H^s(\mathbb{R}^d)$ to itself provided $s > d/2$. The map is uniformly Lipschitzean on bounded subsets of H^s.*

We refer to [Taylor (1997)] or [Rauch (2012)] for the proof of this result, as well as to the following refinement (*tame estimate*):

Lemma 1.24 (Moser's inequality). *Suppose that $G : \mathbb{C} \to \mathbb{C}$ is a smooth function, such that $G(0) = 0$. Then there exists $C : [0, \infty[\to [0, \infty[$ such that for all $u \in H^s(\mathbb{R}^d)$,*

$$\|G(u)\|_{H^s} \leqslant C\left(\|u\|_{L^\infty}\right)\|u\|_{H^s}.$$

For $k \in \mathbb{N}$, denote
$$\Sigma^k = H^k \cap \mathcal{F}(H^k) = \{f \in H^k(\mathbb{R}^d) \; ; \; x \mapsto \langle x \rangle^k f(x) \in L^2(\mathbb{R}^d)\}.$$
When $k = 1$, we simply write Σ.

Proposition 1.25. *Let V satisfy Assumption 1.7, and let $f \in C^\infty(\mathbb{R}_+; \mathbb{R})$. Let $k \in \mathbb{N}$, with $k > d/2$, and fix $\varepsilon \in]0, 1]$.*

- *If $u_0^\varepsilon \in \Sigma^k$, then there exist $T_-^\varepsilon, T_+^\varepsilon > 0$ and a unique maximal solution $u^\varepsilon \in C(]-T_-^\varepsilon, T_+^\varepsilon[; \Sigma^k)$ to (1.1), such that $u_{|t=0}^\varepsilon = u_0^\varepsilon$. It is maximal in the sense that if, say, $T_+^\varepsilon < \infty$, then*

$$\limsup_{t \to T_+^\varepsilon} \|u^\varepsilon(t)\|_{L^\infty(\mathbb{R}^d)} = +\infty. \tag{1.26}$$

- *Assume in addition that V is sub-linear: $\nabla V \in L^\infty_{\mathrm{loc}}(\mathbb{R}; L^\infty(\mathbb{R}^d))$. Let $s > d/2$ (not necessarily an integer). If $u_0^\varepsilon \in H^s(\mathbb{R}^d)$, then there exist $T_-^\varepsilon, T_+^\varepsilon > 0$ and a unique maximal solution $u^\varepsilon \in C(]-T_-^\varepsilon, T_+^\varepsilon[; H^s)$ to (1.1), such that $u_{|t=0}^\varepsilon = u_0^\varepsilon$. It is maximal in the sense that if, say, $T_+^\varepsilon < \infty$, then (1.26) holds. In particular, if $u_0^\varepsilon \in H^\infty$, then $u^\varepsilon \in C^\infty(]-T_-^\varepsilon, T_+^\varepsilon[; H^\infty)$.*

Proof. The proof follows arguments which are classical in the context of semilinear evolution equations. We indicate a few important facts, and refer to [Cazenave and Haraux (1998)] to fill the gaps.

The general idea consists in applying a fixed point argument on the Duhamel's formulation of (1.1) with associated initial datum u_0^ε:

$$u^\varepsilon(t) = U^\varepsilon(t, 0)u_0^\varepsilon - i\varepsilon^{-1} \int_0^t U^\varepsilon(t, \tau)\left(f\left(|u^\varepsilon(\tau)|^2\right)u^\varepsilon(\tau)\right)d\tau. \tag{1.27}$$

We claim that for any $k \in \mathbb{N}$ and any $T > 0$,

$$\sup_{t \in [-T,T]} \|U^\varepsilon(t, 0)u_0^\varepsilon\|_{\Sigma^k} \leqslant C(k, T)\|u_0^\varepsilon\|_{\Sigma^k}. \tag{1.28}$$

For $k = 0$, this is due to the fact that $U^\varepsilon(t, 0)$ is unitary on $L^2(\mathbb{R}^d)$. For $k = 1$, notice the commutator identities

$$\left[\nabla, i\varepsilon\partial_t + \frac{\varepsilon^2}{2}\Delta - V\right] = -\nabla V \quad ; \quad \left[x, i\varepsilon\partial_t + \frac{\varepsilon^2}{2}\Delta - V\right] = -\varepsilon^2\nabla. \tag{1.29}$$

By Assumption 1.7, $|\nabla V(t, x)| \leqslant C(T)\langle x \rangle$ for $|t| \leqslant T$, and (1.28) follows for $k = 1$. For $k \geqslant 2$, the proof follows the same lines.

To estimate the nonlinear term, we can assume without loss of generality that $f(0) = 0$. Indeed, we can replace f with $f - f(0)$ and V with $V + f(0)$. Schauder's lemma shows that

$$u \mapsto f\left(|u|^2\right)u$$

sends $H^s(\mathbb{R}^d)$ (resp. Σ^k) to itself, provided $s > d/2$ (resp. $k > d/2$), and the map is uniformly Lipschitzean on bounded subsets of $H^s(\mathbb{R}^d)$ (resp. Σ^k). The existence and uniqueness of a solution in the first part of the proposition follow easily. The notion of maximality is then a consequence of Lemma 1.24.

When V is sub-linear, notice that in view of the commutator identities (1.29), the estimate (1.28) can be replaced with

$$\sup_{t \in [-T,T]} \|U^\varepsilon(t,0)u_0^\varepsilon\|_{H^s} \leqslant C(s,T)\|u_0^\varepsilon\|_{H^s}.$$

This is straightforward if $s \in \mathbb{N}$, and follows by interpolation for general $s \geqslant 0$. The proof of the second part of the proposition then follows the same lines as the first part. Finally, if $u_0^\varepsilon \in H^\infty$, then u^ε is also smooth with respect to the time variable, $u^\varepsilon \in C^\infty(] - T_-^\varepsilon, T_+^\varepsilon[; H^\infty)$, by a bootstrap argument. $\qquad\square$

Note that the times T_-^ε and T_+^ε may very well go to zero as $\varepsilon \to 0$. The fact that we can bound these two quantities by $T > 0$ independent of $\varepsilon \in]0,1]$ is also a non-trivial information which will be provided by WKB analysis.

1.4.3 *Mild solutions*

Until the end of Sec. 1.4, to simplify the notations, we assume that the nonlinearity is homogeneous:

$$f(y) = \lambda y^\sigma, \quad \lambda \in \mathbb{R}, \sigma > 0.$$

In view of the conservations of mass (1.24) and energy (1.25), it is natural to look for solutions to (1.1) with initial data which are not necessarily as smooth as in Proposition 1.25. Typically, rather that (1.1), we rather consider its Duhamel's formulation, which now reads

$$u^\varepsilon(t) = U^\varepsilon(t,0)u_0^\varepsilon - i\lambda\varepsilon^{-1} \int_0^t U^\varepsilon(t,\tau) \left(|u^\varepsilon(\tau)|^{2\sigma}u^\varepsilon(\tau)\right) d\tau. \qquad (1.30)$$

An extra property of U^ε was proved in [Fujiwara (1979)], which becomes interesting at this stage, that is, a dispersive estimate:

$$\|U^\varepsilon(t,0)U^\varepsilon(s,0)^*\varphi\|_{L^\infty(\mathbb{R}^d)} = \|U^\varepsilon(t,s)\varphi\|_{L^\infty(\mathbb{R}^d)} \leqslant \frac{C}{(\varepsilon|t-s|)^{d/2}}\|\varphi\|_{L^1(\mathbb{R}^d)},$$

provided that $|t - s| \leqslant \delta$, where C and $\delta > 0$ are independent of $\varepsilon \in]0,1]$. As a consequence, Strichartz estimates are available for U^ε (see e.g. [Keel and Tao (1998)]). Note that as $\varepsilon \to 0$, this dispersion estimate becomes worse and worse: the semi-classical limit $\varepsilon \to 0$ is sometimes referred to as *dispersionless limit*. Denoting

$$p = \frac{4\sigma + 4}{d\sigma},$$

(the pair $(p, 2\sigma + 2)$ is *admissible*, see Definition 7.4), we infer:

Proposition 1.26. *Let V satisfying Assumption 1.7.*

- *If $\sigma < 2/d$ and $u_0^\varepsilon \in L^2$, then (1.30) has a unique solution*

$$u^\varepsilon \in C(\mathbb{R}; L^2) \cap L^p_{\text{loc}}(\mathbb{R}; L^{2\sigma+2}),$$

and (1.24) holds for all $t \in \mathbb{R}$.

- *If $u_0^\varepsilon \in \Sigma$ and $\sigma < 2/(d-2)$ when $d \geqslant 3$, then there exist $T_-^\varepsilon, T_+^\varepsilon > 0$ and a unique solution*

$$u^\varepsilon \in C(] - T_-^\varepsilon, T_+^\varepsilon[; \Sigma) \cap L^p_{\text{loc}}(] - T_-^\varepsilon, T_+^\varepsilon[; W^{1,2\sigma+2})$$

to (1.30). Moreover, the mass (1.24) and the energy (1.25) do not depend on $t \in] - T_-^\varepsilon, T_+^\varepsilon[$.

- *If $V = V(x)$ is sub-linear, $u_0^\varepsilon \in H^1$ and $\sigma < 2/(d-2)$ when $d \geqslant 3$, then there exist $T_-^\varepsilon, T_+^\varepsilon > 0$ and a unique solution*

$$u^\varepsilon \in C(] - T_-^\varepsilon, T_+^\varepsilon[; H^1) \cap L^p_{\text{loc}}(] - T_-^\varepsilon, T_+^\varepsilon[; W^{1,2\sigma+2})$$

to (1.30). Moreover, the mass (1.24) does not depend on $t \in] - T_-^\varepsilon, T_+^\varepsilon[$. If the energy (1.25) is finite at time $t = 0$, then it is independent of $t \in] - T_-^\varepsilon, T_+^\varepsilon[$. If $\lambda \geqslant 0$, then we can take $T_-^\varepsilon = T_+^\varepsilon = \infty$, even if the energy is infinite.

- *If $V = 0$, $u_0^\varepsilon \in \Sigma$ and $\sigma < 2/(d-2)$ when $d \geqslant 3$, then the following evolution law holds so long as $u^\varepsilon \in C_t\Sigma$:*

$$\frac{d}{dt}\left(\frac{1}{2}\|(x + i\varepsilon t\nabla)u^\varepsilon\|_{L^2}^2 + \frac{\lambda t^2}{\sigma+1}\|u^\varepsilon\|_{L^{2\sigma+2}}^{2\sigma+2}\right) \tag{1.31}$$
$$= \frac{\lambda t}{\sigma+1}(2 - d\sigma)\|u^\varepsilon\|_{L^{2\sigma+2}}^{2\sigma+2}.$$

In particular, if $\lambda \geqslant 0$, then $T_-^\varepsilon = T_+^\varepsilon = \infty$, and $u^\varepsilon \in C(\mathbb{R}; \Sigma)$.

Proof. The first point follows from the result of Y. Tsutsumi in the case $V = 0$ [Tsutsumi (1987)]. The proof relies on Strichartz estimates. The case $V \neq 0$ proceeds along the same lines, since local in time Strichartz estimates are available thanks to Assumption 1.7: the local in time result is made global thanks to the conservation of mass (1.24), since the local existence time depends only on the L^2 norm of the initial data.

The second point can be found in [Cazenave (2003)] in the case $V = 0$. To adapt it to the case $V \neq 0$, notice that (1.29) show that a closed family of estimates is available for u^ε, ∇u^ε and xu^ε. It is then possible to mimic the proof of the case $V = 0$. For the conservations of mass and energy, we refer to [Cazenave (2003)].

When V is sub-linear, it is possible to work in H^1 only, since

$$\left[\nabla, i\varepsilon\partial_t + \frac{\varepsilon^2}{2}\Delta - V\right] = -\nabla V$$

belongs to $L^\infty_{\text{loc}}(\mathbb{R}; L^\infty(\mathbb{R}^d))$. For the global existence result, rewrite formally the conservation of the energy as

$$\frac{d}{dt}\left(\frac{1}{2}\|\varepsilon\nabla u^\varepsilon(t)\|_{L^2}^2 + \frac{\lambda}{\sigma+1}\|u^\varepsilon(t)\|_{L^{2\sigma+2}}^{2\sigma+2}\right) = -\frac{d}{dt}\int_{\mathbb{R}^d} V(x)|u^\varepsilon(t,x)|^2 dx$$

$$= -2\operatorname{Re}\int_{\mathbb{R}^d} V(x)\overline{u}^\varepsilon\partial_t u^\varepsilon dx = -2\operatorname{Im}\int_{\mathbb{R}^d} V(x)\overline{u}^\varepsilon\left(i\partial_t u^\varepsilon\right)dx$$

$$= \operatorname{Im}\int_{\mathbb{R}^d} V(x)\overline{u}^\varepsilon\varepsilon\Delta u^\varepsilon dx = -\operatorname{Im}\int_{\mathbb{R}^d} \overline{u}^\varepsilon\nabla V(x)\cdot\varepsilon\nabla u^\varepsilon dx.$$

We conclude thanks to Cauchy–Schwarz inequality, the conservation of mass and Gronwall lemma, that $\|\nabla u^\varepsilon(t)\|_{L^2}$ remains bounded on bounded time intervals. Therefore the solution is global in time. See [Carles (2008)] for details.

The identity of the last point follows from the *pseudo-conformal conservation law*, derived by J. Ginibre and G. Velo [Ginibre and Velo (1979)] for $\varepsilon = 1$. The case $\varepsilon \in]0, 1]$ is easily inferred *via* the scaling

$$(t, x) \mapsto \left(\frac{t}{\varepsilon}, \frac{x}{\varepsilon}\right).$$

Since from the previous point, $\varepsilon\nabla u^\varepsilon \in C(\mathbb{R}; L^2)$ and $u^\varepsilon \in C(\mathbb{R}; L^{2\sigma+2})$, this evolution law shows the *a priori* estimate $xu^\varepsilon \in L^\infty_{\text{loc}}(\mathbb{R}; L^2)$. $\square$

1.4.4 *Weak solutions*

We will mention weak solutions only in the case $V = 0$, for a defocusing power-like nonlinearity. We therefore consider

$$i\varepsilon\partial_t u^\varepsilon + \frac{\varepsilon^2}{2}\Delta u^\varepsilon = |u^\varepsilon|^{2\sigma}u^\varepsilon \quad ; \quad u^\varepsilon_{|t=0} = u^\varepsilon_0. \tag{1.32}$$

Definition 1.27 (Weak solution). *Let $u^\varepsilon_0 \in H^1 \cap L^{2\sigma+2}(\mathbb{R}^d)$. A (global) weak solution to (1.32) is a function $u^\varepsilon \in C(\mathbb{R}; \mathcal{D}') \cap L^\infty(\mathbb{R}; H^1 \cap L^{2\sigma+2})$ solving (1.32) in $\mathcal{D}'(\mathbb{R} \times \mathbb{R}^d) \cap C(\mathbb{R}; L^2)$, and such that:*

- $\|u^\varepsilon(t)\|_{L^2} = \|u^\varepsilon_0\|_{L^2}$, $\forall t \in \mathbb{R}$.
- $E^\varepsilon(u^\varepsilon(t)) \leqslant E^\varepsilon(u^\varepsilon_0)$, $\forall t \in \mathbb{R}$.

Essentially, the energy conservation is replaced by an inequality, due to a limiting procedure and the use of Fatou's lemma in the construction of weak solutions.

Proposition 1.28. (**[Ginibre and Velo (1985a)]**) *Let $\sigma > 0$, $\varepsilon \in]0,1]$, and $u_0^\varepsilon \in H^1 \cap L^{2\sigma+2}(\mathbb{R}^d)$. Then (1.32) has a global weak solution. Moreover, if $\sigma < 2/(d-2)_+$, then this weak solution is unique, and coincides with the mild solution of the last point in Proposition 1.26.*

Chapter 2

Weakly Nonlinear Geometric Optics

In this chapter, we consider

$$i\varepsilon\partial_t u^\varepsilon + \frac{\varepsilon^2}{2}\Delta u^\varepsilon = V u^\varepsilon + \varepsilon^\alpha f\left(|u^\varepsilon|^2\right) u^\varepsilon,$$

in the case $\alpha \geqslant 1$; note that α is not necessarily an integer. We first consider the case of a single initial WKB state,

$$i\varepsilon\partial_t u^\varepsilon + \frac{\varepsilon^2}{2}\Delta u^\varepsilon = V u^\varepsilon + \varepsilon^\alpha f\left(|u^\varepsilon|^2\right) u^\varepsilon \quad ; \quad u^\varepsilon_{|t=0} = a_0^\varepsilon e^{i\phi_0/\varepsilon}. \quad (2.1)$$

The formal analysis of Sec. 1.2 suggests that if $\alpha > 1$, then the nonlinearity f is not relevant at leading order in the limit $\varepsilon \to 0$. On the other hand, the value $\alpha = 1$ should be critical, and nonlinear effects are expected to influence the behavior of u^ε at leading order. We prove that this holds true. In the case $\alpha > 1$, this means that at leading order, u^ε is described as in Sec. 1.3. When $\alpha = 1$, we describe precisely the nonlinear effect at leading order: it consists of a *nonlinear phase shift*. In other words, the main nonlinear effect is a phase self-modulation.

In Sec. 2.5, we show a consequence of this analysis on the Cauchy problem without semi-classical parameter.

In Sec. 2.6, we consider the case $\alpha = 1$, with $V = 0$, when the initial datum is the sum of WKB terms, with plane waves only.

In addition to Assumption 1.7, the following assumption is made from Sec. 2.1 to Sec. 2.4 (*single-phase* weakly nonlinear geometric optics):

Assumption 2.1. We assume that the nonlinearity is smooth, and that the initial amplitude is bounded in the following sense:

- $f \in C^\infty(\mathbb{R}; \mathbb{R})$, and $f(0) = 0$.
- There exists $s_0 > d/2$ such that $(a_0^\varepsilon)_\varepsilon$ is bounded in H^{s_0}.

31

As noticed in Sec. 1.4.2, the assumption $f(0) = 0$ comes for free, up to replacing V by $V + \varepsilon^\alpha f(0)$ and f by $f - f(0)$. Note that we could also consider the equation

$$i\varepsilon\partial_t u^\varepsilon + \frac{\varepsilon^2}{2}\Delta u^\varepsilon = Vu^\varepsilon + f\left(\varepsilon^\alpha |u^\varepsilon|^2\right)u^\varepsilon \quad ; \quad u^\varepsilon_{|t=0} = a^\varepsilon_0 e^{i\phi_0/\varepsilon}.$$

Of course, when f is the identity (cubic nonlinearity), this is the same equation as (2.1). Otherwise, it is a different problem. Despite the appearance, this initial value problem is less general than (2.1) from the technical point of view. Indeed, recall that the WKB analysis considers times where the solution u^ε is of order $\mathcal{O}(1)$ (in L^∞) as $\varepsilon \to 0$. Therefore, since $\alpha \geqslant 1$, the Taylor expansion of f yields

$$f\left(\varepsilon^\alpha |u^\varepsilon|^2\right)u^\varepsilon \underset{\varepsilon \to 0}{\sim} f'(0)\varepsilon^\alpha |u^\varepsilon|^2 u^\varepsilon + \frac{f''(0)}{2}\varepsilon^{2\alpha}|u^\varepsilon|^4 u^\varepsilon + \dots$$

Two cases can be distinguished. Either $\alpha > 1$, and the analysis for (2.1) will show that, in this case too, the nonlinearity is negligible at leading order before a caustic is formed (if any); or $\alpha = 1$, and only the cubic term $f'(0)\varepsilon|u^\varepsilon|^2 u^\varepsilon$ is expected to be relevant at leading order. In both cases, we leave it as an exercise to adapt the approach presented below, in order to justify these assertions.

2.1   Improved existence results

If $a^\varepsilon_0 \in \Sigma^k = H^k \cap \mathcal{F}(H^k)$ for some $k > d/2$, Proposition 1.25 shows that (2.1) has a unique solution $u^\varepsilon \in C(] - T^\varepsilon_-, T^\varepsilon_+[; \Sigma^k)$ for some $T^\varepsilon_-, T^\varepsilon_+ > 0$. In this paragraph, we show that we may construct a strong solution u^ε by assuming only $a^\varepsilon_0 \in H^s$ for some $s > d/2$. In addition, we show that u^ε remains bounded on $[-T, T]$, where $T > 0$ is given by Proposition 1.9, provided that a^ε_0 in bounded in H^s as $\varepsilon \to 0$. As a corollary, we show that if a^ε_0 is uniformly in Σ^k for some $k > d/2$, then $T^\varepsilon_\pm \geqslant T$.

As noticed during the proof of Proposition 1.16, it is more natural to work with

$$a^\varepsilon(t, x) = u^\varepsilon(t, x)e^{-i\phi_{\text{eik}}(t,x)/\varepsilon}$$

than with u^ε directly. We check that (2.1) is equivalent to

$$\begin{cases} \partial_t a^\varepsilon + \nabla\phi_{\text{eik}} \cdot \nabla a^\varepsilon + \frac{1}{2}a^\varepsilon\Delta\phi_{\text{eik}} = i\frac{\varepsilon}{2}\Delta a^\varepsilon - i\varepsilon^{\alpha-1}f\left(|a^\varepsilon|^2\right)a^\varepsilon, \\ \\ a^\varepsilon_{|t=0} = a^\varepsilon_0. \end{cases} \qquad (2.2)$$

Proposition 2.2. *Let Assumptions 1.7 and 2.1 be satisfied, and $\alpha \geqslant 1$. There exists $T_0 \in]0, T]$ independent of $\varepsilon \in]0, 1]$, where T is given by Proposition 1.9, such that (2.2) has a unique solution $a^\varepsilon \in C([-T_0, T_0]; H^{s_0})$. Moreover, $(a^\varepsilon)_\varepsilon$ is bounded in $C([-T_0, T_0]; H^{s_0})$. If $(a_0^\varepsilon)_\varepsilon$ is bounded in H^s for some $s \geqslant s_0$, then $(a^\varepsilon)_\varepsilon$ is bounded in $C([-T_0, T_0]; H^s)$.*

Proof. There are at least two procedures to construct a solution to (2.2): an iterative scheme, or Galerkin methods. For the iterative scheme, we solve, for $j \geqslant 0$:

$$\begin{cases} \partial_t a_{j+1}^\varepsilon + \nabla \phi_{\mathrm{eik}} \cdot \nabla a_{j+1}^\varepsilon + \dfrac{1}{2} a_{j+1}^\varepsilon \Delta \phi_{\mathrm{eik}} = i \dfrac{\varepsilon}{2} \Delta a_{j+1}^\varepsilon - i \varepsilon^{\alpha-1} f\left(|a_j^\varepsilon|^2\right) a_j^\varepsilon, \\ a_{j+1|t=0}^\varepsilon = a_0^\varepsilon. \end{cases}$$

This is actually a linear Schrödinger equation: setting $u_j^\varepsilon := a_j^\varepsilon e^{i\phi_{\mathrm{eik}}/\varepsilon}$, we see that the above equation is equivalent to:

$$i\varepsilon \partial_t u_{j+1}^\varepsilon + \frac{\varepsilon^2}{2} \Delta u_{j+1}^\varepsilon = V u_{j+1}^\varepsilon + \varepsilon^\alpha f\left(|u_j^\varepsilon|^2\right) u_j^\varepsilon \quad ; \quad u_{j+1|t=0}^\varepsilon = a_0^\varepsilon e^{i\phi_0/\varepsilon}.$$

Using Galerkin methods, we can mimic the mollification procedure presented for instance in [Alinhac and Gérard (2007); Majda (1984)]; roughly speaking, we solve an ordinary differential equation in H^s along characteristics by considering

$$\begin{cases} \partial_t a_h^\varepsilon + J_h \left(\nabla \phi_{\mathrm{eik}} \cdot \nabla J_h a_h^\varepsilon\right) + \dfrac{1}{2} a_h^\varepsilon \Delta \phi_{\mathrm{eik}} = i \dfrac{\varepsilon}{2} \Delta J_h^2 a_h^\varepsilon - i\varepsilon^{\alpha-1} f\left(|a_h^\varepsilon|^2\right) a_h^\varepsilon, \\ a_{h|t=0}^\varepsilon = a_0^\varepsilon, \end{cases}$$

where $J_h = \jmath(hD)$ is a Fourier multiplier, with $\jmath \in C_0^\infty(\mathbb{R}^d; \mathbb{R})$ equal to one in a neighborhood of the origin.

For both methods, the problem boils down to obtaining energy estimates for (2.2) in H^s, for all $s \geqslant s_0$. Let $s > d/2$. Applying the operator $\Lambda^s = (\mathrm{Id} - \Delta)^{s/2}$ to (2.2), we find:

$$\partial_t \Lambda^s a^\varepsilon + \nabla \phi_{\mathrm{eik}} \cdot \nabla \Lambda^s a^\varepsilon = i \frac{\varepsilon}{2} \Delta \Lambda^s a^\varepsilon - i\varepsilon^{\alpha-1} \Lambda^s \left(f\left(|a^\varepsilon|^2\right) a^\varepsilon\right) + R_s^\varepsilon, \quad (2.3)$$

where

$$R_s^\varepsilon = [\nabla \phi_{\mathrm{eik}} \cdot \nabla, \Lambda^s] a^\varepsilon - \frac{1}{2} \Lambda^s \left(a^\varepsilon \Delta \phi_{\mathrm{eik}}\right).$$

Take the inner product of (2.3) with $\Lambda^s a^\varepsilon$, and consider the real part: the first term of the right-hand side of (2.3) vanishes, since $i\Delta$ is skew-symmetric, and we have:

$$\frac{1}{2} \frac{d}{dt} \|\Lambda^s a^\varepsilon\|_{L^2}^2 + \mathrm{Re} \int_{\mathbb{R}^d} \overline{\Lambda^s a^\varepsilon} \nabla \phi_{\mathrm{eik}} \cdot \nabla \Lambda^s a^\varepsilon \leqslant \varepsilon^{\alpha-1} \left\|f\left(|a^\varepsilon|^2\right) a^\varepsilon\right\|_{H^s} \|a^\varepsilon\|_{H^s}$$

$$+ \|R_s^\varepsilon\|_{L^2} \|a^\varepsilon\|_{H^s}.$$

Notice that we have

$$\left| \text{Re} \int_{\mathbb{R}^d} \overline{\Lambda^s a^\varepsilon} \nabla \phi_{\text{eik}} \cdot \nabla \Lambda^s a^\varepsilon \right| = \frac{1}{2} \left| \int_{\mathbb{R}^d} \nabla \phi_{\text{eik}} \cdot \nabla |\Lambda^s a^\varepsilon|^2 \right|$$

$$= \frac{1}{2} \left| \int_{\mathbb{R}^d} |\Lambda^s a^\varepsilon|^2 \Delta \phi_{\text{eik}} \right| \leqslant C \|a^\varepsilon\|_{H^s}^2,$$

since $\Delta \phi_{\text{eik}} \in L^\infty([-T,T] \times \mathbb{R}^d)$ from Proposition 1.9. Moser's inequality (Lemma 1.24) yields, in view of Assumption 2.1:

$$\left\| f\left(|a^\varepsilon|^2\right) a^\varepsilon \right\|_{H^s} \leqslant C\left(\|a^\varepsilon\|_{L^\infty}\right) \|a^\varepsilon\|_{H^s}.$$

We infer:

$$\frac{d}{dt} \|a^\varepsilon\|_{H^s}^2 \leqslant C\left(\|a^\varepsilon\|_{L^\infty}\right) \|a^\varepsilon\|_{H^s}^2 + \|R_s^\varepsilon\|_{H^s} \|a^\varepsilon\|_{H^s}.$$

Note that the above locally bounded map $C(\cdot)$ can be taken independent of ε if and only if $\alpha \geqslant 1$. To apply Gronwall lemma, we need to estimate the last term: we use the fact that the derivatives of order at least two of ϕ_{eik} are bounded, from Proposition 1.9, to have:

$$\|R_s^\varepsilon\|_{L^2} \leqslant C \|a^\varepsilon\|_{H^s}.$$

We can then conclude by Gronwall lemma and a continuity argument:

$$\|a^\varepsilon\|_{L^\infty([-T,T];H^s)} \leqslant C\left(s, \|a_0^\varepsilon\|_{H^s}\right).$$

This yields boundedness in the "high" norm. Contraction in the "small" norm (that is, contraction in L^2) follows easily. Let a^ε and b^ε be solutions to (2.2), with initial data a_0^ε and b_0^ε respectively. Assume that a^ε and b^ε are bounded in $L^\infty([-T,T];H^s)$ for some $s > n/2$. The difference $w^\varepsilon = a^\varepsilon - b^\varepsilon$ solves

$$\partial_t w^\varepsilon + \nabla \phi_{\text{eik}} \cdot \nabla w^\varepsilon + \frac{1}{2} w^\varepsilon \Delta \phi_{\text{eik}} = i\frac{\varepsilon}{2} \Delta w^\varepsilon - i\varepsilon^{\alpha-1} \left(f\left(|a^\varepsilon|^2\right) a^\varepsilon - f\left(|b^\varepsilon|^2\right) b^\varepsilon \right).$$

The above computations yield

$$\frac{d}{dt} \|w^\varepsilon(t)\|_{L^2}^2 \leqslant C \|w^\varepsilon(t)\|_{L^2}^2 + \left\| f\left(|a^\varepsilon|^2\right) a^\varepsilon - f\left(|b^\varepsilon|^2\right) b^\varepsilon \right\|_{L^2}^2,$$

where we have used Young's inequality

$$xy \leqslant \frac{1}{2}\left(x^2 + y^2\right), \quad \forall x, y \geqslant 0.$$

Denote $g(z) = f(|z|^2)z$. Using Taylor's formula, write

$$f\left(|a^\varepsilon|^2\right) a^\varepsilon - f\left(|b^\varepsilon|^2\right) b^\varepsilon = g\left(w^\varepsilon + b^\varepsilon\right) - g\left(b^\varepsilon\right)$$

$$= w^\varepsilon \int_0^1 \partial_z g\left(b^\varepsilon + \theta w^\varepsilon\right) d\theta + \overline{w}^\varepsilon \int_0^1 \partial_{\overline{z}} g\left(b^\varepsilon + \theta w^\varepsilon\right) d\theta.$$

We infer

$$\left\| f\left(|a^\varepsilon|^2\right) a^\varepsilon - f\left(|b^\varepsilon|^2\right) b^\varepsilon \right\|_{L^2}^2 \leqslant \|w^\varepsilon\|_{L^2}^2 \left\| \int_0^1 g'\left(b^\varepsilon + \theta w^\varepsilon\right) d\theta \right\|_{L^\infty}^2$$

$$\leqslant C\left(\|w^\varepsilon\|_{L^\infty}, \|b^\varepsilon\|_{L^\infty}\right) \|w^\varepsilon\|_{L^2}^2$$

$$\leqslant \widetilde{C}\left(\|a^\varepsilon\|_{L^\infty}, \|b^\varepsilon\|_{L^\infty}\right) \|w^\varepsilon\|_{L^2}^2.$$

This yields the contraction in the L^2 norm on small time intervals, hence the proposition. $\qquad\square$

Remark 2.3. It is not true in general that $u^\varepsilon \in C([-T, T]; H^s)$. Indeed, u^ε is the product of $a^\varepsilon \in C([-T, T]; H^s)$ and $e^{i\phi_{\mathrm{eik}}/\varepsilon}$. If $xa^\varepsilon \notin C([-T, T]; L^2)$ and if ϕ_{eik} grows quadratically in space, then $u^\varepsilon \notin C([-T, T]; H^1)$. This phenomenon is geometric, not nonlinear: if $f = 0$ and if V is an harmonic potential, then u^ε may instantly cease to be in H^s for all $s > 0$ (but not $s = 0$). See Sec. 2.5.

Corollary 2.4. *Let Assumptions 1.7 and 2.1 be satisfied, and $\alpha \geqslant 1$. If in addition $a_0^\varepsilon \in \Sigma^k$ for an integer $k > d/2$, then $a^\varepsilon \in C([-T_0, T_0]; \Sigma^k)$. If $(a_0^\varepsilon)_\varepsilon$ is bounded in Σ^k, then $(a^\varepsilon)_\varepsilon$ is bounded in $C([-T_0, T_0]; \Sigma^k)$.*

Proof. Proposition 2.2 shows that $(a^\varepsilon)_\varepsilon$ is bounded in $C([-T_0, T_0]; H^k)$. By multiplying (2.2) by x^β and using energy estimates as in the proof of Proposition 2.2, induction on $k' = |\beta|$ yields the corollary. $\qquad\square$

2.2 Leading order asymptotic analysis

To pass to the limit in (2.2), we assume:

There exist $s > \dfrac{d}{2} + 2$ and $a_0 \in H^s$, such that $a_0^\varepsilon \underset{\varepsilon \to 0}{\longrightarrow} a_0$ in H^{s-2}. (2.4)

Since we deal with a general $\alpha \geqslant 1$, we keep the last term in (2.2), and consider

$$\begin{cases} \partial_t \widetilde{a}^\varepsilon + \nabla \phi_{\mathrm{eik}} \cdot \nabla \widetilde{a}^\varepsilon + \dfrac{1}{2} \widetilde{a}^\varepsilon \Delta \phi_{\mathrm{eik}} = -i\varepsilon^{\alpha-1} f\left(|\widetilde{a}^\varepsilon|^2\right) \widetilde{a}^\varepsilon, \\[2mm] \widetilde{a}^\varepsilon_{|t=0} = a_0. \end{cases} \quad (2.5)$$

The proof of Proposition 2.2 shows that (2.5) has a unique solution $\widetilde{a}^\varepsilon$, which is bounded in $C([-T_0, T_0]; H^s)$, where s appears in (2.4).

Proposition 2.5. *Let Assumptions 1.7 and 2.1 be satisfied, as well as (2.4), and $\alpha \geqslant 1$. Then there exist $C > 0$ and $\varepsilon_0 \in]0, 1]$ such that*

$$\|a^\varepsilon - \widetilde{a}^\varepsilon\|_{L^\infty([-T,T]; H^{s-2})} \leqslant C\left(\varepsilon + \|a_0^\varepsilon - a_0\|_{H^{s-2}}\right), \quad 0 < \varepsilon \leqslant \varepsilon_0.$$

Proof. Set $w_a^\varepsilon = a^\varepsilon - \widetilde{a}^\varepsilon$. It solves

$$
\begin{cases}
\partial_t w_a^\varepsilon + \nabla \phi_{\text{eik}} \cdot \nabla w_a^\varepsilon + \dfrac{1}{2} w_a^\varepsilon \Delta \phi_{\text{eik}} = i \dfrac{\varepsilon}{2} \Delta w_a^\varepsilon + i \dfrac{\varepsilon}{2} \Delta \widetilde{a}^\varepsilon \\
\hspace{5cm} - i \varepsilon^{\alpha-1} \left(g\left(a^\varepsilon\right) - g\left(\widetilde{a}^\varepsilon\right) \right), \\
\hspace{2cm} w_{a|t=0}^\varepsilon = a_0^\varepsilon - a_0,
\end{cases}
$$

where $g(z) = f(|z|^2)z$. Note that the term $i\Delta w_a^\varepsilon$ vanishes from the energy estimates. The term $i\varepsilon\Delta\widetilde{a}^\varepsilon$ is viewed as a source term of order $\mathcal{O}(\varepsilon)$ in $C([-T_0, T_0]; H^{s-2})$. Since $s - 2 > d/2$, g is uniformly Lipschitzean on bounded sets of H^{s-2} (Lemma 1.23), and the same computations as in the proof of Proposition 2.2 show that we can apply Gronwall lemma, which yields the proposition with T replaced by $T_0 \leqslant T$. The fact that we can actually take $T_0 = T$ will be a consequence of the analysis presented in Sec. 2.3 below, for $\varepsilon > 0$ sufficiently small, thanks to a bootstrap argument using (2.4). □

Proposition 2.5 shows that the leading order asymptotic behavior of u^ε is given by

$$
\left\| u^\varepsilon - \widetilde{a}^\varepsilon e^{i\phi_{\text{eik}}/\varepsilon} \right\|_{L^\infty([-T,T];L^2\cap L^\infty)} \xrightarrow[\varepsilon \to 0]{} 0.
$$

Note also that in the case $\alpha = 1$, $\widetilde{a}^\varepsilon$ *does not* depend on ε. Therefore, the formal computations of Sec. 1.2 are justified at leading order when $\alpha = 1$. It turns out that (2.5) can be solved rather explicitly, in terms of the geometric objects introduced in Sec. 1.3, as shown in the next paragraph, where the case $\alpha > 1$ is also analyzed more precisely.

2.3 Interpretation

Recall from Sec. 1.3 that classical trajectories are given by the Hamiltonian system (1.11):

$$
\begin{cases}
\partial_t x(t,y) = \xi(t,y) & ; \quad x(0,y) = y, \\
\partial_t \xi(t,y) = -\nabla_x V(t, x(t,y)) & ; \quad \xi(0,y) = \nabla \phi_0(y).
\end{cases}
$$

For $t \in [-T, T]$, $y \mapsto x(t, y)$ is a diffeomorphism on $\mathbb{R}^d$, and the Jacobi's determinant

$$
J_t(y) = \det \nabla_y x(t, y)
$$

is uniformly bounded from above and from below:

$$
\exists C > 0, \quad \frac{1}{C} \leqslant J_t(y) \leqslant C, \quad \forall (t, y) \in [-T, T] \times \mathbb{R}^d. \tag{2.6}
$$

Denote

$$\widetilde{A}^{\varepsilon}(t,y) := \widetilde{a}^{\varepsilon}\left(t, x(t,y)\right) \sqrt{J_t(y)}.$$

For $t \in [-T,T]$, (2.5) is equivalent to:

$$\partial_t \widetilde{A}^{\varepsilon} = -i\varepsilon^{\alpha-1} f\left(J_t(y)^{-1}\left|\widetilde{A}^{\varepsilon}\right|^2\right)\widetilde{A}^{\varepsilon} \quad ; \quad \widetilde{A}^{\varepsilon}(0,y) = a_0(y).$$

Despite the appearances, this ordinary differential equation along the rays of geometrical optics is a *linear* equation. Indeed, since f is real-valued, it is of the form

$$\partial_t A = i\mathcal{V}A, \quad \mathcal{V} \in \mathbb{R}.$$

This implies $\partial_t |A|^2 = 0$. In our case, $\partial_t |\widetilde{A}^{\varepsilon}|^2 = 0$, hence

$$\widetilde{A}^{\varepsilon}(t,y) = a_0(y) \exp\left(-i\varepsilon^{\alpha-1}\int_0^t f\left(J_s(y)^{-1}\left|a_0(y)\right|^2\right)ds\right).$$

Back to the function $\widetilde{a}^{\varepsilon}$, we have:

$$\widetilde{a}^{\varepsilon}(t,x) = a(t,x)e^{i\varepsilon^{\alpha-1}G(t,x)},$$

where a is given in the linear case by (1.20), that is

$$a(t,x) = \frac{1}{\sqrt{J_t(y(t,x))}}a_0\left(y(t,x)\right),$$

and the phase shift G is given by:

$$G(t,x) = -\int_0^t f\left(J_s(y(t,x))^{-1}\left|a_0(y(t,x))\right|^2\right)ds. \tag{2.7}$$

This implies that in Proposition 2.2 (and after), we can take $T_0 = T$.

If $\alpha > 1$, then

$$\widetilde{a}^{\varepsilon}(t,x) - a(t,x) = \mathcal{O}\left(\varepsilon^{\alpha-1}\right),$$

and no nonlinear effect is present at leading order. On the other hand, in the case $\alpha = 1$, we see that the leading order nonlinear effect is described by the function G: a phase shift generated by a nonlinear mechanism. In the context of laser physics, this phenomenon is known as *phase self-modulation* (see e.g. [Zakharov and Shabat (1971); Boyd (1992)]). Note that this function G does not affect the convergence of the main two quadratic quantities:

Position density: $\quad \rho^{\varepsilon} = |u^{\varepsilon}|^2 = |a^{\varepsilon}|^2,$

Current density: $\quad J^{\varepsilon} = \varepsilon \operatorname{Im}\left(\overline{u}^{\varepsilon}\nabla u^{\varepsilon}\right) = |a^{\varepsilon}|^2 \nabla \phi_{\text{eik}} + \varepsilon \operatorname{Im}\left(\overline{a}^{\varepsilon}\nabla a^{\varepsilon}\right).$

We refer to Chap. 2.6.4 for further discussions on these quantities.

One may wonder if this approach could be extended to some values $\alpha < 1$. To have a simple ansatz as above, we would like to remove the Laplacian in the limit $\varepsilon \to 0$ in (2.2). We find:

$$\widetilde{a}^{\varepsilon}(t, x) = a(t, x)e^{i\varepsilon^{\alpha-1}G(t,x)},$$

where a and G are given by the same expressions as above. Now recall that in Proposition 2.2, we prove that a^{ε} is bounded in H^s, uniformly for $\varepsilon \in]0, 1]$; this property is used to approximate a^{ε} by $\widetilde{a}^{\varepsilon}$. But when $\alpha < 1$, $\widetilde{a}^{\varepsilon}$ is no longer uniformly bounded in H^s, because what was a phase modulation for $\alpha \geqslant 1$ is now a rapid oscillation. This is another hint that when $\alpha < 1$, the approach must be modified. The analogous study in the case $\alpha = 0$ is presented in Chap. 4. See also Sec. 4.2.3 for the case $0 < \alpha < 1$.

2.4　Higher order asymptotic analysis

To compute the next term in the asymptotic expansion of a^{ε}, we assume that $\alpha \in \mathbb{N} \setminus \{0\}$, and

$$\exists s > \frac{d}{2} + 4, a_0 \in H^s, a_1 \in H^{s-2} / \|a_0^{\varepsilon} - a_0 - \varepsilon a_1\|_{H^{s-4}} = o(\varepsilon). \qquad (2.8)$$

Define $a^{(0)}$ by

$$a^{(0)}(t, x) = \begin{cases} a(t, x) & \text{if } \alpha \geqslant 2, \\ a(t, x)e^{iG(t,x)} & \text{if } \alpha = 1. \end{cases}$$

Define the first corrector $a^{(1)}$ in the WKB analysis as the solution to

$$\partial_t a^{(1)} + \nabla\phi_{\text{eik}} \cdot \nabla a^{(1)} + \frac{1}{2}a^{(1)}\Delta\phi_{\text{eik}} = \frac{i}{2}\Delta a^{(0)} + S_\alpha, \qquad (2.9)$$

with initial datum $a^{(1)}(0, x) = a_1(x)$, where

$$S_\alpha = \begin{cases} 0 & \text{if } \alpha \geqslant 3, \\ -if\left(\left|a^{(0)}\right|^2\right)a^{(0)} & \text{if } \alpha = 2, \\ -if\left(\left|a^{(0)}\right|^2\right)a^{(1)} - 2if'\left(\left|a^{(0)}\right|^2\right)a^{(0)}\,\mathrm{Re}\left(\overline{a}^{(0)}a^{(1)}\right) & \text{if } \alpha = 1. \end{cases}$$

Note that in all the cases, (2.9) is a linear equation. Since ϕ_{eik} is smooth and $a^{(0)} \in C([-T, T]; H^s)$, the regularity of $a^{(1)}$ is given by the regularity of its initial datum and of its source term $i\Delta a^{(0)} \in C([-T, T]; H^{s-2})$. Since $s > d/2$, the term $\|S_\alpha\|_{H^s}$ is controlled by $1 + \|a^{(1)}\|_{H^s}$ in all the cases, so

$a^{(1)} \in C([-T,T]; H^{s-2})$. Using the same approach as above, the following result is left as an exercise:

Proposition 2.6. *Let Assumptions 1.7 and 2.1 be satisfied, as well as (2.9). Then there exists $C > 0$ and $\varepsilon_0 > 0$ such that for $0 < \varepsilon \leqslant \varepsilon_0$,*

$$\left\| a^\varepsilon - a^{(0)} - \varepsilon a^{(1)} \right\|_{L^\infty([-T,T];H^{s-4})} \leqslant C \left(\varepsilon + \left\| a_0^\varepsilon - a_0 - \varepsilon a_1 \right\|_{H^{s-4}} \right).$$

We see that this analysis can be continued to arbitrary order, with essentially the loss of two derivatives at each step, like in the linear case. To justify the asymptotics at order $j \geqslant 0$, we assume $s > d/2 + 2j + 2$. We leave out the computations here, for the higher order analysis bears no new difficulty or interest.

2.5 An application: Cauchy problem in Sobolev spaces for nonlinear Schrödinger equations with potential

In this paragraph, we set $\varepsilon = 1$, and consider the Cauchy problem

$$i\partial_t u + \frac{1}{2}\Delta u = Vu + f\left(|u|^2\right)u \quad ; \quad u_{|t=0} = a_0. \tag{2.10}$$

We assume $a_0 \in H^s$ for $s > d/2$, $f \in C^\infty(\mathbb{R};\mathbb{R})$, and $V = V(x)$ to simplify, satisfying Assumption 1.7. We consider Eq. (2.10) as a weakly nonlinear equation ($\alpha = 1$), even though $\varepsilon = 1$. We have seen in Sec. 1.4.2 that under Assumption 1.7, it is fairly natural to work in the space

$$\Sigma^k = H^k \cap \mathcal{F}(H^k) = \{f \in H^k(\mathbb{R}^d) \; ; \; x \mapsto \langle x \rangle^k f(x) \in L^2(\mathbb{R}^d)\},$$

for some $k > d/2$. When $\nabla V \in L^\infty(\mathbb{R}^d)$, Σ^k can be replaced by H^k. We now address the same question when ∇V is unbounded. The typical example of such an occurrence under Assumption 1.7 is when V is an harmonic potential, say $V(x) = |x|^2$. In the linear case $f = 0$, it is well-known that V acts as a rotation in the phase space. Therefore, a_0 must have similar properties on the x-side and on the ξ-side in order for the Sobolev regularity to be propagated. In particular, if $a_0 \in H^\infty$ with $x \mapsto \langle x \rangle a_0(x) \notin L^2$, then $u(t,\cdot)$ ceases to be in H^s for $s \geqslant 1$ as soon as $t > 0$. It is natural to expect a similar phenomenon in the nonlinear case. We show that this is so, with a proof which is valid both for the linear case and the nonlinear case ([Carles (2008)]).

First, introduce ϕ_{eik} solution to the eikonal equation

$$\partial_t \phi_{\text{eik}} + \frac{1}{2}\left|\nabla \phi_{\text{eik}}\right|^2 + V = 0 \quad ; \quad \phi_{\text{eik}}(0,x) = 0.$$

From Proposition 1.9, there exist $T > 0$ and a unique $\phi_{\text{eik}} \in C^\infty([-T,T] \times \mathbb{R}^d)$ solution to the above equation. Proposition 2.2 shows that there exists $a \in C([-T,T]; H^\infty)$ such that $u = ae^{i\phi_{\text{eik}}}$ solves (2.10). The function a solves:

$$\partial_t a + \nabla\phi_{\text{eik}} \cdot \nabla a + \frac{1}{2}a\Delta\phi_{\text{eik}} = \frac{i}{2}\Delta a - if\left(|a|^2\right)a \quad ; \quad a_{|t=0} = a_0. \quad (2.11)$$

Since in particular, $u \in C([-T,T]; L^2 \cap L^\infty)$, u is the unique such solution: if $v \in C([-T,T]; L^2 \cap L^\infty)$ solves Eq. (2.10), then $w = u - v$ solves

$$i\partial_t w + \frac{1}{2}\Delta w = Vw + f\left(|v+w|^2\right)(v+w) - f\left(|v|^2\right)v \quad ; \quad w_{|t=0} = 0.$$

Lemma 1.2 shows that for $t > 0$,

$$\|w(t)\|_{L^2} \leqslant \int_0^t \left\| f\left(|v+w|^2\right)(v+w) - f\left(|v|^2\right)v \right\|_{L^2} d\tau$$

$$\leqslant C \int_0^t \|w(\tau)\|_{L^2} \, d\tau.$$

Gronwall lemma implies $w \equiv 0$.

Proposition 2.7. *Let $d \geqslant 1$, and f be smooth, $f \in C^\infty(\mathbb{R}_+; \mathbb{R})$. Assume that V is super-linear, and that there exist $0 < k(\leqslant 1)$ and $C > 0$ such that*

$$|\nabla V(x)| \leqslant C\langle x\rangle^k, \quad \forall x \in \mathbb{R}^d,$$

and $\omega, \omega' \in \mathbb{S}^{d-1}$ such that

$$|\omega \cdot \nabla V(x)| \geqslant c|\omega' \cdot x|^k \text{ as } |x| \to \infty, \text{ for some } c > 0. \quad (2.12)$$

Then there exists $a_0 \in H^\infty$ such that for arbitrarily small $t > 0$ and all $s > 0$, the solution $u(t, \cdot)$ to (2.10) fails to be in $H^s(\mathbb{R}^d)$.

Example 2.8. For V, we may consider any non-trivial quadratic form, or $V(x) = \pm\langle x'\rangle^a$, with $1 < a \leqslant 2$, for some decomposition $x = (x', x'') \in \mathbb{R}^d$.

Remark 2.9. No assumption is made on the growth of the nonlinearity at infinity: the above result reveals a *geometric phenomenon*, and not an ill-posedness result like in Chap. 4.5.

To prove the above result, recall that $u = ae^{i\phi_{\text{eik}}}$, where a is given by (2.11). Consider only positive times, and define b as the solution on $[0, T]$ to:

$$\partial_t b + \nabla\phi_{\text{eik}} \cdot \nabla b + \frac{1}{2}b\Delta\phi_{\text{eik}} = -if\left(|b|^2\right)b \quad ; \quad b_{|t=0} = a_0. \quad (2.13)$$

As noticed in Sec. 2.3,

$$b(t,x) = \frac{1}{\sqrt{J_t(y(t,x))}} a_0\left(y(t,x)\right) e^{-i\beta(t,x)},$$

where $\beta(t,x) = \displaystyle\int_0^t f\left(J_s\left(y(t,x)\right)^{-1} |a_0\left(y(t,x)\right)|^2\right) ds,$

$x(t,y)$ is given by Eq. (1.11) with $\phi_0 = 0$, and the Jacobi's determinant is defined by

$$J_t(y) = \det \nabla_y x(t,y).$$

Observe that $b \in C([0,T]; H^\infty)$. Let $r = a - b$: $r \in C([0,T]; H^\infty)$. For $1 \leqslant j \leqslant d$, $x_j r$ solves:

$$\begin{cases} \partial_t(x_j r) + \nabla\phi_{\text{eik}} \cdot \nabla(x_j r) + \dfrac{1}{2} x_j r \Delta\phi_{\text{eik}} = \dfrac{i}{2}\Delta(x_j r) + r\partial_j\phi_{\text{eik}} - i\partial_j r \\[2mm] \qquad\qquad + \dfrac{i}{2} x_j \Delta b - i x_j\left(f\left(|b+r|^2\right)(b+r) - f\left(|b|^2\right) b\right), \\[2mm] \qquad x_j r_{|t=0} = 0. \end{cases}$$

The fundamental theorem of calculus yields:

$$x_j\left(f\left(|a|^2\right) a - f\left(|b|^2\right) b\right) = x_j\left(f\left(|b+r|^2\right)(b+r) - f\left(|b|^2\right) b\right)$$

$$= x_j r \int_0^1 \partial_z g\,(b+sr)\,ds + x_j\bar{r}\int_0^1 \partial_{\bar z} g\,(b+sr)\,ds,$$

where $g(z) = f(|z|^2)z$. In particular, we know that

$$\int_0^1 \partial_z g\,(b+sr)\,ds, \int_0^1 \partial_{\bar z} g\,(b+sr)\,ds \in C \cap L^\infty([0,T] \times \mathbb{R}^d).$$

Energy estimates as in §2.1 show that:

$$\|xr\|_{L^\infty([0,t];L^2)} \leqslant C\left(1 + \|x\Delta b\|_{L^1([0,t];L^2)}\right).$$

We must make sure that the last term is, or can be chosen, finite. We shall demand $x\Delta b \in L^\infty([0,T]; L^2)$. In view of (2.6), this requirement is met as soon as $a_0 \in H^\infty(\mathbb{R}^d)$ is such that $x\Delta a_0, xa_0|\nabla a_0|^2 \in L^2(\mathbb{R}^d)$:

If $a_0 \in H^\infty$ is such that $x\Delta a_0, xa_0|\nabla a_0|^2 \in L^2(\mathbb{R}^d)$, then: $$a = b + r, \text{ with } b,r \in C([0,T]; H^\infty), \text{ and } xr \in C([0,T]; L^2). \tag{2.14}$$

We now prove that for small times, $\nabla\phi_{\text{eik}}(t,x)$ can be approximated by $-t\nabla V(x)$.

Lemma 2.10. *Assume that there exist $0 \leqslant k \leqslant 1$ and $C > 0$ such that*

$$|\nabla V(x)| \leqslant C\langle x\rangle^k, \quad \forall x \in \mathbb{R}^d.$$

Then there exist $T_0, C_0 > 0$ such that

$$|\nabla\phi_{\text{eik}}(t,x) + t\nabla V(x)| \leqslant C_0 t^2 \langle x\rangle^k, \quad \forall t \in [0,T_0].$$

Proof.   We infer from Proposition 1.9 that

$$|\partial_t \nabla \phi_{\text{eik}}(t,x) + \nabla V(x)| \leqslant \|\nabla^2 \phi_{\text{eik}}(t)\|_{L^\infty} |\nabla \phi_{\text{eik}}(t,x)|$$
$$\lesssim |\nabla \phi_{\text{eik}}(t,x)|. \tag{2.15}$$

From Eq. (1.11), we also have

$$|\nabla \phi_{\text{eik}}(t,x)| = |\xi(t, y(t,x))| = \left| \int_0^t \nabla V(x(s, y(t,x)))\, ds \right|$$
$$\lesssim \int_0^t |\nabla V(y(t,x))|\, ds + \int_0^t |x(s, y(t,x)) - y(t,x)|\, ds.$$

We claim that

$$|x(t,y) - y| \lesssim t^2 \langle y \rangle^k. \tag{2.16}$$

Indeed, we have from (1.11),

$$|x(t,y) - y| = \left| \int_0^t \partial_t x(s,y)\, ds \right| = \left| \int_0^t \int_0^s \nabla V(x(s',y))\, ds'\, ds \right|$$
$$= \left| \int_0^t (t - s') \nabla V(x(s',y))\, ds' \right|$$
$$= \left| \int_0^t (t-s) \nabla V(y)\, ds + \int_0^t (t-s)\left(\nabla V(x(s,y)) - \nabla V(y) \right) ds \right|$$
$$\lesssim t^2 \langle y \rangle^k + \int_0^t (t-s)\, |x(s,y) - y|\, ds,$$

and (2.16) follows from Gronwall lemma. We infer that for $t > 0$ sufficiently small,

$$|y(t,x) - x| \lesssim t^2 \langle x \rangle^k,$$

and therefore,

$$|\nabla \phi_{\text{eik}}(t,x)| \lesssim \int_0^t |\nabla V(y(t,x))|\, ds + \int_0^t |x(s, y(t,x)) - y(t,x)|\, ds$$
$$\lesssim \int_0^t |\nabla V(x)|\, ds + \int_0^t |x - y(t,x)|\, ds$$
$$\quad + \int_0^t |x(s, y(t,x)) - y(t,x)|\, ds$$
$$\lesssim t \langle x \rangle^k + t^3 \langle x \rangle^k + \int_0^t s^2 \langle y(t,x) \rangle^k\, ds$$
$$\lesssim t \langle x \rangle^k + t^3 \langle x \rangle^k + t^3 \left(\langle x \rangle^k + t^{2k} \langle x \rangle^{2k} \right).$$

Then (2.15) yields

$$|\partial_t \nabla \phi_{\text{eik}}(t,x) + \nabla V(x)| \lesssim t \langle x \rangle^k,$$

Lemma 2.10 follows by integration in time.   □

We infer that for $t > 0$ small enough,

$$|\omega \cdot \nabla \phi_{\text{eik}}(t, x)| \gtrsim t |\omega \cdot \nabla V(x)|. \tag{2.17}$$

To complete the proof of Proposition 2.7, consider

$$a_0(x) = \frac{1}{\langle x \rangle^{d/2} \log\left(2 + |x|^2\right)}. \tag{2.18}$$

As is easily checked, a_0 meets the requirements of the first line of (2.14). Denote

$$v = b e^{i\phi_{\text{eik}}} \quad ; \quad w = r e^{i\phi_{\text{eik}}}.$$

Obviously, $u = v + w$. From (2.14) and (2.17), $v(t, \cdot) \in L^2(\mathbb{R}^d) \setminus H^1(\mathbb{R}^d)$ for $t > 0$ sufficiently small. On the other hand, $w(t, \cdot) \in H^1(\mathbb{R}^d)$ for all $t \in [0, T]$, hence $u(t, \cdot) \in L^2(\mathbb{R}^d) \setminus H^1(\mathbb{R}^d)$ for $0 < t \ll 1$.

We now just have to see that the same holds if we replace $H^1(\mathbb{R}^d)$ with $H^s(\mathbb{R}^d)$ for $0 < s < 1$. We use the following characterization of $H^s(\mathbb{R}^d)$ (see e.g. [Chemin (1998)]): for $\varphi \in L^2(\mathbb{R}^d)$ and $0 < s < 1$,

$$\varphi \in H^s(\mathbb{R}^d) \iff \iint_{\mathbb{R}^d \times \mathbb{R}^d} \frac{|\varphi(x + y) - \varphi(x)|^2}{|y|^{d+2s}} dx dy < \infty.$$

Since $w(t, \cdot) \in H^1$ for all $t \in [0, T]$, we shall prove that $v(t, \cdot) \in L^2 \setminus H^s$ for t sufficiently small. We prove that for $0 < t \ll 1$ independent of $s \in]0, 1[$,

$$I := \int_{|y| \leqslant 1} \int_{x \in \mathbb{R}^d} \frac{|v(t, x + y) - v(t, x)|^2}{|y|^{d+2s}} dx dy = \infty.$$

To apply a fractional Leibnitz rule, write

$$v(t, x + y) - v(t, x) = (b(t, x + y) - b(t, x)) e^{i\phi_{\text{eik}}(t, x+y)}$$
$$+ \left(e^{i\phi_{\text{eik}}(t, x+y)} - e^{i\phi_{\text{eik}}(t, x)} \right) b(t, x).$$

In view of the inequality $|\alpha - \beta|^2 \geqslant \alpha^2/2 - \beta^2$, we have:

$$|v(t, x + y) - v(t, x)|^2 \geqslant \frac{1}{2} \left| \left(e^{i\phi_{\text{eik}}(t, x+y)} - e^{i\phi_{\text{eik}}(t, x)} \right) b(t, x) \right|^2$$
$$- |b(t, x + y) - b(t, x)|^2.$$

We can leave out the last term, since $b(t, \cdot) \in H^\infty$ for $t \in [0, T]$:

$$\iint_{\mathbb{R}^d \times \mathbb{R}^d} \frac{|b(t, x + y) - b(t, x)|^2}{|y|^{d+2s}} dx dy < \infty, \quad \forall t \in [0, T].$$

We now want to prove

$$\int_{|y| \leqslant 1} \int_{x \in \mathbb{R}^d} |b(t, x)|^2 \frac{\left| \sin\left(\frac{\phi_{\text{eik}}(t, x+y) - \phi_{\text{eik}}(t, x)}{2} \right) \right|^2}{|y|^{d+2s}} dx dy = \infty.$$

Proposition 1.9 yields:

$$(\partial_t + \nabla\phi_{\text{eik}} \cdot \nabla) \nabla^2\phi_{\text{eik}} \in L^\infty \left([0,T] \times \mathbb{R}^d\right)^{d^2} \quad ; \quad \nabla^2\phi_{\text{eik}}|_{t=0} = 0.$$

Therefore,

$$\left\|\nabla^2\phi_{\text{eik}}(t,\cdot)\right\|_{L^\infty(\mathbb{R}^d)^{d^2}} = \mathcal{O}(t) \quad \text{as } t \to 0.$$

We infer:

$$\phi_{\text{eik}}(t, x+y) - \phi_{\text{eik}}(t,x) = y \cdot \nabla\phi_{\text{eik}}(t,x) + \mathcal{O}(t|y|^2), \quad \text{uniformly for } x \in \mathbb{R}^d,$$

and

$$\sin\left(\frac{\phi_{\text{eik}}(t, x+y) - \phi_{\text{eik}}(t,x)}{2}\right) = \sin\left(\frac{y \cdot \nabla\phi_{\text{eik}}(t,x)}{2}\right) \cos\left(\mathcal{O}(t|y|^2)\right)$$

$$+ \cos\left(\frac{y \cdot \nabla\phi_{\text{eik}}(t,x)}{2}\right) \sin\left(\mathcal{O}(t|y|^2)\right).$$

The second term is $\mathcal{O}(t|y|^2)$. Using the estimate $|\alpha - \beta|^2 \geqslant \alpha^2/2 - \beta^2$ again, we see that the integral corresponding to the second term is finite, and can be left out. To prove that

$$I' = \int_{|y| \leqslant 1} \int_{x \in \mathbb{R}^d} |b(t,x)|^2 \frac{\left|\sin\left(\frac{y \cdot \nabla\phi_{\text{eik}}(t,x)}{2}\right)\right|^2}{|y|^{d+2s}} dx dy = \infty \quad \text{for } 0 < t \ll 1,$$

we can localize y in a small conic neighborhood of $\omega\mathbb{R} \cap \{|y| \leqslant 1\}$:

$$\mathcal{V}_\epsilon = \{|y| \leqslant 1 \; ; \; |y - (y \cdot \omega)\omega| \leqslant \epsilon|y|\}, \quad 0 < \epsilon \ll 1.$$

For $0 < \epsilon, t \ll 1$, (2.17) yields:

$$\left|\sin\left(\frac{y \cdot \nabla\phi_{\text{eik}}(t,x)}{2}\right)\right| \gtrsim t|y \cdot \omega| \times |\omega \cdot \nabla V(x)|, \quad y \in \mathcal{V}_\epsilon.$$

Introduce a conic localization for x close to ω', excluding the origin:

$$\mathcal{U}_\epsilon = \{|x| \geqslant 1 \; ; \; |x - (x \cdot \omega')\omega'| \leqslant \epsilon|x|\}.$$

Change the variable in the y-integral: for t and ϵ sufficiently small, and $x \in \mathcal{U}_\epsilon$, set

$$y' = \omega \cdot \nabla\phi_{\text{eik}}(t,x)y.$$

This change of variable is admissible, from (2.12) and (2.17). We infer, for $0 < \epsilon, t \ll 1$:

$$I' \geqslant \int_{y \in \mathcal{V}_\epsilon} \int_{x \in \mathbb{R}^d} |b(t,x)|^2 \frac{\left|\sin\left(\frac{y \cdot \nabla\phi_{\text{eik}}(t,x)}{2}\right)\right|^2}{|y|^{d+2s}} dx dy$$

$$\gtrsim \int_{x \in \mathcal{U}_\epsilon} |b(t,x)|^2 |\omega \cdot \nabla\phi_{\text{eik}}(t,x)|^{2s} \left(\int_{y \in |\omega \cdot \nabla\phi_{\text{eik}}(t,x)|\mathcal{V}_\epsilon} \frac{dy}{|y|^{d+2s-2}}\right) dx$$

$$\gtrsim \int_{x \in \mathcal{U}_\epsilon} |b(t,x)|^2 |\omega \cdot \nabla\phi_{\text{eik}}(t,x)|^{2s} \left(\int_{y \in ct\mathcal{V}_\epsilon} \frac{dy}{|y|^{d+2s-2}}\right) dx.$$

The assumption (2.12), the expression of b and the choice (2.18) for a_0 then show that for $0 < t \ll 1$, $I' = \infty$. This completes the proof of Proposition 2.7.

2.6   Multiphase expansions

In this section, we assume $\alpha = 1$, and we address the case where the initial datum is a sum of terms of the form considered so far,

$$u^\varepsilon(0, x) = \sum_{j \in \mathbb{N}} a_{0j}^\varepsilon(x) e^{i\phi_{0j}(x)/\varepsilon}.$$

First, we may think of the above sum as a finite sum. In the linear case (1.9), the superposition principle shows that we can repeat the analysis presented in Sec. 1.3 to each term of the sum, approximate its evolution under (1.9), and the sum of these approximations will yield an approximation solution to the initial problem, along with error estimates. In the nonlinear case (2.1), nonlinear interactions may force the initially finite sum to become instantaneously infinite, as we will see below, due to the creation of resonant phases by nonlinear interaction. To understand the suitable set of phases, we may think in terms of Picard iterative scheme. Suppose

$$u^\varepsilon(0, x) = \sum_{j=1}^{N} a_{0j}^\varepsilon(x) e^{i\phi_{0j}(x)/\varepsilon},$$

where all the initial phases are at most quadratic, in the sense that Assumption 1.7 is satisfied by all ϕ_{0j}'s. Each initial phase generates a solution ϕ_j to the eikonal equation (1.10). With a Picard iterative scheme in mind, we first solve (1.9) with the above initial datum, and the previous analysis yields

$$u_{(1)}^\varepsilon(t, x) = \sum_{j=1}^{N} a_j^\varepsilon(t, x) e^{i\phi_j(t,x)/\varepsilon},$$

where the subscript indicates that this is the first Picard iterate, and the a_j^ε's are uniformly bounded in Sobolev spaces. For a cubic nonlinearity, the second iterate is given by

$$i\varepsilon \partial_t u_{(2)}^\varepsilon + \frac{\varepsilon^2}{2} \Delta u_{(2)}^\varepsilon = V u_{(2)}^\varepsilon + \varepsilon |u_{(1)}^\varepsilon|^2 u_{(1)}^\varepsilon ; \quad u_{(2)}^\varepsilon(0, x) = \sum_{j=1}^{N} a_{0j}^\varepsilon(x) e^{i\phi_{0j}(x)/\varepsilon}.$$

Expanding

$$|u_{(1)}^\varepsilon|^2 u_{(1)}^\varepsilon = \sum_{1 \leqslant k, \ell, m \leqslant N} a_k^\varepsilon \bar{a}_\ell^\varepsilon a_m^\varepsilon e^{i(\phi_k - \phi_\ell + \phi_m)/\varepsilon},$$

we must examine whether $\phi_k - \phi_\ell + \phi_m$ solves the eikonal equation (1.10) or not. It is possible to create new phases by this nonlinear interaction (as we will see more precisely below). Processing with the Picard iteration, we may create infinitely many resonant phases (in the sense that they solve (1.10)) which were not present initially. In general, nothing guarantees that this infinite set of phases will have a uniform lifespan, and nonlinear interaction might cause instantaneous focusing (a caustic appears instantaneously). Such a phenomenon has been described in the case of hyperbolic systems in [Joly *et al.* (1995a)]. To avoid this, we consider the case $V = 0$, with linear initial phases,

$$\phi_{0j} = \kappa_j \cdot x, \quad \kappa_j \in \mathbb{R}^d.$$

We emphasize that the vectors κ_j need not belong to a network, since on $\mathbb{R}^d$, they do not correspond to Fourier modes, as opposed to the periodic case $x \in \mathbb{T}^d$. The notations J_0 and J below thus only correspond to a labeling of the modes. We consider

$$i\varepsilon\partial_t u^\varepsilon + \frac{\varepsilon^2}{2}\Delta u^\varepsilon = \lambda\varepsilon|u^\varepsilon|^{2\sigma}u^\varepsilon, \quad \lambda \in \mathbb{R} \setminus \{0\}, \tag{2.19}$$

for σ a positive integer. The analysis presented below resumes the main lines from [Carles *et al.* (2010)]. The above preliminaries already suggest that nonlinearities which are multilinear in $(u, \bar{u})$ are easier to analyze, and we stick to this case. In the cubic case $\sigma = 1$, we have a complete characterization of the resonant set. For higher order nonlinearities (quintic and higher), the resonance condition is more intricate, and the complete description of the appearance of new modes (related not only to the resonance condition, but also to the dynamics associated to the system of transport equations, described below in the cubic case) would require some new analysis. We emphasize however that many aspects of the analysis below do not require a precise understanding of the resonance condition, and this is why we may consider $\sigma \in \mathbb{N} \setminus \{0\}$.

As announced above, the initial data that we consider in this section are of the form

$$u^\varepsilon(0, x) = \sum_{j \in J_0} a_{0j}(x)e^{i\kappa_j \cdot x/\varepsilon}, \tag{2.20}$$

where the set $J_0 \subset \mathbb{N}$ is at most countable, the initial amplitudes a_{0j} are chosen independent of ε to simplify discussions, and $\kappa_j \in \mathbb{R}^d$. We seek an approximate solution under the form

$$u_{\text{app}}^\varepsilon(t, x) = \sum_{j \in J} a_j(t, x)e^{i\kappa_j \cdot x/\varepsilon - i|\kappa_j|^2 t/(2\varepsilon)}. \tag{2.21}$$

Several comments are in order regarding this *ansatz*:

- The set J of relevant indices may be (strictly) larger than the initial set of indices J_0, due to the fact that extra phases must be considered.
- We have somehow by-passed the analysis since we directly consider phases solving the eikonal equation, given by Example 1.13.

The main new steps compared to the single phase case considered so far are precisely the description of the set of relevant phases, the analysis of the corresponding family of amplitudes, and the justification of the approximation.

2.6.1 *Resonant phases*

Denote by

$$\phi_j(t, x) = \kappa_j \cdot x - \frac{|\kappa_j|^2}{2} t \tag{2.22}$$

the solutions of the eikonal equation in the present context. We consider the action of the *cubic* nonlinearity on the ansatz (2.21), and more precisely on the rapid oscillations:

$$|u_{\text{app}}^\varepsilon|^2 u_{\text{app}}^\varepsilon = \left(\sum_{k \in J} a_k e^{i\phi_k/\varepsilon} \right) \overline{\left(\sum_{\ell \in J} a_\ell e^{i\phi_\ell/\varepsilon} \right)} \left(\sum_{m \in J} a_m e^{i\phi_m/\varepsilon} \right)$$

$$= \sum_{k, \ell, m \in J} a_k \overline{a}_\ell a_m e^{i(\phi_k - \phi_\ell + \phi_m)/\varepsilon}.$$

For each triplet (k, ℓ, m), the phase $\phi_k - \phi_\ell + \phi_m$ is the sum of a linear function in x and a linear function in t. In view of Example 1.13, this phase solves the eikonal equation if and only if

$$|\kappa_k - \kappa_\ell + \kappa_m|^2 = |\kappa_k|^2 - |\kappa_\ell|^2 + |\kappa_m|^2,$$

a condition which is equivalent to

$$(\kappa_\ell - \kappa_m) \cdot (\kappa_\ell - \kappa_k) = 0. \tag{2.23}$$

We readily see that the one-dimensional case is singular: if $d = 1$, then $\kappa_\ell = \kappa_m$ or $\kappa_\ell = \kappa_k$, and no new phase solving the eikonal equation is created through cubic interaction. On the other hand, if $d \geqslant 2$, we have the following geometric characterization, established initially in [Colliander *et al.* (2010)] on $\mathbb{T}^2$, and easily generalized on $\mathbb{R}^d$ ([Carles *et al.* (2010)]), as a straightforward consequence of (2.23):

Lemma 2.11. *Let $d \geqslant 2$, $\kappa_k, \kappa_\ell, \kappa_m \in \mathbb{R}^d$, and set*

$$\kappa_j := \kappa_k - \kappa_\ell + \kappa_m.$$

Suppose $\kappa_\ell \notin \{\kappa_k, \kappa_m\}$. Then $\phi_k - \phi_\ell - \phi_m = \phi_j$ if and only if the endpoints of the vectors $\kappa_k, \kappa_\ell, \kappa_m, \kappa_j$ form four corners of a non-degenerate rectangle with κ_ℓ and κ_j opposing each other.

In the case $\kappa_\ell \in \{\kappa_k, \kappa_m\}$, we have $\phi_j \in \{\phi_k, \phi_m\}$, and as in the one-dimensional case, the cubic interaction creates no new phase. The above lemma describes the geometric framework which makes the appearance of new characteristic phases possible (in the sense that they solve the eikonal equation). In particular, it takes at least three different initial phases to create a new one. The following example will be crucial in Sec. 5.4.

Example 2.12. Assume $d = 2$, and consider

$$\kappa_1 = (0, 1), \quad \kappa_2 = (1, 1), \quad \kappa_3 = (1, 0).$$

The cubic interaction creates the zero mode $\phi_0 \equiv 0$. If $d \geqslant 3$, the same thing happens by setting the last $d - 2$ coordinates of the κ's to zero.

The set J in (2.21) now corresponds to the set of initial phases J_0, possibly complemented with the new phases created through the above geometric principle. Note however that this principle must be exhausted, in the sense that if we think of the creation of new phases by cubic interaction like in Lemma 2.11 as a discrete dynamical system, we have to apply Lemma 2.11 successively, as phases created at the first step may combine with initial phases to create yet new phases at the second step, and so on. The following illustration is borrowed from [Carles and Faou (2012)], and may be viewed as a particular case of the analysis from [Colliander *et al.* (2010)].

Example 2.13. In dimension 2 (or higher by completing with zero like in Example 2.12), start from a cross (the grid corresponds to $\mathbb{Z}^2$, for instance)

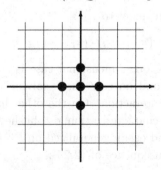

After one iteration of Lemma 2.11, we obtain a full square,

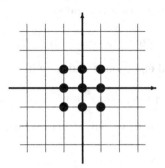

Then by considering the origin and the vertices of this square, we obtain four new points so as to create a larger, rotated, square,

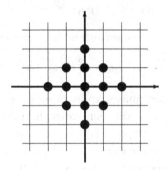

which is the initial cross enlarged by one external layer. Continuing the iteration, we see that the whole network $\mathbb{Z}^2$ is obtained by the exhaustion of Lemma 2.11.

On the other hand, the zero phase is the only generated phase in Example 2.12.

We note that in the case of higher order nonlinearities ($\sigma \geqslant 2$), the cubic mechanism of creation of phases remains, since for instance in the quintic case $\sigma = 2$, we face the linear combination

$$\phi_k - \phi_\ell + \phi_m - \phi_k + \phi_k.$$

We may however have more phases created by higher order interaction. Typically, consider the one-dimensional case $d = 1$. The cubic interaction creates no new phase, but higher order nonlinearities may. To see this, it suffices to show that this is so in the quintic case $\sigma = 2$, since the case

$\sigma \geqslant 3$ follows, in view of the previous remark. Starting from $\phi_1, \ldots, \phi_5$ like in (2.22), the quintic interaction generates

$$\phi_1 - \phi_2 + \phi_3 + \phi_4 - \phi_5,$$

which solves the eikonal equation if and only if

$$(\kappa_1 - \kappa_2 + \kappa_3 - \kappa_4 + \kappa_5)^2 = \kappa_1^2 - \kappa_2^2 + \kappa_3^2 - \kappa_4^2 + \kappa_5^2.$$

Like in Example 2.12, we focus our attention on the creation of the zero mode:

Example 2.14 (From [Carles and Kappeler (2017)]). *For any $p, q \in \mathbb{Z}$ with $p, q \neq 0$ and $p \neq q$, the 5-tuple*

$$(\kappa_1, \kappa_2, \kappa_3, \kappa_4, \kappa_5) = (pq, -q^2, -pq, p^2, p^2 - q^2)$$

creates the zero mode by resonant interaction of nonzero modes, that is

$$\kappa_1 - \kappa_2 + \kappa_3 - \kappa_4 + \kappa_5 = \kappa_1^2 - \kappa_2^2 + \kappa_3^2 - \kappa_4^2 + \kappa_5^2 = 0.$$

The notion of resonant phases appears as rather natural in the present context: the nonlinear interaction generates phases ϕ as linear combinations of the phases initially present, $\phi = \phi_k - \phi_\ell + \phi_m$, and we keep in the analysis only those ϕ's which solve the eikonal equation. The main step in the error estimate consists in checking that the ϕ's which do not solve the eikonal equation lead to a negligible contribution. When no semi-classical analysis is involved, the notion of resonant interactions is central in the study of nonlinear Schrödinger equations when large time dynamics is analyzed, typically with the use of normal forms; see e.g. [Bourgain (1999); Colliander *et al.* (2010); Grébert and Thomann (2012); Hani and Pausader (2014)].

2.6.2　*The set of transport equations*

Proceeding as in Sec. 1.2, the next step in the asymptotic expansion identifies the evolution of the amplitudes a_j. The first difference is that since we now consider several phases, we have to adapt the projection on rapid oscillations. For $j \in J$, cancelling the coefficient of order ε for the term $e^{i\phi_j/\varepsilon}$ yields

$$\partial_t a_j + \kappa_j \cdot \nabla a_j = -i\lambda \sum_{\phi_{j_1} - \phi_{j_2} + \cdots + \phi_{j_{2\sigma+1}} = \phi_j} a_{j_1} \bar{a}_{j_2} \ldots a_{j_{2\sigma+1}}, \qquad (2.24)$$

where we have used $\nabla \phi_j(t,x) = \kappa_j$. Again in view of the specific form of the phase (1.7), the condition on the sum involves a *resonant condition*, $(\kappa_{j_1}, \kappa_{j_2}, \dots, \kappa_{j_{2\sigma+1}}) \in \mathcal{R}_j$, where

$$\mathcal{R}_j = \left\{ (j_\ell)_{1 \leqslant \ell \leqslant 2\sigma+1}, \; \sum_{\ell=1}^{2\sigma+1} (-1)^{\ell+1} \kappa_{j_\ell} = \kappa_j, \; \sum_{\ell=1}^{2\sigma+1} (-1)^{\ell+1} |\kappa_{j_\ell}|^2 = |\kappa_j|^2 \right\}.$$

In the cubic case $\sigma = 1$, the set $\mathcal{R}_j$ has been analyzed in the previous subsection. One has to be cautious regarding multiplicity aspects though.

As noticed above, the one-dimensional cubic case plays a special role, and we discuss it first. In this case, we have seen

$$\mathcal{R}_j = \{ (\ell, \ell, j), \; (j, \ell, \ell), \quad \ell \in J_0 \},$$

and (2.24) boils down to (counting multiplicity)

$$\partial_t a_j + \kappa_j \partial_x a_j = -2i\lambda \sum_{\ell \in J_0} |a_\ell|^2 a_j + i\lambda |a_j|^2 a_j \; ; \quad a_{j|t=0} = a_{0j}. \quad (2.25)$$

In the same fashion as in Sec. 2.3, we note that the modulus of a_j is preserved by the transport equation,

$$\partial_t |a_j|^2 + \kappa_j \partial_x |a_j|^2 = 0,$$

and we obtain

$$|a_j(t,x)|^2 = |a_{0j}(x - t\kappa_j)|^2,$$

so $a_j(t,x) = a_{0j}(x - t\kappa_j) e^{iS_j(t,x)}$ for some (ε-independent) real-valued phase modulation S_j. Plugging this formula into (2.25) yields (after simplification by $a_{0j}(x - t\kappa_j)$)

$$(\partial_t + \kappa_j \partial_x) S_j(t,x) = -2\lambda \sum_{\ell \in J_0} |a_{0\ell}(x - t\kappa_\ell)|^2 + \lambda |a_{0j}(x - t\kappa_j)|^2,$$

hence the explicit formula

$$S_j(t,x) = -2\lambda \sum_{\ell \in J_0 \backslash \{j\}} \int_0^t |a_{0\ell}(x + (\tau - t)\kappa_j - \tau\kappa_\ell)|^2 d\tau - t\lambda |a_{0j}(x - t\kappa_j)|^2.$$

In the multidimensional case $d \geqslant 2$, one must no longer expect the modulus of *each* a_j to be preserved along the linear transport in general, in view of Lemma 2.11.

Example 2.15. In the framework of Example 2.12, supposing $a_{00} \equiv 0$,

$$\partial_t a_{0|t=0} = -2i\lambda a_{01} \bar{a}_{02} a_{03},$$

which may of course be chosen non identically zero, and so a_0 becomes instantaneously non-trivial. For higher order nonlinearities ($\sigma \geqslant 2$), we do not have such a precise description, but we emphasize that if $a_{01} = a_{02} = a_{03} =: \alpha$, then

$$\partial_t a_{0|t=0} = -i\lambda |\alpha|^{2\sigma} \alpha \, \sharp \mathcal{R}_0,$$

and we just notice that $\sharp \mathcal{R}_0 \geqslant 2$, so a_0 becomes instantaneously non-trivial.

Recall that we have seen that it takes at least three initial phases to create new ones. If we start from two modes only,

$$u^\varepsilon(0, x) = a_{01}(x) e^{i\kappa_1 \cdot x / \varepsilon} + a_{02}(x) e^{i\kappa_2 \cdot x / \varepsilon},$$

then the situation is very similar to the one-dimensional case, in the sense that the sets R_1 and R_2 are trivial,

$$\partial_t a_1 + \kappa_1 \cdot \nabla a_1 = -2i\lambda |a_2|^2 a_1 - i\lambda |a_1|^2 a_1; \quad a_{1|t=0} = a_{01},$$

$$\partial_t a_2 + \kappa_2 \cdot \nabla a_2 = -2i\lambda |a_1|^2 a_2 - i\lambda |a_2|^2 a_2; \quad a_{2|t=0} = a_{02},$$

and so we have explicit formulas

$$a_1(t, x) = a_{01}(x - t\kappa_1) e^{-i\lambda \left(2 \int_0^t |a_{02}(x + (\tau - t)\kappa_1 - \tau\kappa_2)|^2 d\tau + t|a_{01}(x - t\kappa_1)|^2 \right)},$$

$$a_2(t, x) = a_{02}(x - t\kappa_2) e^{-i\lambda \left(2 \int_0^t |a_{01}(x + (\tau - t)\kappa_2 - \tau\kappa_1)|^2 d\tau + t|a_{02}(x - t\kappa_2)|^2 \right)}.$$

On the other hand, in the general case, the *global L^2-norm* of the amplitudes is preserved, which may be seen as a remain of the L^2-conservation at the level of the cubic Schrödinger equation:

$$\frac{d}{dt} \left(\sum_{j \in J} \|a_j(t)\|_{L^2(\mathbb{R}^d)}^2 \right) = 0.$$

We emphasize that the set J involved here is the set obtained after the exhaustion of Lemma 2.11, and may be larger than the initial set J_0, as in the case of Examples 2.12 and 2.15. The main step to show the above conservation consists in noticing the identity

$$\sum_{j \in J} (\partial_t + \kappa_j \cdot \nabla) |a_j|^2 = 0,$$

which follows from Lemma 2.11 and symmetry consideration (see [Carles *et al.* (2010)] for more details).

In particular, the creation of new phases is accompanied by a transfer of mass, *in general*. This is not always the case, since for instance in Example 2.15, we may consider a_{01}, a_{02} and a_{03} with disjoint compact supports, such that the transported supports (by $\partial_t + \kappa_j \cdot \nabla$) never all meet at the same time, and so $a_4 \equiv 0$.

2.6.3 Functional setting

In the monokinetic case studied in the beginning of this chapter, it was possible (and convenient) to filter out the rapid oscillations: we changed the unknown function from u^ε to $u^\varepsilon e^{-\phi_{\mathrm{eik}}/\varepsilon}$. This turned a singular limit problem into a perturbative problem, and we could derive some energy estimates in $H^s(\mathbb{R}^d)$ uniformly in ε. In the multiphase case, filtering the rapid oscillations is more delicate, if not hopeless. It turns out that four requirements regarding the function space X in which we want to perform the analysis can be listed:

- X is translation invariant and, denoting by $\tau_k f(x) = f(x - k)$,

$$\|\tau_k f\|_X = \|f\|_X, \quad \forall k \in \mathbb{R}^d, \ \forall f \in X. \tag{2.26}$$

- X is a Banach algebra:

$$\exists C > 0, \quad \|fg\|_X \leqslant C\|f\|_X \|g\|_X, \quad \forall f, g \in X. \tag{2.27}$$

- The norm in X is not affected by the multiplication by plane wave oscillations: for $k \in \mathbb{R}^d$, denote by $e_k(x) = e^{ik \cdot x}$, then

$$\|fe_k\|_X = \|f\|_X, \quad \forall k \in \mathbb{R}^d, \ \forall f \in X. \tag{2.28}$$

- The Schrödinger group acts on X, at least locally in time:

$$\exists T_0, C > 0, \quad \|e^{i\frac{t}{2}\Delta}\|_{\mathcal{L}(X,X)} \leqslant C, \quad \forall t \in [0, T_0]. \tag{2.29}$$

All these properties but the third one are naturally satisfied on $H^s(\mathbb{R}^d)$. Recalling that the multiplication by e_k corresponds to translation on the Fourier side, and in view of Young inequality, a convenient candidate is

$$X = \mathcal{F}L^1(\mathbb{R}^d) = \left\{ f \in \mathcal{S}'(\mathbb{R}^d), \ \widehat{f} \in L^1(\mathbb{R}^d) \right\}.$$

Since $e^{i\frac{t}{2}\Delta}$ is a Fourier multiplier of modulus one, the last property is satisfied as well, with $C = 1$ and $T_0 > 0$ arbitrarily large. This space $\mathcal{F}L^1$, sometimes called Wiener algebra, was considered first in [Joly *et al.* (1994)] for high frequency limits, and has proven very useful in several situations, see also [Colin and Lannes (2009)]. In the context of nonlinear Schrödinger equations, it was used in [Carles *et al.* (2010)]. Note that any space of the form $\mathcal{F}L^1 \cap \mathcal{F}L^p$ can be considered too, as it is an $\mathcal{F}L^1$-module, and the last two requirements on X become trivial on the Fourier side. For instance, the case $X = \mathcal{F}L^1 \cap \mathcal{F}L^2 = \mathcal{F}L^1 \cap L^2$ was considered in [Carles *et al.* (2012); Mouzaoui (2013)] to treat Hartree type nonlinearities in the case of a singular potential.

In the rest of this section, we consider a space X satisfying the above properties.

Lemma 2.16. *Let $d \geqslant 1$. Suppose that $u_0^\varepsilon \in X$. Then there exists $T^\varepsilon > 0$ and a unique solution $u \in C([0, T^\varepsilon]; X)$ to Eq. (2.19) such that $u_{|t=0}^\varepsilon = u_0^\varepsilon$.*

This lemma is a straightforward consequence of the above properties of X, (2.27) and (2.29), which make it possible to use a fixed point argument on Duhamel's formula (1.27). Note that the existence time T^ε may not be uniform in ε, since we did not require a uniform bound on $\|u_0^\varepsilon\|_X$ as $\varepsilon \to 0$. In the case of initial data (2.20), the X-norm of u_0^ε is uniformly bounded, and so we may consider $T^\varepsilon = T_1$ independent of ε. However, in general we cannot claim that the solution u^ε is global in time, with the regularity required by X (in the case $X = \mathcal{F}L^1$, there is no relation with the conservation of mass or energy). The uniform boundedness of u_0^ε is ensured provided that

$$\sum_{j \in J_0} \|a_{0j}\|_X < \infty,$$

which is particular implies that J_0 is at most countable. Up to relabeling phases and amplitudes, we may replace J_0 by $\mathbb{N}$. The appearance of new modes then corresponds to the property: $a_{0j} = 0$, but $a_j(t) \neq 0$ for some $t \neq 0$, due to nonlinear interaction.

We denote by

$$Y = \left\{ (a_j)_{j \in \mathbb{N}}, \ \|a\|_Y := \sum_{j \in \mathbb{N}} \|a_j\|_X < \infty \right\} = \ell^1 X,$$

and

$$Y_2 = \left\{ (a_j)_{j \in \mathbb{N}} \in Y, \ \sum_{j \in \mathbb{N}} \left(\langle \kappa_j \rangle^2 \|a_j\|_X + \langle \kappa_j \rangle \|\nabla a_j\|_X + \|\Delta a_j\|_X \right) < \infty \right\}.$$

Note that the property $(a_{0j})_{j \in \mathbb{N}} \in Y$ ensures $u_0^\varepsilon \in X$.

Lemma 2.17. *Let $d \geqslant 1$.*
- *If $(a_{0j})_{j \in \mathbb{N}} \in Y$, then there exists $T > 0$ independent of $\varepsilon \in [0, 1]$ and a unique solution $(a_j)_{j \in \mathbb{N}} \in C([0, T]; Y)$ to the system (2.24).*
- *If in addition $(a_{0j})_{j \in \mathbb{Z}^d} \in Y_2$, then $(a_j)_{j \in \mathbb{N}} \in C([0, T]; Y_2)$.*

Proof. In the above lemma, it is implicit that new modes may appear, but their number remains at most countable, so we may replace J by $\mathbb{N}$. We remark that (2.24) may be rewritten

$$a_j(t, x) = a_j(0, x - \kappa_j t)$$
$$-i\lambda \sum_{(j_1, j_2, \ldots, j_{2\sigma+1}) \in \mathcal{R}_j} \int_0^t \left(a_{j_1} \bar{a}_{j_2} \ldots a_{j_{2\sigma+1}}\right)(s, x - \kappa_j(t - s))ds. \quad (2.30)$$

At this stage, the assumptions (2.26) and (2.27) make it easy to apply a fixed point argument, and the first point of the lemma follows.

To prove the second point, we first multiply (2.30) by $\langle \kappa_j \rangle^2 = 1 + |\kappa_j|^2$, and remark that for $(j_1, j_2, \ldots, j_{2\sigma+1}) \in \mathcal{R}_j$,

$$1 + |\kappa_j|^2 = 1 + |\kappa_{j_1}|^2 - |\kappa_{j_2}|^2 + \cdots + |\kappa_{j_{2\sigma+1}}|^2 \leqslant \langle \kappa_{j_1} \rangle^2 + \langle \kappa_{j_2} \rangle^2 + \cdots + \langle \kappa_{j_{2\sigma+1}} \rangle^2,$$

so we infer

$$\| \langle \kappa_j \rangle^2 a_j(t) \|_X \leqslant \| \langle \kappa_j \rangle^2 a_{0j} \|_X$$
$$+ C \sum_{\mathcal{R}_j} \int_0^t \prod_{\ell \neq m} \|a_{j_\ell}(s)\|_X \| \langle \kappa_{j_m} \rangle^2 a_{j_m}(s)\|_X ds,$$

where the sum also includes permutations within the $(2\sigma + 1)$-tuples $(j_1, j_2, \ldots, j_{2\sigma+1}) \in \mathcal{R}_j$. Differentiating once (2.30) with respect to the space variable, we obtain similar estimates, hence

$$\sum_{j \in \mathbb{N}} \left(\langle \kappa_j \rangle^2 \|a_j\|_{L^\infty([0,T];X)} + \langle \kappa_j \rangle \|\nabla a_j\|_{L^\infty([0,T];X)} \right) < \infty.$$

Differentiating (2.30) twice, and using the above property, the lemma follows. $\square$

2.6.4 *Error estimate*

At this stage, we have constructed an approximate solution

$$u_{\text{app}}^\varepsilon(t, x) = \sum_{j \in \mathbb{N}} a_j(t, x) e^{i\phi_j(t,x)/\varepsilon},$$

where the profiles are given by Lemma 2.17. By definition, this approximate solution solves

$$i\varepsilon \partial_t u_{\text{app}}^\varepsilon + \frac{\varepsilon^2}{2} \Delta u_{\text{app}}^\varepsilon = \lambda \varepsilon |u_{\text{app}}^\varepsilon|^2 u_{\text{app}}^\varepsilon + r_1^\varepsilon + r_2^\varepsilon, \quad (2.31)$$

where r_1^ε corresponds to the fact that we retained only resonant phases,

$$r_1^\varepsilon = \lambda \varepsilon \sum_{j \in \mathbb{N}} \sum_{(j_1, j_2, \ldots, j_{2\sigma+1}) \notin \mathcal{R}_j} a_{j_1} \bar{a}_{j_2} \ldots a_{j_{2\sigma+1}} e^{i(\phi_{j_1} - \phi_{j_2} + \cdots + \phi_{j_{2\sigma+1}})/\varepsilon},$$

and r_2^ε corresponds to the same $\mathcal{O}(\varepsilon^2)$ terms as in the linear case,

$$r_2^\varepsilon = \frac{\varepsilon^2}{2} \sum_{j\in\mathbb{N}} e^{i\phi_j/\varepsilon} \Delta a_j.$$

The error estimate $w^\varepsilon := u^\varepsilon - u_{\mathrm{app}}^\varepsilon$ thus solves

$$i\varepsilon\partial_t w^\varepsilon + \frac{\varepsilon^2}{2}\Delta w^\varepsilon = \lambda\varepsilon\left(|u^\varepsilon|^2 u^\varepsilon - |u_{\mathrm{app}}|^2 u_{\mathrm{app}}\right) - r_1^\varepsilon - r_2^\varepsilon; \quad w^\varepsilon_{|t=0} = 0.$$

Duhamel's formula reads

$$w^\varepsilon(t) = -i\lambda \int_0^t e^{i\varepsilon\frac{t-s}{2}\Delta}\left(|u^\varepsilon|^2 u^\varepsilon - |u_{\mathrm{app}}|^2 u_{\mathrm{app}}\right)(s)ds$$

$$+ i\varepsilon^{-1}\int_0^t e^{i\varepsilon\frac{t-s}{2}\Delta}r_1^\varepsilon(s)ds + i\varepsilon^{-1}\int_0^t e^{i\varepsilon\frac{t-s}{2}\Delta}r_2^\varepsilon(s)ds.$$

In view of (2.27), (2.29) and Minkowski inequality,

$$\|w^\varepsilon(t)\|_X \leqslant C\int_0^t \left(\|u_{\mathrm{app}}^\varepsilon(s)\|_X^2 + \|w^\varepsilon(s)\|_X^2\right)\|w^\varepsilon(s)\|_X ds$$

$$+ C\varepsilon^{-1}\sum_{j=1,2}\left\|\int_0^t e^{i\varepsilon\frac{t-s}{2}\Delta}r_j^\varepsilon(s)ds\right\|_X,$$

for some C independent of $\varepsilon \in [0,1]$ and $t \in [0,T]$.

In view of the second point of Lemma 2.17, we readily have

$$\sup_{t\in[0,T]}\left\|\int_0^t e^{i\varepsilon\frac{t-s}{2}\Delta}r_2^\varepsilon(s)ds\right\|_X \leqslant \sup_{t\in[0,T]}\int_0^t \left\|e^{i\varepsilon\frac{t-s}{2}\Delta}r_2^\varepsilon(s)\right\|_X ds \lesssim \varepsilon^2.$$

By construction, r_1^ε is the sum of terms of the form $g(t,x)e^{ik\cdot x/\varepsilon - \omega t/(2\varepsilon)}$, and the non-resonance property reads exactly $|k|^2 \neq \omega$.

Lemma 2.18. *Let $k \in \mathbb{R}^d$, $\omega \in \mathbb{R}$, with $|k|^2 \neq \omega$. Define*

$$D^\varepsilon(t,x) = \int_0^t e^{i\varepsilon\frac{t-s}{2}\Delta}\left(g(s,x)e^{ik\cdot x/\varepsilon - i\omega s/(2\varepsilon)}\right)ds.$$

Then we have

$$D^\varepsilon(t,x) = \frac{-2i\varepsilon}{|k|^2 - \omega}e^{i\varepsilon\frac{t-s}{2}\Delta}\left(g(s,x)e^{ik\cdot x/\varepsilon - i\omega s/(2\varepsilon)}\right)\Big|_0^t$$

$$+ \frac{2i\varepsilon}{|k|^2 - \omega}\int_0^t e^{i\varepsilon\frac{t-s}{2}\Delta}\left(e^{ik\cdot x/\varepsilon - i\omega s/(2\varepsilon)}\left(\frac{i}{2}\left(\varepsilon\Delta g + 2k\cdot\nabla g\right) + \partial_t g\right)(s,x)\right)ds.$$

In particular,

$$\|D^\varepsilon(t)\|_X \lesssim \frac{\varepsilon}{||k|^2 - \omega|}\left(\|g\|_{L^\infty([0,t];X)} + \|\Delta g\|_{L^\infty([0,t];X)} + |k|\|\nabla g\|_{L^\infty([0,t];X)}\right.$$

$$\left. + \|\partial_t g\|_{L^\infty([0,t];X)}\right).$$

Proof. The last estimate follows directly from the identity of the lemma, (2.28) and Assumption 2.29, so we only address the identity. Setting $\eta = \xi - k/\varepsilon$, the (spatial) Fourier transform of D is given by

$$\widehat{D}^\varepsilon(t,\xi) = e^{-i\varepsilon t|\eta+k/\varepsilon|^2/2} \int_0^t e^{i\varepsilon s|\eta+k/\varepsilon|^2/2} \, \hat{b}\,(s,\eta)\, e^{-i\omega s/(2\varepsilon)} ds$$

$$= e^{-i\varepsilon t|\eta+k/\varepsilon|^2/2} \int_0^t e^{is\theta/2} \, \hat{b}\,(s,\eta)\, ds$$

$$= e^{-i\varepsilon t|\eta+k/\varepsilon|^2/2} \int_0^t e^{is\theta_2/2}\, e^{is\theta_1/2} \, \hat{b}\,(s,\eta)\, ds,$$

where we have denoted

$$\theta = \varepsilon\left|\eta + \frac{k}{\varepsilon}\right|^2 - \frac{\omega}{\varepsilon} = \underbrace{\varepsilon|\eta|^2 + 2k\cdot\eta}_{\theta_1} + \underbrace{\frac{|k|^2 - \omega}{\varepsilon}}_{\theta_2}.$$

Integrate by parts, by first integrating $e^{is\theta_2/2}$:

$$e^{i\varepsilon\frac{t}{2}|\xi|^2}\widehat{D}^\varepsilon(t,\xi) = -\frac{2i}{\theta_2}e^{is\theta/2}\hat{b}\,(s,\eta)\,\Big|_0^t$$

$$+ \frac{2i}{\theta_2}\int_0^t e^{is\theta/2}\left(i\frac{\theta_1}{2}\widehat{b}\,(s,\eta) + \widehat{\partial_t b}\,(s,\eta)\right)ds.$$

The identity follows by inverting the Fourier transform. $\qquad\square$

We infer:

Proposition 2.19. *Let* $d \geqslant 1$, $\kappa_j \in \mathbb{Z}^d$, *and* $(a_{0j})_{j\in\mathbb{N}} \in Y_2$. *Then for* T *as in Lemma 2.17,*

$$\|u^\varepsilon - u_{\mathrm{app}}^\varepsilon\|_{L^\infty([0,T];X)} \lesssim \varepsilon.$$

Proof. The assumption $\kappa_j \in \mathbb{Z}^d$ ensures that we do not encounter small divisors issues: for all j, $|\kappa_j|^2 \in \mathbb{N}$, and thus in the context of Lemma 2.18, $\big||k|^2 - \omega\big| \geqslant 1$, a uniform lower bound.

First, Lemma 2.17 and (2.24) imply that we also have $(\partial_t a_j)_{j\in\mathbb{N}} \in C([0,T];Y)$. Then, in view of these properties and Lemma 2.18, we have

$$\left\|\int_0^t e^{i\varepsilon\frac{t-s}{2}\Delta}r_1^\varepsilon(s)ds\right\|_X \lesssim \varepsilon^2,$$

where we have used the fact that in the application of Lemma 2.18, $\big||k|^2 - \omega\big| \geqslant 1$, since now $k \in \mathbb{Z}^d$ and $\omega \in \mathbb{Z}$. We infer

$$\|w^\varepsilon(t)\|_X \leqslant C\int_0^t \left(\|u_{\mathrm{app}}^\varepsilon(s)\|_X^2 + \|w^\varepsilon(s)\|_X^2\right)\|w^\varepsilon(s)\|_X ds + C\varepsilon,$$

where C is independent of $\varepsilon \in [0,1]$ and $t \in [0,T]$. Lemmas 2.17 and 2.16 yield $w^\varepsilon \in C([0, \min(T, T^\varepsilon)]; X)$. Since $w^\varepsilon_{|t=0} = 0$, the above inequality and a continuity argument imply that $u^\varepsilon \in C([0,T]; X)$ provided that $\varepsilon > 0$ is sufficiently small, along with the announced error estimate.   □

Remark 2.20 (Periodic setting). *The above analysis is still available in the periodic case $x \in \mathbb{T}^d$, and is actually simpler (see [Carles et al. (2010)]). In the periodic case, the profiles a_j simply correspond to Fourier coefficients, and depend only on time (not on space).*

Chapter 3

Convergence of Quadratic Observables *via* Modulated Energy Functionals

3.1 Presentation

In this chapter, we turn to what has appeared as a supercritical case in Sec. 1.2: the case $\alpha = 0$. We consider the case of a defocusing power-like nonlinearity, with no external potential

$$i\varepsilon\partial_t u^\varepsilon + \frac{\varepsilon^2}{2}\Delta u^\varepsilon = |u^\varepsilon|^{2\sigma}u^\varepsilon \quad ; \quad u^\varepsilon_{|t=0} = a_0^\varepsilon e^{i\phi_0/\varepsilon}. \tag{3.1}$$

We discuss in Chap. 4 how to take the presence of an external potential into account. As noticed in Sec. 1.2, an aspect of supercriticality is that the cascade of equations (1.7), (1.8), etc., is not closed. However, we remark that the transport equation (1.8) reads, in this case:

$$\partial_t a + \nabla\phi \cdot \nabla a + \frac{1}{2}a\Delta\phi = -2i\sigma \operatorname{Re}\left(\bar{a}a^{(1)}\right)|a|^{2\sigma-2}a \text{ if } \alpha = 0.$$

Suppose that we know ϕ (which is not straightforward at all, since (1.7) now contains $|a|^{2\sigma}$). Then the above equation shares an interesting property with (2.5): after changing the space variable to follow the characteristics associated to $\nabla\phi$, this equation is of the form

$$\partial_t A = i\mathcal{V}A, \quad \mathcal{V} \in \mathbb{R}.$$

In particular, $\partial_t |A|^2 = 0$, yielding the identity which can be checked directly:

$$\partial_t |a|^2 + \nabla\phi \cdot \nabla|a|^2 + |a|^2\Delta\phi = 0. \tag{3.2}$$

Setting $(\rho, v) := (|a|^2, \nabla\phi)$, we see that (1.7)–(1.8) *implies*:

$$\begin{cases} \partial_t v + v \cdot \nabla v + \nabla(\rho^\sigma) = 0 \quad ; \quad v_{|t=0} = \nabla\phi_0, \\ \partial_t \rho + \operatorname{div}(\rho v) = 0 \quad ; \quad \rho_{|t=0} = |a_0|^2, \end{cases} \tag{3.3}$$

where we have naturally assumed that $a_0^\varepsilon \to a_0$ as $\varepsilon \to 0$. The above system is an *isentropic, compressible Euler equation*, where the pressure law is such that $\rho\nabla(\rho^\sigma) = \nabla p$, that is

$$p(\rho) = \frac{\sigma}{\sigma + 1}\rho^{\sigma+1}.$$

It is remarkable that this system is *quasi-linear*, while for fixed $\varepsilon > 0$, (3.1) is a *semi-linear* equation. We may say that this increase in the nonlinear aspect of the equations we consider is due to the fact that (3.1) is supercritical as far as WKB analysis is concerned.

In this chapter, we do not prove any asymptotics for the wave function u^ε. This will be done in Chap. 4. We present here an approach that makes it possible to establish the convergence of two quadratic observables which are of particular interest in Physics. We have introduced in Sec. 2.3:

Position density: $\qquad \rho^\varepsilon = |u^\varepsilon|^2,$

Current density: $\qquad J^\varepsilon = \varepsilon \operatorname{Im}(\overline{u}^\varepsilon \nabla u^\varepsilon).$

The above discussion suggests that we should have

$$\rho^\varepsilon \xrightarrow[\varepsilon \to 0]{} \rho \quad ; \quad J^\varepsilon \xrightarrow[\varepsilon \to 0]{} \rho v.$$

A rigorous proof of such convergences is a consequence of the analysis presented in this chapter.

Compare the above convergence to the result of the previous chapter. In (2.1), assume that $V \equiv 0$ and $\alpha \geqslant 1$. Then Proposition 2.5 shows that

$$\rho^\varepsilon \xrightarrow[\varepsilon \to 0]{} \rho \text{ in } L^\infty\left([-T, T]; L^1(\mathbb{R}^d)\right),$$

$$J^\varepsilon \xrightarrow[\varepsilon \to 0]{} \rho v \text{ in } L^\infty\left([-T, T]; L^1_{\text{loc}}(\mathbb{R}^d)\right),$$

where $(\rho, v) = (|a|^2, \nabla\phi_{\text{eik}})$ solves

$$\begin{cases} \partial_t v + v \cdot \nabla v = 0 & ; \quad v_{|t=0} = \nabla\phi_0, \\ \partial_t \rho + \operatorname{div}(\rho v) = 0 & ; \quad \rho_{|t=0} = |a_0|^2. \end{cases}$$

This is a pressure-less Euler system. A non-trivial pressure law appears, when the semi-classical limit is considered, in a supercritical case only.

Note that for linear problems, and the nonlinear problem of the Schrödinger–Poisson system, the convergence of the above two quadratic observables can be determined thanks to the study of the Wigner measure associated to u^ε. See e.g. [Gérard *et al.* (1997); Lions and Paul (1993)], and Sec. 3.4 below. For the nonlinear case (3.1), another tool was introduced by

Y. Brenier [Brenier (2000)], inspired by the notion of dissipative solution in fluid mechanics [Lions (1996)]. In a context very similar to (3.1), this technique has been used initially by F. Lin and P. Zhang [Lin and Zhang (2005)].

The first idea to understand the modulated energy functional is that since the rapid oscillations (at scale ε) in u^ε are expected to be described by ϕ, $u^\varepsilon e^{-i\phi/\varepsilon}$ should not be ε-oscillatory; it should even converge strongly as ε goes to zero, while u^ε does not. Recall that the energy associated to (3.1) is

$$E^\varepsilon = \frac{1}{2}\|\varepsilon\nabla u^\varepsilon\|_{L^2}^2 + \frac{1}{\sigma+1}\|u^\varepsilon\|_{L^{2\sigma+2}}^{2\sigma+2}.$$

It is therefore natural to replace the first part of the energy (*kinetic energy*) with

$$\frac{1}{2}\left\|\varepsilon\nabla\left(u^\varepsilon e^{-i\phi/\varepsilon}\right)\right\|_{L^2}^2 = \frac{1}{2}\left\|(\varepsilon\nabla - iv)u^\varepsilon\right\|_{L^2}^2.$$

To understand how to treat the second part of the energy (*potential energy*), introduce a more general notation, as in Sec. 1.4: for the equation

$$i\varepsilon\partial_t u^\varepsilon + \frac{\varepsilon^2}{2}\Delta u^\varepsilon = f\left(|u^\varepsilon|^2\right)u^\varepsilon, \tag{3.4}$$

setting

$$F(y) = \int_0^y f(\eta)d\eta,$$

the potential energy is

$$\int_{\mathbb{R}^d} F\left(|u^\varepsilon(t,x)|^2\right)dx.$$

The above discussion shows that $|u^\varepsilon|^2$ is expected to be well approximated by ρ, where (ρ, v) solves an Euler equation whose pressure law is related to f. The Taylor expansion of the potential energy yields

$$F\left(|u^\varepsilon|^2\right) = F(\rho) + \left(|u^\varepsilon|^2 - \rho\right)f(\rho) + \mathcal{O}\left(\left(|u^\varepsilon|^2 - \rho\right)^2\right).$$

In the modulated energy functional, we subtract the first two terms of this Taylor expansion. This leads to the definition:

$$\begin{aligned}
H^\varepsilon(t) &= \frac{1}{2}\left\|(\varepsilon\nabla - iv)u^\varepsilon(t)\right\|_{L^2}^2 \\
&+ \int_{\mathbb{R}^d}\left(F\left(|u^\varepsilon|^2\right) - F(\rho) - \left(|u^\varepsilon|^2 - \rho\right)f(\rho)\right)(t,x)dx.
\end{aligned} \tag{3.5}$$

We recover the form suggested in Remark 1, (2) in [Lin and Zhang (2005)].

Note that the energy functional E^ε is non-negative because we consider a defocusing nonlinearity. We will see that thanks to convexity arguments, the modulated energy functional H^ε is non-negative as well. If instead of (3.1), we considered its focusing counterpart

$$i\varepsilon\partial_t u^\varepsilon + \frac{\varepsilon^2}{2}\Delta u^\varepsilon = -|u^\varepsilon|^{2\sigma}u^\varepsilon \quad ; \quad u^\varepsilon_{|t=0} = a_0^\varepsilon e^{i\phi_0/\varepsilon},$$

then the energy functionals would not be signed any more, and the energy estimates presented below would fail. We refer to Sec. 4.5 for the case of focusing nonlinearities.

3.2   Formal computation

Let (ρ, v) solve (3.3) on some time interval $[-T, T]$. Introduce the hydrodynamic variables:

$$\rho^\varepsilon = |u^\varepsilon|^2 \quad ; \quad J^\varepsilon = \operatorname{Im}\left(\varepsilon\overline{u}^\varepsilon\nabla u^\varepsilon\right).$$

For $y \geqslant 0$, denote

$$f(y) = y^\sigma \quad ; \quad F(y) = \int_0^y f(z)dz = \frac{1}{\sigma+1}y^{\sigma+1} \quad ;$$

$$G(y) = \int_0^y zf'(z)dz = yf(y) - F(y) = \frac{\sigma}{\sigma+1}y^{\sigma+1}.$$

We check that $(\rho^\varepsilon, J^\varepsilon)$ satisfies, for $\sigma \geqslant 1$:

$$\begin{cases} \partial_t\rho^\varepsilon + \operatorname{div} J^\varepsilon = 0. \\ \partial_t J_j^\varepsilon + \dfrac{\varepsilon^2}{4}\displaystyle\sum_k \partial_k\left(4\operatorname{Re}\partial_j\overline{u}^\varepsilon\partial_k u^\varepsilon - \partial_{jk}^2\rho^\varepsilon\right) + \partial_j G(\rho^\varepsilon) = 0. \end{cases} \tag{3.6}$$

Split the modulated energy functional (3.5) into two parts, and denote

$$K^\varepsilon(t) = \frac{1}{2}\int_{\mathbb{R}^d}|(\varepsilon\nabla - iv(t,x))\,u^\varepsilon(t,x)|^2\,dx$$

$$= \frac{1}{2}\int|\varepsilon\nabla u^\varepsilon|^2 + \frac{1}{2}\int|v|^2\,|u^\varepsilon|^2 - \int J^\varepsilon \cdot v.$$

Using the conservation of energy, we find

$$\frac{d}{dt}K^\varepsilon = -\frac{d}{dt}\int F(\rho^\varepsilon) + \frac{1}{2}\int|v|^2\partial_t\rho^\varepsilon + \int \rho^\varepsilon v\cdot\partial_t v$$

$$- \int J^\varepsilon\cdot\partial_t v - \int v\cdot\partial_t J^\varepsilon.$$

The first term on the right-hand side is compensated by the first term of the potential energy in the modulated energy functional. In view of (3.6), the second term of the right-hand side is controlled by an integration by parts. The third and fourth are controlled directly, but the last term is not. The advantage of the modulated energy functional is to exactly cancel out this term: integrations by parts, which are studied in more detail below, yield

$$\frac{d}{dt} H^\varepsilon(t) = \mathcal{O}\left(K^\varepsilon + \varepsilon^2\right) - \int_{\mathbb{R}^d} \left(G(\rho^\varepsilon) - G(\rho) - (\rho^\varepsilon - \rho)G'(\rho)\right) \operatorname{div} v \, dx.$$

At first glance, we cannot apply Gronwall lemma directly yet: the right-hand side does not involve H^ε. However, we get by thanks to convexity arguments. We check that there exists $c > 0$ such that

$$H^\varepsilon(t) \geqslant K^\varepsilon(t) + c \int_{\mathbb{R}^d} (\rho^\varepsilon - \rho)^2 \left((\rho^\varepsilon)^{\sigma-1} + \rho^{\sigma-1}\right) dx.$$

On the other hand, we have

$$|G(\rho^\varepsilon) - G(\rho) - (\rho^\varepsilon - \rho)G'(\rho)| \leqslant C(\rho^\varepsilon - \rho)^2 \left((\rho^\varepsilon)^{\sigma-1} + \rho^{\sigma-1}\right).$$

Setting

$$\widetilde{H}^\varepsilon(t) = K^\varepsilon(t) + c \int_{\mathbb{R}^d} (\rho^\varepsilon - \rho)^2 \left((\rho^\varepsilon)^{\sigma-1} + \rho^{\sigma-1}\right) dx,$$

we have therefore:

$$\widetilde{H}^\varepsilon(t) \leqslant \widetilde{H}^\varepsilon(0) + C \int_0^t \left(\widetilde{H}^\varepsilon(s) + \varepsilon^2\right) ds.$$

We check that if $a_0^\varepsilon = a_0 + \mathcal{O}(\varepsilon)$,

$$\widetilde{H}^\varepsilon(0) = \frac{1}{2} \|\varepsilon \nabla a_0^\varepsilon\|_{L^2}^2 + c \int_{\mathbb{R}^d} (|a_0^\varepsilon|^2 - |a_0|^2)^2 \left(|a_0^\varepsilon|^{2\sigma-2} + |a_0|^{2\sigma-2}\right) dx$$
$$= \mathcal{O}(\varepsilon^2).$$

We infer by Gronwall lemma that $\widetilde{H}^\varepsilon(t) = \mathcal{O}(\varepsilon^2)$ so long as it is defined, which can be rewritten:

$$\|(\varepsilon\nabla - iv)u^\varepsilon\|_{L^\infty([-T,T];L^2)}^2 + \left\|(\rho^\varepsilon - \rho)^2 \left((\rho^\varepsilon)^{\sigma-1} + \rho^{\sigma-1}\right)\right\|_{L^\infty([-T,T];L^1)}$$
$$= \mathcal{O}(\varepsilon^2).$$

This suffices to establish the strong convergence of position and current densities.

3.3 Justification

Two aspects have to be pointed out in view of a rigorous justification of the above computations. First, we have to make sure that the limiting Euler equation possesses a unique, sufficiently smooth, solution. We will see that this is rather straightforward in the cubic case $\sigma = 1$, but demands a non-trivial argument when $\sigma \geqslant 2$. Next, the above integrations by part require sufficient regularity on the solution u^ε. Even if we know from Proposition 1.25 that u^ε remains smooth on some time interval $] - T_-^\varepsilon, T_+^\varepsilon [$, it may happen that T_-^ε or T_+^ε goes to zero as $\varepsilon \to 0$. In general, we only know that u^ε exists as a global weak solution, from Proposition 1.28.

3.3.1 *The Cauchy problem for* (3.3)

For a more general nonlinearity like in (3.4), the same arguments as above lead to the following Euler equation:

$$\begin{cases} \partial_t v + v \cdot \nabla v + \nabla \left(f\left(\rho \right) \right) = 0 & ; \quad v_{|t=0} = \nabla \phi_0, \\ \partial_t \rho + \operatorname{div}\left(\rho v \right) = 0 & ; \quad \rho_{|t=0} = |a_0|^2. \end{cases} \tag{3.7}$$

When $f' > 0$, this system enters the framework of symmetric, hyperbolic, quasi-linear equations. It is classical that if the initial data are in $H^s(\mathbb{R}^d)$ for some $s > d/2 + 1$, then there exists $T > 0$ such that (3.7) has a unique solution $(\rho, v) \in C([-T, T]; H^s)^2$. We refer for instance to [Majda (1984)] or [Taylor (1997)]. Moreover, tame estimates show that the time of existence $T > 0$ can be chosen independent of $s > d/2 + 1$.

In the homogeneous case $f(y) = y^\sigma$, $\sigma \in \mathbb{N} \setminus \{0\}$, which we have in mind, $f' > 0$ only in the cubic case $\sigma = 1$. For $\sigma \geqslant 2$, f' possesses zeroes, and this causes a lack of (strict) hyperbolicity in (3.7). This corresponds to the presence of vacuum in fluid dynamics. From the analytical point of view, such lack of hyperbolicity may cause a loss of regularity in energy estimates; see e.g. [Cicognani and Colombini (2006a,b)] and references therein. We overcome this issue, and in view of Chap. 4, we prove:

Lemma 3.1. *Let $\sigma \in \mathbb{N} \setminus \{0\}$, $s > d/2 + 1$, $\phi_0 \in H^{s+1}$ and $a_0 \in H^s$. Consider*

$$\begin{cases} \partial_t v + v \cdot \nabla v + \nabla \left(|a|^{2\sigma} \right) = 0 & ; \quad v_{|t=0} = \nabla \phi_0. \\ \partial_t a + v \cdot \nabla a + \dfrac{1}{2} a \operatorname{div} v = 0 & ; \quad a_{|t=0} = a_0. \end{cases} \tag{3.8}$$

There exists $T_-, T_+ > 0$ such that Eq. (3.8) has a unique maximal solution $(v, a) \in C(] - T_-, T_+[; H^s \times H^{s-1})$. It is maximal in the sense that if, say,

$T_+ < \infty$, *then*

$$\|(v,a)(t)\|_{W^{1,\infty}(\mathbb{R}^d)} \underset{t \to T_+}{\longrightarrow} \infty.$$

In addition, if $\phi_0, a_0 \in H^\infty$, *then* $v, a \in C^\infty(]-T_-, T_+[; H^\infty)$. *Setting* $\rho = |a|^2$, *we infer that* (3.3) *has a unique solution* $(v, \rho) \in C(]-T_-, T_+[; H^s \times H^{s-1})$.

Remark 3.2. This shows that the possible loss of regularity due to the lack of hyperbolicity for Eq. (3.3) remains limited.

Remark 3.3. The backward and forward lifespans T_- and T_+ are finite for all compactly supported initial data, as shown in [Makino *et al.* (1986)] (see also [Chemin (1990)]): singularities appear in finite time.

Proof. Adapting the idea of [Makino *et al.* (1986)], consider the unknown $(v, u) = (v, a^\sigma)$. Even though the map $a \mapsto a^\sigma$ is not bijective, this will suffice to prove the lemma. The pair (v, u) solves:

$$\begin{cases} \partial_t v + v \cdot \nabla v + \nabla\left(|u|^2\right) = 0 & ; \quad v_{|t=0} = \nabla \phi_0 \in H^s(\mathbb{R}^d), \\ \partial_t u + v \cdot \nabla u + \dfrac{\sigma}{2} u \operatorname{div} v = 0 & ; \quad u_{|t=0} = a_0^\sigma \in H^s(\mathbb{R}^d). \end{cases} \tag{3.9}$$

This system is hyperbolic symmetric, with a constant symmetrizer. Therefore, there exist $T_-, T_+ > 0$ and a unique maximal solution $(v, u) \in C(]-T_-, T_+[; H^s)^2$. The notion of maximality follows from Moser's inequality (Lemma 1.24). Now that v is known, we define a as the solution of the linear transport equation

$$\partial_t a + v \cdot \nabla a + \frac{1}{2} a \operatorname{div} v = 0 \quad ; \quad a_{|t=0} = a_0.$$

The function a has the regularity announced in Lemma 3.1. We check that a^σ solves the second equation in (3.9). Since v is a smooth coefficient, by uniqueness for this linear equation, we have $u = a^\sigma$. Therefore, (v, a) solves Eq. (3.8). Note that the local existence times T_-, T_+ may be chosen independent of $s > d/2 + 1$, thanks to tame estimates. $\qquad \square$

3.3.2 *Rigorous estimates for the modulated energy*

Fix $T > 0$ such that $T < \min(T_-, T_+)$, where T_- and T_+ are given by Lemma 3.1. We stop counting the derivatives, and assume that the initial data are in H^∞:

Theorem 3.4. *Let* $d \geqslant 1$, $\sigma \geqslant 1$ *be an integer, and* $\phi_0, a_0 \in H^\infty$. *Let* $(v, \rho) \in C([-T, T]; H^\infty)^2$ *given by Lemma 3.1. Assume that there exists*

$s > d/2$ such that

$$\|a_0^\varepsilon - a_0\|_{H^s} = \mathcal{O}(\varepsilon).$$

Denote $\rho^\varepsilon = |u^\varepsilon|^2$. Then we have the following estimate:

$$\|(\varepsilon\nabla - iv)u^\varepsilon\|^2_{L^\infty([-T,T];L^2)} + \left\|(\rho^\varepsilon - \rho)^2\left((\rho^\varepsilon)^{\sigma-1} + \rho^{\sigma-1}\right)\right\|_{L^\infty([-T,T];L^1)}$$
$$= \mathcal{O}(\varepsilon^2).$$

Remark 3.5. The above quantities are well-defined for weak solutions, so the above quantities are well-defined, since u^ε is (at least) a global weak solution.

Proof.   Recall that in general, the integrations by parts mentioned in Sec. 3.2 do not make sense for all $t \in [-T, T]$, since we consider weak solutions only. To make the above approach rigorous, we work on a sequence of global strong solutions, converging to a weak solution. For $(\delta_m)_m$ a sequence of positive numbers going to zero, introduce the saturated nonlinearity, defined for $y \geqslant 0$:

$$f_m(y) = \frac{y^\sigma}{1 + (\delta_m y)^\sigma}.$$

Note that f_m is a symbol of degree 0. For fixed m and $\varepsilon > 0$, we have a global mild solution $u_m^\varepsilon \in C(\mathbb{R}; H^1)$ to:

$$i\varepsilon\partial_t u_m^\varepsilon + \frac{\varepsilon^2}{2}\Delta u_m^\varepsilon = f_m\left(|u_m^\varepsilon|^2\right)u_m^\varepsilon \quad ; \quad u_m^\varepsilon(0,x) = a_0^\varepsilon(x)e^{i\phi_0(x)/\varepsilon}. \quad (3.10)$$

As $m \to \infty$, the sequence $(u_m^\varepsilon)_m$ converges to a weak solution of (3.1) (see [Ginibre and Velo (1985a); Lebeau (2005)]). For $y \geqslant 0$, introduce also

$$F_m(y) = \int_0^y f_m(z)dz \quad ; \quad G_m(y) = \int_0^y zf_m'(z)dz = yf_m(y) - F_m(y).$$

The mass and energy associated to u_m^ε are conserved:

$$M_m^\varepsilon(t) = \int |u_m^\varepsilon(t,x)|^2 dx \equiv \|a_0\|_{L^2}^2.$$

$$E_m^\varepsilon(t) = \frac{1}{2}\|\varepsilon\nabla u_m^\varepsilon(t)\|_{L^2}^2 + \int_{\mathbb{R}^d} F_m\left(|u_m^\varepsilon(t,x)|^2\right)dx \equiv E_m^\varepsilon(0).$$

Moreover, the solution is in $H^2(\mathbb{R}^d)$ for all time: $u_m^\varepsilon \in C(\mathbb{R}; H^2)$. To see this, we use an idea due to T. Kato [Kato (1987, 1989)], and consider $\partial_t u_m^\varepsilon$. It solves

$$\left(i\varepsilon\partial_t + \frac{\varepsilon^2}{2}\Delta\right)\partial_t u_m^\varepsilon = 2f_m'\left(|u_m^\varepsilon|^2\right)\mathrm{Re}\left(\overline{u_m^\varepsilon}\partial_t u_m^\varepsilon\right)u_m^\varepsilon + f_m\left(|u_m^\varepsilon|^2\right)\partial_t u_m^\varepsilon.$$

To compute the initial data for $\partial_t u_m^\varepsilon$, we use (3.10):

$$\partial_t u_m^\varepsilon{}_{|t=0} = \frac{i}{\varepsilon}\left(\frac{\varepsilon^2}{2}\Delta u_m^\varepsilon - f_m\left(|u_m^\varepsilon|^2\right)u_m^\varepsilon\right)\Big|_{t=0} \in H^\infty.$$

Note that since f_m is a symbol of degree 0, there exists $C_m > 0$ independent of u_m^ε such that

$$|u_m^\varepsilon|^2 f_m'\left(|u_m^\varepsilon|^2\right) + f_m\left(|u_m^\varepsilon|^2\right) \leqslant C_m.$$

Energy estimates (see Lemma 1.2) then show that $\partial_t u_m^\varepsilon \in C(\mathbb{R}; L^2)$. Using (3.10) and the boundedness of f_m, we infer $\Delta u_m^\varepsilon \in C(\mathbb{R}; L^2)$.

We consider the hydrodynamic variables:

$$\rho_m^\varepsilon = |u_m^\varepsilon|^2 \quad ; \quad J_m^\varepsilon = \mathrm{Im}\left(\varepsilon \overline{u}_m^\varepsilon \nabla u_m^\varepsilon\right).$$

From the above discussion, we have:

$$\rho_m^\varepsilon(t) \in W^{2,1}(\mathbb{R}^d) \text{ and } J_m^\varepsilon(t) \in W^{1,1}(\mathbb{R}^d), \quad \forall t \in \mathbb{R}. \tag{3.11}$$

The analogue of (3.6) is:

$$\begin{cases} \partial_t \rho_m^\varepsilon + \mathrm{div}\, J_m^\varepsilon = 0. \\ \partial_t (J_m^\varepsilon)_j + \dfrac{\varepsilon^2}{4}\sum_k \partial_k (4\,\mathrm{Re}\,\partial_j \overline{u}_m^\varepsilon \partial_k u_m^\varepsilon - \partial_{jk}^2 \rho_m^\varepsilon) + \partial_j G_m(\rho_m^\varepsilon) = 0. \end{cases} \tag{3.12}$$

Introduce the modulated energy functional "adapted to (3.10)":

$$H_m^\varepsilon(t) = \frac{1}{2}\int_{\mathbb{R}^d} |(\varepsilon\nabla - iv)u_m^\varepsilon|^2\, dx$$

$$+ \int_{\mathbb{R}^d} \left(F_m(\rho_m^\varepsilon) - F_m(\rho) - (\rho_m^\varepsilon - \rho)f_m(\rho)\right) dx.$$

Notice that this functional is not exactly adapted to (3.10), since the limiting quantities (as $\varepsilon \to 0$) ρ and v are constructed with the nonlinearity f and not the nonlinearity f_m. We also distinguish the kinetic part:

$$K_m^\varepsilon(t) = \frac{1}{2}\int_{\mathbb{R}^d} |(\varepsilon\nabla - iv)u_m^\varepsilon|^2\, dx.$$

Thanks to the conservation of energy for u_m^ε, we have:

$$\frac{d}{dt}K_m^\varepsilon = -\frac{d}{dt}\int F_m(\rho_m^\varepsilon) + \frac{1}{2}\int |v|^2 \partial_t \rho_m^\varepsilon + \int \rho_m^\varepsilon v \cdot \partial_t v$$

$$- \int J_m^\varepsilon \cdot \partial_t v - \int v \cdot \partial_t J_m^\varepsilon.$$

Using Lemma 3.1, Eqs. (3.11) and (3.12), (licit) integrations by parts yield:

$$\frac{d}{dt} K_m^\varepsilon = -\frac{d}{dt} \int F_m(\rho_m^\varepsilon) - \frac{1}{2} \int |v|^2 \operatorname{div} J_m^\varepsilon - \sum_{j,k} \rho_m^\varepsilon v_j v_k \partial_j v_k$$

$$- \int \rho_m^\varepsilon \nabla f(\rho) \cdot v + \int (v \cdot \nabla v) \cdot J_m^\varepsilon + \int \nabla f(\rho) \cdot J_m^\varepsilon$$

$$- \sum_{j,k} \int \partial_k v_j \operatorname{Re} \left(\varepsilon \partial_j \overline{u}_m^\varepsilon \varepsilon \partial_k u_m^\varepsilon \right) - \frac{\varepsilon^2}{4} \int \nabla \left(\operatorname{div} v \right) \cdot \nabla \rho_m^\varepsilon + \int \rho_m^\varepsilon v \cdot \nabla f_m(\rho^\varepsilon).$$

Proceeding as in [Lin and Zhang (2005)], we have:

$$\varepsilon^2 \int \operatorname{div}(\nabla v) \cdot \nabla \rho_m^\varepsilon = \varepsilon \int \operatorname{div}(\nabla v) \cdot (\overline{u}_m^\varepsilon \varepsilon \nabla u_m^\varepsilon + u_m^\varepsilon \varepsilon \nabla \overline{u}_m^\varepsilon)$$

$$= \varepsilon \int \operatorname{div}(\nabla v) \cdot \left(\overline{u}_m^\varepsilon (\varepsilon \nabla - iv) u_m^\varepsilon + u_m^\varepsilon \overline{(\varepsilon \nabla - iv) u^\varepsilon}_m \right)$$

$$= \mathcal{O} \left(K_m^\varepsilon + \varepsilon^2 \right),$$

where we have used the conservation of mass and Young inequality

$$ab \leqslant \frac{1}{2} \left(a^2 + b^2 \right), \quad \forall a, b \geqslant 0.$$

From now on, we use the convention that the constant associated to the notation $\mathcal{O}$ is independent of m and ε. Treating the term involving $\partial_k v_j \operatorname{Re} \left(\varepsilon \partial_j \overline{u}_m^\varepsilon \varepsilon \partial_k u_m^\varepsilon \right)$ in a similar fashion, simplifications yield:

$$\frac{d}{dt} K_m^\varepsilon = \mathcal{O} \left(K_m^\varepsilon + \varepsilon^2 \right) - \frac{d}{dt} \int F_m(\rho_m^\varepsilon) + \int \nabla f_m(\rho_m^\varepsilon) \rho_m^\varepsilon v$$

$$- \int \nabla f(\rho) \cdot (\rho_m^\varepsilon v - J_m^\varepsilon).$$

Similar computations for $H_m^\varepsilon - K_m^\varepsilon$ yield:

$$\frac{d}{dt} H_m^\varepsilon = \mathcal{O} \left(K_m^\varepsilon + \varepsilon^2 \right) - \int (G_m(\rho_m^\varepsilon) - G_m(\rho) - (\rho_m^\varepsilon - \rho) G_m'(\rho)) \operatorname{div} v$$

$$+ \int \nabla (f(\rho) - f_m(\rho)) \cdot (J_m^\varepsilon - \rho_m^\varepsilon v).$$

Note that $f(\rho) - f_m(\rho) \to 0$ in $L^\infty([0, T]; W^{1,\infty})$ as $m \to \infty$. We can thus write:

$$\frac{d}{dt} H_m^\varepsilon = \mathcal{O} \left(K_m^\varepsilon + \varepsilon^2 \right) + o_{m \to \infty}(1)$$

$$- \int (G_m(\rho_m^\varepsilon) - G_m(\rho) - (\rho_m^\varepsilon - \rho) G_m'(\rho)) \operatorname{div} v. \tag{3.13}$$

We check that there exists C independent of m such that

$$|G_m(\rho_m^\varepsilon) - G_m(\rho) - (\rho_m^\varepsilon - \rho) G_m'(\rho)| \leqslant C(\rho_m^\varepsilon - \rho)^2 \left(\theta_m(\rho_m^\varepsilon) + \theta_m(\rho) \right),$$

where we have set, for $y \geqslant 0$,

$$\theta(y) = \frac{y^{\sigma-1}}{1+y^\sigma} \quad ; \quad \theta_m(y) = \frac{y^{\sigma-1}}{1+(\delta_m y)^\sigma}.$$

Lemma 3.6. *There exists $K > 0$ independent of m such that for all $\rho', \rho \geqslant 0$,*

$$|G_m(\rho') - G_m(\rho) - (\rho' - \rho)G_m'(\rho)| \leqslant K\,|F_m(\rho') - F_m(\rho) - (\rho' - \rho)F_m'(\rho)|\,.$$

Proof. This lemma stems from Taylor formula with an integral remainder, and the identities

$$F_m''(y) = f_m'(y) \quad ; \quad G_m''(y) = f_m'(y) + y f_m''(y).$$

Setting, for $y \geqslant 0$, $h(y) = y^\sigma/(1+y^\sigma)$ we have:

$$f_m'(y) = \delta_m^{1-\sigma} h'(\delta_m y) \quad ; \quad f_m''(y) = \delta_m^{2-\sigma} h''(\delta_m y).$$

Moreover,

$$h'(y) = \frac{\sigma y^{\sigma-1}}{(1+y^\sigma)^2} \geqslant 0.$$

Therefore, to prove the lemma, it suffices to note that

$$y h''(y) = h'(y) \times \frac{\sigma - 1 - (\sigma+1)y^\sigma}{1+y^\sigma},$$

hence for all $y \geqslant 0$,

$$|y h''(y)| \leqslant C h'(y). \qquad \square$$

The lemma implies

$$\frac{d}{dt} H_m^\varepsilon \leqslant C\left(H_m^\varepsilon + \varepsilon^2\right) + o_{m\to\infty}(1),$$

for some C independent of m. Using Gronwall lemma, we infer

$$\sup_{t\in[-T,T]} H_m^\varepsilon(t) \leqslant C\varepsilon^2 + o_{m\to\infty}(1),$$

for some constant C independent of m. Letting $m \to \infty$, Fatou's lemma yields

$$\|(\varepsilon\nabla - iv)\,u^\varepsilon(t)\|_{L^2}^2 + \int_{\mathbb{R}^d} (\rho^\varepsilon - \rho)^2 \left((\rho^\varepsilon)^{\sigma-1} + \rho^{\sigma-1}\right) dx = \mathcal{O}\left(\varepsilon^2\right),$$

uniformly for $t \in [-T, T]$. This completes the proof of Theorem 3.4. $\qquad \square$

3.4   Convergence of quadratic observables

We infer the convergence of quadratic observables from Theorem 3.4. Recall a few basic facts about the Wigner measures. For $(x, \xi) \in \mathbb{R}^d \times \mathbb{R}^d$, the Wigner transform of $u^\varepsilon \in L^\infty_t L^2_x$ is defined by

$$w^\varepsilon(t, x, \xi) = (2\pi)^{-d} \int_{\mathbb{R}^d} u^\varepsilon \left(t, x - \varepsilon \frac{\eta}{2}\right) \overline{u}^\varepsilon \left(t, x + \varepsilon \frac{\eta}{2}\right) e^{i\eta \cdot \xi} d\eta.$$

The position and current densities can be recovered from w^ε, by

$$\rho^\varepsilon(t, x) = |u^\varepsilon(t, x)|^2 = \int_{\mathbb{R}^d} w^\varepsilon(t, x, \xi) d\xi,$$

$$J^\varepsilon(t, x) = \operatorname{Im} \left(\varepsilon \overline{u}^\varepsilon \nabla u^\varepsilon\right)(t, x) = \int_{\mathbb{R}^d} \xi w^\varepsilon(t, x, \xi) d\xi.$$

A measure μ is a Wigner measure associated to u^ε (there is no uniqueness in general) if, up to extracting a subsequence, w^ε converges to μ as $\varepsilon \to 0$. Note that μ is a non-negative measure on the phase space. We refer to [Burq (1997)] and references therein for various results on Wigner measures, as well as applications to several problems.

Recall that if X and Y are two Banach spaces, $X + Y$ is equipped with the norm

$$\|u\|_{X+Y} = \inf \left\{\|u_1\|_X + \|u_2\|_Y \quad ; \quad u = u_1 + u_2, \ u_1 \in X, \ u_2 \in Y\right\}.$$

Corollary 3.7. *Under the assumptions of Theorem 3.4, the position and current densities converge strongly on $[-T, T]$ as $\varepsilon \to 0$:*

$$|u^\varepsilon|^2 \xrightarrow[\varepsilon \to 0]{} |a|^2 \qquad \qquad \text{in } C\left([-T, T]; L^{\sigma+1}(\mathbb{R}^d)\right).$$

$$\operatorname{Im}\left(\varepsilon \overline{u}^\varepsilon \nabla u^\varepsilon\right) \xrightarrow[\varepsilon \to 0]{} |a|^2 v \qquad \text{in } C\left([-T, T]; L^{\sigma+1}(\mathbb{R}^d) + L^1(\mathbb{R}^d)\right).$$

In particular, there is only one Wigner measure associated to $(u^\varepsilon)_\varepsilon$, and it is given by

$$\mu(t, dx, d\xi) = |a(t, x)|^2 dx \otimes \delta\left(\xi - v(t, x)\right).$$

Remark 3.8. The above convergences imply the following local L^1 convergences:

$$|u^\varepsilon|^2 \xrightarrow[\varepsilon \to 0]{} |a|^2 \qquad \qquad \text{in } C\left([-T, T]; L^1(|x| \leqslant R)\right),$$

$$\operatorname{Im}\left(\varepsilon \overline{u}^\varepsilon \nabla u^\varepsilon\right) \xrightarrow[\varepsilon \to 0]{} |a|^2 v \qquad \text{in } C\left([-T, T]; L^1(|x| \leqslant R)\right), \quad \forall R \geqslant 0.$$

Proof. The second part of the estimate in Theorem 3.4 yields

$$\sup_{t \in [-T,T]} \int_{\mathbb{R}^d} \left(|u^\varepsilon(t,x)|^2 - |a(t,x)|^2 \right)^2 \left(|u^\varepsilon(t,x)|^{2\sigma-2} + |a(t,x)|^{2\sigma-2} \right)^2 dx$$

$$= \mathcal{O}\left(\varepsilon^2 \right).$$

Therefore, since there exists C_σ such that

$$|\alpha - \beta|^{\sigma+1} \leqslant C_\sigma (\alpha - \beta)^2 \left(\alpha^{\sigma-1} + \beta^{\sigma-1} \right), \quad \forall \alpha, \beta \geqslant 0,$$

we infer:

$$\sup_{t \in [T,T]} \int_{\mathbb{R}^d} \left| |u^\varepsilon(t,x)|^2 - |a(t,x)|^2 \right|^{\sigma+1} dx = \mathcal{O}\left(\varepsilon^2 \right).$$

This yields the first part of the corollary, along with a bound on the rate of convergence as $\varepsilon \to 0$. For the current density, write

$$\mathrm{Im}\left(\varepsilon \overline{u}^\varepsilon \nabla u^\varepsilon \right) = \mathrm{Im}\left(\overline{u}^\varepsilon \left(\varepsilon \nabla - iv \right) u^\varepsilon \right) + |u^\varepsilon|^2 v.$$

Since $v \in L^\infty([-T,T] \times \mathbb{R}^d)$, we have

$$|u^\varepsilon|^2 v \xrightarrow[\varepsilon \to 0]{} |a|^2 v \quad \text{in } C([-T,T]; L^{\sigma+1}).$$

On the other hand, Cauchy–Schwarz inequality and Theorem 3.4 yield

$$\mathrm{Im}\left(\overline{u}^\varepsilon \left(\varepsilon \nabla - iv \right) u^\varepsilon \right) = \mathcal{O}(\varepsilon) \quad \text{in } C([-T,T]; L^1).$$

This completes the proof of the corollary. $\qquad \square$

To conclude this paragraph, we point out a phenomenon which is typical of the supercritical WKB régime, and which is the key point at the origin of the instability mechanisms presented in Chap. 4.5. Suppose that no rapid oscillation is present in the initial datum for u^ε:

$$\phi_0 = 0.$$

Then $v_{|t=0} = 0$ in (3.8). The equation for v at time $t = 0$ then yields:

$$\partial_t v_{|t=0} = -\nabla \left(|a_0|^{2\sigma} \right).$$

Thus, at least for $t > 0$ independent of ε but sufficiently small,

$$v(t, \cdot) = -t \nabla \left(|a_0|^{2\sigma} \right) + \mathcal{O}\left(t^2 \right).$$

Note also that for a non-trivial $a_0 \in L^2(\mathbb{R}^d)$, $\nabla \left(|a_0|^{2\sigma} \right) \neq 0$. Therefore, even if no rapid oscillation is present initially, u^ε becomes *instantaneously* ε-oscillatory.

Chapter 4

Pointwise Description of the Wave Function

In the previous chapter, we have established the convergence of quadratic observables thanks to a modulated energy functional, in the case of a defocusing nonlinearity, with no external potential. In this chapter, we study the asymptotic behavior of the wave function itself, as $\varepsilon \to 0$, in the same supercritical WKB régime: we consider

$$i\varepsilon\partial_t u^\varepsilon + \frac{\varepsilon^2}{2}\Delta u^\varepsilon = V u^\varepsilon + f\left(|u^\varepsilon|^2\right) u^\varepsilon \quad ; \quad u^\varepsilon_{|t=0} = a_0^\varepsilon e^{i\phi_0/\varepsilon}. \tag{4.1}$$

Recall that the standard WKB approach meets the problem of closing the cascade of equations. This problem was eluded in the previous section, since we have noticed that the system relating the phase ϕ (or equivalently, its gradient v) and the *modulus* of the leading order profile a is closed (Euler equation). To study the wave function u^ε itself, this does not suffice: we will see in particular that $\mathcal{O}(\varepsilon)$ perturbations of the initial profile a_0^ε affect the wave function u^ε at *leading order*, through a term of modulus one (phase modulation).

We first discuss several possibilities to adapt the WKB method to this case. In particular, we explain why the case of a focusing nonlinearity is extremely different (Sec. 4.5). For instance, the framework of Sobolev spaces is not well adapted to this case. For defocusing nonlinearities, we provide a pointwise description of u^ε as $\varepsilon \to 0$ in two cases:

- Defocusing nonlinearity, which is *cubic at the origin*: $f' > 0$.
- Smooth, *homogeneous*, defocusing nonlinearity: $f(y) = y^\sigma$, $\sigma \in \mathbb{N}$.

4.1   Several possible approaches

To overcome the absence of closure in the regular WKB analysis, a possibility consists in trying to write the exact solution u^ε as the product of an amplitude and of a rapidly oscillatory factor:

$$u^\varepsilon = a^\varepsilon e^{i\Phi^\varepsilon/\varepsilon}, \qquad (4.2)$$

where a^ε and Φ^ε depend on ε. If one can construct a^ε and Φ^ε such that u^ε can be represented as above, then we recover a WKB-like expansion as soon as a^ε and Φ^ε have asymptotic expansions as $\varepsilon \to 0$.

A standard approach is to assume that a_0^ε is real-valued, and to seek a real-valued amplitude a^ε; see e.g. [Landau and Lifschitz (1967)]. The phase Φ^ε is real-valued too. Plugging (4.2) into Eq. (4.1), and separating real and imaginary parts, we find (adding tildes to avoid confusions):

$$\begin{cases} \widetilde{a}^\varepsilon \left(\partial_t \widetilde{\Phi}^\varepsilon + \dfrac{1}{2}|\nabla\widetilde{\Phi}^\varepsilon|^2 + V + f\left(|\widetilde{a}^\varepsilon|^2\right) \right) = \dfrac{\varepsilon^2}{2}\Delta\widetilde{a}^\varepsilon \quad ; \quad \widetilde{\Phi}^\varepsilon_{|t=0} = \phi_0, \\[2mm] \partial_t \widetilde{a}^\varepsilon + \nabla\widetilde{\Phi}^\varepsilon \cdot \widetilde{\nabla}a^\varepsilon + \dfrac{1}{2}\widetilde{a}^\varepsilon\Delta\widetilde{\Phi}^\varepsilon = 0 \qquad\qquad\quad ; \quad \widetilde{a}^\varepsilon_{|t=0} = a_0^\varepsilon. \end{cases} \qquad (4.3)$$

The problem in seeking a solution $(\widetilde{\Phi}^\varepsilon, \widetilde{a}^\varepsilon)$ to the above system, with $\widetilde{a}^\varepsilon(t,\cdot) \in L^2(\mathbb{R}^d)$, is the meaning of the first equation at the zeroes of $\widetilde{a}^\varepsilon$ (that is, the zeroes of u^ε).

Another possibility consists in allowing a^ε to be complex-valued. In that case, we have an extra degree of freedom in imposing the system solved by $(\Phi^\varepsilon, a^\varepsilon)$. The choice proposed by E. Grenier [Grenier (1998)] (in the case $V \equiv 0$) consists in considering

$$\begin{cases} \partial_t \Phi^\varepsilon + \dfrac{1}{2}|\nabla\Phi^\varepsilon|^2 + V + f\left(|a^\varepsilon|^2\right) = 0 \quad ; \quad \Phi^\varepsilon_{|t=0} = \phi_0, \\[2mm] \partial_t a^\varepsilon + \nabla\Phi^\varepsilon \cdot \nabla a^\varepsilon + \dfrac{1}{2}a^\varepsilon\Delta\Phi^\varepsilon = i\dfrac{\varepsilon}{2}\Delta a^\varepsilon \quad ; \quad a^\varepsilon_{|t=0} = a_0^\varepsilon. \end{cases} \qquad (4.4)$$

We will see in the next section that this approach is very efficient when $f' > 0$, that is, for a defocusing nonlinearity which is cubic at the origin.

In Eq. (4.3) as well as in Eq. (4.4), assuming that Φ^ε and a^ε are bounded in, say, $C([-T, T]; H^s)$ for some sufficiently large s, it is natural to expect

$$\Phi^\varepsilon \underset{\varepsilon\to 0}{\longrightarrow} \Phi, \quad a^\varepsilon \underset{\varepsilon\to 0}{\longrightarrow} a,$$

where (Φ, a) solves

$$\begin{cases} \partial_t \Phi + \dfrac{1}{2}|\nabla\Phi|^2 + V + f\left(|a|^2\right) = 0 \quad ; \quad \Phi_{|t=0} = \phi_0, \\[2mm] \partial_t a + \nabla\Phi \cdot \nabla a + \dfrac{1}{2}a\Delta\Phi = 0 \qquad\qquad ; \quad a_{|t=0} = a_0. \end{cases} \qquad (4.5)$$

Note that when $V = 0$, we see that $(\nabla\Phi, a)$ solves the Euler type system (3.8). This leads to the second approach which we present here. This approach makes it possible to treat the case of defocusing nonlinearities which are not cubic at the origin, but homogeneous of degree $2\sigma + 1$, with $\sigma \in \mathbb{N} \setminus \{0\}$. The idea, already present in Chap. 2.6.4, is to say that if Eq. (4.5) has some rigorous meaning, then at least the rapid oscillations of u^ε should be described by Φ. We point out at this stage that this is the most reasonable thing to expect. In general, a does not suffice to describe the (complex-valued) amplitude of the wave function u^ε, unless, for instance, $a_0 \in \mathbb{R}$ and $a_1 \in i\mathbb{R}$ (e.g. $a_1 = 0$), where

$$a_0^\varepsilon = a_0 + \varepsilon a_1 + o(\varepsilon) \quad \text{in } H^s(\mathbb{R}^d).$$

See Sec. 4.2.1 below. Once Φ is determined, change the unknown function u^ε — and the notation (4.2) — to

$$a^\varepsilon := u^\varepsilon e^{-i\Phi/\varepsilon}.$$

The idea is that this process should filter out all the rapid oscillations, so that a^ε is bounded in Sobolev spaces, and converges strongly (while u^ε does not, as soon as Φ is not trivial). With this definition for a^ε, Eq. (4.1) is *equivalent* to

$$\partial_t a^\varepsilon + \nabla\Phi \cdot \nabla a^\varepsilon + \frac{1}{2} a^\varepsilon \Delta\Phi = i\frac{\varepsilon}{2}\Delta a^\varepsilon - \frac{i}{\varepsilon}\left(f\left(|a^\varepsilon|^2\right) - f\left(|a|^2\right)\right)a^\varepsilon.$$

The major difficulty to prove that a^ε is bounded and converges strongly in Sobolev spaces, is the singular factor $1/\varepsilon$ in front of the last term. Note already that it is reasonable to hope that this singularity is "artificial", since we expect $|a^\varepsilon|^2 = |a|^2 + \mathcal{O}(\varepsilon)$ (this is already suggested by the results of Chap. 2.6.4). We prove that this is so in the homogeneous case $f(y) = y^\sigma$, $\sigma \in \mathbb{N}$, in Sec. 4.3. Note that the results in the case $f' > 0$ as well as in the case $f(y) = y^\sigma$, $\sigma \in \mathbb{N}$, show that the presence of vacuum (zeroes of u^ε) is not a real problem, but barely a technical difficulty (a non-trivial one, though).

Finally, we discuss the case of a focusing nonlinearity $(f' < 0)$ in Sec. 4.5.

4.2 E. Grenier's idea

In this section, we explain the approach based on Eq. (4.4). We first consider the case $V = 0$ for some initial phase $\phi_0 \in H^s$ for large s, in Sec. 4.2.1. We then show how to adapt the approach to the case where V and ϕ_0 are smooth and subquadratic, that is, under Assumption 1.7, in Sec. 4.2.2.

4.2.1 *Without external potential*

In this paragraph, we assume $V = 0$, and we recall the approach of Grenier (1998). To study Eq. (4.4), we introduce an intermediary system, in terms of the amplitude a^ε and the "velocity" $v^\varepsilon := \nabla \Phi^\varepsilon$. The second equation in Eq. (4.4) can directly be expressed in terms of a^ε and v^ε. Differentiating the first equation in (4.4) with respect to x, we find:

$$\begin{cases} \partial_t v^\varepsilon + v^\varepsilon \cdot \nabla v^\varepsilon + 2f'\left(|a^\varepsilon|^2\right)\operatorname{Re}\left(\overline{a}^\varepsilon \nabla a^\varepsilon\right) = 0 & ; \ v^\varepsilon_{|t=0} = \nabla\phi_0, \\ \partial_t a^\varepsilon + v^\varepsilon \cdot \nabla a^\varepsilon + \dfrac{1}{2}a^\varepsilon \operatorname{div} v^\varepsilon = i\dfrac{\varepsilon}{2}\Delta a^\varepsilon & ; \ a^\varepsilon_{|t=0} = a^\varepsilon_0. \end{cases} \quad (4.6)$$

The important remark made by E. Grenier is to notice that if $f' > 0$, the above system is hyperbolic symmetric, perturbed by a skew-symmetric term. To make this fact more explicit, separate the real and imaginary parts of a^ε, to consider the unknown

$$\mathbf{u}^\varepsilon = \begin{pmatrix} \operatorname{Re} a^\varepsilon \\ \operatorname{Im} a^\varepsilon \\ v^\varepsilon_1 \\ \vdots \\ v^\varepsilon_d \end{pmatrix} = \begin{pmatrix} a^\varepsilon_1 \\ a^\varepsilon_2 \\ v^\varepsilon_1 \\ \vdots \\ v^\varepsilon_d \end{pmatrix} \in \mathbb{R}^{d+2}.$$

In terms of this unknown function, Eq. (4.6) reads

$$\partial_t \mathbf{u}^\varepsilon + \sum_{j=1}^d A_j(\mathbf{u}^\varepsilon)\partial_j \mathbf{u}^\varepsilon = \frac{\varepsilon}{2}L\mathbf{u}^\varepsilon, \quad (4.7)$$

where the matrices $A_j \in \mathcal{M}_{d+2}(\mathbb{R})$ are given by:

$$A(\mathbf{u},\xi) = \sum_{j=1}^d A_j(\mathbf{u})\xi_j = \begin{pmatrix} v\cdot\xi & 0 & \frac{1}{2}a_1\,{}^t\xi \\ 0 & v\cdot\xi & \frac{1}{2}a_2\,{}^t\xi \\ 2f'a_1\,\xi & 2f'a_2\,\xi & v\cdot\xi I_d \end{pmatrix},$$

where f' stands for $f'(|a_1|^2 + |a_2|^2)$. The linear operator L is given by

$$L = \begin{pmatrix} 0 & -\Delta & 0 & \dots & 0 \\ \Delta & 0 & 0 & \dots & 0 \\ 0 & 0 & & 0_{d\times d} & \end{pmatrix}.$$

The important remark is that even though L is a differential operator of order two, it causes no loss of regularity in the energy estimates, since it is skew-symmetric. The other important fact is that the left-hand side of Eq. (4.7) is hyperbolic symmetric (or symmetrizable), provided $f' > 0$. Let

$$S = \begin{pmatrix} I_2 & 0 \\ 0 & \frac{1}{4f'}I_d \end{pmatrix}. \quad (4.8)$$

This matrix is symmetric and positive if (and only if) $f' > 0$, and SA is symmetric,

$$SA(\mathbf{u}, \xi) \in \mathcal{S}_{d+2}(\mathbb{R}), \quad \forall (\mathbf{u}, \xi) \in \mathbb{R}^{d+2} \times \mathbb{R}^d.$$

Theorem 4.1 ([Grenier (1998)]). *Let $f \in C^\infty(\mathbb{R}_+; \mathbb{R})$ with $f(0) = 0$ and $f' > 0$. Let $s > 2 + d/2$. Assume that $\phi_0 \in H^{s+1}$, and that a_0^ε is uniformly bounded in H^s for $\varepsilon \in]0, 1]$. There exist $T > 0$ independent of $\varepsilon \in]0, 1]$ and $s > d/2 + 2$, and $u^\varepsilon = a^\varepsilon e^{i\Phi^\varepsilon/\varepsilon}$ solution to (4.1) on $[-T, T]$. Moreover, a^ε and Φ^ε are bounded in $C([-T, T]; H^s)$ and $C([-T, T]; H^{s+1})$ respectively, uniformly in $\varepsilon \in]0, 1]$.*

Remark 4.2. The assumption $f(0) = 0$ is not really one. Indeed, considering $u^\varepsilon e^{itf(0)/\varepsilon}$ instead of u^ε turns f into $f - f(0)$ in Eq. (4.1).

Proof. We first prove that (4.6) has a unique solution $(v^\varepsilon, a^\varepsilon)$ in $C([-T, T]; H^s)^2$, uniformly in $\varepsilon \in]0, 1]$. The main step to prove this fact consists in obtaining *a priori* estimates, so we shall detail this part only. For $s > d/2 + 2$, we bound

$$\langle S\Lambda^s \mathbf{u}^\varepsilon, \Lambda^s \mathbf{u}^\varepsilon \rangle,$$

(scalar product in $L^2(\mathbb{R}^{d+2})$) by computing its time derivative:

$$\frac{d}{dt} \langle S\Lambda^s \mathbf{u}^\varepsilon, \Lambda^s \mathbf{u}^\varepsilon \rangle = \langle \partial_t S\Lambda^s \mathbf{u}^\varepsilon, \Lambda^s \mathbf{u}^\varepsilon \rangle + 2 \langle S\partial_t \Lambda^s \mathbf{u}^\varepsilon, \Lambda^s \mathbf{u}^\varepsilon \rangle,$$

since S is symmetric. For the first term, we consider the lower $d \times d$ block:

$$\langle \partial_t S\Lambda^s \mathbf{u}^\varepsilon, \Lambda^s \mathbf{u}^\varepsilon \rangle \leqslant \left\| \frac{1}{f'} \partial_t \left(f' \left(|a_1^\varepsilon|^2 + a_2^\varepsilon|^2 \right) \right) \right\|_{L^\infty} \langle S\Lambda^s \mathbf{u}^\varepsilon, \Lambda^s \mathbf{u}^\varepsilon \rangle.$$

Since a_0^ε is bounded in $H^s(\mathbb{R}^d) \subset L^\infty(\mathbb{R}^d)$, there exists C_0 independent of $\varepsilon \in]0, 1]$ such that

$$\|a_0^\varepsilon\|_{L^\infty} \leqslant C_0.$$

So long as $\|\mathbf{u}^\varepsilon\|_{L^\infty} \leqslant 2C_0$, we have:

$$f' \left(|a_1^\varepsilon|^2 + |a_2^\varepsilon|^2 \right) \geqslant \inf \left\{ f'(y) \; ; \; 0 \leqslant y \leqslant 4C_0^2 \right\} = \delta_d > 0,$$

where δ_d is now fixed, since f' is continuous with $f' > 0$. Note that this property implies that there exists $C > 0$ such that

$$\frac{1}{C} I_{d+2} \leqslant S \leqslant C I_{d+2}, \tag{4.9}$$

in the sense of symmetric matrices. We infer,

$$\left\| \frac{1}{f'} \partial_t \left(f' \left(|a_1^\varepsilon|^2 + |a_2^\varepsilon|^2 \right) \right) \right\|_{L^\infty} \leqslant C \left\| \partial_t \left(|a_1^\varepsilon|^2 + |a_2^\varepsilon|^2 \right) \right\|_{L^\infty}$$

$$\leqslant C \| \partial_t a^\varepsilon \|_{L^\infty} \leqslant C \| \mathbf{u}^\varepsilon \|_{H^s},$$

where we have used (4.7), the assumption $s > d/2 + 2$, and Sobolev embeddings. Note that we need to assume $s > d/2 + 2$ instead of the more standard assumption $s > d/2 + 1$ for quasi-linear systems, because the operator L is of second order. If the symmetrizer S is constant (in the case of an exactly cubic, defocusing nonlinearity), we can assume simply $s > d/2 + 1$.

For the second term we use

$$\langle S \partial_t \Lambda^s \mathbf{u}^\varepsilon, \Lambda^s \mathbf{u}^\varepsilon \rangle = \frac{\varepsilon}{2} \langle SL(\Lambda^s \mathbf{u}^\varepsilon), \Lambda^s \mathbf{u}^\varepsilon \rangle - \Big\langle S\Lambda^s \Big(\sum_{j=1}^n A_j(\mathbf{u}^\varepsilon) \partial_j \mathbf{u}^\varepsilon \Big), \Lambda^s \mathbf{u}^\varepsilon \Big\rangle.$$

We notice that SL is a skew-symmetric second order operator, so the first term is zero. For the second term, write

$$\Big\langle S\Lambda^s \Big(\sum_{j=1}^n A_j(\mathbf{u}^\varepsilon) \partial_j \mathbf{u}^\varepsilon \Big), \Lambda^s \mathbf{u}^\varepsilon \Big\rangle = \sum_{j=1}^n \Big\langle SA_j(\mathbf{u}^\varepsilon) \partial_j \Lambda^s \mathbf{u}^\varepsilon, \Lambda^s \mathbf{u}^\varepsilon \Big\rangle$$

$$+ \Big\langle S\Big(\sum_{j=1}^n [\Lambda^s, A_j(\mathbf{u}^\varepsilon) \partial_j] \mathbf{u}^\varepsilon \Big), \Lambda^s \mathbf{u}^\varepsilon \Big\rangle. \quad (4.10)$$

Since the matrices $SA_j(\mathbf{u}^\varepsilon)$ are symmetric, we have

$$\Big\langle SA_j(\mathbf{u}^\varepsilon) \partial_j \Lambda^s \mathbf{u}^\varepsilon, \Lambda^s \mathbf{u}^\varepsilon \Big\rangle = \Big\langle \partial_j \Lambda^s \mathbf{u}^\varepsilon, SA_j(\mathbf{u}^\varepsilon) \Lambda^s \mathbf{u}^\varepsilon \Big\rangle$$

$$= - \Big\langle \Lambda^s \mathbf{u}^\varepsilon, \partial_j \left(SA_j(\mathbf{u}^\varepsilon) \right) \Lambda^s \mathbf{u}^\varepsilon \Big\rangle$$

$$- \Big\langle \Lambda^s \mathbf{u}^\varepsilon, SA_j(\mathbf{u}^\varepsilon) \Lambda^s \partial_j \mathbf{u}^\varepsilon \Big\rangle$$

$$= - \frac{1}{2} \Big\langle \Lambda^s \mathbf{u}^\varepsilon, \partial_j \left(SA_j(\mathbf{u}^\varepsilon) \right) \Lambda^s \mathbf{u}^\varepsilon \Big\rangle.$$

This yields

$$\Big| \Big\langle SA_j(\mathbf{u}^\varepsilon) \partial_j \Lambda^s \mathbf{u}^\varepsilon, \Lambda^s \mathbf{u}^\varepsilon \Big\rangle \Big| \leqslant \| \partial_j \left(SA_j(\mathbf{u}^\varepsilon) \right) \|_{L^\infty} \| \mathbf{u}^\varepsilon \|_{H^s}^2$$

$$\leqslant C \left(\| \mathbf{u}^\varepsilon \|_{H^s} \right) \| \mathbf{u}^\varepsilon \|_{H^s}^2,$$

where we have used Schauder's lemma (Lemma 1.23) and the assumption $s > d/2 + 1$. Usual estimates on commutators (see e.g. [Majda (1984); Taylor (1997)]), and Eq. (4.9), yield finally:

$$\frac{d}{dt} \langle S\Lambda^s \mathbf{u}^\varepsilon, \Lambda^s \mathbf{u}^\varepsilon \rangle \leqslant C \left(\| \mathbf{u}^\varepsilon \|_{H^s} \right) \langle S\Lambda^s \mathbf{u}^\varepsilon, \Lambda^s \mathbf{u}^\varepsilon \rangle,$$

for $s > d/2 + 2$. Gronwall lemma along with a continuity argument (to make sure that $\|\mathbf{u}^\varepsilon\|_{L^\infty} \leqslant 2C_0$) show that we can find $T > 0$ independent of ε, such that Eq. (4.4) has a unique solution $(v^\varepsilon, a^\varepsilon)$ in $C([-T,T]; H^s)^2$, uniformly in $\varepsilon \in]0, 1]$.

The fact that T can be chosen independent of $s > d/2 + 2$ follows from tame estimates (see Lemma 1.24).

Finally, once v^ε is known, we can proceed in two ways to conclude. Either remark that v^ε is irrotational ($\nabla \times v^\varepsilon \equiv 0$), so there exists $\widetilde{\Phi}^\varepsilon$ such that $v^\varepsilon = \nabla\widetilde{\Phi}^\varepsilon$; up to adding a function $F = F(t)$ of time only, $\Phi^\varepsilon = \widetilde{\Phi}^\varepsilon + F$ solves the first equation in (4.4). The other possibility is to define directly Φ^ε as

$$\Phi^\varepsilon(t, x) = \phi_0(x) - \int_0^t \left(\frac{1}{2}|v^\varepsilon(\tau, x)|^2 + f\left(|a^\varepsilon(\tau, x)|^2\right) \right) d\tau.$$

We check

$$\partial_t \left(\nabla\Phi^\varepsilon - v^\varepsilon \right) = \nabla\partial_t\Phi^\varepsilon - \partial_t v^\varepsilon = 0.$$

Since $\nabla\Phi^\varepsilon$ and v^ε have the same initial data, we infer that $\nabla\Phi^\varepsilon = v^\varepsilon$, and $(\Phi^\varepsilon, a^\varepsilon)$ solves Eq. (4.4). Since $v^\varepsilon, a^\varepsilon \in C([-T,T]; H^s)$ and $s > d/2$, we have directly $\Phi^\varepsilon \in C([-T,T]; L^2)$ (this is where we need $f(0) = 0$), and we conclude $\Phi^\varepsilon \in C([-T,T]; H^{s+1})$. $\qquad\square$

Once we have constructed the solution to Eq. (4.4), the next step is to study the asymptotic behavior of $(\Phi^\varepsilon, a^\varepsilon)$ as $\varepsilon \to 0$. In view of Eq. (4.4), it is natural to consider Eq. (4.5) in the case $V = 0$:

$$\begin{cases} \partial_t\Phi + \dfrac{1}{2}|\nabla\Phi|^2 + f\left(|a|^2\right) = 0 \quad ; \quad \Phi_{|t=0} = \phi_0, \\[2mm] \partial_t a + \nabla\Phi \cdot \nabla a + \dfrac{1}{2}a\Delta\Phi = 0 \quad ; \quad a_{|t=0} = a_0. \end{cases} \tag{4.11}$$

The proof of Theorem 4.1 shows that if $a_0 \in H^s$ (and $\phi_0 \in H^{s+1}$) for some $s > d/2 + 1$ (there is no second order operator in the analogue of Eq. (4.6) with $\varepsilon = 0$), then Eq. (4.11) has a unique solution

$$(\Phi, a) \in C([-T,T]; H^{s+1} \times H^s).$$

The error estimate between $(\Phi^\varepsilon, a^\varepsilon)$ and (Φ, a) is given by:

Proposition 4.3. *Let* $f \in C^\infty(\mathbb{R}_+; \mathbb{R})$ *with* $f(0) = 0$ *and* $f' > 0$. *Let* $s > 2 + d/2$. *Assume that* $\phi_0 \in H^{s+3}$, $a_0 \in H^{s+2}$ *and*

$$\|a_0^\varepsilon - a_0\|_{H^s} \xrightarrow[\varepsilon \to 0]{} 0.$$

Then for $T > 0$ *given by Theorem 4.1, there exists* $C > 0$ *independent of* ε *such that*

$$\|\Phi^\varepsilon - \Phi\|_{L^\infty([-T,T]; H^{s+1})} + \|a^\varepsilon - a\|_{L^\infty([-T,T]; H^s)} \leqslant C\left(\varepsilon + \|a_0^\varepsilon - a_0\|_{H^s}\right).$$

Proof.   Resume the above notation $\mathbf{u}^\varepsilon$, and introduce $\mathbf{u}$, its counterpart associated to (Φ, a). We know that $\mathbf{u} \in C([-T, T]; H^{s+2})$. Consider the error $\mathbf{w}^\varepsilon = \mathbf{u}^\varepsilon - \mathbf{u}$. It solves

$$\partial_t \mathbf{w}^\varepsilon + \sum_{j=1}^d \left(A_j(\mathbf{u}^\varepsilon)\partial_j \mathbf{u}^\varepsilon - A_j(\mathbf{u})\partial_j \mathbf{u} \right) = \frac{\varepsilon}{2} L \mathbf{u}^\varepsilon.$$

Rewrite this equation as:

$$\partial_t \mathbf{w}^\varepsilon + \sum_{j=1}^d A_j(\mathbf{u}^\varepsilon)\partial_j \mathbf{w}^\varepsilon = -\sum_{j=1}^d \left(A_j(\mathbf{u}^\varepsilon) - A_j(\mathbf{u}) \right) \partial_j \mathbf{u} + \frac{\varepsilon}{2} L \mathbf{w}^\varepsilon + \frac{\varepsilon}{2} L \mathbf{u}.$$

The operator on the left-hand side is the same operator as in Eq. (4.7). It is symmetrized by S, defined in Eq. (4.8). This means that we keep the symmetrizer associated to $\mathbf{u}^\varepsilon$. We do not consider the symmetrizer associated to $\mathbf{u}$. The term $L\mathbf{w}^\varepsilon$ is not present in the energy estimates, since it is skew-symmetric. The term $\varepsilon L \mathbf{u}$ is considered as a source term: it is of order ε, uniformly in $C([-T, T]; H^s)$. Finally, the first term on the right-hand side is a semi-linear perturbation:

$$\begin{aligned}
\left\| \left(A_j(\mathbf{u}^\varepsilon) - A_j(\mathbf{u}) \right) \partial_j \mathbf{u} \right\|_{H^s} &\leqslant \left\| \left(A_j(\mathbf{u}^\varepsilon) - A_j(\mathbf{u}) \right) \right\|_{H^s} \| \mathbf{u} \|_{H^{s+1}} \\
&\leqslant C \left\| \left(A_j(\mathbf{w}^\varepsilon + \mathbf{u}) - A_j(\mathbf{u}) \right) \right\|_{H^s} \\
&\leqslant C \left(\| \mathbf{w}^\varepsilon \|_{L^\infty}, \| \mathbf{u} \|_{L^\infty} \right) \| \mathbf{w}^\varepsilon \|_{H^s},
\end{aligned}$$

where we have used Moser's inequality. Finally, we know that $\mathbf{w}^\varepsilon$ is bounded in $L^\infty([-T, T] \times \mathbb{R}^d)$, as the difference of two bounded terms. With the same approach as in the proof of Theorem 4.1, and thanks to Cauchy–Schwarz and Young inequalities $(2ab \leqslant a^2 + b^2)$, we end up with:

$$\frac{d}{dt} \langle S\Lambda^s \mathbf{w}^\varepsilon, \Lambda^s \mathbf{w}^\varepsilon \rangle \leqslant C \left(\varepsilon^2 + \| \mathbf{w}^\varepsilon \|_{H^s}^2 \right) \leqslant C \left(\varepsilon^2 + \langle S\Lambda^s \mathbf{w}^\varepsilon, \Lambda^s \mathbf{w}^\varepsilon \rangle \right).$$

Gronwall lemma yields:

$$\| \nabla \Phi^\varepsilon - \nabla \Phi \|_{L^\infty([-T,T];H^s)} + \| a^\varepsilon - a \|_{L^\infty([-T,T];H^s)} \leqslant C \left(\varepsilon + \| a_0^\varepsilon - a_0 \|_{H^s} \right).$$

We infer

$$\| \partial_t (\Phi^\varepsilon - \Phi) \|_{H^s} \leqslant C \left(\varepsilon + \| a_0^\varepsilon - a_0 \|_{H^s} \right),$$

and since Φ^ε and Φ coincide at time $t = 0$,

$$\| \Phi^\varepsilon(t) - \Phi(t) \|_{H^{s+1}} \leqslant Ct \left(\varepsilon + \| a_0^\varepsilon - a_0 \|_{H^s} \right), \quad t \in [-T, T]. \tag{4.12}$$

This completes the proof of the proposition. $\square$

At this stage of the study, it is tempting to consider $ae^{i\Phi/\varepsilon}$ as a decent approximation for u^ε. In general, this approximation is interesting for very small time only:

$$
\begin{aligned}
\left| u^\varepsilon - ae^{i\Phi/\varepsilon} \right| &= \left| a^\varepsilon e^{i\Phi^\varepsilon/\varepsilon} - ae^{i\Phi/\varepsilon} \right| \\
&\leqslant \left| a^\varepsilon - a \right| + |a| \left| e^{i\Phi^\varepsilon/\varepsilon} - e^{i\Phi/\varepsilon} \right| \\
&\leqslant C\left(\varepsilon + \|a_0^\varepsilon - a_0\|_{H^s}\right) + C\left| \sin\left(\frac{\Phi^\varepsilon - \Phi}{2\varepsilon}\right) \right|.
\end{aligned}
$$

In view of (4.12), we infer:

$$
\left\| u^\varepsilon(t) - a(t)e^{i\Phi(t)/\varepsilon} \right\|_{L^\infty} \leqslant o_{\varepsilon\to0}(1) + \mathcal{O}(t) + \mathcal{O}\left(\frac{t\|a_0^\varepsilon - a_0\|_{H^s}}{\varepsilon}\right).
$$

The best we can expect in general is, provided $\|a_0^\varepsilon - a_0\|_{H^s} = \mathcal{O}(\varepsilon)$,

$$
\left\| u^\varepsilon(t) - a(t)e^{i\Phi(t)/\varepsilon} \right\|_{L^\infty} = o_{\varepsilon\to0}(1) + \mathcal{O}(t). \tag{4.13}
$$

This shows that because the phase is divided by ε, we will obtain a good approximation for u^ε only if we know the asymptotic behavior of Φ^ε as $\varepsilon \to 0$ up to a remainder which is at least $o(\varepsilon)$. The above computation shows that it is reasonable to require the same thing about a_0^ε, because of the coupling because phase and amplitude.

We therefore seek an asymptotic expansion for $(\Phi^\varepsilon, a^\varepsilon)$. The formal approach is the same as the one presented in Sec. 1.2: we plug an asymptotic expansion of the form

$$
(\Phi^\varepsilon, a^\varepsilon) = (\Phi, a) + \varepsilon\left(\Phi^{(1)}, a^{(1)}\right) + \varepsilon^2\left(\Phi^{(2)}, a^{(2)}\right) + \dots
$$

into Eq. (4.4), and we identify the powers of ε. Of course, we also assume

$$
a_0^\varepsilon = a_0 + \varepsilon a_1 + \varepsilon^2 a_2 + \dots
$$

The term in ε^0 yields Eq. (4.5). For the term in ε^1, we obtain:

$$
\begin{cases}
\partial_t \Phi^{(1)} + \nabla\Phi \cdot \nabla\Phi^{(1)} + 2f'\left(|a|^2\right) \operatorname{Re}\left(\overline{a}a^{(1)}\right) = 0, \\[2mm]
\partial_t a^{(1)} + \nabla\Phi \cdot \nabla a^{(1)} + \nabla\Phi^{(1)} \cdot \nabla a + \dfrac{1}{2}a^{(1)}\Delta\Phi + \dfrac{1}{2}a\Delta\Phi^{(1)} = \dfrac{i}{2}\Delta a, \\[2mm]
\Phi^{(1)}_{|t=0} = 0 \quad ; \quad a^{(1)}_{|t=0} = a_1.
\end{cases}
$$

To solve this system, introduce $v^{(1)} = \nabla\Phi^{(1)}$ (and $v = \nabla\Phi$):

$$
\begin{cases}
\partial_t v^{(1)} + v \cdot \nabla v^{(1)} + 2\nabla\left(f'\left(|a|^2\right)\operatorname{Re}\left(\overline{a}a^{(1)}\right)\right) = -v^{(1)} \cdot \nabla v, \\[2mm]
\partial_t a^{(1)} + v \cdot \nabla a^{(1)} + \dfrac{1}{2}a\operatorname{div}v^{(1)} = -v^{(1)} \cdot \nabla a - \dfrac{1}{2}a^{(1)}\Delta\Phi + \dfrac{i}{2}\Delta a, \tag{4.14} \\[2mm]
v^{(1)}_{|t=0} = 0 \quad ; \quad a^{(1)}_{|t=0} = a_1.
\end{cases}
$$

The left-hand side is a *linear* hyperbolic symmetric operator, applied to $(v^{(1)}, a^{(1)})$. The right-hand side consists of terms which are linear in $(v^{(1)}, a^{(1)})$, plus the source $i\Delta a$. We infer the following existence lemma, whose easy proof is left out.

Lemma 4.4. *Let $s > d/2 + 2$. Assume that $\phi_0 \in H^{s+3}$, $a_0 \in H^{s+2}$ and $a_1 \in H^s$. Then Eq. (4.14) has a unique solution*

$$(v^{(1)}, a^{(1)}) \in C([-T; T]; H^s)^2.$$

With this lemma, we can find $(\Phi^{(1)}, a^{(1)})$ as the second term of the asymptotic expansion for $(\Phi^\varepsilon, a^\varepsilon)$. Set

$$(\Phi_1^\varepsilon, a_1^\varepsilon) = (\Phi, a) + \varepsilon\left(\Phi^{(1)}, a^{(1)}\right).$$

By construction, it solves Eq. (4.4), up to a source term of order $\mathcal{O}(\varepsilon^2)$:

$$\begin{cases} \partial_t \Phi_1^\varepsilon + \dfrac{1}{2}|\nabla\Phi_1^\varepsilon|^2 + f\left(|a_1^\varepsilon|^2\right) = -\dfrac{\varepsilon^2}{2}|\nabla\Phi^{(1)}|^2 \\ \qquad\qquad\qquad - \varepsilon^2 \displaystyle\int_0^1 h''\left(a + \theta\varepsilon a^{(1)}\right) \cdot a^{(1)} \cdot a^{(1)} d\theta, \\ \partial_t a_1^\varepsilon + \nabla\Phi_1^\varepsilon \cdot \nabla a_1^\varepsilon + \dfrac{1}{2}a_1^\varepsilon \Delta\Phi_1^\varepsilon = i\dfrac{\varepsilon}{2}\Delta a_1^\varepsilon - \varepsilon^2 \nabla\Phi^{(1)} \cdot \nabla a^{(1)} \\ \qquad\qquad\qquad - \dfrac{\varepsilon^2}{2}a^{(1)}\Delta\Phi^{(1)} - i\dfrac{\varepsilon^2}{2}\Delta a^{(1)}, \\ \Phi_{1|t=0}^\varepsilon = \phi_0 \quad ; \quad a_{1|t=0}^\varepsilon = a_0 + \varepsilon a_1. \end{cases}$$

We have used the notation $h(z) = f(|z|^2)$, and the last term in the equation for Φ_1^ε is an obvious formal notation. Mimicking the proof of Proposition 4.3, the following result is left as an exercise:

Proposition 4.5. *Let $f \in C^\infty(\mathbb{R}_+; \mathbb{R})$ with $f(0) = 0$ and $f' > 0$. Let $s > 2 + d/2$. Assume that $\phi_0 \in H^{s+5}$, $a_0 \in H^{s+4}$, $a_1 \in H^{s+2}$, and*

$$\|a_0^\varepsilon - a_0 - \varepsilon a_1\|_{H^s} = o(\varepsilon) \quad \text{as } \varepsilon \to 0.$$

Then for $T > 0$ given by Theorem 4.1, there exists $C > 0$ independent of ε such that

$$\left\|\Phi^\varepsilon - \Phi - \varepsilon\Phi^{(1)}\right\|_{L^\infty([-T,T];H^{s+1})} + \left\|a^\varepsilon - a - \varepsilon a^{(1)}\right\|_{L^\infty([-T,T];H^s)}$$

$$\leqslant C\left(\varepsilon^2 + \|a_0^\varepsilon - a_0 - \varepsilon a^{(1)}\|_{H^s}\right).$$

Despite the notations, it seems unadapted to consider $\Phi^{(1)}$ as being part of the phase. Indeed, we infer from the above proposition

$$\left\| u^\varepsilon - a e^{i\Phi^{(1)}} e^{i\Phi/\varepsilon} \right\|_{L^\infty([-T,T];L^2 \cap L^\infty)} \leqslant \left\| a^\varepsilon - a \right\|_{L^\infty([-T,T];L^2 \cap L^\infty)}$$

$$+ \left\| a \right\|_{L^\infty([-T,T];L^2 \cap L^\infty)} \left\| e^{i\Phi^\varepsilon/\varepsilon} - e^{i\Phi^{(1)}} e^{i\Phi/\varepsilon} \right\|_{L^\infty([-T,T] \times \mathbb{R}^d)}$$

$$\leqslant \mathcal{O}(\varepsilon) + \mathcal{O}\left(\frac{\left\| a_0^\varepsilon - a_0 - \varepsilon a^{(1)} \right\|_{H^s}}{\varepsilon} \right)$$

$$\leqslant o(1).$$

If in addition $\left\| a_0^\varepsilon - a_0 - \varepsilon a^{(1)} \right\|_{H^s} = \mathcal{O}(\varepsilon^2)$ (as is usual in WKB analysis), we find

$$\left\| u^\varepsilon - a e^{i\Phi^{(1)}} e^{i\Phi/\varepsilon} \right\|_{L^\infty([-T,T];L^2 \cap L^\infty)} = \mathcal{O}(\varepsilon).$$

Since $\Phi^{(1)}$ depends on a_1 while a does not, we retrieve the fact that in supercritical régimes, the leading order amplitude in WKB methods depends on the initial first corrector a_1. This phenomenon was called *ghost effect* in the context of gas dynamics [Sone *et al.* (1996)]: the corrector a_1 vanishes in the limit $\varepsilon \to 0$ at time $t = 0$, but plays a non-negligible role for $t > 0$.

Remark 4.6. The term $e^{i\Phi^{(1)}}$ does not appear in the Wigner measure of $a e^{i\Phi^{(1)}} e^{i\Phi/\varepsilon}$. Thus, from the point of view of Wigner measures, the asymptotic behavior of the exact solution is described by the Euler-type system (3.7). We also recover the result of Chap. 2.6.4

The above procedure can be pursued to arbitrary order, and we leave it at this stage. To conclude this paragraph, we examine more closely the relevance of the term $\Phi^{(1)}$. From the equation, we find

$$\Phi^{(1)}_{|t=0} = 0 \quad ; \quad \partial_t \Phi^{(1)}_{|t=0} = -2 f' \left(|a_0|^2 \right) \operatorname{Re}\left(\overline{a}_0 a_1 \right).$$

So if $\operatorname{Re}\left(\overline{a}_0 a_1 \right) \neq 0$, $\Phi^{(1)}$ is non-trivial for $t > 0$. Note that even if $a_1 = 0$, then in general, $\Phi^{(1)}$ is non-trivial for $t > 0$. Indeed, if $a_1 = 0$, we have

$$a^{(1)}_{|t=0} = 0 \quad ; \quad \partial_t a^{(1)}_{|t=0} = \frac{i}{2} \Delta a_0,$$

and therefore

$$\Phi^{(1)}_{|t=0} = \partial_t \Phi^{(1)}_{|t=0} = 0 \quad ; \quad \partial_t^2 \Phi^{(1)}_{|t=0} = -f' \left(|a_0|^2 \right) \operatorname{Im}\left(\overline{a}_0 \Delta a_0 \right).$$

If a_0 has a constant argument, then $\Phi^{(1)} \equiv 0$, as shown below.

With the case of Eq. (4.3) in mind, assume that $a_0 e^{i\theta}$ is real-valued some some *constant* $\theta \in \mathbb{R}$; then so is $ae^{i\theta}$, from Eq. (4.5). In that case, we check that $\left(\Phi^{(1)}, \text{Re}\left(\bar{a}a^{(1)}\right)\right)$ solves an *homogeneous* linear system, since then

$$\partial_t \text{Re}\left(\bar{a}a^{(1)}\right) + \nabla\Phi \cdot \nabla \text{Re}\left(\bar{a}a^{(1)}\right) = -\frac{1}{2}\text{div}\left(|a|^2\nabla\Phi^{(1)}\right) - \text{Re}\left(\bar{a}a^{(1)}\right)\Delta\Phi.$$

By uniqueness, if $a_0 e^{i\theta}$ is real-valued and $a_1 e^{i\theta}$ is purely imaginary (e.g. $a_1 = 0$), then $\Phi^{(1)} \equiv 0$. Note however that if $a_1 e^{i\theta} \notin i\mathbb{R}$ is non-trivial (e.g. $a_1 e^{i\theta} \in \mathbb{R}$), then $\Phi^{(1)}$ is non-trivial.

Remark 4.7. The proof of Theorem 4.1 relies on techniques from quasi-linear hyperbolic systems, ignoring the influence of the term $i\varepsilon\Delta a^\varepsilon$ on the right-hand side of the second equation in (4.6), since it has no influence in L^2-based energy estimates, being skew-symmetric. One might hope that using smoothing effects related to Schrödinger equation (this term corresponds to the Schrödinger dispersion), as first presented in [Constantin and Saut (1988)], this term (for $\varepsilon > 0$) could make the solution to (4.6) smooth for all time, in sharp constrast with the case $\varepsilon = 0$, see Remark 3.3. However, numerical evidence from [Besse *et al.* (2013)] suggests that this is not the case: the Schrödinger smoothing effect does not seem sufficient to allow $T = \infty$ in Theorem 4.1.

4.2.2 *With an external potential*

When $V \neq 0$, the first idea consists in trying the same arguments as above. Obviously, if $V \in H^\infty$, then the previous approach can be repeated. Even if we assume only $\nabla V \in H^\infty$, we can construct $(v^\varepsilon, a^\varepsilon)$ in Sobolev spaces, and then Φ^ε is not necessarily in L^2, which is not a big issue.

This approach is essentially perturbative. Its main drawback is that it is incompatible with the case when V is an harmonic potential for instance, a case motivated by Physics. We seek a solution to Eq. (4.4), with

$$\Phi^\varepsilon = \phi_{\text{eik}} + \phi^\varepsilon,$$

where ϕ_{eik} was constructed in Sec. 1.3.1, and ϕ^ε belongs to some Sobolev space. This idea is very naïve, since the equations at stake are nonlinear (note that even when $f = 0$, the eikonal equation is a nonlinear equation). However, it turns out to be fruitful, essentially because ϕ_{eik} is subquadratic with respect to the space variable. For the sake of readability, we rewrite Assumption 1.7 and the main result of Sec. 1.3.1:

Assumption 4.8 (Geometric assumption). *We assume that the potential and the initial phase are smooth, real-valued, and subquadratic:*

- $V \in C^\infty(\mathbb{R} \times \mathbb{R}^d)$, *and* $\partial_x^\alpha V \in C(\mathbb{R}; L^\infty(\mathbb{R}^d))$ *as soon as* $|\alpha| \geqslant 2$.
- $\phi_0 \in C^\infty(\mathbb{R}^d)$, *and* $\partial^\alpha \phi_0 \in L^\infty(\mathbb{R}^d)$ *as soon as* $|\alpha| \geqslant 2$.

Proposition 4.9. *Under Assumption 4.8, there exists $T > 0$ and a unique solution* $\phi_{\mathrm{eik}} \in C^\infty\left([-T, T] \times \mathbb{R}^d\right)$ *to*

$$\partial_t \phi_{\mathrm{eik}} + \frac{1}{2} \left|\nabla \phi_{\mathrm{eik}}\right|^2 + V = 0 \quad ; \quad \phi_{\mathrm{eik}}(0, x) = \phi_0(x).$$

In addition, this solution is subquadratic: $\partial_x^\alpha \phi_{\mathrm{eik}} \in L^\infty([-T, T] \times \mathbb{R}^d)$ *as soon as* $|\alpha| \geqslant 2$.

In terms of the unknown function $(\phi^\varepsilon, a^\varepsilon)$, Eq. (4.4) reads:

$$\begin{cases} \partial_t \phi^\varepsilon + \nabla \phi_{\mathrm{eik}} \cdot \nabla \phi^\varepsilon + \dfrac{1}{2}|\nabla \phi^\varepsilon|^2 + f\left(|a^\varepsilon|^2\right) = 0, \\[2mm] \partial_t a^\varepsilon + \nabla \phi_{\mathrm{eik}} \cdot \nabla a^\varepsilon + \nabla \phi^\varepsilon \cdot \nabla a^\varepsilon + \dfrac{1}{2} a^\varepsilon \Delta \phi_{\mathrm{eik}} + \dfrac{1}{2} a^\varepsilon \Delta \phi^\varepsilon = i\dfrac{\varepsilon}{2} \Delta a^\varepsilon, \\[2mm] \phi^\varepsilon_{|t=0} = 0 \quad ; \quad a^\varepsilon_{|t=0} = a_0^\varepsilon. \end{cases}$$

By construction, the potential V has disappeared from the equation. To prove the analogue of Theorem 4.1, we stop counting the derivatives, and we distinguish two cases, whether the nonlinearity is exactly cubic or not:

Assumption 4.10 (Analytical assumption). *We assume that a_0^ε is bounded in H^s for all $s \geqslant 0$, and $f \in C^\infty(\mathbb{R}_+; \mathbb{R})$ with $f(0) = 0$ and $f' > 0$. Moreover,*

- *The nonlinearity is exactly cubic, $f(y) = \lambda y$ for some $\lambda > 0$, or*
- *The first momentum of a_0^ε is bounded in Sobolev spaces: xa_0^ε is bounded in H^s for all $s \geqslant 0$.*

The above distinction makes it possible to refine a result in [Carles (2007c)]:

Theorem 4.11. *Let Assumptions 4.8 and 4.10 be satisfied. There exist $T_* > 0$ independent of $\varepsilon \in]0, 1]$, and $u^\varepsilon = a^\varepsilon e^{i(\phi_{\mathrm{eik}} + \phi^\varepsilon)/\varepsilon}$ solution to (4.1) on $[-T_*, T_*]$. Moreover, a^ε and ϕ^ε are bounded in $C([-T_*, T_*]; H^\infty)$. In addition, in the second case of Assumption 4.10, xa^ε and $x\nabla \phi^\varepsilon$ are bounded in $L^\infty([-T_*, T_*]; H^s)$ for all $s \geqslant 0$.*

Proof.   The proof proceeds along the same lines as the proof of Theorem 4.1, so we point out the main differences. We only work with times such that $|t| \leqslant T$, so that ϕ_{eik} remains smooth. Like in the previous paragraph, we introduce $v^\varepsilon = \nabla \phi^\varepsilon$, and the notations

$$\mathbf{u}^\varepsilon = \begin{pmatrix} \operatorname{Re} a^\varepsilon \\ \operatorname{Im} a^\varepsilon \\ v_1^\varepsilon \\ \vdots \\ v_d^\varepsilon \end{pmatrix} = \begin{pmatrix} a_1^\varepsilon \\ a_2^\varepsilon \\ v_1^\varepsilon \\ \vdots \\ v_d^\varepsilon \end{pmatrix} \quad , \quad L = \begin{pmatrix} 0 & -\Delta & 0 & \dots & 0 \\ \Delta & 0 & 0 & \dots & 0 \\ 0 & 0 & & 0_{d \times d} & \end{pmatrix},$$

and $\quad A(\mathbf{u}, \xi) = \displaystyle\sum_{j=1}^d A_j(\mathbf{u})\xi_j = \begin{pmatrix} v \cdot \xi & 0 & \frac{a_1}{2}{}^t\xi \\ 0 & v \cdot \xi & \frac{a_2}{2}{}^t\xi \\ 2f'a_1 \xi & 2f'a_2 \xi & v \cdot \xi I_d \end{pmatrix},$

where f' stands for $f'(|a_1|^2 + |a_2|^2)$. Instead of (4.7), we now have a system of the form

$$\partial_t \mathbf{u}^\varepsilon + \sum_{j=1}^d A_j(\mathbf{u}^\varepsilon)\partial_j \mathbf{u}^\varepsilon + \nabla\phi_{\text{eik}} \cdot \nabla \mathbf{u}^\varepsilon + M\left(\nabla^2 \phi_{\text{eik}}\right)\mathbf{u}^\varepsilon = \frac{\varepsilon}{2}L\mathbf{u}^\varepsilon, \quad (4.15)$$

where the matrix M is smooth and locally bounded. The quasi-linear part of the above equation is the same as in Eq. (4.7), and involves the matrices A_j. In particular, we keep the same symmetrizer S given by (4.8). The term $\nabla\phi_{\text{eik}} \cdot \nabla\mathbf{u}^\varepsilon$ has a semi-linear contribution, as we see below. The term corresponding to the matrix M can obviously be considered as a source term, since ϕ_{eik} is subquadratic.

For $s > d/2 + 2$, we still have

$$\frac{d}{dt}\langle S\Lambda^s\mathbf{u}^\varepsilon, \Lambda^s\mathbf{u}^\varepsilon\rangle = \langle \partial_t S\Lambda^s\mathbf{u}^\varepsilon, \Lambda^s\mathbf{u}^\varepsilon\rangle + 2\langle S\partial_t\Lambda^s\mathbf{u}^\varepsilon, \Lambda^s\mathbf{u}^\varepsilon\rangle.$$

Two cases must then be distinguished, which explain the two cases in Assumption 4.10: if f' is constant, then so is the symmetrizer S. Otherwise, we have

$$\langle \partial_t S\Lambda^s\mathbf{u}^\varepsilon, \Lambda^s\mathbf{u}^\varepsilon\rangle \leqslant \left\|\frac{1}{f'}\partial_t\left(f'\left(|a_1^\varepsilon|^2 + |a_2^\varepsilon|^2\right)\right)\right\|_{L^\infty} \langle S\Lambda^s\mathbf{u}^\varepsilon, \Lambda^s\mathbf{u}^\varepsilon\rangle.$$

Since a_0^ε is bounded in $H^s(\mathbb{R}^d) \subset L^\infty(\mathbb{R}^d)$, there exists C_0 independent of $\varepsilon \in]0,1]$ such that

$$\|a_0^\varepsilon\|_{L^\infty} \leqslant C_0.$$

So long as $\|\mathbf{u}^\varepsilon\|_{L^\infty} \leqslant 2C_0$, we have:

$$f'\left(|a_1^\varepsilon|^2 + |a_2^\varepsilon|^2\right) \geqslant \inf\left\{f'(y) \; ; \; 0 \leqslant y \leqslant 4C_0^2\right\} = \delta_d > 0,$$

where δ_d is now fixed, since f' is continuous with $f' > 0$. We infer,

$$\left\|\frac{1}{f'}\partial_t\left(f'\left(|a_1^\varepsilon|^2 + |a_2^\varepsilon|^2\right)\right)\right\|_{L^\infty} \leqslant C\left\|\partial_t\left(|a_1^\varepsilon|^2 + |a_2^\varepsilon|^2\right)\right\|_{L^\infty}.$$

Using the equation for a^ε, we see that to estimate $\partial_t a^\varepsilon$, new terms appear, compared to the proof of Theorem 4.1: $\nabla\phi_{\text{eik}} \cdot \nabla a^\varepsilon$ and $a^\varepsilon\Delta\phi_{\text{eik}}$. Since ϕ_{eik} is subquadratic, we have, thanks to Sobolev embeddings:

$$\left\|\frac{1}{f'}\partial_t\left(f'\left(|a_1^\varepsilon|^2 + |a_2^\varepsilon|^2\right)\right)\right\|_{L^\infty} \leqslant C\left(\|\mathbf{u}^\varepsilon\|_{H^s} + \|x\mathbf{u}^\varepsilon\|_{H^{s-1}}\right).$$

For the clarity of the proof, we distinguish the cases for the rest of the computations.

First case: exactly cubic nonlinearity. In the first case of Assumption 4.10, we have noticed that the symmetrizer S is constant. For the quasi-linear estimates involving the matrices A_j, we can mimic the proof of Theorem 4.1:

$$\frac{d}{dt}\langle S\Lambda^s\mathbf{u}^\varepsilon, \Lambda^s\mathbf{u}^\varepsilon\rangle \leqslant C\left(\|\mathbf{u}^\varepsilon\|_{H^s}\right)\langle S\Lambda^s\mathbf{u}^\varepsilon, \Lambda^s\mathbf{u}^\varepsilon\rangle$$
$$+ \left|\left\langle S\Lambda^s\left(\nabla\phi_{\text{eik}} \cdot \nabla\mathbf{u}^\varepsilon\right), \Lambda^s\mathbf{u}^\varepsilon\right\rangle\right|$$
$$+ \left|\left\langle S\Lambda^s\left(M(\nabla^2\phi_{\text{eik}})\mathbf{u}^\varepsilon\right), \Lambda^s\mathbf{u}^\varepsilon\right\rangle\right|.$$

For the second term of the right-hand side, write:

$$\left\langle S\Lambda^s\left(\partial_j\phi_{\text{eik}}\partial_j\mathbf{u}^\varepsilon\right), \Lambda^s\mathbf{u}^\varepsilon\right\rangle = \left\langle S\partial_j\phi_{\text{eik}}\partial_j\Lambda^s\mathbf{u}^\varepsilon, \Lambda^s\mathbf{u}^\varepsilon\right\rangle$$
$$+ \left\langle S\left[\Lambda^s, \partial_j\phi_{\text{eik}}\partial_j\right]\mathbf{u}^\varepsilon, \Lambda^s\mathbf{u}^\varepsilon\right\rangle.$$

For the first term of the right-hand side, an integration by parts yields:

$$\left|\left\langle S\partial_j\phi_{\text{eik}}\partial_j\Lambda^s\mathbf{u}^\varepsilon, \Lambda^s\mathbf{u}^\varepsilon\right\rangle\right| \leqslant \|\partial_j\left(S\partial_j\phi_{\text{eik}}\right)\|_{L^\infty}\|\mathbf{u}^\varepsilon\|_{H^s}^2 \qquad (4.16)$$
$$\leqslant C\|\mathbf{u}^\varepsilon\|_{H^s}^2,$$

where we have used the fact that S is constant and ϕ_{eik} is subquadratic. This also shows that the commutator

$$[\Lambda^s, \partial_j\phi_{\text{eik}}\partial_j]$$

is a pseudo-differential operator of degree $\leqslant s$, with bounded coefficients. We infer:

$$\left|\left\langle S\Lambda^s\left(\nabla\phi_{\text{eik}} \cdot \nabla\mathbf{u}^\varepsilon\right), \Lambda^s\mathbf{u}^\varepsilon\right\rangle\right| \leqslant C\|\mathbf{u}^\varepsilon\|_{H^s}^2.$$

We have obviously

$$\left| \left\langle S\Lambda^s \left(M(\nabla^2 \phi_{\text{eik}}) \mathbf{u}^\varepsilon \right), \Lambda^s \mathbf{u}^\varepsilon \right\rangle \right| \leqslant C \|\mathbf{u}^\varepsilon\|_{H^s}^2.$$

This yields:

$$\frac{d}{dt} \left\langle S\Lambda^s \mathbf{u}^\varepsilon, \Lambda^s \mathbf{u}^\varepsilon \right\rangle \leqslant C \left(\|\mathbf{u}^\varepsilon\|_{H^s} \right) \left\langle S\Lambda^s \mathbf{u}^\varepsilon, \Lambda^s \mathbf{u}^\varepsilon \right\rangle, \qquad (4.17)$$

and we conclude like in Theorem 4.1.

Second case. If we assume only $f' > 0$, the assumption $x a_0^\varepsilon \in H^s$ makes it possible to conclude in a similar fashion. We have:

$$\frac{d}{dt} \left\langle S\Lambda^s \mathbf{u}^\varepsilon, \Lambda^s \mathbf{u}^\varepsilon \right\rangle \leqslant C \left(\|\mathbf{u}^\varepsilon\|_{H^s} + \|x\mathbf{u}^\varepsilon\|_{H^{s-1}} \right) \left\langle S\Lambda^s \mathbf{u}^\varepsilon, \Lambda^s \mathbf{u}^\varepsilon \right\rangle$$
$$+ \left| \left\langle S\Lambda^s \left(\nabla\phi_{\text{eik}} \cdot \nabla \mathbf{u}^\varepsilon \right), \Lambda^s \mathbf{u}^\varepsilon \right\rangle \right|$$
$$+ \left| \left\langle S\Lambda^s \left(M(\nabla^2 \phi_{\text{eik}}) \mathbf{u}^\varepsilon \right), \Lambda^s \mathbf{u}^\varepsilon \right\rangle \right|.$$

The last term is obviously controlled by $\|\mathbf{u}^\varepsilon\|_{H^s}^2$. For the second term, resume the estimate (4.16). Using the definition of S, we find

$$\left\| \partial_j \left(S\partial_j \phi_{\text{eik}} \right) \right\|_{L^\infty} \leqslant C \left(\|\mathbf{u}^\varepsilon\|_{L^\infty} \right) \left\| \langle x \rangle \, a^\varepsilon \right\|_{L^\infty} \left\| \partial_j a^\varepsilon \right\|_{L^\infty}$$
$$\leqslant C \left(\|\mathbf{u}^\varepsilon\|_{L^\infty} \right) \left\| \langle x \rangle \, a^\varepsilon \right\|_{H^{s-1}} \|a^\varepsilon\|_{H^s}.$$

We infer:

$$\frac{d}{dt} \left\langle S\Lambda^s \mathbf{u}^\varepsilon, \Lambda^s \mathbf{u}^\varepsilon \right\rangle \leqslant F \left(\|\mathbf{u}^\varepsilon\|_{H^s} + \|x\mathbf{u}^\varepsilon\|_{H^{s-1}} \right) \left\langle S\Lambda^s \mathbf{u}^\varepsilon, \Lambda^s \mathbf{u}^\varepsilon \right\rangle.$$

To close the family of estimates, we show that

$$\frac{d}{dt} \left\langle S\Lambda^{s-1} \left(x\mathbf{u}^\varepsilon \right), \Lambda^{s-1} \left(x\mathbf{u}^\varepsilon \right) \right\rangle$$

can be bounded in a similar fashion. Let $1 \leqslant k \leqslant d$:

$$\partial_t(x_k \mathbf{u}^\varepsilon) + \sum_{j=1}^d A_j(\mathbf{u}^\varepsilon) \partial_j(x_k \mathbf{u}^\varepsilon) + \nabla\phi_{\text{eik}} \cdot \nabla(x_k \mathbf{u}^\varepsilon)$$

$$+ M \left(\nabla^2 \phi_{\text{eik}} \right) x_k \mathbf{u}^\varepsilon = \frac{\varepsilon}{2} L(x_k \mathbf{u}^\varepsilon) + A_k(\mathbf{u}^\varepsilon) \mathbf{u}^\varepsilon + \partial_k \phi_{\text{eik}} \mathbf{u}^\varepsilon + \frac{\varepsilon}{2} [x_k, L] \mathbf{u}^\varepsilon.$$

The quasi-linear part and the term $\nabla\phi_{\text{eik}} \cdot \nabla(x_k \mathbf{u}^\varepsilon)$ are estimated like before. The terms $M \left(\nabla^2 \phi_{\text{eik}} \right) x_k \mathbf{u}^\varepsilon$ and $A_k(\mathbf{u}^\varepsilon) \mathbf{u}^\varepsilon$ are controlled in an obvious way. The term $\partial_k \phi_{\text{eik}} \mathbf{u}^\varepsilon$ is controlled by $\langle x \rangle \, \mathbf{u}^\varepsilon$, since ϕ_{eik} is subquadratic: this is a linear perturbation. Finally,

$$[x_k, L] = \begin{pmatrix} 0 & 2\partial_k & 0 & \dots & 0 \\ -2\partial_k & 0 & 0 & \dots & 0 \\ 0 & 0 & & 0_{d \times d} & \end{pmatrix}.$$

Therefore,

$$\left\| \Lambda^{s-1}[x_k, L]\mathbf{u}^\varepsilon \right\|_{L^2} \leqslant 2\|\mathbf{u}^\varepsilon\|_{H^s}.$$

This shows that we obtain a closed family of estimates. Gronwall lemma and a continuity argument yield existence and uniqueness for $\mathbf{u}^\varepsilon$, like for Theorem 4.1. Note that thanks to tame estimates, T_* does not depend on $s > d/2 + 2$. Finally, we define ϕ^ε by

$$\phi^\varepsilon(t) = -\int_0^t \left(\nabla\phi_{\text{eik}} \cdot v^\varepsilon + \frac{1}{2}|v^\varepsilon|^2 + f\left(|a_0^\varepsilon|^2\right) \right)(\tau)d\tau.$$

We check that $\phi^\varepsilon \in C([-T_*, T_*]; L^2)$ and $\partial_t (\nabla\phi^\varepsilon - v^\varepsilon) = 0$, and the proof of the theorem is complete. $\square$

Note that the phase $\phi_{\text{eik}} + \phi^\varepsilon$ belongs to a somewhat non-standard space: it is the sum of a subquadratic function and an H^∞ function. Even if ϕ^ε goes to zero at infinity, it should not be considered as a small perturbation of ϕ_{eik} in L^∞. Indeed, like for the case $V \equiv 0$, the coupling between the amplitude a^ε and ϕ^ε is so strong that even though $\phi^\varepsilon_{|t=0} = 0$,

$$\partial_t \phi^\varepsilon_{|t=0} = -f\left(|a_0^\varepsilon|^2\right),$$

so ϕ^ε is of order $\mathcal{O}(1)$ in L^∞ for $t > 0$.

The next step in the semi-classical analysis is to study the asymptotic expansion of $(\phi^\varepsilon, a^\varepsilon)$ as $\varepsilon \to 0$. It proceeds along the same lines as in the case $V = 0$ (Sec. 4.2.1), up to the adaptations pointed out in the above proof. We leave out the discussion at this stage, since all the tools have been given, and the conclusion is essentially the same as in Sec. 4.2.1.

4.2.3 The case $0 < \alpha < 1$

So far, we have addressed the case of weakly nonlinear geometric optics (Chap. 2), and the supercritical case (4.1). In this paragraph, we discuss the intermediary case of

$$i\varepsilon\partial_t u^\varepsilon + \frac{\varepsilon^2}{2}\Delta u^\varepsilon = Vu^\varepsilon + \varepsilon^\alpha f\left(|u^\varepsilon|^2\right) u^\varepsilon \quad ; \quad u^\varepsilon_{|t=0} = a_0^\varepsilon e^{i\phi_0/\varepsilon}, \quad (4.18)$$

in the case $0 < \alpha < 1$. We have seen in Chap. 2 that the nonlinear term is so strong that we should not expect $u^\varepsilon e^{-i\phi_{\text{eik}}/\varepsilon}$ to be bounded in H^s ($s > 0$) as $\varepsilon \to 0$. So we adapt the point of view of the previous paragraph, that is, Eq. (4.18) with $\alpha = 0$. Again, we write the exact solution as

$$u^\varepsilon = a^\varepsilon e^{i\Phi^\varepsilon/\varepsilon}, \quad \text{with } \Phi^\varepsilon = \phi_{\text{eik}} + \phi^\varepsilon.$$

Let Assumptions 4.8 and 4.10 be satisfied. To simplify the discussion, suppose that we are in the second case of Assumption 4.10 (which is not incompatible with the first case!). The unknown function is the pair $(a^\varepsilon, \phi^\varepsilon)$. We have two unknown functions to solve a single equation, (1.1). We can choose how to balance the terms: we resume the approach followed when $\alpha = 0$. Note that this approach would also be efficient for the case $\alpha \geqslant 1$, with the serious drawback that we still assume $f' > 0$, an assumption proven to be unnecessary when $\alpha \geqslant 1$ (see Chap. 2). We impose:

$$\begin{cases} \partial_t \phi^\varepsilon + \dfrac{1}{2} |\nabla \phi^\varepsilon|^2 + \nabla \phi_{\text{eik}} \cdot \nabla \phi^\varepsilon + \varepsilon^\alpha f\left(|a^\varepsilon|^2\right) = 0, \\[2mm] \partial_t a^\varepsilon + \nabla \phi^\varepsilon \cdot \nabla a^\varepsilon + \nabla \phi_{\text{eik}} \cdot \nabla a^\varepsilon + \dfrac{1}{2} a^\varepsilon \Delta \phi^\varepsilon + \dfrac{1}{2} a^\varepsilon \Delta \phi_{\text{eik}} = i \dfrac{\varepsilon}{2} \Delta a^\varepsilon, \\[2mm] \phi^\varepsilon_{|t=0} = 0 \;\; ; \;\; a^\varepsilon_{|t=0} = a^\varepsilon_0. \end{cases}$$

This is the same system as before, with only f replaced by $\varepsilon^\alpha f$. Mimicking the analysis of the previous paragraph, we work with the unknown $\mathbf{u}^\varepsilon$ given by the same definition: it solves the system (4.15), where only the matrices A_j have changed, and now depend on ε. The symmetrizer is the same as before, with f' replaced by $\varepsilon^\alpha f'$: the matrix $S = S^\varepsilon$ is not bounded as $\varepsilon \to 0$, but its inverse is. We claim that we can still proceed as before, thanks to this remark and the following reasons:

- The matrix S^ε is diagonal.
- The matrix M is block diagonal.
- The matrices $S^\varepsilon A_j^\varepsilon$ are independent of $\varepsilon \in]0, 1]$.
- The inverse of S^ε is uniformly bounded on compact sets, as $\varepsilon \to 0$.

Gronwall lemma then implies the analogue of Theorem 4.11: in particular, $\mathbf{u}^\varepsilon$ exists locally in time, with H^s-norm uniformly bounded as $\varepsilon \to 0$. Note that since $\phi^\varepsilon\big|_{t=0} = 0$, we have:

$$\left(S^\varepsilon \Lambda^s \mathbf{u}^\varepsilon, \Lambda^s \mathbf{u}^\varepsilon\right)\big|_{t=0} = \mathcal{O}(1),$$

and we infer more precisely:

$$\|a^\varepsilon\|_{L^\infty([-T_*, T_*]; H^s)} + \|x a^\varepsilon\|_{L^\infty([-T_*, T_*]; H^s)} = \mathcal{O}(1),$$

$$\|\phi^\varepsilon\|_{L^\infty([-T_*, T_*]; H^s)} + \|x \nabla \phi^\varepsilon\|_{L^\infty([-T_*, T_*]; H^s)} = \mathcal{O}\left(\varepsilon^\alpha\right).$$

It seems natural to change unknown functions, and work with $\widetilde{\phi}^\varepsilon = \varepsilon^{-\alpha} \phi^\varepsilon$ instead of ϕ^ε. With this, we somehow correct the shift in the cascade of equations caused by the factor ε^α in front of the nonlinearity. We find

$$\begin{cases} \partial_t\widetilde{\phi}^\varepsilon + \dfrac{\varepsilon^\alpha}{2}\left|\nabla\widetilde{\phi}^\varepsilon\right|^2 + \nabla\phi_{\text{eik}}\cdot\nabla\widetilde{\phi}^\varepsilon + f\left(|a^\varepsilon|^2\right) = 0, \\[2mm] \partial_t a^\varepsilon + \varepsilon^\alpha\nabla\widetilde{\phi}^\varepsilon\cdot\nabla a^\varepsilon + \nabla\phi_{\text{eik}}\cdot\nabla a^\varepsilon + \dfrac{\varepsilon^\alpha}{2}a^\varepsilon\Delta\widetilde{\phi}^\varepsilon + \dfrac{1}{2}a^\varepsilon\Delta\phi_{\text{eik}} = i\dfrac{\varepsilon}{2}\Delta a^\varepsilon, \\[2mm] \widetilde{\phi}^\varepsilon_{|t=0} = 0 \quad ; \quad a^\varepsilon_{|t=0} = a_0^\varepsilon. \end{cases}$$

The pairs $(\widetilde{\phi}^\varepsilon, a^\varepsilon)$ and $(\partial_t\widetilde{\phi}^\varepsilon, \partial_t a^\varepsilon)$ are bounded in $C([-T_*,T_*];H^s)$. Therefore, Arzela–Ascoli's theorem shows that a subsequence is convergent, and the limit is given by:

$$\begin{cases} \partial_t\widetilde{\phi} + \nabla\phi_{\text{eik}}\cdot\nabla\widetilde{\phi} + f\left(|a|^2\right) = 0 \ ; \quad \widetilde{\phi}_{|t=0} = 0, \\[2mm] \partial_t a + \nabla\phi_{\text{eik}}\cdot\nabla a + \dfrac{1}{2}a\Delta\phi_{\text{eik}} = 0 \ ; \quad a_{|t=0} = a_0. \end{cases}$$

We see that a solves the same transport equation as in the linear case, Eq. (1.19); $\widetilde{\phi}$ is given by an ordinary differential equation along the rays associated to ϕ_{eik}, with a source term showing nonlinear effect: $f\left(|a|^2\right)$. By uniqueness, the whole sequence is convergent. Roughly speaking, we see that if

$$\mathbf{w}^\varepsilon = {}^t\left(\nabla\left(\widetilde{\phi}^\varepsilon - \widetilde{\phi}\right), a^\varepsilon - a\right),$$

then Gronwall lemma yields:

$$(S^\varepsilon\partial_x^\alpha\mathbf{w}^\varepsilon, \partial_x^\alpha\mathbf{w}^\varepsilon) \leqslant C\left(\varepsilon + \varepsilon^\alpha\right) \leqslant 2C\varepsilon^\alpha.$$

We infer, for $|t| \leqslant T_*$:

$$\|a^\varepsilon(t) - a(t)\|_{H^s} \leqslant C_s\varepsilon^\alpha \quad ; \quad \|\phi^\varepsilon(t) - \varepsilon^\alpha\widetilde{\phi}\|_{H^s} \leqslant C_s\varepsilon^{2\alpha}t.$$

Three cases must be distinguished:

- If $1/2 < \alpha < 1$, then we can infer

$$\left\|u^\varepsilon - ae^{i\phi_{\text{eik}}/\varepsilon}e^{i\widetilde{\phi}/\varepsilon^{1-\alpha}}\right\|_{L^\infty([-T_*,T_*];L^2\cap L^\infty)} \xrightarrow[\varepsilon\to 0]{} 0.$$

- If $\alpha = 1/2$, then we can infer a similar result for small time only:

$$\left\|u^\varepsilon - ae^{i\phi_{\text{eik}}/\varepsilon}e^{i\widetilde{\phi}/\varepsilon^{1-\alpha}}\right\|_{L^\infty([-t,t];L^2\cap L^\infty)} \to 0 \quad \text{as } \varepsilon \text{ and } t \to 0.$$

- If $0 < \alpha < 1/2$, then we must pursue the analysis, and compute a corrector of order $\varepsilon^{2\alpha}$.

We shall not go further into detailed computations, but instead, discuss the whole analysis in a rather loose fashion. However, we note that all the ingredients have been given for a complete justification.

Let $N = [1/\alpha]$, where $[r]$ is the largest integer not larger than $r > 0$. We construct $a^{(1)}, \ldots, a^{(N)}$ and $\widetilde{\phi}^{(1)}, \ldots, \widetilde{\phi}^{(N)}$ such that:

$$\left\| a^{\varepsilon} - a - \varepsilon^{\alpha} a^{(1)} - \ldots - \varepsilon^{N\alpha} a^{(N)} \right\|_{L^{\infty}([-T_*, T_*]; H^s)}$$
$$+ \left\| \widetilde{\phi}^{\varepsilon} - \widetilde{\phi} - \varepsilon^{\alpha} \widetilde{\phi}^{(1)} - \ldots - \varepsilon^{N\alpha} \widetilde{\phi}^{(N)} \right\|_{L^{\infty}([-T_*, T_*]; H^s)} = o\left(\varepsilon^{N\alpha}\right).$$

But since $N + 1 > 1/\alpha$, we have:

$$\left\| \widetilde{\phi}^{\varepsilon} - \varepsilon^{\alpha} \widetilde{\phi} - \varepsilon^{2\alpha} \widetilde{\phi}^{(1)} - \ldots - \varepsilon^{N\alpha} \widetilde{\phi}^{(N-1)} \right\|_{L^{\infty}([-T_*, T_*]; H^s)} = \mathcal{O}\left(\varepsilon^{(N+1)\alpha}\right)$$
$$= o(\varepsilon).$$

We infer:

$$\left\| u^{\varepsilon} - a e^{i\phi_{\mathrm{eik}}/\varepsilon + i\phi_{\mathrm{app}}^{\varepsilon}} \right\|_{L^{\infty}([-T_*, T_*]; L^2 \cap L^{\infty})} = o(1),$$

where

$$\phi_{\mathrm{app}}^{\varepsilon} = \frac{\widetilde{\phi}}{\varepsilon^{1-\alpha}} + \frac{\widetilde{\phi}^{(1)}}{\varepsilon^{1-2\alpha}} + \ldots + \frac{\widetilde{\phi}^{(N-1)}}{\varepsilon^{1-N\alpha}}.$$

Remark 4.12. In the case $\alpha = 1$, $N = 1$, and the above analysis shows that one phase shift factor appears: we retrieve the result of Chap. 2, and $\widetilde{\phi}$ coincides with the function G of Eq. (2.7) (under the unnecessary assumption $f' > 0$). If $\alpha > 1$, then $N = 0$, and we see that $a e^{i\phi_{\mathrm{eik}}/\varepsilon}$ is a good approximation for u^{ε}.

To conclude this paragraph, we consider the convergence of quadratic observables. It follows from the pointwise description of the wave function u^{ε}. Since the nonlinear effects are present at leading order only through $\phi_{\mathrm{app}}^{\varepsilon}$, the quadratic observables converge to the same quantities as in the linear case (just like for the case $\alpha = 1$):

$$|u^{\varepsilon}|^2 = \rho^{\varepsilon} \xrightarrow[\varepsilon \to 0]{} \rho \quad ; \quad \varepsilon \, \mathrm{Im}\left(\overline{u}^{\varepsilon} \nabla u^{\varepsilon}\right) = J^{\varepsilon} \xrightarrow[\varepsilon \to 0]{} \rho \underline{v},$$

where $(\rho, \underline{v}) = (|a|^2, \nabla\phi_{\mathrm{eik}})$ solves the Euler equation

$$\begin{cases} \partial_t \underline{v} + \underline{v} \cdot \nabla \underline{v} + \nabla V = 0 \quad ; \quad \underline{v}_{|t=0} = \nabla\phi_0, \\ \partial_t \rho + \mathrm{div}\left(\rho \underline{v}\right) = 0 \quad ; \quad \rho_{|t=0} = |a_0|^2. \end{cases}$$

The pressure is given by $p = 0$, and the external force is ∇V. We see that even if $\alpha = 1$ is the critical threshold as far as WKB analysis is concerned, when it turns to quadratic observables, the critical threshold becomes $\alpha = 0$ (when the pressure law in Euler equations depends on f).

4.3 Higher order homogeneous nonlinearities

If we consider a quintic nonlinearity in (4.1), $f(y) = y^2$, the previous approach fails. Essentially, the symmetrizer

$$S = \begin{pmatrix} I_2 & 0 \\ 0 & \frac{1}{4f'(|u^\varepsilon|^2)} I_d \end{pmatrix}.$$

becomes singular at the zeroes of u^ε. Since we have no control on the zeroes of u^ε, the approach must be modified. We present the result of [Alazard and Carles (2009b)], for the case of

$$i\varepsilon \partial_t u^\varepsilon + \frac{\varepsilon^2}{2} \Delta u^\varepsilon = |u^\varepsilon|^{2\sigma} u^\varepsilon \quad ; \quad u^\varepsilon_{|t=0} = a_0^\varepsilon e^{i\phi_0/\varepsilon}. \tag{4.19}$$

An alternative approach appears in [Chiron and Rousset (2009)], allowing nonlinearities which are more general than those considered here, based on the linear stability of oscillating solutions of the form $a e^{i\phi/\varepsilon}$ directly on the Schrödinger equation (as opposed to a phase-amplitude system).

We assume $V = 0$: in the spirit of Sec. 4.2.2, inserting an external potential adds no technical difficulty, but makes the presentation heavier. We also assume that the nonlinearity is homogeneous, and smooth: $\sigma \in \mathbb{N} \setminus \{0\}$.

Several ideas are natural in view of the results of Sec. 4.2. First, assuming that we could construct $(\Phi^\varepsilon, a^\varepsilon)$ solution to Eq. (4.4), bounded in Sobolev spaces, then passing to the limit, we expect $(\Phi^\varepsilon, a^\varepsilon)$ to converge to (Φ, a) given by Eq. (4.5). Write

$$u^\varepsilon \underset{\varepsilon \to 0}{\sim} a e^{i\Phi/\varepsilon} e^{i(\Phi^\varepsilon - \Phi)/\varepsilon}.$$

The only ε-oscillatory factor is measured by Φ: set $a^\varepsilon := u^\varepsilon e^{-i\Phi/\varepsilon}$. For a^ε to be bounded in H^s, we need another information: we need a rate of convergence of Φ^ε towards Φ,

$$\Phi^\varepsilon - \Phi = \mathcal{O}(\varepsilon).$$

Otherwise, ∇a^ε may not be bounded as $\varepsilon \to 0$. The study led in Sec. 4.2.1 shows that for this property to be satisfied, we have to know a_0^ε up to an error of order $\mathcal{O}(\varepsilon)$. We refuse to count the derivatives when not necessary:

Assumption 4.13. There exists $a_0 \in H^\infty$ such that for all $s \geqslant 0$,

$$\|a_0^\varepsilon - a_0\|_{H^s} = \mathcal{O}(\varepsilon).$$

Lemma 4.14. *Let $\sigma \in \mathbb{N}$ and $\phi_0, a_0 \in H^\infty$. There exists $T^* > 0$ such that*

$$\begin{cases} \partial_t \phi + \dfrac{1}{2}|\nabla\phi|^2 + |a|^{2\sigma} = 0 & ; \quad \phi_{|t=0} = \phi_0, \\[2mm] \partial_t a + \nabla\phi \cdot \nabla a + \dfrac{1}{2}a\Delta\phi = 0 & ; \quad a_{|t=0} = a_0 \end{cases} \tag{4.20}$$

has a unique solution $(\phi, a) \in C^\infty([-T^*, T^*]; H^\infty(\mathbb{R}^d))^2$.

Proof. In Chap. 2.6.4, we have considered the system

$$\begin{cases} \partial_t v + v \cdot \nabla v + \nabla\left(|a|^{2\sigma}\right) = 0 & ; \quad v_{|t=0} = \nabla\phi_0. \\[2mm] \partial_t a + v \cdot \nabla a + \dfrac{1}{2}a\,\mathrm{div}\,v = 0 & ; \quad a_{|t=0} = a_0. \end{cases}$$

Thanks to the idea of [Makino *et al.* (1986)], we have proved that it possesses a unique solution $(v, a) \in C^\infty([-T^*, T^*]; H^\infty(\mathbb{R}^d))^2$. Now set

$$\phi(t, x) = \phi_0(x) - \int_0^t \left(\frac{1}{2}|v(\tau, x)|^2 + |a(\tau, x)|^{2\sigma}\right) d\tau.$$

We check that $\partial_t(\nabla\phi - v) = \nabla\partial_t\phi - \partial_t v = 0$, and $\phi \in C([-T^*, T^*]; L^2)$. Hence the lemma. $\qquad\square$

Define

$$a^\varepsilon := u^\varepsilon e^{-i\phi/\varepsilon}.$$

Equation (4.19) is *equivalent* to

$$\begin{cases} \partial_t a^\varepsilon + \nabla\phi \cdot \nabla a^\varepsilon + \dfrac{1}{2}a^\varepsilon \Delta\phi = i\dfrac{\varepsilon}{2}\Delta a^\varepsilon - \dfrac{i}{\varepsilon}\left(|a^\varepsilon|^{2\sigma} - |a|^{2\sigma}\right)a^\varepsilon, \\[2mm] a^\varepsilon_{|t=0} = a_0^\varepsilon. \end{cases} \tag{4.21}$$

The main problem to prove that a^ε is bounded in Sobolev spaces is the presence of the singular factor $1/\varepsilon$ on the right-hand side. On the other hand, the analysis of Sec. 4.2 shows that it is natural to expect

$$a^\varepsilon = a + \mathcal{O}(\varepsilon).$$

In this case, the singular factor $1/\varepsilon$ is compensated. Nevertheless, this argument does not seem closed: apparently, to prove that a^ε is bounded in Sobolev spaces, we need a more precise information. The idea in [Alazard and Carles (2009b)] consists in introducing an extra unknown function in order to obtain a closed system of estimates. The approach of considering more unknown functions that in the initial problem has proven very efficient in several contexts. We can mention the study of blow-up for the nonlinear wave equation, [Alinhac (1995b)] (see also [Alinhac (1995a, 2002)]), low Mach number limit of the full Navier–Stokes equations [Alazard (2006)], or geometric optics for the incompressible Euler or Navier–Stokes equations [Cheverry (2004, 2006); Cheverry and Guès (2008)].

Inspired by the analysis of E. Grenier, the idea is to symmetrize the equations, and to obtain a system for the family of unknown functions which is hyperbolic symmetric. Split the term $|a^\varepsilon|^{2\sigma} - |a|^{2\sigma}$ as a product

$$|a^\varepsilon|^2 - |a|^{2\sigma} = \varepsilon g^\varepsilon q^\varepsilon = \varepsilon(G\,Q)(|a^\varepsilon|^2, |a|^2)$$

$$= \varepsilon G(r_1, r_2) Q(r_1, r_2)\big|_{(r_1, r_2) = (|a^\varepsilon|^2, |a|^2)},$$

where q^ε satisfies an equation of the form

$$\partial_t q^\varepsilon + L(a, \phi, \partial_x) q^\varepsilon + g^\varepsilon \,\mathrm{div}\,(\mathrm{Im}(\overline{a}^\varepsilon \nabla a^\varepsilon)) = 0, \qquad (4.22)$$

and L is a first order differential operator. Introduce the position densities

$$\rho := |a|^2 \quad ; \quad \rho^\varepsilon := |a^\varepsilon|^2 = |u^\varepsilon|^2.$$

Recall that $v = \nabla\phi$. Elementary computations show that:

$$\partial_t \rho + \mathrm{div}(\rho v) = 0, \qquad (4.23)$$

$$\partial_t \rho^\varepsilon + \mathrm{div}\,\mathrm{Im}\,(\varepsilon \overline{u}^\varepsilon \nabla u^\varepsilon) = 0, \qquad (4.24)$$

$$\partial_t \rho^\varepsilon + \mathrm{div}\,(\mathrm{Im}(\varepsilon \overline{a}^\varepsilon \nabla a^\varepsilon) + \rho^\varepsilon v) = 0. \qquad (4.25)$$

Denote

$$\beta^\varepsilon = \varepsilon q^\varepsilon = B(r_1, r_2)\big|_{(r_1, r_2) = (|a^\varepsilon|^2, |a|^2)} \quad ; \quad J^\varepsilon := \varepsilon\,\mathrm{Im}(\overline{a}^\varepsilon \nabla a^\varepsilon).$$

By writing

$$\partial_t \beta^\varepsilon = (\partial_{r_1} B)(\rho^\varepsilon, \rho)\partial_t \rho^\varepsilon + (\partial_{r_2} B)(\rho^\varepsilon, \rho)\partial_t \rho,$$

we compute, from (4.23) and (4.25):

$$\partial_t \beta^\varepsilon + (\partial_{r_1} B)(\rho^\varepsilon, \rho)\,\mathrm{div}(J^\varepsilon + \rho^\varepsilon v) + (\partial_{r_2} B)(\rho^\varepsilon, \rho)\,\mathrm{div}(\rho v) = 0.$$

Hence, in order to have an equation of the desired form (4.22), we impose

$$\partial_{r_1} B(r_1, r_2) = G(r_1, r_2).$$

Since on the other hand,

$$G(r_1, r_2) B(r_1, r_2) = r_1^\sigma - r_2^\sigma,$$

this suggests to choose β^ε such that

$$(\beta^\varepsilon)^2 = \frac{2}{\sigma + 1}(\rho^\varepsilon)^{\sigma+1} - 2\rho^\sigma \rho^\varepsilon + f(\rho). \qquad (4.26)$$

To obtain an operator L which is linear with respect to β^ε we choose

$$(\beta^\varepsilon)^2 = \frac{2}{\sigma + 1}(\rho^\varepsilon)^{\sigma+1} - \frac{2}{\sigma + 1}\rho^{\sigma+1} - 2\rho^\sigma(\rho^\varepsilon - \rho). \qquad (4.27)$$

With this choice, we formally compute:

$$\partial_t \beta^\varepsilon + \varepsilon g^\varepsilon \operatorname{div}(\operatorname{Im}(\overline{a}^\varepsilon \nabla a^\varepsilon)) + v \cdot \nabla \beta^\varepsilon + \frac{\sigma+1}{2} \beta^\varepsilon \operatorname{div} v = 0.$$

This equation is derived rigorously in [Alazard and Carles (2009b)], and we refer to the paper for the complete proof. Examine the right-hand side of (4.27). Taylor's formula yields

$$\frac{2}{\sigma+1}(\rho^\varepsilon)^{\sigma+1} - \frac{2}{\sigma+1}\rho^{\sigma+1} - 2\rho^\sigma(\rho^\varepsilon - \rho) = (\rho^\varepsilon - \rho)^2 Q_\sigma(\rho^\varepsilon, \rho),$$

where Q_σ is given by:

$$Q_\sigma(r_1, r_2) := 2\sigma \int_0^1 (1-s)\,(r_2 + s(r_1 - r_2))^{\sigma-1}\,ds. \tag{4.28}$$

Note that there exists C_σ such that:

$$Q_\sigma(r_1, r_2) \geqslant C_\sigma \left(r_1^{\sigma-1} + r_2^{\sigma-1} \right). \tag{4.29}$$

This is the same convexity inequality as the one we have used in Sec. 3.2.

Notation 4.15. Let $\sigma \in \mathbb{N}$. Introduce

$$G_\sigma(r_1, r_2) = \frac{P_\sigma(r_1, r_2)}{\sqrt{Q_\sigma(r_1, r_2)}} \quad ; \quad B_\sigma(r_1, r_2) := (r_1 - r_2)\sqrt{Q_\sigma(r_1, r_2)},$$

where Q_σ is given by (4.28) and

$$P_\sigma(r_1, r_2) = \frac{r_1^\sigma - r_2^\sigma}{r_1 - r_2} = \sum_{\ell=0}^{\sigma-1} r_1^{\sigma-1-\ell} r_2^\ell.$$

Example 4.16. For $\sigma = 1, 2, 3$, we compute

$$G_1 = 1, \qquad\qquad\qquad B_1 = r_1 - r_2.$$

$$G_2 = \sqrt{\frac{3}{2}} \frac{r_1 + r_2}{\sqrt{r_1 + 2r_2}}, \qquad B_2 = \sqrt{\frac{2}{3}}(r_1 - r_2)\sqrt{r_1 + 2r_2}.$$

$$G_3 = \sqrt{2} \frac{r_1^2 + r_1 r_2 + r_2^2}{\sqrt{(r_1 - r_2)^2 + 2r_2^2}}, \qquad B_3 = \frac{1}{\sqrt{2}}(r_1 - r_2)\sqrt{(r_1 - r_2)^2 + 2r_2^2}.$$

A remarkable fact is that, although the functions G_σ and B_σ are not smooth for $\sigma \geqslant 2$, one can compute an evolution equation for the unknown β^ε, as mentioned above.

Denote $\psi^\varepsilon = \nabla a^\varepsilon$. Noticing that

$$g^\varepsilon \operatorname{div}(\operatorname{Im}(\overline{a}^\varepsilon \psi^\varepsilon)) = \operatorname{Im}(g^\varepsilon \overline{a}^\varepsilon \operatorname{div} \psi^\varepsilon),$$

we find that the unknown $(a^\varepsilon, \psi^\varepsilon, q^\varepsilon)$ solves

$$
\begin{cases}
\partial_t a^\varepsilon + v \cdot \nabla a^\varepsilon - i\dfrac{\varepsilon}{2}\Delta a^\varepsilon = -\dfrac{1}{2}a^\varepsilon \operatorname{div} v - ig^\varepsilon q^\varepsilon a^\varepsilon. \\[2mm]
\partial_t \psi^\varepsilon + v \cdot \nabla \psi^\varepsilon + ia^\varepsilon g^\varepsilon \nabla q^\varepsilon - i\dfrac{\varepsilon}{2}\Delta \psi^\varepsilon \\[2mm]
\quad = -\dfrac{1}{2}\psi^\varepsilon \operatorname{div} v - \psi^\varepsilon \cdot \nabla v - \dfrac{1}{2}a^\varepsilon \nabla \operatorname{div} v - iq^\varepsilon \nabla (a^\varepsilon g^\varepsilon), \\[2mm]
\partial_t q^\varepsilon + v \cdot \nabla q^\varepsilon + \operatorname{Im}(g^\varepsilon \overline{a}^\varepsilon \operatorname{div} \psi^\varepsilon) = -\dfrac{\sigma+1}{2}q^\varepsilon \operatorname{div} v.
\end{cases}
\tag{4.30}
$$

Note that by assumption,

$$
\|a^\varepsilon_{|t=0}\|_{H^s(\mathbb{R}^d)} + \|\psi^\varepsilon_{|t=0}\|_{H^s(\mathbb{R}^d)} = \mathcal{O}(1), \quad \forall s \geqslant 0.
\tag{4.31}
$$

A similar estimate for the initial data of q^ε is a more delicate issue, since B_σ is not a smooth function. An important remark is that q^ε, viewed as a function of a^ε and a, is an homogeneous function of degree $\sigma + 1$. We then use the following lemma, which can be proved by induction on m:

Lemma 4.17. *Let $p \geqslant 1$ and $m \geqslant 2$ be integers and consider $F \colon \mathbb{R}^p \to \mathbb{C}$. Assume that $F \in C^\infty(\mathbb{R}^p \setminus \{0\})$ is homogeneous of degree m, that is:*

$$
F(\lambda y) = \lambda^m F(y), \qquad \forall \lambda \geqslant 0, \forall y \in \mathbb{R}^p.
$$

Then, for $d \leqslant 3$, there exists $K > 0$ such that, for all $u \in H^m(\mathbb{R}^d)$ with values in $\mathbb{R}^p$, $F(u) \in H^m(\mathbb{R}^d)$ and

$$
\|F(u)\|_{H^m} \leqslant K\|u\|_{H^m}^m.
$$

The same is true when $m = 1$ and $d \in \mathbb{N}$.

Remark 4.18. Note that the result is false for $d \geqslant 4$ and $m \geqslant 2$. Also, one must not expect $F(u) \in H^{m+1}(\mathbb{R}^d)$, even for $u \in H^\infty(\mathbb{R}^d)$. For instance, if

$$
d = 1 = p, \quad m = 2, \quad F(y) = y|y|, \quad u(x) = xe^{-x^2},
$$

then $F(u) \in H^2(\mathbb{R})$ and $F(u) \notin H^3(\mathbb{R})$. Similarly, in general, one must not expect $F_\sigma(u,v) \in H^{\sigma+1}(\mathbb{R}^d)$, even for $(u,v) \in H^\infty(\mathbb{R}^d)^2$.

The left-hand side of (4.30) is a first order quasi-linear symmetric hyperbolic system, plus a second order skew-symmetric term. The right-hand side can be viewed as a semi-linear source term. Denoting $U^\varepsilon := (2q^\varepsilon, a^\varepsilon, \overline{a}^\varepsilon, \psi^\varepsilon, \overline{\psi}^\varepsilon)$, we see that

$$
\partial_t U^\varepsilon + \sum_{1 \leqslant j \leqslant n} A_j(v, a^\varepsilon g^\varepsilon, \overline{a}^\varepsilon g^\varepsilon)\partial_j U^\varepsilon + \varepsilon\mathcal{L}(\partial_x)U^\varepsilon = E(\Phi, U^\varepsilon, a^\varepsilon g^\varepsilon, \nabla(a^\varepsilon g^\varepsilon)),
$$

where $\Phi = (\nabla\phi, \nabla^2\phi, \nabla^3\phi)$, the A_j's are Hermitian matrices linear in their arguments, $\mathcal{L}(\partial_x) = \sum L_{jk}\partial_j\partial_k$ is a skew-symmetric second order differential operator with constant coefficients, and E is a C^∞ function of its arguments, vanishing at the origin.

Using the above structure, quasi-linear analysis for hyperbolic systems, and estimates for non-smooth homogeneous functions, the main result in [Alazard and Carles (2009b)] follows:

Theorem 4.19. *Let $d \leqslant 3$, $\phi_0 \in H^\infty$ and let Assumption 4.13 be satisfied. There exists $T \in]0, T^*[$, where T^* is given by Lemma 4.14, such that the following holds. For all $\varepsilon \in]0, 1]$, the Cauchy problem (1.1) has a unique solution $u^\varepsilon \in C([-T, T]; H^\infty(\mathbb{R}^d))$. Moreover,*

$$\sup_{\varepsilon \in]0,1]} \left(\left\| a^\varepsilon \right\|_{L^\infty([-T,T];H^k(\mathbb{R}^d))} + \left\| q^\varepsilon \right\|_{L^\infty([-T,T];H^{k-1}(\mathbb{R}^d))} \right) < +\infty, \quad (4.32)$$

where the index k is as follows:

- *If $\sigma = 1$, then $k \in \mathbb{N}$ is arbitrary.*
- *If $\sigma = 2$ and $d = 1$, then we can take $k = 2$.*
- *If $\sigma = 2$ and $2 \leqslant d \leqslant 3$, then we can take $k = 1$.*
- *If $\sigma \geqslant 3$, then we can take $k = \sigma$.*

The assumption $d \leqslant 3$ appears when estimating non-smooth homogeneous functions. It could be removed, up to considering sufficiently large σ. The restriction for k when $\sigma \geqslant 2$ follows from Lemma 4.17. This argument is used not only to estimate q^ε at time $t = 0$, but also the factor $a^\varepsilon g^\varepsilon$ in terms of a^ε, in Eq. (4.30).

Note that for $k = 1$, Eq. (4.32) is exactly Theorem 3.4. From this point of view, Eq. (4.30) can be interpreted as a local form for the modulated energy. This is even more explicit when considering

$$e^\varepsilon := |a^\varepsilon|^2 + |\psi^\varepsilon|^2 + |q^\varepsilon|^2.$$

It satisfies an equation of the form $\partial_t e^\varepsilon + \text{div}(\eta^\varepsilon) + \flat^\varepsilon = \mathcal{O}(e^\varepsilon)$, where $\int \flat^\varepsilon = 0$. Indeed, directly from Eq. (4.30), we compute

$$\partial_t e^\varepsilon + \text{div}(v e^\varepsilon) + 2 \, \text{div}\big(\text{Im}(g^\varepsilon q^\varepsilon \overline{a}^\varepsilon \psi^\varepsilon)\big) + \varepsilon \, \text{Im}\left(\overline{a}^\varepsilon \Delta a^\varepsilon + \overline{\psi^\varepsilon} \Delta \psi^\varepsilon \right)$$
$$= -\sigma |q^\varepsilon|^2 \, \text{div}\, v - \text{Re}\left((2\psi^\varepsilon \cdot \nabla v + a^\varepsilon \nabla \, \text{div}\, v) \overline{\psi^\varepsilon} \right).$$

We have thus obtained an evolution equation for a local modulated energy. Gronwall lemma yields

$$\|e^\varepsilon(t)\|_{L^1(\mathbb{R}^d)} \leqslant \|e^\varepsilon(0)\|_{L^1(\mathbb{R}^d)} \exp\left(Ct \right).$$

Finally, we check that $(e^\varepsilon(0))_\varepsilon$ is bounded in $L^1(\mathbb{R}^d)$. Therefore, this argument yields Eq. (4.32) for $k = 1$.

To obtain the pointwise asymptotics of the wave function u^ε, we can proceed in a similar spirit, guided by the results of Sec. 4.2.1. We make the following assumption on the initial amplitude:

Assumption 4.20. There exist $a_0, a_1 \in H^\infty$ such that for all $s \geqslant 0$,

$$\|a_0^\varepsilon - a_0 - \varepsilon a_1\|_{H^s} = \mathcal{O}\left(\varepsilon^2\right).$$

Introduce the system

$$\begin{cases} \partial_t \phi^{(1)} + \nabla\phi \cdot \nabla\phi^{(1)} + 2\sigma \operatorname{Re}\left(\overline{a}a^{(1)}\right) |a|^{2\sigma-2} = 0, \\ \partial_t a^{(1)} + \nabla\phi \cdot \nabla a^{(1)} + \nabla\phi^{(1)} \cdot \nabla a + \frac{1}{2}a^{(1)}\Delta\phi + \frac{1}{2}a\Delta\phi^{(1)} = \frac{i}{2}\Delta a, \\ \left.\phi^{(1)}\right|_{t=0} = 0 \quad ; \quad \left.a^{(1)}\right|_{t=0} = a_1. \end{cases}$$

$$(4.33)$$

Again, at the zeroes of a, Eq. (4.33) ceases to be strictly hyperbolic, and we cannot solve the Cauchy problem by a standard argument. Yet, we can prove:

Lemma 4.21. *Let $n \geqslant 1$, $\phi_0 \in H^\infty$, and let Assumption 4.20 be satisfied. Then (4.33) has a unique solution $(\phi^{(1)}, a^{(1)})$ in $C([-T^*, T^*]; H^\infty)^2$, where T^* is given by Lemma 4.14.*

Proof. [Sketch of the proof] We transform the equations so as to obtain an auxiliary hyperbolic system for $(\nabla\phi^{(1)}, A_1)$ for some unknown A_1, depending linearly upon $a^{(1)}$. The definition of A_1 depends on the parity of σ. This allows to determine a function $\phi^{(1)}$ and, next, to define a function $a^{(1)}$ by solving the second equation in (4.33). We conclude the proof by checking that $(\phi^{(1)}, a^{(1)})$ does solve (4.33). The first change of unknown consists in considering $v_1 := \nabla\phi^{(1)}$. The first equation in (4.33) yields:

$$\partial_t v_1 + v \cdot \nabla v_1 + 2\sigma\nabla \operatorname{Re}\left(|a|^{2\sigma-2}\overline{a}a^{(1)}\right) = -v_1 \cdot \nabla v.$$

First case: $\sigma \geqslant 2$ **is even.** Consider the new unknown

$$A_1 := |a|^{\sigma-2} \operatorname{Re}\left(\overline{a}a^{(1)}\right).$$

We check that, if $(\phi^{(1)}, a^{(1)})$ solves (4.33), then

$$\begin{cases} \partial_t v_1 + v \cdot \nabla v_1 + 2\sigma|a|^\sigma \nabla A_1 = -v_1 \cdot \nabla v - 2\sigma A_1 \nabla\left(|a|^\sigma\right), \\ \partial_t A_1 + v \cdot \nabla A_1 + \frac{1}{2}|a|^\sigma \operatorname{div} v_1 = -\frac{1}{\sigma}\nabla\left(|a|^\sigma\right) \cdot v_1 - \frac{\sigma}{2}A_1 \operatorname{div} v \\ \hphantom{\partial_t A_1 + v \cdot \nabla A_1 + \frac{1}{2}|a|^\sigma \operatorname{div} v_1 = } + \frac{i}{2}\operatorname{Re}\left(|a|^{\sigma-2}\overline{a}\Delta a\right). \end{cases}$$

$$(4.34)$$

This linear system is hyperbolic symmetric, and its coefficients are smooth. In particular, uniqueness for (4.33) follows from the uniqueness for (4.34). Equation (4.34) possesses a unique solution in $C^\infty([-T^*, T^*]; H^\infty)$. We next define $\phi^{(1)}$ and $a^{(1)}$ as announced above.

Second case: σ is odd. In this case, $\sigma = 2m + 1$, for some $m \in \mathbb{N}$. We consider the new unknown

$$A_1 := |a|^{\sigma-1} a^{(1)} = |a|^{2m} a^{(1)}.$$

We check that (v_1, A_1) must solve

$$\begin{cases} \partial_t v_1 + v \cdot \nabla v_1 + 2\sigma \operatorname{Re}\left(|a|^{2m} \overline{a} \nabla A_1\right) = -v_1 \cdot \nabla v - 2\sigma \operatorname{Re}\left(A_1 \nabla \left(|a|^{2m} \overline{a}\right)\right), \\ \partial_t A_1 + v \cdot \nabla A_1 + \dfrac{1}{2} |a|^{2m} a \operatorname{div} v_1 = -\dfrac{\sigma}{2} A_1 \operatorname{div} v - |a|^{2m} \nabla a \cdot v_1 \\ \qquad\qquad\qquad\qquad + \dfrac{i}{2} |a|^{2m} \Delta a. \end{cases}$$

We can then conclude as in the first case. $\qquad\qquad\qquad\qquad\qquad\square$

We can now describe the wave function u^ε at leading order as $\varepsilon \to 0$:

Proposition 4.22. *Let $d \leqslant 3$, $\phi_0 \in H^\infty$, and let Assumption 4.20 be satisfied. Set $\widetilde{a} := a e^{i\phi^{(1)}}$. Then there exists $\varepsilon_0 > 0$ such that $a^\varepsilon \in C([-T^*, T^*]; H^\infty)$ for $\varepsilon \in]0, \varepsilon_0]$, and*

$$\|a^\varepsilon - \widetilde{a}\|_{L^\infty([-T^*, T^*]; H^k)} = \mathcal{O}(\varepsilon),$$

where k is as in Theorem 4.19.

Proof. We indicate the main steps of the proof only. Denote

$$r^\varepsilon = a^\varepsilon - \widetilde{a} \quad ; \quad \widetilde{a}^{(1)} = a^{(1)} e^{i\phi^{(1)}}.$$

From (4.20), (4.21) and (4.33), we see that r^ε solves

$$\begin{cases} \partial_t r^\varepsilon + v \cdot \nabla r^\varepsilon + \dfrac{1}{2} r^\varepsilon \operatorname{div} v - i \dfrac{\varepsilon}{2} \Delta r^\varepsilon = i \dfrac{\varepsilon}{2} \Delta \widetilde{a} - i S^\varepsilon, \\ r^\varepsilon_{|t=0} = a_0^\varepsilon - a_0 = \varepsilon a_1 + \mathcal{O}\left(\varepsilon^2\right), \end{cases}$$

where the term S^ε is given by:

$$S^\varepsilon = \dfrac{1}{\varepsilon} \left(|a^\varepsilon|^{2\sigma} - |\widetilde{a}|^{2\sigma}\right) a^\varepsilon - 2\sigma \widetilde{a} |\widetilde{a}|^{2\sigma-2} \operatorname{Re}\left(\overline{\widetilde{a}} \widetilde{a}^{(1)}\right).$$

We check that for all $s \geqslant 0$, we have, in $H^s(\mathbb{R}^d)$:

$$S^\varepsilon = \dfrac{1}{\varepsilon} \left(|a^\varepsilon|^{2\sigma} - |\widetilde{a} + \varepsilon \widetilde{a}^{(1)}|^{2\sigma}\right) a^\varepsilon + 2\sigma r^\varepsilon |\widetilde{a}|^{2\sigma-2} \operatorname{Re}\left(\overline{\widetilde{a}} \widetilde{a}^{(1)}\right) + \mathcal{O}(\varepsilon).$$

The last term should be viewed as a small source term. The second one is linear in r^ε, and is suitable in view of an application of the Gronwall lemma. There remains to handle the first term. At this stage, we can mimic the previous approach. Introduce the nonlinear change of unknown:

$$\widetilde{q}^\varepsilon = \frac{1}{\varepsilon} B_\sigma \left(|a^\varepsilon|^2, |\widetilde{a} + \varepsilon \widetilde{a}^{(1)}|^2 \right) \quad ; \quad \widetilde{g}^\varepsilon = G_\sigma \left(|a^\varepsilon|^2, |\widetilde{a} + \varepsilon \widetilde{a}^{(1)}|^2 \right),$$

where B_σ and G_σ are defined in Notation 4.15. We check that $(r^\varepsilon, \nabla r^\varepsilon, \widetilde{q}^\varepsilon)$ solves a system of the form (4.30), plus some extra source terms of order $\mathcal{O}(\varepsilon)$ in $H^s(\mathbb{R}^d)$. We also note that the initial data are of order $\mathcal{O}(\varepsilon)$, from Assumption 4.20:

$$\left\| (r^\varepsilon, \nabla r^\varepsilon) \big|_{t=0} \right\|_{H^s} = \mathcal{O}(\varepsilon), \ \forall s \geqslant 0.$$

We also have

$$\left\| \widetilde{q}^\varepsilon \big|_{t=0} \right\|_{H^{k-1}} = \mathcal{O}(\varepsilon),$$

where k is as Theorem 4.19.

The proposition then stems from Gronwall lemma and a standard continuity argument, which we sketch for the convenience of the reader (see also [Rauch (2012)]). Proceeding like in the proof of Theorem 4.19, Gronwall lemma shows that there exists $T > 0$ and $C > 0$ independent of ε, such that

$$\|r^\varepsilon\|_{L^\infty([-T,T];H^k)} + \|\widetilde{q}^\varepsilon\|_{L^\infty([-T,T];H^{k-1})} \leqslant C\varepsilon.$$

The rate $\mathcal{O}(\varepsilon)$ follows from the fact that the initial data are of order $\mathcal{O}(\varepsilon)$, as well as the source term in the system for $(r^\varepsilon, \nabla r^\varepsilon, \widetilde{q}^\varepsilon)$. If a^ε were not smooth on $[0, T^*]$, then there would exist $t^\varepsilon > 0$ such that

$$\|r^\varepsilon(t^\varepsilon)\|_{H^k} + \|\widetilde{q}^\varepsilon(t^\varepsilon)\|_{H^{k-1}} = 1.$$

Let $\underline{t}^\varepsilon$ the smallest such $t^\varepsilon > 0$:

$$\|r^\varepsilon\|_{L^\infty([0,\underline{t}^\varepsilon];H^k)} + \|\widetilde{q}^\varepsilon\|_{L^\infty([0,\underline{t}^\varepsilon];H^{k-1})} \leqslant 1.$$

Using this estimate, and Gronwall lemma, we infer

$$\|r^\varepsilon\|_{L^\infty([0,\underline{t}^\varepsilon];H^k)} + \|\widetilde{q}^\varepsilon\|_{L^\infty([0,\underline{t}^\varepsilon];H^{k-1})} \leqslant \underline{C}\varepsilon,$$

for some $\underline{C} > 0$ independent of ε. For $\varepsilon > 0$ sufficiently small, $\underline{C}\varepsilon \leqslant 1/2$. This contradicts the definition of t^ε. Therefore, a^ε is smooth on $[0, T^*]$ (and on $[-T^*, 0]$ by the same argument) provided $\varepsilon \leqslant \varepsilon_0$ for some $\varepsilon_0 > 0$ sufficiently small, and the error estimate follows. $\square$

4.4   On conservation laws

Recall some important evolution laws for (4.19):

$$\text{Mass: } \frac{d}{dt}\|u^\varepsilon(t)\|_{L^2} = 0\,.$$

$$\text{Energy: } \frac{d}{dt}\left(\frac{1}{2}\|\varepsilon\nabla_x u^\varepsilon\|_{L^2}^2 + \frac{1}{\sigma+1}\|u^\varepsilon\|_{L^{2\sigma+2}}^{2\sigma+2}\right) = 0\,.$$

$$\text{Momentum: } \frac{d}{dt}\,\text{Im}\int \overline{u}^\varepsilon(t,x)\varepsilon\nabla_x u^\varepsilon(t,x)dx = 0\,.$$

$$\text{Pseudo-conformal law: } \frac{d}{dt}\left(\frac{1}{2}\|J^\varepsilon(t)u^\varepsilon\|_{L^2}^2 + \frac{t^2}{\sigma+1}\|u^\varepsilon\|_{L^{2\sigma+2}}^{2\sigma+2}\right)$$

$$= \frac{t}{\sigma+1}(2 - d\sigma)\|u^\varepsilon\|_{L^{2\sigma+2}}^{2\sigma+2}\,,$$

where $J^\varepsilon(t) = x + i\varepsilon t\nabla$. These evolutions are deduced from the usual ones ($\varepsilon = 1$, see e.g. [Cazenave (2003); Sulem and Sulem (1999)]) via the scaling $\psi(t,x) = u(\varepsilon t, \varepsilon x)$. Writing $u^\varepsilon = a^\varepsilon e^{i\phi/\varepsilon}$, and passing to the limit formally in the above formulae yields:

$$\frac{d}{dt}\|a(t)\|_{L^2} = 0\,.$$

$$\frac{d}{dt}\int\left(\frac{1}{2}|a(t,x)|^2|\nabla\phi(t,x)|^2 + \frac{1}{\sigma+1}|a(t,x)|^{2\sigma+2}\right)dx = 0\,.$$

$$\frac{d}{dt}\int |a(t,x)|^2\nabla\phi(t,x)dx = 0\,.$$

$$\frac{d}{dt}\int\left(\frac{1}{2}|(x - t\nabla\phi(t,x))\,a(t,x)|^2 + \frac{t^2}{\sigma+1}|a(t,x)|^{2\sigma+2}\right)dx$$

$$= \frac{t}{\sigma+1}(2 - d\sigma)\int |a(t,x)|^{2\sigma+2}dx\,.$$

Note that we also have the conservation ([Carles and Nakamura (2004)]):

$$\frac{d}{dt}\,\text{Re}\int \overline{u}^\varepsilon(t,x)J^\varepsilon(t)u^\varepsilon(t,x)dx = 0\,,$$

which yields:

$$\frac{d}{dt}\int (x - t\nabla\phi(t,x))\,|a(t,x)|^2dx = 0\,.$$

All these expressions involve only $(|a|^2, \nabla\phi) = (|\widetilde{a}|^2, \nabla\phi)$. Recall that if we set $(\rho, v) = (|a|^2, \nabla\phi)$,

$$\begin{cases} \partial_t v + v \cdot \nabla v + \nabla(\rho^\sigma) = 0 &;\quad v_{|t=0} = \nabla\phi_0, \\ \partial_t \rho + \text{div}(\rho v) = 0 &;\quad \rho_{|t=0} = |a_0|^2. \end{cases} \tag{4.35}$$

Rewriting the above evolution laws, we get:

$$\frac{d}{dt}\int_{\mathbb{R}^d}\rho(t,x)dx = 0.$$

$$\frac{d}{dt}\int\left(\frac{1}{2}\rho(t,x)|v(t,x)|^2 + \frac{1}{\sigma+1}\rho(t,x)^{\sigma+1}\right)dx = 0.$$

$$\frac{d}{dt}\int\rho(t,x)v(t,x)dx = 0.$$

$$\frac{d}{dt}\int\left(\frac{1}{2}\left|(x-tv(t,x))\right|^2\rho(t,x) + \frac{t^2}{\sigma+1}\rho(t,x)^{\sigma+1}\right)dx$$

$$= \frac{t}{\sigma+1}(2-d\sigma)\int\rho(t,x)^{\sigma+1}dx.$$

$$\frac{d}{dt}\int\left(x-tv(t,x)\right)\rho(t,x)dx = 0.$$

We thus retrieve formally some evolution laws for the compressible Euler equation (4.35) (see e.g. [Serre (1997); Xin (1998)]), with the pressure law $p(\rho) = c\rho^{\sigma+1}$.

4.5 Focusing nonlinearities

The main feature of the limit system we used is that it enters, up to a change of unknowns, into the framework of quasi-linear hyperbolic systems. This comes from the fact that we consider the defocusing case. Had we worked instead with the focusing case, where $+|u|^{2\sigma}u$ is replaced with $-|u|^{2\sigma}u$, the corresponding limit system would have been ill-posed. We refer to [Métivier (2005)], in which G. Métivier establishes Hadamard's instabilities for non-hyperbolic nonlinear equations.

As an example, consider the Cauchy problem in space dimension $n = 1$

$$\begin{cases}\partial_t\phi + \frac{1}{2}|\partial_x\phi|^2 - |a|^{2\sigma} = 0 & ; \quad \phi_{|t=0} = \phi_0, \\ \partial_t a + \partial_x\phi\partial_x a + \frac{1}{2}a\partial_x^2\phi = 0 & ; \quad a_{|t=0} = a_0.\end{cases} \tag{4.36}$$

As opposed to the previous analysis, this system is not hyperbolic, but elliptic. The following result follows from Hadamard's argument (see [Métivier (2005)]).

Proposition 4.23. *Suppose that (ϕ, a) in $C^2([0,T]\times\mathbb{R})^2$ solves (4.36). If $\phi_0(x)$ is real analytic near $\underline{x}$ and if $a_0(\underline{x}) > 0$, then $a_0(x)$ is real analytic near $\underline{x}$. Consequently, there are smooth initial data for which the Cauchy problem has no solution.*

This result was extended to the multi-dimensional framework in [Lerner *et al.* (2018)].

This shows that to study the semi-classical limit for the focusing analogue of (1.1), working with analytic data, is not only convenient: it is necessary.

On the other hand, data and solutions with analytic regularity seem appropriate. In [Gérard (1993)], P. Gérard works with the analytic regularity, when the space variable x belongs to the torus $\mathbb{T}^d$, without external potential ($V \equiv 0$). Note that the only assumption needed on the nonlinearity f is analyticity near the range of $\left| a_0^{(0)} \right|^2$ (see below for the definition of $a_0^{(0)}$). This includes the case $f(y) = \lambda y^\sigma$, $\lambda \in \mathbb{R}$, $\sigma \in \mathbb{N}$. This result was extended to the case of the whole Euclidean space $\mathbb{R}^d$ in [Thomann (2008)].

The initial phase ϕ_0 is supposed real analytic, and the initial amplitude is analytic in the following sense (see [Sjöstrand (1982)]):

$$a_0^\varepsilon(z) = \sum_{j \geqslant 0} \varepsilon^j a_0^{(j)}(z),$$

where the functions $a_0^{(j)}$ satisfy the following properties. There exist $\ell > 0$, $A > 0$, $B > 0$ such that, for all $j \geqslant 0$, $a_0^{(j)}$ is holomorphic in $\{|\operatorname{Im} z| < \ell\}$ ($z = (z_1, \ldots, z_d) \in \mathbb{C}^d$), and

$$\left| a_0^{(j)}(z) \right| \leqslant AB^j j! \quad \text{on } \{|\operatorname{Im} z| < \ell\}.$$

To consider functions which decay sufficiently at infinity to be in $H^s(\mathbb{R}^d)$, L. Thomann introduces the weight

$$W(z) = \exp\left(1 + z^2\right)^{1/2}, \quad z^2 = z_1^2 + \ldots + z_n^2.$$

The condition on the coefficients $a_0^{(j)}$ becomes:

$$\left| W(z) a_0^{(j)}(z) \right| \leqslant AB^j j! \quad \text{on } \{|\operatorname{Im} z| < \ell\}.$$

Denoting $\overline{a}(t, z)$ the complex conjugate of $a(\overline{t}, \overline{z})$, P. Gérard constructs a formal solution of the form

$$u^\varepsilon = a^\varepsilon e^{i\phi/\varepsilon}, \quad a^\varepsilon(t, z) = \sum_{j \geqslant 0} \varepsilon^j a^{(j)}(t, z),$$

which satisfies:

$$\begin{cases} \partial_t v = -v \cdot \nabla v - \nabla f\left(a^{(0)} \overline{a}^{(0)}\right), \\ \partial_t a^\varepsilon = -v \cdot \nabla a^\varepsilon - \dfrac{1}{2} a^\varepsilon \operatorname{div} v + i\dfrac{\varepsilon}{2} \Delta a^\varepsilon - \dfrac{i a^\varepsilon}{\varepsilon}\left(f\left(a^\varepsilon \overline{a}^\varepsilon\right) - f\left(a^{(0)} \overline{a}^{(0)}\right)\right). \end{cases}$$

The sum is defined in the sense of J. Sjöstrand: there exists $t_0 > 0$ such that, for all $j \geqslant 0$, $a^{(j)}$ is holomorphic in $\{|t| < t_0\} \times \{|\operatorname{Im} z| < \ell\}$, and

$$\left| W(z) a^{(j)}(t, z) \right| \leqslant AB^j j! \quad \text{on } \{|t| \leqslant t_0\} \times \{|\operatorname{Im} z| < \ell\},$$

with the convention $W \equiv 1$ in the periodic case. To make the argument complete, one has to truncate the formal series, in such a way that the resulting function solves the nonlinear Schrödinger, up to an error term as small as possible. Setting

$$v^\varepsilon = e^{i\phi/\varepsilon} \sum_{j \leqslant 1/(C_0 \varepsilon)} \varepsilon^j a^{(j)}$$

for C_0 sufficiently large, the approximate solution v^ε satisfies:

$$i\varepsilon \partial_t v^\varepsilon + \frac{\varepsilon^2}{2} \Delta v^\varepsilon = f\left(|v^\varepsilon|^2\right) v^\varepsilon + \mathcal{O}\left(e^{-\delta/\varepsilon}\right),$$

for some $\delta > 0$. Essentially, this source term is sufficiently small to overcome the difficulty pointed out at the end of Sec. 1.2: for small time independent of ε, the exponential growth provided by Gronwall lemma is more than compensated by the term $e^{-\delta/\varepsilon}$, so it is possible to justify nonlinear geometric optics by semi-linear arguments. We refer to [Gérard (1993)] and [Thomann (2008)] for precise statements and complete proofs.

An alternative technique to work at the analytic level is inspired by the approach of [Ginibre and Velo (2001)], where one of the ingredients consists in allowing a time-dependent (nonincreasing) radius of analyticity. We simply outline the main steps of the argument in this setting. Introducing the exponential weight, for $\varrho > 0$, $\mathtt{w}(\xi) = e^{\varrho \langle \xi \rangle}$, consider the family of norms

$$\|\psi\|_{\mathcal{H}_\varrho^\ell}^2 = \int_{\mathbb{R}^d} \langle \xi \rangle^{2\ell} e^{2\varrho \langle \xi \rangle} |\widehat{\psi}(\xi)|^2 d\xi, \quad \ell \geqslant 0.$$

Allowing ϱ to depend on time, we compute

$$\frac{d}{dt} \|\psi\|_{\mathcal{H}_\varrho^\ell}^2 = 2\dot{\varrho} \|\psi\|_{\mathcal{H}_\varrho^{\ell+1/2}}^2 + 2\operatorname{Re} \langle \psi, \partial_t \psi \rangle_{\mathcal{H}_\varrho^\ell},$$

so a decreasing ϱ provides regularizing estimates in the same fashion as for parabolic equations, and makes it possible to consider (4.4) (we assume $V = 0$ and f is a polynomial), thanks to the technical lemma:

Lemma 4.24. *Let $m \geqslant 0$. Then,*
1. *For $k + s > m + d/2 + 2$, and $k, s \geqslant m + 1$,*

$$\|\nabla \phi \cdot \nabla a\|_{\mathcal{H}_\varrho^m} \leqslant C \|\phi\|_{\mathcal{H}_\varrho^s} \|a\|_{\mathcal{H}_\varrho^k}.$$

2. *For* $k + s > m + 2 + d/2$, $k \geqslant m$ *and* $s \geqslant m + 2$,

$$\|a\Delta\phi\|_{\mathcal{H}_\varrho^m} \leqslant C\|\phi\|_{\mathcal{H}_\varrho^s}\|a\|_{\mathcal{H}_\varrho^k}.$$

3. *For* $s > d/2$,

$$\|\psi_1\psi_2\|_{\mathcal{H}_\varrho^m} \leqslant C\left(\|\psi_1\|_{\mathcal{H}_\varrho^m}\|\psi_2\|_{\mathcal{H}_\varrho^s} + \|\psi_1\|_{\mathcal{H}_\varrho^s}\|\psi_2\|_{\mathcal{H}_\varrho^m}\right). \tag{4.37}$$

The various constants C *are independent of* ϱ.

In this L^2-based setting, like in the proof of Theorem 4.1, the term $i\varepsilon\Delta a^\varepsilon$ vanishes in the energy estimates. More precisely, set

$$\|\psi\|_{X_\varrho^\ell(T)} := \sup_{0\leqslant t\leqslant T}\left\|\mathcal{F}^{-1}\left(\mathbf{w}\hat{\psi}\right)\right\|_{H^\ell(\mathbb{R}^d)} + \|\psi\|_{L^2(0,T;\mathcal{H}_\varrho^{\ell+1/2})},$$

where ϱ (hence $\mathbf{w}$) is a decreasing function of time, we can construct a solution to (4.4),

$$(\phi^\varepsilon, a^\varepsilon) \in X_\varrho^{\ell+1}(T) \times X_\varrho^\ell(T),$$

for $\ell > d/2+1$ and ϱ linear in t, $\varrho(t) = M_0 - Mt$ ($T < M_0/M$). The fact that ϕ^ε has not the same regularity as a^ε in the above statement is reminiscent of the fact that in the hyperbolic case, the right unknown is $(\nabla\phi^\varepsilon, a^\varepsilon)$, and not directly $(\phi^\varepsilon, a^\varepsilon)$. This approach was introduced in [Ginibre and Velo (2001)] to construct long range wave operators for Hartree type equations, by using a phase-amplitude decomposition, and adapted to the semi-classical régime in [Carles and Gallo (2017)] to provide some error estimates in the time discretization of (4.1) known as time splitting method.

Chapter 5

Some Instability Phenomena

In this chapter, we show some instability phenomena which are related to WKB analysis. The results concern the regularity of the flow map for nonlinear Schrödinger equations in Sobolev spaces $H^s(\mathbb{R}^d)$.

In the case of positive regularity, $s > 0$, the analysis relies on the supercritical régime in WKB analysis. We first present some ill-posedness results for the "usual" nonlinear Schrödinger equation, that is, with $\varepsilon = 1$ in Eq. (4.19). Roughly speaking, justification of nonlinear geometric optics on very small time intervals yields ill-posedness, and a justification on a longer time interval yields a worse phenomenon, indicating a loss of regularity for the flow map associated to the nonlinear Schrödinger equation.

In Sec. 5.3, we show that even at the semi-classical level, some instabilities occur. These are strong instabilities in L^2, but which do not affect the quadratic observables, or the Wigner measures.

In the case of negative regularity, $s < 0$, instability results are established by using the multiphase weakly nonlinear analysis of Sec. 2.6, and we sketch some results in this direction in Sec. 5.4.

5.1 Ill-posedness for nonlinear Schrödinger equations

In this paragraph, we consider the Cauchy problem

$$i\partial_t \psi + \frac{1}{2}\Delta\psi = \omega|\psi|^{2\sigma}\psi \quad ; \quad \psi_{|t=0} = \varphi, \tag{5.1}$$

where $x \in \mathbb{R}^d$, $\sigma \in \mathbb{N} \setminus \{0\}$ and $\omega \in \mathbb{R} \setminus \{0\}$. There is only one β such that the map

$$\psi(t, x) \mapsto \lambda^\beta \psi(\lambda^2 t, \lambda x) \tag{5.2}$$

leaves the equation in (5.1) (but not the initial data) unchanged for all $\lambda > 0$. It is given by

$$\beta = \frac{1}{\sigma}.$$

Recall that for $0 < s < d/2$, the homogeneous Sobolev space $\dot{H}^s$ is defined as

$$\dot{H}^s(\mathbb{R}^d) = \left\{ u \in L^{\frac{2d}{d-2s}}(\mathbb{R}^d) \quad ; \quad \exists v \in L^2(\mathbb{R}^d), \; \widehat{u}(\xi) = \frac{\widehat{v}(\xi)}{|\xi|^s} \right\}.$$

It is equipped with the norm

$$\|u\|_{\dot{H}^s} = \left(\int_{\mathbb{R}^d} |\xi|^{2s} \left| \widehat{u}(\xi) \right|^2 d\xi \right)^{1/2}.$$

There is only one s such that the $\dot{H}^s$-norm remains unchanged by the spatial scaling of Eq. (5.2) associated to Eq. (5.1), that is

$$f(x) \mapsto \lambda^{1/\sigma} f(\lambda x).$$

It is given by

$$s_c = \frac{d}{2} - \frac{1}{\sigma}.$$

In Sec. 5.1 and Sec. 5.2, we assume $s_c > 0$.

Definition 5.1. Let $s \geqslant k \geqslant 0$. The Cauchy problem for (5.1) is locally well-posed from $H^s(\mathbb{R}^d)$ to $H^k(\mathbb{R}^d)$ if, for all bounded subset $B \subset H^s(\mathbb{R}^d)$, there exist $T > 0$ and a Banach space $X_T \hookrightarrow C([-T, T]; H^k(\mathbb{R}^d))$ such that:
(1) For all $\varphi \in B \cap H^\infty$, (1.32) has a unique solution $\psi \in C([-T, T]; H^\infty)$.
(2) The mapping $\varphi \in (H^\infty, \| \cdot \|_B) \mapsto \psi \in X_T$ is continuous.

It has been established in [Cazenave and Weissler (1990)] that the Cauchy problem for (5.1) is locally well-posed from H^s to H^s as soon as $s \geqslant s_c$. The situation is different when $0 < s < s_c$ (recall that the notations below are defined at the beginning of the book):

Theorem 5.2 ([Christ *et al.* (2003b)]). *Let $d \geqslant 1$, $\sigma \in \mathbb{N} \setminus \{0\}$ and $\omega \in \mathbb{R} \setminus \{0\}$. Assume that $0 < s < s_c$.*
(1) *Ill-posedness. The Cauchy problem for (5.1) is not locally well-posed from H^s to H^s: for any $\delta > 0$, we can find families $(\varphi_1^h)_{0 < h \leqslant 1}$ and $(\varphi_2^h)_{0 < h \leqslant 1}$ with*

$$\varphi_1^h, \varphi_2^h \in \mathcal{S}(\mathbb{R}^d) \; ; \; \|\varphi_1^h\|_{H^s}, \|\varphi_2^h\|_{H^s} \leqslant \delta, \; \|\varphi_1^h - \varphi_2^h\|_{H^s} \xrightarrow[h \to 0]{} 0,$$

such that if ψ_1^h *and* ψ_2^h *denote the solutions to* (5.1) *with these initial data, there exists* $0 < t^h \ll 1$ *such that*

$$\liminf_{h \to 0} \left\| \psi_1^h \left(t^h \right) - \psi_2^h \left(t^h \right) \right\|_{H^s} > 0.$$

(2) Norm inflation. *We can find* $(\psi^h)_{0 < h \leqslant 1}$ *solving* (5.1), *such that*

$$\varphi^h \in \mathcal{S}(\mathbb{R}^d), \quad \left\| \varphi^h \right\|_{H^s} \ll 1 \; ; \quad \exists t^h \ll 1, \quad \left\| \psi^h \left(t^h \right) \right\|_{H^s} \gg 1.$$

The proof we give below relies on WKB analysis for very small time. Such an approach to prove ill-posedness results is due to G. Lebeau for the case of the nonlinear wave equation [Lebeau (2001)] (see also [Métivier (2004a)]). For the case of nonlinear Schrödinger equations, a proof in this spirit appears in the appendix of [Burq *et al.* (2005)]. We present the proof given in Appendix B of [Carles (2007b)]. The main idea of the proof is that in a semi-classical régime, if the initial data do not oscillate rapidly, then for very small time, the Laplacian is negligible, and the nonlinear Schrödinger equation can be approximated by an explicitly solvable ordinary differential equation (recall that the semi-classical limit is sometimes referred to as dispersionless limit). Such an idea appears in [Kuksin (1995)], in the context of the nonlinear wave equation.

Proposition 5.3. *Let* $d \geqslant 1$, $\sigma \in \mathbb{N} \setminus \{0\}$, $\omega \in \mathbb{R} \setminus \{0\}$ *and* $a_0^\varepsilon \in \mathcal{S}(\mathbb{R}^d)$. *Fix* $k > d/2$, *and assume that* $(a_0^\varepsilon)_{0 < \varepsilon \leqslant 1}$ *is bounded in* H^k. *Consider the initial value problems:*

$$\begin{cases} i\varepsilon \partial_t u^\varepsilon + \dfrac{\varepsilon^2}{2} \Delta u^\varepsilon = \omega |u^\varepsilon|^{2\sigma} u^\varepsilon, \\[2mm] i\varepsilon \partial_t v^\varepsilon = \omega |v^\varepsilon|^{2\sigma} v^\varepsilon, \\[2mm] u^\varepsilon_{|t=0} = v^\varepsilon_{|t=0} = a_0^\varepsilon. \end{cases}$$

We can find $c_0, c_1, \theta > 0$ *independent of* $\varepsilon \in]0, 1]$ *such that*

$$\| u^\varepsilon - v^\varepsilon \|_{L^\infty([-c_0 \varepsilon| \log \varepsilon|^\theta, c_0 \varepsilon| \log \varepsilon|^\theta]; H^k)} \lesssim \varepsilon \, \langle \log \varepsilon \rangle^{c_1}.$$

Before proving the above estimate, we show that neglecting the Laplacian for very small time is rather natural with WKB analysis in mind. Suppose for instance that $a_0^\varepsilon \to a_0$ as $\varepsilon \to 0$. We have seen in Chap. 4 that, at least in the case $\omega > 0$,

$$\left\| u^\varepsilon(t) - a(t) e^{i\phi(t)/\varepsilon} \right\|_{L^\infty} = o_{\varepsilon \to 0}(1) + \mathcal{O}(t),$$

where (ϕ, a) is given by

$$\begin{cases} \partial_t \phi + \dfrac{1}{2} |\nabla \phi|^2 + \omega |a|^{2\sigma} = 0 \; ; \quad \phi_{|t=0} = 0, \\[2mm] \partial_t a + \nabla \phi \cdot \nabla a + \dfrac{1}{2} a \Delta \phi = 0 \; ; \quad a_{|t=0} = a_0. \end{cases}$$

Since the above approximation is interesting for very small time only, examine the Taylor expansion for ϕ and a as $t \to 0$. For instance, plug formal series of the form

$$a(t,x) \sim \sum_{j \geqslant 0} t^j a_j(x) \quad ; \quad \phi(t,x) \sim \sum_{j \geqslant 0} t^j \phi_j(x)$$

into the above system, and identify the powers of t. Formally, we find

$$a(t,x) = a_0(x) + \mathcal{O}\left(t^2\right) \quad ; \quad \phi(t,x) = -t\omega|a_0(x)|^{2\sigma} + \mathcal{O}\left(t^3\right).$$

Since the phase ϕ is divided by ε, the function

$$a_0(x) \exp\left(-i\omega\frac{t}{\varepsilon}|a_0(x)|^{2\sigma}\right)$$

is expected to be a decent approximation of u^ε for $|t|^3 \ll \varepsilon$. This issue is discussed more precisely in Sec. 5.3. Note that the above approximation was derived without assessing any spatial derivative: it is not surprising that, when $a_0^\varepsilon = a_0$, it coincides with the solution v^ε of the above ordinary differential equation. Note also that in the statement of Proposition 5.3, the error estimate is described for $|t| \lesssim \varepsilon|\log\varepsilon|^\theta$, which is still very far from the borderline $|t| \ll \varepsilon^{1/3}$. The technical reason is that for $|t| \lesssim \varepsilon|\log\varepsilon|^\theta$, a semi-linear analysis is sufficient (see below), while for larger time (even for $\varepsilon|\log\varepsilon|^\theta \ll |t| \ll \varepsilon^{1/3}$), it seems that a quasi-linear analysis (in the spirit of Chaps. 2.6.4 and 4) is needed.

Proof. [Proof of Proposition 5.3] Let $w^\varepsilon = u^\varepsilon - v^\varepsilon$. It solves:

$$i\varepsilon\partial_t w^\varepsilon + \frac{\varepsilon^2}{2}\Delta w^\varepsilon = (g(w^\varepsilon + v^\varepsilon) - g(v^\varepsilon)) - \frac{\varepsilon^2}{2}\Delta v^\varepsilon, \tag{5.3}$$

with $w^\varepsilon_{|t=0} = 0$, where we have set $g(z) = \omega|z|^{2\sigma}z$. For $k \geqslant 0$, we have, from Lemma 1.2:

$$\|w^\varepsilon\|_{L^\infty([-t,t];H^k)} \lesssim \frac{1}{\varepsilon}\|g(w^\varepsilon + v^\varepsilon) - g(v^\varepsilon)\|_{L^1([-t,t];H^k)}$$
$$+ \varepsilon\|\Delta v^\varepsilon\|_{L^1([-t,t];H^k)}.$$

Using Taylor formula, and the fact that H^k is an algebra since $k > d/2$, we have:

$$\|g(w^\varepsilon(t) + v^\varepsilon(t)) - g(v^\varepsilon(t))\|_{H^k} \lesssim \left(\|w^\varepsilon(t)\|_{H^k}^{2\sigma} + \|v^\varepsilon(t)\|_{H^k}^{2\sigma}\right)\|w^\varepsilon(t)\|_{H^k}.$$

Since we have

$$v^\varepsilon(t,x) = a_0^\varepsilon(x) \exp\left(-i\omega\frac{t}{\varepsilon}|a_0^\varepsilon|^{2\sigma}\right),$$

we check, for all $s \geqslant 0$:

$$\|v^\varepsilon(t)\|_{H^s} \lesssim \left\langle \frac{t}{\varepsilon} \right\rangle^s.$$

On any time interval where we have, say, $\|w^\varepsilon\|_{H^k} \leqslant 1$, we infer:

$$\|w^\varepsilon\|_{L^\infty([-t,t];H^k)} \leqslant \frac{C}{\varepsilon} \int_{-t}^t \left\langle \frac{\tau}{\varepsilon} \right\rangle^{2\sigma k} \|w^\varepsilon(\tau)\|_{H^k} d\tau + C_1 \varepsilon \int_{-t}^t \left\langle \frac{\tau}{\varepsilon} \right\rangle^{k+2} d\tau.$$

Gronwall lemma yields:

$$\|w^\varepsilon\|_{L^\infty([-t,t];H^k)} \lesssim \varepsilon \int_{-t}^t \left\langle \frac{\tau}{\varepsilon} \right\rangle^{k+2} \exp\left(\frac{C}{\varepsilon} \int_\tau^t \left\langle \frac{\tau'}{\varepsilon} \right\rangle^{2\sigma k} d\tau' \right) d\tau.$$

Let $t^\varepsilon = c_0 \varepsilon |\log \varepsilon|^\theta$:

$$\|w^\varepsilon\|_{L^\infty([-t^\varepsilon,t^\varepsilon];H^k)} \lesssim \varepsilon \exp\left(2Cc_0 |\log \varepsilon|^\theta \langle c_0 \log \varepsilon \rangle^{2\sigma k\theta} \right) \int_{-t^\varepsilon}^{t^\varepsilon} \left\langle \frac{\tau}{\varepsilon} \right\rangle^{k+2} d\tau$$

$$\lesssim \exp\left(2Cc_0 |\log \varepsilon|^\theta \langle c_0 \log \varepsilon \rangle^{2\sigma k\theta} \right) \varepsilon^2 \langle \log \varepsilon \rangle^{(k+3)\theta}.$$

For $\theta = (1 + 2\sigma k)^{-1}$ and c_0 sufficiently small, this yields:

$$\|w^\varepsilon\|_{L^\infty([-t^\varepsilon,t^\varepsilon];H^k)} \lesssim \varepsilon \langle \log \varepsilon \rangle^{\frac{3+k}{1+2\sigma k}}.$$

Up to choosing ε sufficiently small, a continuity argument shows that $\|w^\varepsilon\|_{H^k}$ remains bounded by 1 on this time interval, and the proposition follows. □

As mentioned already, the above proof relies on a perturbative analysis (semi-linear approach). The main remark is that the exponential amplification factor $e^{Ct/\varepsilon}$, already pointed out at the end of Sec. 1.2, can be controlled on very small time intervals, of the form given above.

Proof. [Proof of Theorem 5.2] The result is a straightforward consequence of Proposition 5.3 and explicit computations on ordinary differential equations.

For $a_0 \in \mathcal{S}(\mathbb{R}^d)$ with $\|a_0\|_{H^s} \leqslant \delta/2$, and $h > 0$, consider ψ_1^h solving (5.1) with:

$$\varphi_1^h(x) = h^{-\frac{n}{2}+s} a_0 \left(\frac{x}{h} \right).$$

Using the parabolic scaling and the scaling of $\dot{H}^s$, define $\mathbf{u}^h$ by:

$$\mathbf{u}^h(t,x) = h^{\frac{n}{2}-s} \psi_1^h \left(h^2 t, hx \right).$$

It solves:

$$i\partial_t \mathbf{u}^h + \frac{1}{2}\Delta \mathbf{u}^h = \omega h^{2-d\sigma+2\sigma s}|\mathbf{u}^h|^{2\sigma}\mathbf{u}^h \quad ; \quad \mathbf{u}^h_{|t=0} = a_0.$$

Let $\varepsilon = h^{\frac{d\sigma}{2}-1-s\sigma} = h^{\sigma(s_c-s)}$: ε and h go to zero simultaneously since $s < s_c$. With this relation between ε and h, we denote the dependence upon one or the other according to the more natural context. Define

$$u_1^\varepsilon(t,x) = \mathbf{u}^h(ht,x) = h^{\frac{d}{2}-s}\psi_1^h\left(h^{\frac{d\sigma}{2}+1-s\sigma}t, hx\right).$$

It solves:

$$i\varepsilon\partial_t u_1^\varepsilon + \frac{\varepsilon^2}{2}\Delta u_1^\varepsilon = \omega|u_1^\varepsilon|^{2\sigma}u_1^\varepsilon \quad ; \quad u_1^\varepsilon{}_{|t=0} = a_0.$$

We go back to ψ_1^h via the formula:

$$\psi_1^h(t,x) = h^{-\frac{d}{2}+s}u_1^\varepsilon\left(\frac{t}{h^{\frac{d\sigma}{2}+1-s\sigma}}, \frac{x}{h}\right).$$

In particular, ψ_1^h and u_1^ε have the same $\dot{H}^s$ norm.

(1) *Ill-posedness.* Define u_2^ε solution to the same equation, but with initial datum

$$u_2^\varepsilon{}_{|t=0} = (1+\delta^\varepsilon)a_0,$$

with $\delta^\varepsilon = |\log\varepsilon|^{-\theta} \ll 1$, where $\theta > 0$ stems from Proposition 5.3. We define ψ_2^h by the same scaling as above, and it is straightforward to check, for h sufficiently small:

$$\varphi_1^h, \varphi_2^h \in \mathcal{S}(\mathbb{R}^d) \; ; \; \|\varphi_1^h\|_{H^s}, \|\varphi_2^h\|_{H^s} \leqslant \delta, \quad \|\varphi_1^h - \varphi_2^h\|_{H^s} \xrightarrow[h\to 0]{} 0.$$

From Proposition 5.3, we have, for $k > d/2$ and $t^\varepsilon = c\varepsilon|\log\varepsilon|^\theta$, with $0 < c \leqslant c_0$,

$$\left\|u_j^\varepsilon(t^\varepsilon) - v_j^\varepsilon(t^\varepsilon)\right\|_{H^k} \ll 1, \quad j = 1, 2,$$

where

$$v_1^\varepsilon(t,x) = a_0(x)\exp\left(-i\omega\frac{t}{\varepsilon}|a_0(x)|^{2\sigma}\right),$$

$$v_2^\varepsilon(t,x) = (1+\delta^\varepsilon)a_0(x)\exp\left(-i\omega\frac{t}{\varepsilon}|(1+\delta^\varepsilon)a_0(x)|^{2\sigma}\right).$$

We infer:

$$\left\|u_1^\varepsilon(t^\varepsilon) - u_2^\varepsilon(t^\varepsilon)\right\|_{\dot{H}^s} \gtrsim \left\|v_1^\varepsilon(t^\varepsilon) - v_2^\varepsilon(t^\varepsilon)\right\|_{\dot{H}^s}.$$

On the other hand, we have, from the explicit form of v_j^ε:

$$\left|v_1^\varepsilon(t,x) - v_2^\varepsilon(t,x)\right| \underset{\varepsilon\to 0}{\sim} |a_0(x)| \left|e^{-i\omega t|a_0(x)|^{2\sigma}/\varepsilon} - e^{-i\omega t|(1+\delta^\varepsilon)a_0(x)|^{2\sigma}/\varepsilon}\right|$$

$$\underset{\varepsilon\to 0}{\sim} 2\,|a_0(x)| \left|\sin\left(\frac{\omega t\left((1+\delta^\varepsilon)^{2\sigma}-1\right)|a_0(x)|^{2\sigma}}{2\varepsilon}\right)\right|$$

$$\underset{\varepsilon\to 0}{\sim} 2|a_0(x)| \left|\sin\left(\omega\sigma\frac{t}{\varepsilon}\delta^\varepsilon|a_0(x)|^{2\sigma}\right)\right|.$$

Hence, from the definition of t^ε and δ^ε:

$$\left|v_1^\varepsilon(t^\varepsilon,x) - v_2^\varepsilon(t^\varepsilon,x)\right| \underset{\varepsilon\to 0}{\sim} 2|a_0(x)| \left|\sin\left(\omega\sigma c|a_0(x)|^{2\sigma}\right)\right|.$$

Up to adjusting $c \in\,]0, c_0]$, we infer:

$$\liminf_{\varepsilon\to 0} \left\|u_1^\varepsilon\left(t^\varepsilon\right) - u_2^\varepsilon\left(t^\varepsilon\right)\right\|_{\dot H^s} > 0.$$

Since u_j^ε and ψ_j^h have the same $\dot H^s$ norm, the first part of Theorem 5.2 follows.

(2) *Norm inflation.* In [Christ *et al.* (2003b)], this phenomenon appears as a transfer of energy from low to high Fourier modes. It corresponds to the appearance of rapid oscillations in a supercritical WKB régime: even though u^ε is not ε-oscillatory initially, rapid oscillations appear instantaneously. Still from Proposition 5.3, with $t^\varepsilon = c\varepsilon|\log\varepsilon|^\theta$, we have

$$u_1^\varepsilon(t^\varepsilon,x) \underset{\varepsilon\to 0}{\sim} a_0(x)e^{-i\omega\frac{t^\varepsilon}{\varepsilon}|a_0(x)|^{2\sigma}} = a_0(x)e^{-i\omega|a_0(x)|^{2\sigma}\left(\log\frac{1}{\varepsilon}\right)^\theta}.$$

Even though u^ε is not yet ε-oscillatory, "rapid" oscillations have appeared already. If we replace a_0 with $|\log h|^{-\theta'}a_0$, this proves the second part of the theorem. $\qquad\square$

5.2 Loss of regularity for nonlinear Schrödinger equations

The end of the proof of Theorem 5.2 suggests that there is some room left: if we can show that u^ε becomes exactly ε-oscillatory, then we should obtain a stronger result. On the other hand, to prove that u^ε becomes exactly ε-oscillatory, we have to justify some asymptotics on a time interval which is independent of ε. This was achieved in Chap. 4. However, note that it is not necessary to know the pointwise asymptotic behavior of u^ε to know that it is ε-oscillatory: it turns out that the analysis developed in Chap. 2.6.4 suffices to do so (for defocusing nonlinearities, $\omega > 0$), and we have:

Theorem 5.4 ([Alazard and Carles (2009a)]). *Let* $\omega > 0$, $\sigma \in \mathbb{N} \backslash \{0\}$, *and* $0 < s < s_c = d/2 - 1/\sigma$. *There exists a family* $(\varphi^h)_{0 < h \leqslant 1}$ *in* $\mathcal{S}(\mathbb{R}^d)$ *with*

$$\|\varphi^h\|_{H^s(\mathbb{R}^n)} \to 0 \text{ as } h \to 0,$$

a solution ψ^h *to* (5.1) *and* $0 < t^h \to 0$, *such that:*

$$\|\psi^h(t^h)\|_{H^k(\mathbb{R}^d)} \to +\infty \text{ as } h \to 0, \ \forall k > \frac{s}{1 + \sigma(s_c - s)}.$$

In particular, (5.1) *is not locally well-posed from* H^s *to* H^k.

Note that in general, the solutions of the above theorem must be understood as the weak solutions given by Proposition 1.28.

From now on, we assume $\omega > 0$, and fix $\omega = +1$ for simplicity.

This result is to be compared with the main result in [Lebeau (2005)], which we recall with notations adapted to make the comparison with the Schrödinger case easier. For $d \geqslant 3$ and energy-supercritical wave equations

$$\left(\partial_t^2 - \Delta\right) u + u^{2\sigma+1} = 0, \quad \sigma \in \mathbb{N}, \ \sigma > \frac{2}{d-2},$$

G. Lebeau shows that one can find a *fixed* initial datum in H^s, $s > 1$, and a sequence of times $0 < t^h \to 0$, such that the H^k norms of the solution are unbounded along the sequence t^h, for $k \in]I(s), s]$. The expression for $I(s)$ is related to the critical Sobolev exponent

$$s_{\text{sob}} = \frac{d}{2} \frac{\sigma}{\sigma + 1},$$

which corresponds to the embedding $H^{s_{\text{sob}}}(\mathbb{R}^d) \subset L^{2\sigma+2}(\mathbb{R}^d)$. In [Lebeau (2005)], we find:

$$I(s) = 1 \text{ if } 1 < s \leqslant s_{\text{sob}} \quad ; \quad I(s) = \frac{s}{1 + \sigma(s_c - s)} \text{ if } s_{\text{sob}} \leqslant s < s_c. \quad (5.4)$$

Note that we have

$$\frac{s_{\text{sob}}}{1 + \sigma(s_c - s_{\text{sob}})} = 1. \quad (5.5)$$

The approach in [Lebeau (2005)] consists in using an *anisotropic* scaling, as opposed to the isotropic scaling used in [Lebeau (2001); Christ *et al.* (2003b)]. Compare Theorem 5.4 with the approach of [Lebeau (2005)]. Recall that (5.1) has two important (formally) conserved quantities: mass and energy,

$$M(t) = \int_{\mathbb{R}^d} |\psi(t, x)|^2 dx \equiv M(0),$$

$$E(\psi(t)) = \frac{1}{2} \int_{\mathbb{R}^d} |\nabla \psi(t, x)|^2 dx + \frac{1}{\sigma + 1} \int_{\mathbb{R}^d} |\psi(t, x)|^{2\sigma+2} dx \equiv E(\varphi). \quad (5.6)$$

In view of (5.5), we obtain, for H^1-supercritical nonlinearities:

Corollary 5.5. *Let $d \geqslant 3$ and $\sigma > \frac{2}{d-2}$. There exists a family $(\varphi^h)_{0 < h \leqslant 1}$ in $\mathcal{S}(\mathbb{R}^d)$ with*

$$\|\varphi^h\|_{H^1} + \|\varphi^h\|_{L^{2\sigma+2}} \to 0 \text{ as } h \to 0,$$

a solution ψ^h to

$$i\partial_t \psi^h + \frac{1}{2}\Delta\psi^h = |\psi^h|^{2\sigma}\psi^h \quad ; \quad \psi^h_{|t=0} = \varphi^h,$$

and $0 < t^h \to 0$, such that:

$$\|\psi^h(t^h)\|_{H^k(\mathbb{R}^d)} \to +\infty \text{ as } h \to 0, \; \forall k > 1.$$

We thus get the analogue of the result of G. Lebeau when $I(s) = 1$, with the drawback that we consider a *sequence* of initial data only. The information that we do not have for Schrödinger equations, and which is available for wave equations, is the finite speed of propagation, that is used in [Lebeau (2005)] to construct a fixed initial datum. On the other hand, the range for k in Theorem 5.4 is broader when $1 < s < s_{\text{sob}}$, and also, we allow the range $0 < s \leqslant 1$, for which no analogous result is available for the wave equation. However, we choose to present the proof of Theorem 5.4 for $k \geqslant 1$ only, and refer to [Alazard and Carles (2009a)] for the remaining cases. Note that $k \geqslant 1$ suffices to establish Corollary 5.5.

The proof starts like the proof of Theorem 5.2: for $a_0 \in \mathcal{S}(\mathbb{R}^d)$ and $h > 0$, consider ψ^h solving (5.1) with:

$$\varphi^h(x) = h^{-\frac{d}{2}+s} a_0\left(\frac{x}{h}\right).$$

Let $\varepsilon = h^{\sigma(s_c - s)}$: ε and h go to zero simultaneously since $s < s_c$. Define

$$u^\varepsilon(t, x) = h^{\frac{d}{2}-s}\psi^h\left(h^{\frac{d\sigma}{2}+1-s\sigma}t, hx\right).$$

It solves:

$$i\varepsilon\partial_t u^\varepsilon + \frac{\varepsilon^2}{2}\Delta u^\varepsilon = \omega|u^\varepsilon|^{2\sigma}u^\varepsilon \quad ; \quad u^\varepsilon_{|t=0} = a_0.$$

The approach consists in showing that for some $\tau > 0$ independent of ε,

$$\liminf_{\varepsilon \to 0} \varepsilon^k \|u^\varepsilon(\tau)\|_{\dot{H}^k} > 0, \quad \forall k \geqslant 1. \tag{5.7}$$

Back to ψ, this will yield $t^h = \tau h^2 \varepsilon$ and

$$\|\psi^h(t^h)\|_{\dot{H}^k} \gtrsim h^{s-k}\varepsilon^{-k} = h^{s-k(1+\sigma(s_c - s))}.$$

To complete the above reduction, note that we only have to prove (5.7) for $k = 1$. Indeed, for $k > 1$, there exists $C_k > 0$ such that

$$\|f\|_{\dot{H}^1} \leqslant C_k \|f\|_{L^2}^{1-1/k} \|f\|_{\dot{H}^k}^{1/k}, \quad \forall f \in H^k(\mathbb{R}^d).$$

This inequality is straightforward thanks to Fourier analysis. Note also that thanks to the conservation of mass for u^ε, we have:

$$\|u^\varepsilon(t)\|_{\dot{H}^1} \leqslant C_k \|a_0\|_{L^2}^{1-1/k} \|u^\varepsilon(t)\|_{\dot{H}^k}^{1/k}.$$

Up to replacing a_0 with $|\log h|^{-1} a_0$, the result then follows from:

Proposition 5.6. *Let $n \geqslant 1$, $a_0 \in \mathcal{S}(\mathbb{R}^d)$ be non-trivial, and $\sigma \geqslant 1$. There exists $\tau > 0$ such that*

$$\liminf_{\varepsilon \to 0} \|\varepsilon \nabla u^\varepsilon(\tau)\|_{L^2} > 0.$$

Proof. The result is a consequence of Theorem 3.4, whose proof is the core of [Alazard and Carles (2009a)]. Indeed, Theorem 3.4 shows that for all $\tau \in [-T, T]$,

$$\liminf_{\varepsilon \to 0} \|\varepsilon \nabla u^\varepsilon(\tau)\|_{L^2} \geqslant \liminf_{\varepsilon \to 0} \|v(\tau) u^\varepsilon(\tau)\|_{L^2},$$

where $(v, a) \in C^\infty([-T, T]; H^\infty)^2$ solves

$$\begin{cases} \partial_t v + v \cdot \nabla v + \nabla \left(|a|^{2\sigma}\right) = 0 & ; \quad v_{|t=0} = 0. \\ \partial_t a + v \cdot \nabla a + \dfrac{1}{2} a \operatorname{div} v = 0 & ; \quad a_{|t=0} = a_0. \end{cases}$$

Theorem 3.4 also yields:

$$\lim_{\varepsilon \to 0} \left\| |u^\varepsilon|^2 - |a|^2 \right\|_{L^\infty([-T,T];L^{\sigma+1})} = 0.$$

Using Hölder's inequality, we find

$$\|v(\tau) a(\tau)\|_{L^2}^2 = \left\| |v(\tau)|^2 |a(\tau)|^2 \right\|_{L^1}$$

$$\leqslant \left\| |v(\tau)|^2 |u^\varepsilon(\tau)|^2 \right\|_{L^1} + \left\| |v(\tau)|^2 \left(|a(\tau)|^2 - |u^\varepsilon(\tau)|^2\right) \right\|_{L^1}$$

$$\leqslant \left\| |v(\tau)|^2 |u^\varepsilon(\tau)|^2 \right\|_{L^1} + \left\| |v(\tau)|^2 \right\|_{L^{\frac{\sigma+1}{\sigma}}} \left\| |a(\tau)|^2 - |u^\varepsilon(\tau)|^2 \right\|_{L^{\sigma+1}}.$$

Therefore, for all $\tau \in [-T, T]$,

$$\liminf_{\varepsilon \to 0} \|\varepsilon \nabla u^\varepsilon(\tau)\|_{L^2} \geqslant \|v(\tau) a(\tau)\|_{L^2}.$$

To show that there exists $\tau > 0$ such that $\|v(\tau) a(\tau)\|_{L^2} > 0$, we argue by continuity. We use the identities, in H^s for all $s \geqslant 0$, and as $t \to 0$:

$$v(t, x) = -t \nabla \left(|a_0(x)|^{2\sigma}\right) + \mathcal{O}\left(t^3\right) \quad ; \quad a(t, x) = a_0(x) + \mathcal{O}\left(t^2\right).$$

These identities follow directly from the equation satisfied by (v, a). $\square$

As discussed in Sec. 4.5, for focusing nonlinearities ($\omega < 0$), the above analysis fails. On the other hand, working in an analytic setting is possible, and we have:

Theorem 5.7 ([Thomann (2008)]). *The conclusions of Theorem 5.4 still hold if we assume $\omega < 0$.*

5.3 Instability at the semi-classical level

We have seen in Chap. 4 that a perturbation of the initial amplitude at order ε (that is, the presence of a non-trivial corrector a_1) alters the leading order amplitude for times of order $\mathcal{O}(1)$; see Proposition 4.5 and the discussion below, as well as Proposition 4.22. It seems natural to guess that a perturbation of the initial amplitude at order $\varepsilon^{1-\delta}$ for some $\delta \in]0,1[$ will alter the leading order amplitude for times of order $o(1)$, which can be understood as "instantaneously". In this section, we show that this is the case, and that the above mentioned perturbation becomes visible for times of order $\mathcal{O}(\varepsilon^\delta)$.

For the sake of readability, we detail the approach in the case of a cubic nonlinearity, in the absence of external potential:

$$i\varepsilon\partial_t u^\varepsilon + \frac{\varepsilon^2}{2}\Delta u^\varepsilon = |u^\varepsilon|^2 u^\varepsilon. \tag{5.8}$$

We present only formal computations here, which can be justified thanks to the method detailed in Chap. 4. In particular, we could repeat the argument in the presence of a sub-quadratic external potential, or for higher order defocusing, smooth, nonlinearity in space dimension $d \leqslant 3$. Recall that the notations in the following result have been defined in the beginning of these notes.

Theorem 5.8. *Let* $d \geqslant 1$, $a_0, \widetilde{a}_0^\varepsilon \in \mathcal{S}(\mathbb{R}^d)$, $\phi_0 \in C^\infty(\mathbb{R}^d; \mathbb{R})$, *where* a_0 *and* ϕ_0 *are independent of* ε, *and* $\nabla\phi_0 \in H^\infty$. *Let* u^ε *and* v^ε *solve the initial value problems:*

$$i\varepsilon\partial_t u^\varepsilon + \frac{\varepsilon^2}{2}\Delta u^\varepsilon = |u^\varepsilon|^2 u^\varepsilon \;;\; u^\varepsilon\big|_{t=0} = a_0 e^{i\phi_0/\varepsilon}.$$

$$i\varepsilon\partial_t v^\varepsilon + \frac{\varepsilon^2}{2}\Delta v^\varepsilon = |v^\varepsilon|^2 v^\varepsilon \;;\; v^\varepsilon\big|_{t=0} = \widetilde{a}_0^\varepsilon e^{i\phi_0/\varepsilon}.$$

Assume that there exists $N \in \mathbb{N}$ *and* $\varepsilon^{1-\frac{1}{N}} \ll \delta^\varepsilon \ll 1$ *such that:*

$$\|a_0 - \widetilde{a}_0^\varepsilon\|_{H^s} \approx \delta^\varepsilon, \; \forall s \geqslant 0 \;;\; \limsup_{\varepsilon\to 0} \left\|\frac{\operatorname{Re}(a_0 - \widetilde{a}_0^\varepsilon)\overline{a}_0}{\delta^\varepsilon}\right\|_{L^\infty(\mathbb{R}^d)} \neq 0. \tag{5.9}$$

Then we can find $0 < t^\varepsilon \ll 1$ *such that:* $\|u^\varepsilon(t^\varepsilon) - v^\varepsilon(t^\varepsilon)\|_{L^2} \gtrsim 1$, *and* $\|u^\varepsilon(t^\varepsilon) - v^\varepsilon(t^\varepsilon)\|_{L^\infty} \gtrsim 1$. *More precisely, this occurs for* $t^\varepsilon\delta^\varepsilon \approx \varepsilon$. *In particular,*

$$\frac{\|u^\varepsilon - v^\varepsilon\|_{L^\infty([0,t^\varepsilon];L^2\cup L^\infty)}}{\left\|u^\varepsilon_{|t=0} - v^\varepsilon_{|t=0}\right\|_{L^2\cap L^\infty}} \to +\infty \quad as \; \varepsilon \to 0.$$

Remark 5.9. From the conservation of the L^2 norm for u^ε and v^ε (Eq. (1.24)), the instability cannot be much stronger, at least in L^2, since

$$\|u^\varepsilon(t) - v^\varepsilon(t)\|_{L^2} \leqslant \|u^\varepsilon(t)\|_{L^2} + \|v^\varepsilon(t)\|_{L^2} \lesssim 1.$$

Remark 5.10. The second part of the assumption (5.9) can be viewed as a polarization condition. We could remove it with essentially the same approach as below, up to demanding $\varepsilon^{1/2-1/N} \ll \delta^\varepsilon \ll 1$.

Example 5.11. Consider $a_0, b_0 \in \mathcal{S}(\mathbb{R}^d)$ independent of h, such that $\mathrm{Re}(\bar{a}_0 b_0) \not\equiv 0$, and take $\tilde{a}_0^\varepsilon = a_0 + \delta^\varepsilon b_0$.

Example 5.12. Consider $a_0 \in \mathcal{S}(\mathbb{R}^d)$ independent of ε and $x^\varepsilon \in \mathbb{R}^d$. We can take $\tilde{a}_0^\varepsilon(x) = a_0(x - x^\varepsilon)$, provided that $|x^\varepsilon| = \delta^\varepsilon$ and

$$\limsup_{\varepsilon \to 0} \left\| \frac{x^\varepsilon}{|x^\varepsilon|} \cdot \nabla \left(|a_0|^2 \right) \right\|_{L^\infty} \neq 0.$$

Typically, we can think of $x^\varepsilon = \delta^\varepsilon e_j$, for $1 \leqslant j \leqslant n$, where $(e_j)_{1 \leqslant j \leqslant n}$ denotes the canonical basis of $\mathbb{R}^d$, and $\partial_j a_0 \not\equiv 0$ (the latter is merely an assumption, since $a_0 \in \mathcal{S}(\mathbb{R}^d)$).

Remark 5.13. The last example is motivated by the result of [Burq and Zworski (2005)]. There, instability is established for

$$\begin{cases} i\varepsilon \partial_t u^\varepsilon + \dfrac{\varepsilon^2}{2} \Delta u^\varepsilon = \dfrac{|x|^2}{2} u^\varepsilon + \varepsilon^2 |u^\varepsilon|^2 u^\varepsilon, \quad x \in \mathbb{R}^3, \\ u^\varepsilon(0, x) = \dfrac{1}{\varepsilon^{3/2}} \Phi \left(\dfrac{x - x_0}{\varepsilon} \right). \end{cases} \tag{5.10}$$

As above, a small perturbation of x_0 yields an instability phenomenon in the limit $\varepsilon \to 0$. We will come back to this framework at the end of this paragraph.

The above result addresses perturbations which satisfy in particular $\delta^\varepsilon \gg \varepsilon$. This excludes the standard WKB data of the form

$$u^\varepsilon(0, x) = a_0^\varepsilon(x) e^{i\phi_0(x)/\varepsilon}, \quad \text{where } a_0^\varepsilon \sim a_0 + \varepsilon a_1 + \varepsilon^2 a_2 + \dots$$

In that case, a perturbation of a_1 is relevant at time $t^\varepsilon \approx 1$, and the previous result is essentially sharp:

Proposition 5.14. *Let* $d \geqslant 1$, $a_0, a_1 \in \mathcal{S}(\mathbb{R}^d)$, $\phi_0 \in C^\infty(\mathbb{R}^d; \mathbb{R})$ *independent of* ε, *with* $\nabla \phi_0 \in H^\infty$. *Assume that* $\mathrm{Re}(a_0 \overline{a_1}) \not\equiv 0$. *Let* u^ε *and* v^ε *solve the initial value problems:*

$$i\varepsilon \partial_t u^\varepsilon + \frac{\varepsilon^2}{2} \Delta u^\varepsilon = |u^\varepsilon|^2 u^\varepsilon \; ; \; u^\varepsilon \big|_{t=0} = a_0 e^{i\phi_0/\varepsilon}.$$

$$i\varepsilon \partial_t v^\varepsilon + \frac{\varepsilon^2}{2} \Delta v^\varepsilon = |v^\varepsilon|^2 v^\varepsilon \; ; \; v^\varepsilon \big|_{t=0} = (a_0 + \varepsilon a_1) e^{i\phi_0/\varepsilon}.$$

Then for any $\tau^\varepsilon \ll 1$, $\|u^\varepsilon - v^\varepsilon\|_{L^\infty([0,\tau^\varepsilon];L^2)} \ll 1$, and for $t > 0$ independent of ε, and arbitrarily small: $\|u^\varepsilon(t) - v^\varepsilon(t)\|_{L^2} \gtrsim 1$.

This result is a direct consequence of the analysis presented in Chap. 4. Essentially, the reason why u^ε and v^ε diverge from each other for $t > 0$ is the presence of the phase corrector $\Phi^{(1)}$ for v^ε; we have already seen that under the above assumptions, there is no such phase corrector for u^ε. We then remark that $\Phi^{(1)}\big|_{t=0} = 0$ and $\partial_t \Phi^{(1)}\big|_{t=0} \neq 0$, so $\|\Phi^{(1)}(t)\|_{L^\infty} \approx t$ as $t \to 0$. Hence the proposition.

As announced above, we give only the formal aspect of the proof of Theorem 5.8 here, which originates from [Carles (2007b)]. Since the statement of the result involves times which go to zero as $\varepsilon \to 0$, we can use the approximation, see (4.13):

$$\limsup_{\varepsilon \to 0} \left\| u^\varepsilon(t, \cdot) - a(t, \cdot)e^{i\phi(t,\cdot)/\varepsilon} \right\|_{L^2 \cap L^\infty} = \mathcal{O}(t),$$

where (ϕ, a) is given by:

$$\begin{cases} \partial_t \phi + \dfrac{1}{2}|\nabla\phi|^2 + |a|^2 = 0 & ;\ \phi_{|t=0} = \phi_0, \\ \partial_t a + \nabla\phi \cdot \nabla a + \dfrac{1}{2}a\Delta\phi = 0 & ;\ a_{|t=0} = a_0. \end{cases}$$

To approximate v^ε, we use the same system. However, we do not pass to the limit in the initial data. We have

$$\limsup_{\varepsilon \to 0} \left\| v^\varepsilon(t, \cdot) - \widetilde{a}^\varepsilon(t, \cdot)e^{i\widetilde{\phi}^\varepsilon(t,\cdot)/\varepsilon} \right\|_{L^2 \cap L^\infty} = \mathcal{O}(t),$$

where $(\widetilde{\phi}^\varepsilon, \widetilde{a}^\varepsilon)$ is given by:

$$\begin{cases} \partial_t \widetilde{\phi}^\varepsilon + \dfrac{1}{2}\left|\nabla\widetilde{\phi}^\varepsilon\right|^2 + |\widetilde{a}^\varepsilon|^2 = 0 & ;\ \widetilde{\phi}^\varepsilon_{|t=0} = \phi_0, \\ \partial_t \widetilde{a}^\varepsilon + \nabla\widetilde{\phi}^\varepsilon \cdot \nabla\widetilde{a}^\varepsilon + \dfrac{1}{2}\widetilde{a}^\varepsilon\Delta\widetilde{\phi}^\varepsilon = 0 & ;\ \widetilde{a}^\varepsilon_{|t=0} = \widetilde{a}^\varepsilon_0. \end{cases}$$

In particular,

$$u^\varepsilon(t, x) - v^\varepsilon(t, x) \approx a(t, x)e^{i\phi(t,x)/\varepsilon} - \widetilde{a}^\varepsilon(t, x)e^{i\widetilde{\phi}^\varepsilon(t,x)/\varepsilon}.$$

The stability analysis of Chap. 4 shows that

$$\|a - \widetilde{a}^\varepsilon\|_{L^\infty([0,T];H^s)} = \mathcal{O}\left(\delta^\varepsilon\right),$$

for some time $T > 0$ independent of ε. We infer:

$$u^\varepsilon(t, x) - v^\varepsilon(t, x) \approx a(t, x)\left(e^{i\phi(t,x)/\varepsilon} - e^{i\widetilde{\phi}^\varepsilon(t,x)/\varepsilon}\right),$$

and
$$|u^\varepsilon(t,x) - v^\varepsilon(t,x)| \approx 2|a(t,x)| \left| \sin\left(\frac{\phi(t,x) - \widetilde{\phi}^\varepsilon(t,x)}{2\varepsilon} \right) \right|.$$

The idea is then to consider the Taylor expansion for the phases with respect to the time variable:
$$\phi(t,x) \approx \sum_{j \geqslant 0} t^j \phi_j(x) \quad ; \quad \widetilde{\phi}^\varepsilon(t,x) \approx \sum_{j \geqslant 0} t^j \widetilde{\phi}_j^\varepsilon(x).$$

The notations are consistent when $j = 0$: $\phi_0 = \widetilde{\phi}_0^\varepsilon$. We have already computed:
$$\phi_1(x) = -|a_0(x)|^2 \quad ; \quad \widetilde{\phi}_1^\varepsilon(x) = -|\widetilde{a}_0^\varepsilon(x)|^2.$$

To compute the higher order terms, we see that we must also consider the Taylor expansion in time for a and $\widetilde{a}^\varepsilon$:
$$a(t,x) \approx \sum_{j \geqslant 0} t^j a_j(x) \quad ; \quad \widetilde{a}^\varepsilon(t,x) \approx \sum_{j \geqslant 0} t^j \widetilde{a}_j^\varepsilon(x).$$

Here again, the notations are consistent when $j = 0$. We have
$$a_1 = -\nabla\phi_0 \cdot \nabla a_0 - \frac{1}{2}a_0\Delta\phi_0 \quad ; \quad \widetilde{a}_1^\varepsilon = -\nabla\phi_0 \cdot \nabla\widetilde{a}_0^\varepsilon - \frac{1}{2}\widetilde{a}_0^\varepsilon\Delta\phi_0.$$

We check that for $j \geqslant 1$, (ϕ_j, a_j) is determined by $(\phi_k, a_k)_{0 \leqslant k \leqslant j-1}$. Also, since ϕ and $\widetilde{\phi}^\varepsilon$ have the same initial datum, we have
$$\phi(t,x) - \widetilde{\phi}^\varepsilon(t,x) \approx \sum_{j \geqslant 1} t^j \left(\phi_j(x) - \widetilde{\phi}_j^\varepsilon(x) \right).$$

The first term of this series is given by
$$\phi_1 - \widetilde{\phi}_1^\varepsilon = |\widetilde{a}_0^\varepsilon|^2 - |a_0|^2 = |a_0 + \widetilde{a}_0^\varepsilon - a_0|^2 - |a_0|^2$$
$$= 2\,\mathrm{Re}\left(\overline{a}_0\left(\widetilde{a}_0^\varepsilon - a_0\right)\right) + |\widetilde{a}_0^\varepsilon - a_0|^2.$$

Note that by assumption, the first term on the last line is *exactly* of order δ^ε (on the support of a_0). The last term is smaller, controlled by $(\delta^\varepsilon)^2$:
$$\phi_1 - \widetilde{\phi}_1^\varepsilon = 2\,\mathrm{Re}\left(\overline{a}_0\left(\widetilde{a}_0^\varepsilon - a_0\right)\right) + \mathcal{O}\left((\delta^\varepsilon)^2\right).$$

By induction, it is easy to check
$$\phi_j - \widetilde{\phi}_j^\varepsilon = \mathcal{O}(\delta^\varepsilon), \quad \forall j \geqslant 2.$$

Therefore, for any $K \in \mathbb{N} \setminus \{0\}$,
$$\phi(t,x) - \widetilde{\phi}^\varepsilon(t,x) = \sum_{j=1}^{K} t^j \left(\phi_j(x) - \widetilde{\phi}_j^\varepsilon(x) \right) + \mathcal{O}\left(t^{K+1}\delta^\varepsilon\right) \approx t\left(\phi_1(x) - \widetilde{\phi}_1^\varepsilon(x) \right).$$

We infer as $\varepsilon \to 0$ and $t \to 0$:

$$|u^\varepsilon(t,x) - v^\varepsilon(t,x)| \approx 2|a(t,x)| \left| \sin\left(\frac{\phi(t,x) - \widetilde{\phi}^\varepsilon(t,x)}{2\varepsilon} \right) \right|$$

$$\approx 2|a_0(x)| \left| \sin\left(\frac{t}{2\varepsilon} \left(\phi_1(x) - \widetilde{\phi}_1^\varepsilon(x) \right) \right) \right|.$$

The argument of the sine function is of order exactly $t\delta^\varepsilon/\varepsilon$. Therefore, we can find $t = t^\varepsilon$ of order $\varepsilon/\delta^\varepsilon$, with:

- $t^\varepsilon \to 0$ as $\varepsilon \to 0$.
- $\liminf\limits_{\varepsilon \to 0} \|u^\varepsilon(t^\varepsilon) - v^\varepsilon(t^\varepsilon)\|_{L^2 \cap L^\infty} > 0.$

This completes the proof of Theorem 5.8.

Remark 5.15. In view of [Burq and Zworski (2005)], introduce the complex projective distance:

$$u_j \in L^2(\mathbb{R}^d), \quad d_{\mathrm{pr}}(u_1, u_2) := \arccos\left(\frac{|\langle u_1, u_2 \rangle|}{\|u_1\|_{L^2} \|u_2\|_{L^2}} \right).$$

Then we can check that, up to demanding $\varepsilon/\delta^\varepsilon \ll t^\varepsilon \ll 1$,

$$\frac{d_{\mathrm{pr}}\left(u^\varepsilon(t^\varepsilon), v^\varepsilon(t^\varepsilon) \right)}{d_{\mathrm{pr}}\left(u^\varepsilon(0), v^\varepsilon(0) \right)} \to +\infty \quad \text{as } \varepsilon \to 0.$$

Essentially, Theorem 5.8 uses the fact that oscillations of order $\mathcal{O}(1)$ appear for time of order $\varepsilon/\delta^\varepsilon$. The above result uses the fact that for larger time, these oscillations are rapid as $\varepsilon \to 0$ (but not of order $\mathcal{O}(1/\varepsilon)$).

We now turn our attention to the special case where the initial data contain no rapid oscillation: $\phi_0 = 0$. An easy induction shows that in the Taylor expansion for ϕ (resp. a), all the even (resp. odd) powers of t vanish:

$$\phi(t,x) \approx \sum_{j \geqslant 0} t^{2j+1} \phi_{2j+1}(x) \quad ; \quad a(t,x) \approx \sum_{j \geqslant 0} t^{2j} a_{2j}(x).$$

The same holds for $\widetilde{\phi}^\varepsilon$ and $\widetilde{a}^\varepsilon$, since at time $t = 0$, $\widetilde{\phi}^\varepsilon = \phi_0 = 0$. In particular,

$$\phi(t,x) = t\phi_1(x) + \mathcal{O}\left(t^3\right) = -t|a_0(x)|^2 + \mathcal{O}\left(t^3\right).$$

As $\varepsilon \to 0$ and $t \to 0$, we infer

$$u^\varepsilon(t,x) \approx a(t,x)e^{i\phi(t,x)/\varepsilon} \approx a_0(x)e^{-it|a_0(x)|^2/\varepsilon}e^{i\mathcal{O}(t^3)/\varepsilon}.$$

We have therefore:

$$u^\varepsilon(t,x) \approx a_0(x)e^{-it|a_0(x)|^2/\varepsilon} \quad \text{for } |t| \ll \varepsilon^{1/3}.$$

Note that the function of the right-hand side involves no spatial derivative/integration. In other words, it is constructed without considering the Laplacian in the nonlinear Schrödinger equation: we recover the approximate solution considered in Proposition 5.3. The above approximation for $|t| \ll \varepsilon^{1/3}$ is expected to remain valid for any smooth nonlinearity, that is even in cases where the rigorous justification of WKB analysis is not known. An advantage of replacing the assumption $|t| \leqslant c\varepsilon |\log \varepsilon|^{\theta}$ with $|t| \ll \varepsilon^{1/3}$ would be to infer a loss of regularity phenomenon, as in Sec. 5.2, even though the range for the Sobolev index k should be decreased compared to Theorem 5.4. In particular, this would not suffice to prove Corollary 5.5.

Finally, we discuss more precisely the link between this approach and the result of [Burq and Zworski (2005)] mentioned in Remark 5.13. The initial data in Eq. (5.10) are not of WKB type. As we will see in the second part of this book, they correspond to a focusing phenomenon (a caustic reduced to a point). The important aspect is that the nonlinearity is cubic, and in space dimension three, the size of the coupling constant, ε^2 in Eq. (5.10), is *supercritical* as far as nonlinear effects near the caustic are concerned; see Sec. 6.3 and Chap. 7. Therefore, supercriticality is the main feature shared by Eq. (5.8) and Eq. (5.10), and is the reason why instability occurs in the limit $\varepsilon \to 0$. Without entering into the details of the proof of [Burq and Zworski (2005)], we mention that the main idea consists in neglecting the Laplacian for small times, in order to reduce the problem to an ordinary differential equation mechanism, like in §5.1.

In the particular case of Eq. (5.10) (where the harmonic potential is isotropic), this link can be made more explicit. Consider u^ε solution to a generalization of Eq. (5.10):

$$i\varepsilon \partial_t u^\varepsilon + \frac{\varepsilon^2}{2} \Delta u^\varepsilon = \frac{|x|^2}{2} u^\varepsilon + \varepsilon^k |u^\varepsilon|^2 u^\varepsilon, \quad x \in \mathbb{R}^d.$$

Assume that $1 < k < d$ (hence $d \geqslant 2$). First, introduce U^ε given by a lens transform (such transforms are discussed in more details in Sec. 11.2.1):

$$U^\varepsilon(t, x) = \frac{1}{(1+t^2)^{d/4}} e^{i\frac{t}{1+t^2}\frac{|x|^2}{2\varepsilon}} u^\varepsilon \left(\arctan t, \frac{x}{\sqrt{1+t^2}} \right).$$

It solves:

$$\begin{cases} i\varepsilon \partial_t U^\varepsilon + \frac{\varepsilon^2}{2} \Delta U^\varepsilon = \varepsilon^k \left(1+t^2\right)^{d/2-1} |U^\varepsilon|^2 U^\varepsilon, \\ \quad U^\varepsilon(0, x) = u^\varepsilon(0, x). \end{cases}$$

Introduce $\gamma = k/d < 1$, $h = \varepsilon^{1-\gamma}$, $t_0^h = h^{\gamma/(1-\gamma)}$ and $\psi = \psi^h$ given by

$$\psi^h(t, x) = U^\varepsilon \left(\frac{t}{\varepsilon^\gamma} - 1, x \right).$$

It solves:

$$\begin{cases} ih\partial_t \psi^h + \dfrac{h^2}{2} \Delta \psi^h = \left(\left(t_0^h \right)^2 + \left(t - t_0^h \right)^2 \right)^{d/2 - 1} \left| \psi^h \right|^2 \psi^h, \\ \psi^h \left(t_0^h, x \right) = u^\varepsilon(0, x). \end{cases}$$

The above equation is closely akin to the cubic nonlinear Schrödinger equation in a supercritical WKB régime (think of the coupling constant in front of the nonlinearity as t^{d-2}, and recall that $d \geqslant 2$). Consider the case where $u^\varepsilon(0, x) = a_0(x)$ is independent of ε. Instabilities as $h \to 0$ can be established in the same spirit as above (with a slightly different scaling, because of the factor in front of the cubic nonlinearity for ψ^h), yielding instabilities for u^ε. More precisely, we can prove

$$\left\| \psi^h(t) - a(t)e^{i\phi(t)/h} \right\|_{L^2 \cap L^\infty} = \mathcal{O} \left(t + h^{\frac{d(k-1)}{d-k}} \right),$$

where (ϕ, a) is given by:

$$\begin{cases} \partial_t \phi + \dfrac{1}{2} |\nabla \phi|^2 + t^{d-2} |a|^2 = 0 \quad ; \quad \phi_{|t=0} = 0, \\ \partial_t a + \nabla \phi \cdot \nabla a + \dfrac{1}{2} a \Delta \phi = 0 \quad ; \quad a_{|t=0} = a_0. \end{cases}$$

The Taylor expansion for ϕ yields:

$$\phi(t, x) = -\frac{t^{d-1}}{d-1} |a_0(x)|^2 + \mathcal{O} \left(t^{2d-1} \right).$$

Therefore, a perturbation of a_0 of order δ^h, with $h^{1-1/N} \ll \delta^h \ll 1$ (as in Theorem 5.8) becomes visible on ψ^h for times t^h such that $(t^h)^{d-1}\delta^h \approx h$. In the case of Eq. (5.10), $1 < k = 2 < d = 3$, and this yields $t^h \approx (h/\delta^h)^{1/2}$. This corresponds to a time

$$\frac{1}{\sqrt{\varepsilon\delta}} - 1$$

for U^ε, that is,

$$\arctan \left(\frac{1}{\sqrt{\varepsilon\delta}} - 1 \right) = \frac{\pi}{2} - \arctan \left(\frac{\sqrt{\varepsilon\delta}}{1 - \sqrt{\varepsilon\delta}} \right) \approx \frac{\pi}{2} - \arctan \left(\sqrt{\varepsilon\delta} \right)$$

for u^ε. Note that instabilities occur for small time for ψ^h, corresponding to time of order $\pi/2$ for u^ε. This is due to the fact that the harmonic potential causes focusing at time $\pi/2$ (see Examples 1.12 and 1.18, and §8.1 below).

Note also that U^ε is of order $\mathcal{O}(1)$ in L^∞ when the instability occurs, which implies that u^ε is of order

$$|u^\varepsilon| \approx (\varepsilon\delta)^{-3/4} \ll \varepsilon^{-1+1/(4N)} \ll \varepsilon^{-3/2},$$

since $\varepsilon^{1/3-1/(3N)} = h^{1-1/N} \ll \delta \ll 1$. When the instability occurs, the wave function is not as concentrated as in [Burq and Zworski (2005)]; see Eq. (5.10). Heuristically, it is not surprising that instability occurs also for more concentrated data.

Finally, we point out that the instabilities at the semi-classical level affect the wave function, but not the usual quadratic observables. The instability mechanism is due to a phase modulation, and the creation of oscillations whose period is of order $\varepsilon/\delta^\varepsilon$. By assumption,

$$\varepsilon \ll \varepsilon/\delta^\varepsilon \ll 1.$$

Therefore, this scale of oscillation is not detected by the Wigner measure, which accounts for phenomena at scales 1 and ε only. Similarly, the instability does not affect the convergence of the position and current densities for small time. Indeed, these quadratic observables, as well as the Wigner measure, are described by an Euler equation, which is stable on $[-T, T]$, for some $T > 0$. On the other hand, nothing seems to be known as for what happens when the solution to the Euler equation develops singularities. The example developed in Sec. 7.4.3 suggests that the instabilities at the semi-classical level may affect also the Wigner measures after the solution to the Euler equation has become singular.

5.4   Negative order and infinite loss of regularity

In Secs. 5.1 and 5.2, instability phenomena were due to a transfer from low Fourier modes to high Fourier modes: the nonlinearity forces the appearance of rapid oscillations. Conversely, nonlinear interaction may cause the transfer of energy from high to low Fourier modes. Positive Sobolev norms (H^s for $s > 0$) are barely affected by this transfer, but on the other hand, negative Sobolev norms (H^s for $s < 0$) are much more sensitive to low modes.

So far in this chapter, we have only mentioned the invariance by scaling, which led to a notion of critical Sobolev regularity, $s_c = d/2 - 1/\sigma$. Another important invariance has repercussions on the notion of well-posedness, which is the Galilean invariance: if $\psi(t, x)$ solves

$$i\partial_t \psi + \frac{1}{2}\Delta\psi = \omega|\psi|^{2\sigma}\psi,$$

then so does $\psi(t, x - vt)e^{iv \cdot x - i|v|^2 t/2}$, for any constant vector $v \in \mathbb{R}^d$. This change of unknown function leaves the L^2-norm invariant, so it is natural to expect that well-posedness in H^s for $s \geqslant \max(0, s_c)$. On the other hand, instability may occur for $s < \max(0, s_c)$. We have seen in Sec. 5.1 and Sec. 5.2 that this is indeed the case when $s_c > 0$, and we now address the case $s < 0$, even though our presentation concerns situations where $s_c \geqslant 0$.

We note that there are several results addressing instability at negative regularity, in the same spirit as Theorem 5.2, starting with [Kenig *et al.* (2001)]; see e.g. [Christ *et al.* (2003a,b); Oh (2017); Kishimoto (2019)]. There are even more results when the space variable is periodic, $x \in \mathbb{T}^d$; see e.g. [Christ *et al.* (2003c); Carles *et al.* (2010); Oh and Wang (2018)].

In this section, we only focus on a result analogous to Theorem 5.4. The case $d \geqslant 2$ was established in [Carles *et al.* (2012)].

Theorem 5.16. *Let* $\omega \in \mathbb{R} \setminus \{0\}$, $\sigma \in \mathbb{N} \setminus \{0\}$, $d \geqslant 1$, *with* $d\sigma \geqslant 2$, *and*

$$s < -\frac{1}{2\sigma + 1}.$$

There exists a family $(\varphi^h)_{0 < h \leqslant 1}$ *in* $\mathcal{S}(\mathbb{R}^d)$ *with*

$$\|\varphi^h\|_{H^s(\mathbb{R}^d)} \to 0 \ as \ h \to 0,$$

a solution ψ^h *to*

$$i\partial_t \psi + \frac{1}{2}\Delta\psi = \omega|\psi|^{2\sigma}\psi,$$

and $0 < t^h \to 0$, *such that:*

$$\left\|\psi^h(t^h)\right\|_{H^k(\mathbb{R}^d)} \to +\infty \ as \ h \to 0, \ \forall k \in \mathbb{R}.$$

In particular, (5.1) *is not locally well-posed from* H^s *to* H^k, *regardless how small* k *is.*

The somehow surprising aspect of this result, in comparison with Theorem 5.4, is that *all* the Sobolev norms become infinite at the same time.

Proof. We proceed in three steps:

- We first consider the multidimensional case $d \geqslant 2$, under the assumption $s < -1/(2\sigma)$.
- We refine the approach to relax the assumption to $s < -1/(2\sigma+1)$.
- Finally, we address the one-dimensional case.

First case: $d \geqslant 2$. Consider initial data

$$\varphi^\varepsilon(x) = \varepsilon^{-1/(2\sigma)} e^{-|x|^2} \sum_{j=1}^{3} e^{i\kappa_j \cdot x / \varepsilon},$$

where $(\kappa_1, \kappa_2, \kappa_3)$ is like in Example 2.12. We readily check, for $s \in \mathbb{R}$,

$$\|\varphi^\varepsilon\|_{H^s(\mathbb{R}^d)} \lesssim \varepsilon^{-1/(2\sigma)} \times \varepsilon^{-s}.$$

The right-hand side is small precisely for $s < -1/(2\sigma)$. Rescale ψ^ε through

$$u^\varepsilon(t, x) = \varepsilon^{1/(2\sigma)} \psi^\varepsilon(\varepsilon t, x).$$

This new unknown solves

$$i\varepsilon \partial_t u^\varepsilon + \frac{\varepsilon^2}{2} \Delta u^\varepsilon = \omega \varepsilon |u^\varepsilon|^{2\sigma} u^\varepsilon \quad ; \quad u^\varepsilon(0, x) = e^{-|x|^2} \sum_{j=1}^{3} e^{i\kappa_j \cdot x / \varepsilon}.$$

In view of Example 2.15, the approximate solution is given by

$$u_{\text{app}}^\varepsilon(t, x) = a_0(t, x) + \sum_{j=1}^{3} a_j(t, x) e^{i\phi_j(t, x)/\varepsilon},$$

where ϕ_j is given by (2.22), $a_0(0, x) = 0$ and $\partial_t a_0(0, x) \neq 0$, so a_0 is instantaneously non-trivial:

$$\exists \tau > 0, \quad a_0(\tau, \cdot) \not\equiv 0,$$

and therefore

$$\|u_{\text{app}}^\varepsilon(\tau)\|_{H^k(\mathbb{R}^d)} \gtrsim 1, \quad \forall k \in \mathbb{R}.$$

Now the analysis of Sec. 2.6 with $X = \mathcal{F}L^1 \cap L^2$ yields

$$\|u^\varepsilon(\tau) - u_{\text{app}}^\varepsilon(\tau)\|_{\mathcal{F}L^1 \cap L^2} \lesssim \varepsilon,$$

and thus, for any $k \leqslant 0$,

$$\|u^\varepsilon(\tau) - u_{\text{app}}^\varepsilon(\tau)\|_{H^k} \leqslant \|u^\varepsilon(\tau) - u_{\text{app}}^\varepsilon(\tau)\|_{L^2} \lesssim \varepsilon.$$

We infer, for $k \leqslant 0$,

$$\|u^\varepsilon(\tau)\|_{H^k} \gtrsim 1, \quad \text{hence} \quad \|\psi^\varepsilon(\varepsilon\tau)\|_{H^k} \gtrsim \varepsilon^{-1/(2\sigma)},$$

and the result follows, in the case $d \geqslant 2$ with $s < -1/(2\sigma)$.

To relax the assumption on s, the idea from [Carles *et al.* (2012)] consists in allowing a "more weakly nonlinear régime" for u^ε. Introduce a parameter $1 < J < 2$, and consider

$$\varphi^\varepsilon(x) = \varepsilon^{(J-2)/(2\sigma)} e^{-|x|^2} \sum_{j=1}^{3} e^{i\kappa_j \cdot x/\varepsilon},$$

where $(\kappa_1, \kappa_2, \kappa_3)$ is like in Example 2.12. For $s \in \mathbb{R}$,

$$\|\varphi^\varepsilon\|_{H^s(\mathbb{R}^d)} \lesssim \varepsilon^{(J-2)/(2\sigma)} \times \varepsilon^{-s}.$$

Rescale ψ^ε through

$$\tilde{u}^\varepsilon(t,x) = \varepsilon^{(2-J)/(2\sigma)} \psi^\varepsilon(\varepsilon t, x).$$

This new unknown solves

$$i\varepsilon \partial_t \tilde{u}^\varepsilon + \frac{\varepsilon^2}{2} \Delta \tilde{u}^\varepsilon = \omega \varepsilon^J |\tilde{u}^\varepsilon|^{2\sigma} \tilde{u}^\varepsilon \quad ; \quad \tilde{u}^\varepsilon(0,x) = e^{-|x|^2} \sum_{j=1}^{3} e^{i\kappa_j \cdot x/\varepsilon}.$$

With $J > 1$, the natural approximate solution is the linear solution, like in Sec. 1.3. However, we may reproduce the analysis of Sec. 2.6, by simply introducing an extra factor ε^{J-1} on the right-hand side of (2.24). This actually induces very few modifications in the analysis. In particular, Proposition 2.19 remains. The only novelty that we must keep in mind is that in Example 2.15, we now have $\partial_t a_0(0,x) = \varepsilon^{J-1} e^{-(2\sigma+1)|x|^2} \sharp \mathcal{R}_0$, and thus the above lower bound becomes

$$\|\tilde{u}_{\text{app}}^\varepsilon(\tau)\|_{H^k(\mathbb{R}^d)} \gtrsim \varepsilon^{J-1}, \quad \forall k \in \mathbb{R}.$$

Note that the rapidly oscillatory part of $\tilde{u}_{\text{app}}^\varepsilon$ is much smaller, when measured in sufficiently negative Sobolev spaces,

$$\|\tilde{u}_{\text{app}}^\varepsilon(\tau) - \varepsilon^{J-1} a_0(\tau)\|_{H^k(\mathbb{R}^d)} \lesssim \varepsilon^{-k}, \quad \forall k \leqslant 0.$$

We still have

$$\|\tilde{u}^\varepsilon(\tau) - \tilde{u}_{\text{app}}^\varepsilon(\tau)\|_{\mathcal{F}L^1 \cap L^2} \lesssim \varepsilon,$$

and thus, for any $k \leqslant 0$,

$$\|\tilde{u}^\varepsilon(\tau) - \tilde{u}_{\text{app}}^\varepsilon(\tau)\|_{H^k} \leqslant \|\tilde{u}^\varepsilon(\tau) - \tilde{u}_{\text{app}}^\varepsilon(\tau)\|_{L^2} \lesssim \varepsilon.$$

We infer, for $k \leqslant 0$,

$$\|u^\varepsilon(\tau)\|_{H^k} \gtrsim \varepsilon^{J-1}, \quad \text{hence} \quad \|\psi^\varepsilon(\varepsilon\tau)\|_{H^k} \gtrsim \varepsilon^{J-1} \varepsilon^{(J-2)/(2\sigma)}.$$

The result follows, provided that the two constraints

$$\varepsilon^{(J-2)/(2\sigma)} \times \varepsilon^{-s} \to 0, \quad \varepsilon^{J-1} \varepsilon^{(J-2)/(2\sigma)} \to \infty,$$

are satisfied, which boils down to:

$$s < \frac{J-2}{2\sigma}, \quad J < \frac{2\sigma+2}{2\sigma+1}.$$

Now given $s < -1/(2\sigma + 1)$, we can always find $J > 1$ satisfying the above constraints, hence the theorem in the case $d \geqslant 2$.

Second case: $d = 1$, and $\sigma \geqslant 2$. We can reproduce the same approach as above, by replacing Example 2.12 with Example 2.14. The zero mode is created by (at least) quintic interaction, out of non-zero modes. Example 2.15 is readily adapted, to check that we can start from initial amplitudes all equal (to a Gaussian, like above, for instance), to create an instantaneously non-trivial zero mode. □

Remark 5.17. In the periodic case $x \in \mathbb{T}^d$, similar "infinite loss of regularity" results were proven in [Carles and Kappeler (2017)], removing the assumption $s < -1/(2\sigma + 1)$: we only require $s < 0$, provided that $d\sigma \geqslant 2$. In the one-dimensional cubic case, $d = \sigma = 1$, the same phenomenon is proved for $s < -2/3$. In [Oh and Wang (2018)], for $d = \sigma = 1$, norm inflation from $H^s(\mathbb{T})$ to $H^s(\mathbb{T})$ is established for $s \leqslant -1/2$.

To conclude this chapter, we resume a result from Appendix B in [Carles *et al.* (2012)], to show that measuring oscillatory functions in negative order Sobolev spaces may yield counter-intuitive results: the H^s-norm of an ε-oscillatory function is not always of order ε^{-s}. Indeed, in the case of plane waves,

$$f^\varepsilon(x) = a(x)e^{ik\cdot x/\varepsilon}, \quad a \in \mathcal{S}(\mathbb{R}^d), \ k \in \mathbb{R}^d \setminus \{0\}, \, , x \in \mathbb{R}^d,$$

we compute, for $s \in \mathbb{R}$,

$$\|f^\varepsilon\|_{H^s(\mathbb{R}^d)}^2 = \int_{\mathbb{R}^d} \langle\xi\rangle^{2s} |\widehat{f^\varepsilon}(\xi)|^2 d\xi = \int_{\mathbb{R}^d} \langle\xi\rangle^{2s} |\widehat{a}(\xi - k/\varepsilon)|^2 d\xi$$
$$\leqslant \sup_{\xi \in \mathbb{R}^d} \left(\langle\xi\rangle^{2s} \langle\xi - k/\varepsilon\rangle^{-2s} \right) \|a\|_{H^s(\mathbb{R}^d)}^2,$$

and, for $|a| \gg 1$,

$$\frac{1}{|a|} \lesssim \inf_{\eta \in \mathbb{R}^d} \frac{\langle\eta + a\rangle}{\langle\eta\rangle} = \inf_{\xi \in \mathbb{R}^d} \frac{\langle\xi\rangle}{\langle\xi - a\rangle} \leqslant \sup_{\xi \in \mathbb{R}^d} \frac{\langle\xi\rangle}{\langle\xi - a\rangle} \lesssim |a|,$$

hence, using either the lower bound or the upper bound above, according to the sign of s,

$$\|f^\varepsilon\|_{H^s(\mathbb{R}^d)} \lesssim \varepsilon^{-s}, \quad s \in \mathbb{R}.$$

It is easy to check that this order of magnitude is sharp. The picture is different in the case of nonlinear oscillations:

Lemma 5.18. *Let $d \geqslant 1$, and set*

$$g^\varepsilon(x) = e^{-|x|^2/2} e^{-i|x|^2/(2\varepsilon)}, \quad x \in \mathbb{R}^d.$$

Then as $\varepsilon \to 0$,

$$\|g^\varepsilon\|_{H^s(\mathbb{R}^d)} \approx \begin{cases} \varepsilon^{-s} & \text{if } s > -d/2, \\ \varepsilon^{d/2} \log \dfrac{1}{\varepsilon} & \text{if } s = -d/2, \\ \varepsilon^{d/2} & \text{if } s < -d/2. \end{cases}$$

Remark 5.19. The same asymptotic result is true in homogeneous Sobolev spaces (it will be clear from the proof), but the result in the inhomogeneous case is perhaps more striking.

Proof. Consider more generally, for $z \in \mathbb{C}$ with $\operatorname{Re} z > 0$,

$$g_z(x) = e^{-z|x|^2/2}.$$

We compute:

$$\mathcal{F} g_z(\xi) = z^{-d/2} e^{-|\xi|^2/(2z)}.$$

For $s \in \mathbb{R}$, we have, if $z = a + ib$, $a, b \in \mathbb{R}$, $a > 0$:

$$\|g_z\|_{H^s(\mathbb{R}^d)}^2 = \int_{\mathbb{R}^d} \langle \xi \rangle^{2s} |\mathcal{F} g_z(\xi)|^2 \, d\xi = \frac{1}{|z|^d} \int_{\mathbb{R}^d} \langle \xi \rangle^{2s} e^{-\frac{a}{a^2+b^2}|\xi|^2} \, d\xi$$

$$= \frac{1}{a^{d/2}} \int_{\mathbb{R}^d} \left\langle \left(\frac{a^2+b^2}{a} \right)^{1/2} \eta \right\rangle^{2s} e^{-|\eta|^2} \, d\eta. \tag{5.11}$$

In the present case, $z = 1 + i/\varepsilon$:

$$\|g^\varepsilon\|_{H^s(\mathbb{R}^d)}^2 = \int_{\mathbb{R}^d} \left\langle \left(1 + \frac{1}{\varepsilon^2} \right)^{1/2} \eta \right\rangle^{2s} e^{-|\eta|^2} \, d\eta$$

$$\approx \int_{\mathbb{R}^d} \left\langle \frac{\eta}{\varepsilon} \right\rangle^{2s} e^{-|\eta|^2} \, d\eta = c(d) \int_0^{+\infty} \left(1 + \frac{r^2}{\varepsilon^2} \right)^s e^{-r^2} r^{d-1} \, dr.$$

We split the last integral into $\int_0^\varepsilon + \int_\varepsilon^{+\infty}$. Then, we have for all s, since $1 \leqslant 1 + r^2/\varepsilon^2 \leqslant 2$ in the first integral,

$$\int_0^\varepsilon \left(1 + \frac{r^2}{\varepsilon^2} \right)^s e^{-r^2} r^{d-1} \, dr \approx \int_0^\varepsilon e^{-r^2} r^{d-1} \, dr \approx \varepsilon^d.$$

For the other term, since $r^2/\varepsilon^2 \leqslant 1 + r^2/\varepsilon^2 \leqslant 2r^2/\varepsilon^2$, we have

$$I(\varepsilon) := \int_\varepsilon^{+\infty} \left(1 + \frac{r^2}{\varepsilon^2}\right)^s e^{-r^2} r^{d-1} dr \approx \varepsilon^{-2s} \int_\varepsilon^{+\infty} e^{-r^2} r^{2s+d-1} dr.$$

Therefore, if $d + 2s > 0$, then $I(\varepsilon) \approx \varepsilon^{-2s}$. On the other hand, if $d + 2s < 0$, the lack of integrability near the origin yields

$$I(\varepsilon) \approx \varepsilon^d,$$

and if $d + 2s = 0$, we have, for the same reason,

$$I(\varepsilon) \approx \varepsilon^{-2s} \log \frac{1}{\varepsilon} = \varepsilon^d \log \frac{1}{\varepsilon}.$$

The lemma follows. $\square$

PART 2

Caustic Crossing: the Case of Focal Points

Chapter 6

Caustic Crossing: Formal Analysis

6.1 Presentation

In this second part, we consider the régime where the WKB analysis breaks down. This analysis is essentially independent of the first part, which may be considered as a motivation only. Roughly speaking, in the linear case, many results are available, while in the nonlinear case, very few phenomena have been identified.

Resume one of the examples given in Chap. 1:

$$i\varepsilon\partial_t u^\varepsilon_{\text{lin}} + \frac{\varepsilon^2}{2}\Delta u^\varepsilon_{\text{lin}} = 0 \quad ; \quad u^\varepsilon_{\text{lin}}(0, x) = a_0(x)e^{-i|x|^2/(2\varepsilon)}. \tag{6.1}$$

We have seen that for $t < 1$, the solution to the eikonal equation, and to the leading order transport equation respectively, are given by:

$$\phi_{\text{eik}}(t, x) = \frac{|x|^2}{2(t-1)} \quad ; \quad a(t, x) = \frac{1}{(1-t)^{d/2}}a_0\left(\frac{x}{1-t}\right).$$

As already noticed, these terms (as well as all the others involved in the WKB analysis) become singular as $t \to 1$. On the other hand, if, say, $a_0 \in \mathcal{S}(\mathbb{R}^d)$, then for every fixed ε, Eq. (6.1) has a unique global solution $u^\varepsilon_{\text{lin}} \in C^\infty\left(\mathbb{R}; \mathcal{S}(\mathbb{R}^d)\right)$. Since we consider the free Schrödinger equation (by "free", we mean linear, and without external potential), $u^\varepsilon_{\text{lin}}$ is given explicitly by the integral

$$u^\varepsilon_{\text{lin}}(t, x) = \frac{1}{(2i\pi t)^{d/2}}\int_{\mathbb{R}^d} e^{i\frac{|x-y|^2}{2\varepsilon t}}u^\varepsilon_{\text{lin}}(0, y)dy$$

$$= \frac{1}{(2i\pi t)^{d/2}}\int_{\mathbb{R}^d} e^{i\frac{|x-y|^2}{2\varepsilon t} - i\frac{|y|^2}{2\varepsilon}}a_0(y)dy.$$

This formula also yields the expression for ϕ_{eik} and a very easily, thanks to the stationary phase formula (which can be found for instance in [Alinhac and Gérard (2007); Grigis and Sjöstrand (1994); Hörmander (1994)]).

Indeed, the last expression can be viewed as an oscillatory integral, with phase

$$\varphi(t, x, y) = \frac{|x - y|^2}{2t} - \frac{|y|^2}{2}.$$

For $t \neq 1$, the critical points of φ, considered as a function of y, are given by:

$$\nabla_y \varphi(t, x, y_c) = 0 \iff \frac{y_c - x}{t} = y_c \iff y_c = \frac{x}{1 - t}.$$

The Hessian of φ is

$$\nabla_y^2 \varphi(t, x, y) = \left(\frac{1}{t} - 1 \right) I_d.$$

Therefore, if $t \neq 1$, φ has a unique critical point, which is non-degenerate. Stationary phase formula then yields:

$$u_{\text{lin}}^\varepsilon(t, x) \underset{\varepsilon \to 0}{\sim} \begin{cases} \dfrac{1}{(1 - t)^{d/2}} a_0 \left(\dfrac{x}{1 - t} \right) e^{i \frac{|x|^2}{2\varepsilon(t-1)}} & \text{if } t < 1, \\[3mm] \dfrac{e^{-id\pi/2}}{(t - 1)^{d/2}} a_0 \left(\dfrac{x}{1 - t} \right) e^{i \frac{|x|^2}{2\varepsilon(t-1)}} & \text{if } t > 1. \end{cases}$$

From this point of view, the main difference between the asymptotics for $t < 1$ and for $t > 1$ is the phase shift $e^{-id\pi/2}$, which is due to the sign change of the Hessian of φ as $t - 1$ changes signs.

For $t = 1$, we can no longer apply stationary phase formula, but we have directly from the integral expression of $u_{\text{lin}}^\varepsilon$:

$$u_{\text{lin}}^\varepsilon(1, x) = \frac{1}{(2i\pi)^{d/2}} \int_{\mathbb{R}^d} e^{i \frac{|x - y|^2}{2\varepsilon} - i \frac{|y|^2}{2\varepsilon}} a_0(y) dy = \frac{e^{i \frac{|x|^2}{2\varepsilon}}}{(i\varepsilon)^{d/2}} \widehat{a_0} \left(\frac{x}{\varepsilon} \right).$$

By writing

$$u_{\text{lin}}^\varepsilon(1, x) = \frac{1}{(i\varepsilon)^{d/2}} \widehat{a_0} \left(\frac{x}{\varepsilon} \right) \exp \left(i \frac{\varepsilon}{2} \left| \frac{x}{\varepsilon} \right|^2 \right),$$

we see that

$$u_{\text{lin}}^\varepsilon(1, x) = \frac{1}{(i\varepsilon)^{d/2}} \widehat{a_0} \left(\frac{x}{\varepsilon} \right) + \mathcal{O}(\varepsilon) \quad \text{in } L^2(\mathbb{R}^d).$$

To summarize, we have:

$$u_{\text{lin}}^\varepsilon(t, x) \underset{\varepsilon \to 0}{\sim} \begin{cases} \dfrac{1}{(1 - t)^{d/2}} a_0 \left(\dfrac{x}{1 - t} \right) e^{i \frac{|x|^2}{2\varepsilon(t-1)}} & \text{if } t < 1, \\[3mm] \dfrac{e^{-id\pi/4}}{\varepsilon^{d/2}} \widehat{a_0} \left(\dfrac{x}{\varepsilon} \right) & \text{if } t = 1, \\[3mm] \dfrac{e^{-id\pi/2}}{(t - 1)^{d/2}} a_0 \left(\dfrac{x}{1 - t} \right) e^{i \frac{|x|^2}{2\varepsilon(t-1)}} & \text{if } t > 1. \end{cases} \qquad (6.2)$$

We see that with initial data of order $\mathcal{O}(1)$ as $\varepsilon \to 0$, the size of $u_{\text{lin}}^\varepsilon$ grows as t approaches one. As $t = 1$, the amplitude is saturated, of order $\mathcal{O}(\varepsilon^{-d/2})$. After that critical time, the amplitude decreases, and is of order $\mathcal{O}(1)$ again. In particular, the family $(\|u^\varepsilon(1,\cdot)\|_{L^\infty})_{0<\varepsilon\leqslant 1}$ is unbounded: if $|u^\varepsilon|^2$ is viewed as the intensity of a laser beam, it becomes extremely high for $t \approx 1$. This is why the phenomenon is called *caustic* (from the Greek *kaustikos*, after the verb *katein* = to burn).

Recall that in Chap. 1, the caustic was defined as the set where the solution to the eikonal equation becomes singular. This is a geometrical definition, which is equivalent to saying that the caustic is the locus where rays of geometric optics form an envelope. This may seem different from the etymological definition, which refers to an analytical phenomenon (growth of the amplitude). Yet, since in the semi-classical limit, the energy is carried by the rays of geometric optics, these points of view agree. On the other hand, we will see in Sec. 9.2 that in some nonlinear cases, it might be sensible to distinguish these two notions.

It turns out that the result outlined above is fairly general. To study the high frequency limit of linear equations, the use of oscillatory integrals has proven very efficient to describe the wave function itself. Working in the phase space (that is, $(t, x, \tau, \xi) \in T^*\mathbb{R}^{1+n}$) instead of the physical space $((t, x) \in \mathbb{R}^{1+n})$, the singularity corresponding to the caustic disappears. In other words, the caustic phenomenon appears when projecting from the phase space to the physical space. In the case of Eq. (6.1), this approach leads to the following Lagrangian integral:

$$u_{\text{lin}}^\varepsilon(t, x) = \frac{1}{(2\pi\varepsilon)^{d/2}} \int_{\mathbb{R}^d} e^{-i\frac{t-1}{2\varepsilon}|\xi|^2 + i\frac{x\cdot\xi}{\varepsilon}} A^\varepsilon(\xi) d\xi. \tag{6.3}$$

Obviously, the Lagrangian symbol A^ε is given by the formula

$$A^\varepsilon(\xi) = \frac{1}{\varepsilon^{d/2}} e^{i\frac{t-1}{2\varepsilon}|\xi|^2} \widehat{u}_{\text{lin}}^\varepsilon\left(t, \frac{\xi}{\varepsilon}\right).$$

Note that the right-hand side is indeed independent of time, since $u_{\text{lin}}^\varepsilon$ solves the free Schrödinger equation. We check that A^ε converges as $\varepsilon \to 0$:

$$A^\varepsilon(\xi) = \frac{1}{(2\pi\varepsilon)^{d/2}} e^{-i\frac{|\xi|^2}{2\varepsilon}} \int_{\mathbb{R}^d} e^{-i\frac{x\cdot\xi}{\varepsilon}} u_{\text{lin}}^\varepsilon(0, x) dx$$

$$= \frac{1}{(2\pi\varepsilon)^{d/2}} \int_{\mathbb{R}^d} e^{-i\frac{|x+\xi|^2}{2\varepsilon}} a_0(x) dx = e^{-id\frac{\pi}{4}} a_0(-\xi) + \mathcal{O}(\varepsilon).$$

Roughly speaking, in general two phenomena occur at leading order in the high frequency limit for linear equations. First, a phase shift appears after the caustic crossing, in the same fashion as for $t > 1$ in Eq. (6.2). This phase shift is called the Maslov index, and is related to the geometry of the caustic; see e.g. [Duistermaat (1974); Maslov and Fedoriuk (1981); Yajima (1979)]. From a technical point of view, it corresponds to a signature change of the Hessian of the phase involved in the oscillatory integral approach: in Eq. (6.3), the Hessian of the phase is also $(1-t)I_d$, and its signature changes from $+d$ to $-d$ as $t-1$ changes signs, hence a phase shift of $(-d-(+d))\pi/4 = -d\pi/2$. The second phenomenon, which is not present in the example of Eq. (6.1), is the creation of new phases. Typically, when applying a stationary phase argument beyond the caustic, it may happen that the phase in the oscillatory integral has several non-degenerate critical points, even if it has only one before the caustic. In that case, we need a description of the form:

$$u^\varepsilon_{\text{lin}}(t, x) \approx \sum_{\text{critical points}} a_j(t, x)e^{i\phi_j(t,x)/\varepsilon}.$$

All the phases ϕ_j solve the eikonal equation (Eq. (1.10) in the case of Schrödinger equations). Note that in order to describe the wave function, it is not sufficient to consider only the viscosity solution to this Hamilton–Jacobi equation: by selecting only one of the above phases, the error we make is of the same order as the wave function $u^\varepsilon_{\text{lin}}$ itself, in L^∞; see [Kossioris (1993)]. We refer to [Duistermaat (1974); Maslov and Fedoriuk (1981)] and references therein for a more precise description of the phenomena mentioned above.

Note that if one is interested in quadratic observables rather than in the wave function itself, the Wigner measures share an important feature with the Lagrangian integral approach mentioned above: they unfold the singularities. In other words, the caustic phenomenon does not exist on Wigner measures; see e.g. [Gérard et al. (1997); Lions and Paul (1993); Sparber et al. (2003)]. On the other hand, the Maslov index is lost when working with Wigner measures.

For nonlinear equations, far less is known. As pointed out in Chap. 1, a new parameter appears for nonlinear problems: the size of the data. Moreover, the nature of the nonlinearity (dissipative, accretive, conservative, Lipschitzean...) is crucial, as discussed below. So potentially, a huge variety of phenomena is likely to happen. Let us mention the most important results on the subject, concerning hyperbolic equations: [Joly et al. (1995b, 1996a, 2000)] (see also [Joly et al. (1997a)]). As a typical striking result of this work, consider the following semi-linear wave equation, with highly oscillatory initial data:

$$\begin{cases} \left(\partial_t^2 - \Delta\right) u^\varepsilon + |\partial_t u^\varepsilon|^{p-1} \partial_t u^\varepsilon = 0. \\ u^\varepsilon(0, x) = \varepsilon U_0\left(x, \dfrac{\phi_0(x)}{\varepsilon}\right) \quad ; \quad \partial_t u^\varepsilon(0, x) = U_1\left(x, \dfrac{\phi_0(x)}{\varepsilon}\right), \end{cases} \quad (6.4)$$

where the functions $U_j(x, \cdot)$ are 2π-periodic for all $x \in \mathbb{R}^d$, and $p > 1$ (not necessarily an integer). One can check that in a WKB régime, the above equation is of weakly nonlinear type: the eikonal equation is the same as in the linear case, and the nonlinearity is present in the leading order transport equation. Consider the case where the initial phase ϕ_0 is spherically symmetric, with

$$\phi_0(x) = |x| =: r.$$

As mentioned in Chap. 1, the eikonal equation associated to the wave equation is

$$\left(\partial_t \phi\right)^2 = |\nabla \phi|^2.$$

In the particular case under consideration, we find the two solutions

$$\phi_\pm = \phi_\pm(t, r) = r \pm t.$$

We see that the rays associated to ϕ_- meet at the origin for positive time: as in the case of Eq. (6.1), a caustic reduced to a point is formed. Note that Eq. (6.4) is a dissipative equation: the energy associated to the linear equation is a non-increasing function of time,

$$\frac{d}{dt} \int_{\mathbb{R}^d} \left(\left(\partial_t u^\varepsilon(t, x)\right)^2 + |\nabla u^\varepsilon(t, x)|^2\right) dx \leqslant 0.$$

To give a flavor of the results of J.-L. Joly, G. Métivier and J. Rauch, let us describe qualitatively the main result of [Joly et al. (1995b)], generalized in [Joly et al. (2000)]. Suppose that $p > 2$. As the wave focuses at the origin, its amplitude grows, and the dissipative mechanism becomes very strong. The part of the wave carrying the most energy is highly dissipated. Since the highest energy terms are those which are highly oscillatory, the rapid oscillations of u^ε are absorbed at the focus: past the focus, u^ε is no longer ε-oscillatory at leading order.

Of course, the above mechanism is due to the nature of the nonlinearity: it is dissipative, and the phenomenon sketched here is due to this aspect. On the other hand, the nonlinear Schrödinger equations which we consider here, of the form

$$i\varepsilon \partial_t u^\varepsilon + \frac{\varepsilon^2}{2} \Delta u^\varepsilon = |u^\varepsilon|^{2\sigma} u^\varepsilon,$$

are conservative and Hamiltonian: the L^2-norm of u^ε is conserved, and the L^2-norm of $\varepsilon \nabla u^\varepsilon$ is bounded; see Eq. (1.24) and Eq. (1.25). So the above dissipation mechanism is less likely to occur. Before describing some caustic phenomena for this nonlinear Schrödinger equation, we present an idea due to J. Hunter and J. Keller, which suggests that the discussion presented in §1.2 for relevant phenomena in a WKB régime, should be repeated, with some differences though, near a caustic.

6.2 The idea of J. Hunter and J. Keller

The idea presented in [Hunter and Keller (1987)] for conservation laws is essentially the following. In a WKB régime, according to the amplitude of a wave, the nonlinearity can influence the geometry of the propagation (supercritical case, where the nonlinearity is present in the eikonal equation), or simply the leading order amplitude (weakly nonlinear régime), or can even be negligible at leading order. In the linear case, near a caustic, the amplitude of the wave is altered, as we have seen on the example of Eq. (6.1). The amplification factor depends on the geometry of the caustic, and the maximal amplification corresponds to a focal point: the energy concentrates on a single point rather than on, say, a curve or a surface. Therefore, we can resume a similar discussion: according to the amplitude of the wave near the caustic, the nonlinearity should have several possible effects at leading order. Note that the amplification is localized near the caustic, in some boundary layer. The idea in [Hunter and Keller (1987)] consists in saying that the important quantity to consider is the *average* nonlinear effect near the caustic.

For instance, in the case of Eq. (6.1), Eq. (6.2) shows that the amplification near the caustic is of order $\varepsilon^{-d/2}$, and that the boundary layer associated to the focal point is of order ε.

Before describing more precisely this idea for solutions to the nonlinear Schrödinger equation, resume the case of the semi-linear wave equation mentioned in the previous paragraph, and consider spherically symmetric initial data:

$$\begin{cases} \left(\partial_t^2 - \Delta \right) u^\varepsilon + |\partial_t u^\varepsilon|^{p-1} \, \partial_t u^\varepsilon = 0, & (t,x) \in [0,T] \times \mathbb{R}^3, \\ u^\varepsilon(0,x) = \varepsilon^{J+1} U_0 \left(r, \dfrac{r-r_0}{\varepsilon} \right) \; ; \; \partial_t u^\varepsilon(0,x) = \varepsilon^J U_1 \left(r, \dfrac{r-r_0}{\varepsilon} \right). \end{cases} \quad (6.5)$$

The new parameter $J \geqslant 0$ measures the size of the initial data. For $r_0 = 0$, and functions $U_j(r, \cdot)$ which are 2π-periodic for all $r \geqslant 0$, we recover the framework of [Joly *et al.* (1995b)]. We consider rather the case $r_0 > 0$ and $\operatorname{supp} U_j(r, .) \subset [-z_0, z_0]$ for some $z_0 > 0$ independent of $r \geqslant 0$ (the initial data are *short pulses*, as opposed to the previous wave trains). In the linear case, the solution to

$$\begin{cases} \left(\partial_t^2 - \Delta\right) v^\varepsilon = 0, & (t,x) \in [0,T] \times \mathbb{R}^3, \\ v^\varepsilon(0,x) = \varepsilon^{J+1} U_0\left(r, \dfrac{r-r_0}{\varepsilon}\right) \ ; \quad \partial_t v^\varepsilon(0,x) = \varepsilon^J U_1\left(r, \dfrac{r-r_0}{\varepsilon}\right), \end{cases}$$

is given by, since we consider a three-dimensional setting:

$$v^\varepsilon(t,r) = \begin{cases} \varepsilon^{J+1} \dfrac{g^\varepsilon(t+r) - g^\varepsilon(t-r)}{2r} & \text{for } r > 0, \\ \varepsilon^{J+1} \left(g^\varepsilon\right)'(t) & \text{for } r = 0, \end{cases}$$

where $g^\varepsilon(r)$ is essentially $r\left(U_0 + V_1\right)\left(r, \frac{r-r_0}{\varepsilon}\right)$, where V_1 is an antiderivative of U_1 with respect to its second variable. We see that $\partial_t v^\varepsilon$ is of order ε^J outside the origin, and of order ε^{J-1} near the origin. The characteristic boundary layer about the origin is of order ε, since the pulses are supported on a domain of thickness of order ε, and propagate along the characteristic directions, $r \pm t = $ Const., see Fig. 6.1. Plugging this estimate in the nonlinear potential, we have, for $r \approx 0$:

$$\left|\partial_t v^\varepsilon\right|^{p-1} \approx \varepsilon^{(p-1)(J-1)}.$$

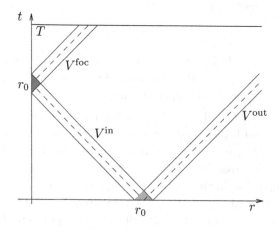

Fig. 6.1   Propagation of short pulses.

Integrated on the focusing region, we find:

$$\int_{|r|\lesssim\varepsilon} |\partial_t v^\varepsilon|^{p-1} \approx \varepsilon^{(p-1)(J-1)+1}.$$

This suggests that if $(p-1)(J-1)+1 > 0$, then in the nonlinear case (6.5), nonlinear effects are negligible at the focus, while they become relevant for $(p-1)(J-1)+1 = 0$. To resume the terminology of [Hunter and Keller (1987)], the former case is referred to as "linear caustic" while the latter is called "nonlinear caustic". We can also check that the WKB régime is linear if $J > 0$ ("linear geometric optics", or "linear propagation"), weakly nonlinear if $J = 0$, as in the previous paragraph ("nonlinear propagation"). This leads to the following distinctions:

	$J > 0$	$J = 0$
$J > \frac{p-2}{p-1}$	linear propagation linear caustic	nonlinear propagation linear caustic
$J = \frac{p-2}{p-1}$	linear propagation nonlinear caustic	nonlinear propagation nonlinear caustic
$J < \frac{p-2}{p-1}$	linear propagation supercritical caustic	nonlinear propagation supercritical caustic

The above line of reasoning is slightly different from that of [Hunter and Keller (1987)]: we have not considered the whole nonlinear term $|\partial_t v^\varepsilon|^{p-1}\partial_t v^\varepsilon$, but only the "potential" $|\partial_t v^\varepsilon|^{p-1}$. The reason is that energy estimates are available for the equation

$$\left(\partial_t^2 - \Delta\right) u^\varepsilon + V^\varepsilon \partial_t u^\varepsilon = 0,$$

and if V^ε is small in $L_t^1 L_x^\infty$ for instance, then u^ε and v^ε are close to each other in the energy space. We give more details on this approach in the case of nonlinear Schrödinger equations below.

For short pulses, the above table has been justified in a series of three articles, [Carles and Rauch (2002, 2004a,b)]. In the case of a slowly varying envelopes (when the profile U_j are periodic with respect to their second variable, and not compactly supported), the same distinctions are expected, but a justification for all the cases is not available so far.

Notice that assuming that the above table remains valid for slowly varying envelopes, the result of [Joly et al. (1995b)] corresponds to $J = 0 < \frac{p-2}{p-1}$: the supercritical effect near the caustic is the absorption of oscillations. On the other hand, the analogous phenomenon for the pulses is the absorption of the pulse at the focal point [Carles and Rauch (2004b)]. Indeed, formally, a pulse has zero mean value:

$$\frac{1}{2M}\int_{-M}^{M} f(z)dz \xrightarrow[M\to\infty]{} 0 \quad \text{if } f \text{ is compactly supported.}$$

So in a way, like for the case of wave trains, only the mean value of the wave remains past the focus. We leave out the discussion for the semi-linear wave equation at this stage.

Back to the case of nonlinear Schrödinger equations, consider the initial value problem

$$i\varepsilon\partial_t u^\varepsilon + \frac{\varepsilon^2}{2}\Delta u^\varepsilon = \varepsilon^\alpha |u^\varepsilon|^{2\sigma} u^\varepsilon \quad ; \quad u^\varepsilon(0,x) = a_0(x)e^{i\phi_0(x)/\varepsilon}. \tag{6.6}$$

Recall that the factor ε^α can be viewed as a scaling factor for the size of the initial data; see the end of §1.1. We have seen in the first part of these notes that the value $\alpha = 1$ is critical for the WKB methods: if $\alpha > 1$, then the nonlinearity does not affect the transport equation (1.19), while if $\alpha = 1$, then the nonlinearity appears in the right-hand side of (1.19). To resume the terminology of [Hunter and Keller (1987)], we say that $\alpha > 1$ corresponds to a "linear geometric optics", or "linear propagation", while $\alpha = 1$ corresponds to a "nonlinear geometric optics", or "nonlinear propagation". The idea presented in [Hunter and Keller (1987)] consists in saying that according to the geometry of the caustic $\mathcal{C}$, different notions of criticality exist, as far as α is concerned, near the caustic. In the linear setting, the influence of the caustic is relevant only in a neighborhood of this set (essentially, in a boundary layer whose size depends on ε and the geometry of $\mathcal{C}$). View the nonlinearity in Eq. (6.6) as a potential, and assume that the nonlinear effects are negligible near the caustic: then $u^\varepsilon \sim v^\varepsilon$ near $\mathcal{C}$, where v^ε solves the free Schrödinger equation. Consider the term $\varepsilon^\alpha |u^\varepsilon|^{2\sigma}$ as a (nonlinear) potential. The average nonlinear effect near $\mathcal{C}$ is expected to be:

$$\varepsilon^{-1}\int_{\mathcal{C}(\varepsilon)} \varepsilon^\alpha |u^\varepsilon|^{2\sigma} \sim \varepsilon^{-1}\int_{\mathcal{C}(\varepsilon)} \varepsilon^\alpha |v^\varepsilon|^{2\sigma},$$

where $\mathcal{C}(\varepsilon)$ is the region where caustic effects are relevant, and the factor ε^{-1} is due to the integration in time (recall that there is an ε in front of the time derivative in Eq. (6.6)). The idea of this heuristic argument is that when the nonlinear effects are negligible near $\mathcal{C}$ (in the sense that the uniform norm of $u^\varepsilon - v^\varepsilon$ is small compared to that of v^ε near $\mathcal{C}$), the above approximation should be valid. On the other hand, it is expected that it ceases to be valid precisely when nonlinear effects can no longer be neglected near the caustic: $u^\varepsilon - v^\varepsilon$ is (at least) of the same order of magnitude as v^ε in $L^\infty(\mathcal{C}(\varepsilon))$. Like for the case of the semi-linear wave equation (6.5), we have not considered the whole nonlinearity $|u^\varepsilon|^{2\sigma} u^\varepsilon$, but only the nonlinear potential $|u^\varepsilon|^{2\sigma}$. Like for the wave equation, this is due to the fact that energy estimates are available for Schrödinger equations, showing that the norm of $|u^\varepsilon|^{2\sigma}$ in $L_t^1 L_x^\infty$ measures the influence of the whole nonlinear term in Eq. (6.6). Indeed, write

$$\left(i\varepsilon\partial_t + \frac{\varepsilon^2}{2}\Delta\right)(u^\varepsilon - v^\varepsilon) = \varepsilon^\alpha |u^\varepsilon|^{2\sigma} u^\varepsilon = \varepsilon^\alpha |u^\varepsilon|^{2\sigma}(u^\varepsilon - v^\varepsilon) + \varepsilon^\alpha |u^\varepsilon|^{2\sigma} v^\varepsilon.$$

Lemma 1.2 yields:

$$\begin{aligned}
\|u^\varepsilon - v^\varepsilon\|_{L^\infty(I;L^2)} &\leqslant \varepsilon^{\alpha-1} \int_I \left\| |u^\varepsilon|^{2\sigma} v^\varepsilon \right\|_{L^2} \\
&\leqslant \varepsilon^{\alpha-1} \|u^\varepsilon\|_{L^1(I;L^\infty)}^{2\sigma} \|v^\varepsilon\|_{L^\infty(I;L^2)} \\
&\leqslant C\varepsilon^{\alpha-1} \|u^\varepsilon\|_{L^1(I;L^\infty)}^{2\sigma},
\end{aligned}$$

where C does not depend on ε, since the L^2 norm of v^ε is conserved, and independent of ε. In the case of a linear caustic, we can replace u^ε by v^ε in the last term of the above inequality, hence the above discussion.

Practically, assume that in the linear case, v^ε has an amplitude $\varepsilon^{-\ell}$ in a boundary layer of size ε^k; then the above quantity is

$$\varepsilon^{-1} \int_{\mathcal{C}(\varepsilon)} \varepsilon^\alpha |v^\varepsilon|^{2\sigma} \sim \varepsilon^{-1} \varepsilon^\alpha \left| \varepsilon^{-\ell} \right|^{2\sigma} \varepsilon^k.$$

The value α is then critical when the above cumulated effects are not negligible:

$$\alpha_c = 1 + 2\ell\sigma - k.$$

When $\alpha > \alpha_c$, the nonlinear effects are expected to be negligible near the caustic: "linear caustic". The case $\alpha = \alpha_c$ corresponds to a "nonlinear caustic". We illustrate this discussion in the case of two geometries below:

- When the caustic is reduced to a single point.
- When the caustic is a cusp.

6.3   The case of a focal point

In the case of a focal point, which corresponds to a quadratic initial phase (e.g. $\phi_0(x) = -|x|^2/2$), we have seen that $k = 1$ and $\ell = d/2$. See Eq. (6.2)

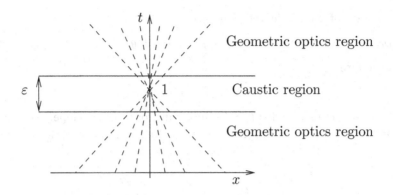

Fig. 6.2   Rays of geometric optics: focal point.

and Fig. 6.2. This leads us to the value: $\alpha_c(\text{focal point}) = d\sigma$. Therefore, the following distinctions are expected:

	$\alpha > 1$	$\alpha = 1$
$\alpha > d\sigma$	linear propagation linear caustic	nonlinear propagation linear caustic
$\alpha = d\sigma$	linear propagation, nonlinear caustic	nonlinear propagation nonlinear caustic

Note that as in the case of the semi-linear wave equation, the entries for rows and columns are fairly independent, as suggested in [Hunter and Keller (1987)]. In Chap. 7, we show that the above distinctions are relevant, and we describe the asymptotics of the wave functions in each of the four cases. Unlike for the semi-linear wave equation, we have note mentioned the case of a "supercritical caustic". A complete description in that case is not available yet, and we give a very partial answer in Chap. 9.

6.4   The case of a cusp

It seems that there are no rigorous results concerning nonlinear effects near a cusped caustic for nonlinear Schrödinger equations. Therefore, the discussion in this paragraph is only formal.

In space dimension $d = 1$, a cusped caustic appears for instance if the initial phase is given by

$$\phi_0(x) = \cos x.$$

The rays of geometric optics are given

$$\partial_t x(t,y) = \xi(t,y) \quad ; \quad x(0,y) = y \quad ; \quad \partial_t \xi(t,y) = 0 \quad ; \quad \xi(0,y) = -\sin y.$$

Therefore,

$$x(t,y) = y - t\sin y.$$

A caustic is present for $t \geqslant 1$, since rays form an envelope, see Fig. 6.3. At $t = 1$, there is a cusp at the origin. For $t > 1$, the caustic is a smooth

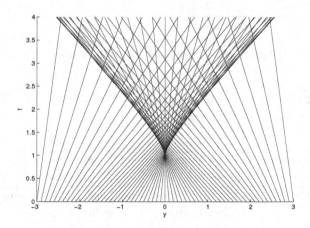

Fig. 6.3   Rays of geometric optics: cusped caustic.

curve. Moreover, for $t > 1$, three phases are necessary to describe the wave function inside the caustic region. To see this, recall that the linear solution is given by

$$u_{\text{lin}}^\varepsilon(t,x) = \frac{1}{(2i\pi t)^{1/2}} \int_{\mathbb{R}} e^{i\frac{|x-y|^2}{2\varepsilon t} + i\frac{\cos y}{\varepsilon}} a_0(y) dy.$$

The critical points y_c associated to the phase of this integral are such that

$$\frac{y_c - x}{t} = \sin y_c,$$

that is $x = y_c - t\sin y_c$. To compute the number of critical points, we map $y \mapsto y - t\sin y$ for various values of t. At time $t = 0.5$ (Fig. 6.4), there is only one critical point: the caustic is not formed yet. At time $t = 1$ (Fig. 6.5), we see that the tangent is vertical at $(x,y) = (0,0)$: this corresponds to the cusp. At time $t = 1.5$ (Fig. 6.6), there are three critical points for $-x_0(1.5) < x < x_0(1.5)$, for some $x_0(1.5) \approx 0.3$. This corresponds

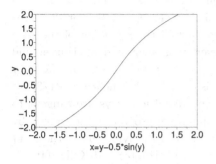

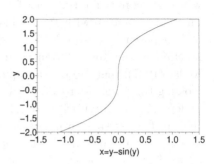

Fig. 6.4 Time $t = 0.5$. Fig. 6.5 Time $t = 1$.

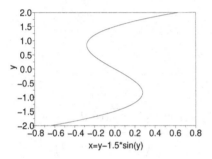

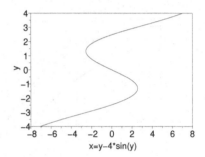

Fig. 6.6 Time $t = 1.5$. Fig. 6.7 Time $t = 4$.

to the three phases inside the caustic region. Note that $|x| = x_0(1.5)$ corresponds to the caustic itself. For $t = 4$ (Fig. 6.7), the caustic is broader, and the corresponding $x_0(4)$ is larger: $x_0(4) > 2$.

To apply the argument of J. Hunter and J. Keller, we must distinguish two families of points: the cusp (at $(t, x) = (1, 0)$), and smooth points of the caustic (for $t > 1$ and $|x| = x_0(t)$).

At smooth points, the wave function is described thanks to the Airy function: on the caustic, it is not possible to apply the usual stationary phase formula, because the critical points are degenerate, but the third order derivative is not zero. This was first noticed in [Ludwig (1966)]. See also [Duistermaat (1974); Hörmander (1994); Hunter and Keller (1987)]. Typically, near the caustic, the linear solution can be approximated by

$$\varepsilon^{-1/6} \left(\alpha Ai \left(\frac{\psi(t, x)}{\varepsilon^{2/3}} \right) + \beta \varepsilon^{1/3} Ai' \left(\frac{\psi(t, x)}{\varepsilon^{2/3}} \right) \right) e^{i\rho(t, x)/\varepsilon},$$

where Ai stands for the Airy function, for $\alpha, \beta \in \mathbb{C}$ and some smooth functions ψ and ρ, with $\psi = 0$ on $\mathcal{C}$. This yields $\ell = 1/6$. For the value of k, it is tempting to take $k = 2/3$, in view of the argument of the Airy function in the above formula. However, it is suggested in [Hunter and Keller (1987)] that the relevant quantity to consider is the length of a ray crossing this layer of order $\varepsilon^{2/3}$, which is of order $\varepsilon^{1/3}$, hence $k = 1/3$. To understand this derivation, recall that by definition, rays are tangent to the caustic. At smooth points of the caustic, approximate the caustic by a circle. Figure 6.8 shows why the length of a ray lying inside the layer of order $\varepsilon^{2/3}$ is of order $\varepsilon^{1/3}$. The approach of [Hunter and Keller (1987)] therefore suggests:

$$\alpha_c(\text{smooth point in 1D}) = \frac{\sigma + 2}{3}.$$

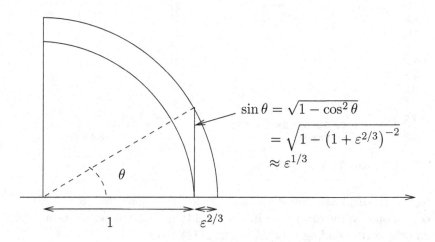

Fig. 6.8 Length of a ray inside a typical boundary layer.

At the cusp, the wave function grows like $\varepsilon^{-1/4}$, in a region of order $\varepsilon^{1/2}$ (see [Duistermaat (1974); Hunter and Keller (1987)]), hence

$$\alpha_c(\text{cusp in 1D}) = 1 - \frac{2\sigma}{4} - \frac{1}{2} = \frac{\sigma + 1}{2}.$$

We see that we have

$$\alpha_c(\text{cusp in 1D}) > \alpha_c(\text{smooth point in 1D}) \iff \sigma > 1.$$

Therefore, nonlinear effects might be stronger either at the cusp, or at smooth points of the caustic, according to the power of the nonlinearity.

This may seem paradoxical, since the cusp is expected to concentrate more energy than a smooth point on the caustic. However, we invite the reader to consult [Joly *et al.* (2000)] (or [Joly *et al.* (1997b)]): the analytical results (to prove estimates in Lebesgue spaces for oscillatory integrals) differ from the topological results of [Duistermaat (1974)]. In the case studied by J.-L. Joly, G. Métivier and J. Rauch, the critical index for L^p estimates is given by smooth points of the caustic, and the influence of the cusp is negligible. However, the approach of [Joly *et al.* (2000)] does not seem to yield directly estimates which can be used in the case of Schrödinger equations. Even in the apparently simple case considered in this paragraph, the following two questions remain open so far:

- Find α_c such that nonlinear effects are negligible near the caustic for $\alpha > \alpha_c$, but not for $\alpha < \alpha_c$.
- For $\alpha = \alpha_c$, describe the (possible) leading order nonlinear effects at the caustic.

Chapter 7

Focal Point without External Potential

7.1 Presentation

In this chapter, we consider the initial value problem

$$i\varepsilon\partial_t u^\varepsilon + \frac{\varepsilon^2}{2}\Delta u^\varepsilon = \varepsilon^\alpha |u^\varepsilon|^{2\sigma} u^\varepsilon \quad ; \quad u^\varepsilon(0,x) = a_0(x)e^{-i|x|^2/(2\varepsilon)}. \tag{7.1}$$

To simplify the notations and the discussions, we consider only non-negative time: $t \geqslant 0$. Recall that we have derived formally the following distinctions in the previous chapter:

	$\alpha > 1$	$\alpha = 1$
$\alpha > d\sigma$	linear propagation linear caustic	nonlinear propagation linear caustic
$\alpha = d\sigma$	linear propagation, nonlinear caustic	nonlinear propagation nonlinear caustic

Consider the solution to the free Schrödinger equation which coincides with u^ε at some fixed time τ:

$$i\varepsilon\partial_t v_\tau^\varepsilon + \frac{\varepsilon^2}{2}\Delta v_\tau^\varepsilon = 0 \quad ; \quad v_\tau^\varepsilon(\tau,x) = u^\varepsilon(\tau,x). \tag{7.2}$$

We can then give a more precise interpretation of the above table, where the expression $f^\varepsilon - g^\varepsilon = o(h^\varepsilon)$ stands for $\|f^\varepsilon - g^\varepsilon\|_{L^2} = o(\|h^\varepsilon\|_{L^2})$ as $\varepsilon \to 0$:

- If $\alpha > \max(1, d\sigma)$, then we expect $u^\varepsilon(t) - v_0^\varepsilon(t) = o(v_0^\varepsilon(t))$ for all t.
- If $\alpha = 1 > d\sigma$, then outside the focal point at $t = 1$, we already know that this is a weakly nonlinear régime: the leading order nonlinear effect consists of a self-phase modulation, and u^ε is not comparable to v_0^ε. On the other hand, the nonlinearity is negligible near the focal point: $u^\varepsilon(t) - v_1^\varepsilon(t) = o(v_1^\varepsilon(t))$ for $|t-1| \leqslant C\varepsilon$. Recall that ε also measures the influence zone of the focal point.

- If $\alpha = d\sigma > 1$, then we expect $u^\varepsilon(t) - v_0^\varepsilon(t) = o(v_0^\varepsilon(t))$ for $t < 1$ and $u^\varepsilon(t) - v_\tau^\varepsilon(t) = o(v_\tau^\varepsilon(t))$ for $t > 1$, for (any) $\tau > 1$ independent of ε. These two relations express the fact that the nonlinearity in Eq. (7.1) is negligible off $t = 1$. On the other hand, it should not be negligible near $t = 1$, and therefore,

$$\liminf_{\varepsilon \to 0} \|v_0^\varepsilon(t) - v_\tau^\varepsilon(t)\|_{L^2} > 0, \forall t \geqslant 0, \ \forall \tau > 1, \quad \text{in general.}$$

We will see that at leading order, v_0^ε and v_τ^ε differ in terms of a scattering operator associated to the nonlinear Schrödinger equation.

- If $\alpha = 1 = d\sigma$, then we don't expect $u^\varepsilon(t) - v_\tau^\varepsilon(t) = o(v_\tau^\varepsilon(t))$ for any $t \neq \tau$: the solution u^ε never behaves like a free solution.

These four assertions have been justified in [Carles (2000b)], thus proving that the previous table is correct.

For technical reasons, we will always assume that the nonlinearity is H^1-subcritical (see Sec. 1.4), and some lower bounds on σ:

$$0 < \sigma < \infty \text{ if } d = 1; \ \frac{1}{2} < \sigma < \infty \text{ if } d = 2; \ \frac{2}{d+2} < \sigma < \frac{2}{d-2} \text{ if } d \geqslant 3.$$

We also assume that the initial amplitude a_0 is in $\Sigma(1)$, defined in §1.4.2. For the sake of readability, we drop the index 1, and consider therefore:

$$a_0 \in \Sigma := \left\{ f \in H^1(\mathbb{R}^d) \ ; \ x \mapsto \langle x \rangle f(x) \in L^2(\mathbb{R}^d) \right\}.$$

The space Σ is equipped with the norm

$$\|f\|_\Sigma = \|f\|_{L^2} + \|\nabla f\|_{L^2} + \|xf\|_{L^2}.$$

Obviously, for fixed ε, $u^\varepsilon(0, \cdot) \in \Sigma$.

There are at least two ways to present the results. First, we may write the asymptotics for the function u^ε itself. Second, we can proceed as in [Joly et al. (1996a, 2000)], and write the solution as a modified Lagrangian integral. This approach was followed in [Carles (2000b)]. We will present the results from both points of view here. In our case, the generalization of the representation (6.3) is:

$$u^\varepsilon(t, x) = \frac{1}{(2\pi\varepsilon)^{d/2}} \int_{\mathbb{R}^d} e^{-i\frac{t-1}{2\varepsilon}|\xi|^2 + i\frac{x \cdot \xi}{\varepsilon}} A^\varepsilon(t, \xi) d\xi. \tag{7.3}$$

Note that unlike in the linear case, the Lagrangian symbol A^ε depends on time. It is still given by

$$A^\varepsilon(t, \xi) = \frac{1}{\varepsilon^{d/2}} e^{i\frac{t-1}{2\varepsilon}|\xi|^2} \widehat{u^\varepsilon}\left(t, \frac{\xi}{\varepsilon}\right).$$

We have the following preliminary result:

Lemma 7.1. *Let $d \geqslant 1$ and $a_0 \in \Sigma$. The initial Lagrangian symbol converges in Σ:*

$$A^\varepsilon(0, \xi) = e^{-id\pi/4} a_0(-\xi) + o(1) \quad in \ \Sigma, \ as \ \varepsilon \to 0.$$

Moreover, the Lagrangian symbol satisfies the equation:

$$i\partial_t A^\varepsilon(t, \xi) = \varepsilon^{\alpha - d/2 - 1} e^{i\frac{t-1}{2\varepsilon}|\xi|^2} \mathcal{F}\left(|u^\varepsilon|^{2\sigma} u^\varepsilon\right)\left(t, \frac{\xi}{\varepsilon}\right). \tag{7.4}$$

Proof. As in the linear case, we have

$$A^\varepsilon(0, \xi) = \frac{1}{(2\pi\varepsilon)^{d/2}} \int_{\mathbb{R}^d} e^{-i\frac{|x+\xi|^2}{2\varepsilon}} a_0(x) dx$$

$$= \frac{1}{(2\pi\varepsilon)^{d/2}} \int_{\mathbb{R}^d} e^{-i\frac{|x|^2}{2\varepsilon}} a_0(x - \xi) dx.$$

Recalling that

$$\mathcal{F}\left(e^{-i|x|^2/(2\varepsilon)}\right)(\eta) = \left(\frac{\varepsilon}{i}\right)^{d/2} e^{i\varepsilon|\eta|^2/2},$$

$$\mathcal{F}_{x \to \eta}\left(a_0(x - \xi)\right)(\eta) = e^{i\eta \cdot \xi} \widehat{a}_0(\eta),$$

Parseval formula yields:

$$A^\varepsilon(0, \xi) = \frac{1}{(2i\pi)^{d/2}} \int e^{i\eta \cdot \xi} e^{-i\varepsilon|\eta|^2/2} \widehat{a}_0(\eta) d\eta.$$

Then

$$A^\varepsilon(0, \xi) - e^{-id\pi/4} a_0(-\xi) = \frac{1}{(2i\pi)^{d/2}} \int e^{i\eta \cdot \xi} \left(e^{-i\varepsilon|\eta|^2/2} - 1\right) \widehat{a}_0(\eta) d\eta.$$

By Plancherel equality, we infer

$$\left\| A^\varepsilon(0, \cdot) - e^{-id\pi/4} a_0(-\cdot) \right\|_{L^2} = \left\| \left(e^{-i\varepsilon|\cdot|^2/2} - 1\right) \widehat{a}_0(\cdot) \right\|_{L^2}.$$

Now since

$$\left| e^{-i\varepsilon|\eta|^2/2} - 1 \right| = 2 \left| \sin\left(\varepsilon \frac{|\eta|^2}{4}\right) \right|,$$

the Dominated Convergence Theorem yields:

$$\lim_{\varepsilon \to 0} \left\| A^\varepsilon(0, \cdot) - e^{-id\pi/4} a_0(-\cdot) \right\|_{L^2} = 0.$$

Noting that

$$\nabla\left(A^\varepsilon(0, \xi) - e^{-id\pi/4} a_0(-\xi)\right) = \frac{i}{(2i\pi)^{d/2}} \int e^{i\eta \cdot \xi} \left(e^{-i\varepsilon|\eta|^2/2} - 1\right) \eta \widehat{a}_0(\eta) d\eta,$$

$$\xi\left(A^\varepsilon(0, \xi) - e^{-id\pi/4} a_0(-\xi)\right)$$

$$= \frac{i}{(2i\pi)^{d/2}} \int e^{i\eta \cdot \xi} \partial_\eta\left(\left(e^{-i\varepsilon|\eta|^2/2} - 1\right) \widehat{a}_0(\eta)\right) d\eta,$$

and since $a_0 \in \Sigma$, the estimate of the lemma follows by the same argument. The second part of the lemma is straightforward. $\qquad\square$

For $t \geqslant 0$, we check the following identities:

$$\|A^\varepsilon(t)\|_{L^2} = \|u^\varepsilon(t)\|_{L^2} \quad ; \quad \|\nabla A^\varepsilon(t)\|_{L^2} = \|J^\varepsilon(t)u^\varepsilon(t)\|_{L^2}$$
$$\|\xi A^\varepsilon(t)\|_{L^2} = \|\varepsilon \nabla u^\varepsilon(t)\|_{L^2}, \tag{7.5}$$

where the operator J^ε is given by:

$$J^\varepsilon(t) = \frac{x}{\varepsilon} + i(t-1)\nabla.$$

Note that we have

$$J^\varepsilon(t) = X + iT\nabla_X\big|_{(T,X)=((t-1)/\varepsilon, x/\varepsilon)}.$$

We recover the operator $X + iT\nabla$, introduced in [Ginibre and Velo (1979)], which is classical in the scattering theory for nonlinear Schrödinger equation; see also e.g. [Tsutsumi and Yajima (1984); Hayashi and Tsutsumi (1987); Cazenave and Weissler (1992)].

Let us mention another point of view, from which the introduction of this operator is also very natural. Recall that in Chap. 2 (and also in §4.3), we have performed H^s estimates not directly on u^ε, but on $u^\varepsilon e^{-i\phi_{\mathrm{eik}}/\varepsilon}$. The idea is that the H^s norm of the latter is bounded uniformly in $\varepsilon \in]0,1]$, while the former is not. In the present case, we have

$$\phi_{\mathrm{eik}}(t,x) = \frac{|x|^2}{2(t-1)},$$

so it is natural to consider

$$e^{i\phi_{\mathrm{eik}}(t,x)/\varepsilon}\nabla\left(e^{-i\phi_{\mathrm{eik}}(t,x)/\varepsilon}\cdot\right) = -i\frac{x}{\varepsilon(t-1)} + \nabla.$$

We can be more precise by recalling that the concentration rate for the approximate solution in the linear case is exactly $1-t$, see (6.2). Therefore, up to an irrelevant factor i, we retrieve

$$i(t-1)e^{i\phi_{\mathrm{eik}}(t,x)/\varepsilon}\nabla\left(e^{-i\phi_{\mathrm{eik}}(t,x)/\varepsilon}\cdot\right) = J^\varepsilon(t).$$

This formula has two interesting consequences from a technical point of view:

- The operator acts on gauge invariant nonlinearities like a derivative: if $G(z) = F\left(|z|^2\right)z$ is C^1, then

$$J^\varepsilon(t)G(u) = \partial_z G(u)J^\varepsilon(t)u - \partial_{\bar{z}}G(u)\overline{J^\varepsilon(t)u}.$$

- Weighted Gagliardo–Nirenberg inequalities are available:

$$\|u\|_{L^r} \leqslant \frac{C_r}{|1-t|^{\delta(r)}} \|u\|_{L^2}^{1-\delta(r)} \|J^{\varepsilon}(t)u\|_{L^2}^{\delta(r)},$$

$$\text{where } \delta(r) := d\left(\frac{1}{2} - \frac{1}{r}\right) \in [0,1[, \tag{7.6}$$

and C_r depends only on r and d. This is a direct consequence of the standard inequality (without the weight $|1-t|$), where $J^{\varepsilon}(t)$ is replaced by ∇.

Example 7.2. Apply the operator $J^{\varepsilon}(t)$ to the approximate solution of the linear case, that is, the right-hand side of (6.2). For $t < 1$, we find:

$$J^{\varepsilon}(t)\left(\frac{1}{(1-t)^{d/2}}a_0\left(\frac{x}{1-t}\right)e^{i\frac{|x|^2}{2\varepsilon(t-1)}}\right) = \frac{i}{(1-t)^{d/2}}\nabla a_0\left(\frac{x}{1-t}\right)e^{i\frac{|x|^2}{2\varepsilon(t-1)}}.$$

When applying the weighted Gagliardo–Nirenberg inequality (7.6) to this function, we see that both the left-hand side and the right-hand side are of order

$$|1-t|^{-\delta(r)}.$$

This suggests that for $t \neq 1$, in the semi-classical régime, the operator J^{ε} yields sharp estimates, as far as the parameters t and ε are concerned.

For $t = 1$, the inequality (7.6) becomes singular. Instead, resume the standard inequality, rescaled by ε:

$$\|u\|_{L^r} \leqslant \frac{C_r}{\varepsilon^{\delta(r)}} \|u\|_{L^2}^{1-\delta(r)} \|\varepsilon\nabla u\|_{L^2}^{\delta(r)}.$$

Then again, the power $\varepsilon^{-\delta(r)}$ is such that when applied to

$$\frac{1}{(i\varepsilon)^{d/2}}\widehat{a}_0\left(\frac{x}{\varepsilon}\right),$$

both sides of the above estimates are of order $\varepsilon^{-\delta(r)}$.

The conclusion suggested by this example, and which turns out to be useful in the nonlinear estimates, is that the operator J^{ε} yields good estimates off $t = 1$, while near $t = 1$, the natural operator is $\varepsilon\nabla$.

Before describing more precisely the results, we mention a third important property of the operator J^{ε}: it commutes with the linear Schrödinger operator,

$$\left[i\varepsilon\partial_t + \frac{\varepsilon^2}{2}\Delta, J^{\varepsilon}(t)\right] = 0.$$

This property, classical in the case $\varepsilon = 1$ [Hayashi and Tsutsumi (1987)], stems from the fact that J^ε can be factorized in a different way. Let

$$U^\varepsilon(t) = \exp\left(i\varepsilon\frac{t}{2}\Delta\right)$$

denote the group associated to the free semi-classical Schrödinger equation. We have:

$$J^\varepsilon(t) = U^\varepsilon(t-1)\frac{x}{\varepsilon}U^\varepsilon(1-t).$$

This expression implies the above commutation with the linear Schrödinger operator. We will see in Sec. 8.3 that the existence of such an operator with nice properties both for nonlinear estimates and for linear commutators does not seem to be generic, in the presence of an external potential.

7.2   Linear propagation, linear caustic

In view of Lemma 7.1, denote

$$A_0(\xi) = e^{-id\pi/4}a_0(-\xi).$$

Proposition 7.3. *Assume $\alpha > \max(1, d\sigma)$. Then*

$$A^\varepsilon \xrightarrow[\varepsilon \to 0]{} A_0 \quad in \ L^\infty_{\mathrm{loc}}(\mathbb{R}; \Sigma).$$

Equivalently,

$$\|\mathcal{B}^\varepsilon\left(u^\varepsilon - v_0^\varepsilon\right)\|_{L^\infty_{\mathrm{loc}}(\mathbb{R};L^2)} \xrightarrow[\varepsilon \to 0]{} 0, \quad for \ all \ \mathcal{B}^\varepsilon \in \{\mathrm{Id}, \varepsilon\nabla, J^\varepsilon\}.$$

To make the proof more intuitive, we distinguish the special case $d = 1$ from the general case $d \geqslant 1$.

The case $d = 1$. In space dimension $d = 1$, we can use the Sobolev embedding $H^1(\mathbb{R}) \hookrightarrow L^\infty(\mathbb{R})$. More precisely, Gagliardo–Nirenberg inequality shows that there exists C independent of ε and t such that for all $u \in \Sigma$:

$$\|u\|_{L^\infty} \leqslant \frac{C}{(\varepsilon + |t-1|)^{1/2}} \|u\|_{L^2}^{1/2}\left(\|\varepsilon\partial_x u\|_{L^2} + \|J^\varepsilon(t)u\|_{L^2}\right)^{1/2}. \tag{7.7}$$

We also note that (for any $d \geqslant 1$)

$$\begin{aligned}
\|v_0^\varepsilon(t)\|_{L^2} &= \|v_0^\varepsilon(0)\|_{L^2} = \|a_0\|_{L^2}, \\
\|\varepsilon\nabla v_0^\varepsilon(t)\|_{L^2} &= \|\varepsilon\nabla v_0^\varepsilon(0)\|_{L^2} = \mathcal{O}(1), \\
\|J^\varepsilon(t)v_0^\varepsilon(t)\|_{L^2} &= \|J^\varepsilon(0)v_0^\varepsilon(0)\|_{L^2} = \|\nabla a_0\|_{L^2}.
\end{aligned} \tag{7.8}$$

Consider the error term $w^\varepsilon = u^\varepsilon - v_0^\varepsilon$. It solves:

$$i\varepsilon\partial_t w^\varepsilon + \frac{\varepsilon^2}{2}\Delta w^\varepsilon = \varepsilon^\alpha |u^\varepsilon|^{2\sigma} u^\varepsilon \quad ; \quad w^\varepsilon_{|t=0} = 0. \tag{7.9}$$

Lemma 1.2 yields, for $t > 0$:

$$\|w^\varepsilon\|_{L^\infty([0,t];L^2)} \leqslant \varepsilon^{\alpha-1} \int_0^t \left\| |u^\varepsilon(\tau)|^{2\sigma} u^\varepsilon(\tau) \right\|_{L^2} d\tau$$

$$\leqslant \varepsilon^{\alpha-1} \|a_0\|_{L^2} \int_0^t \|u^\varepsilon(\tau)\|_{L^\infty}^{2\sigma} d\tau,$$

where we have used the conservation of the L^2-norm of u^ε. Recalling that $u^\varepsilon = w^\varepsilon + v_0^\varepsilon$, (7.7) and Eq. (7.8) yield:

$$\|u^\varepsilon(\tau)\|_{L^\infty}^{2\sigma} \leqslant C_\sigma \left(\|w^\varepsilon(\tau)\|_{L^\infty}^{2\sigma} + \|v_0^\varepsilon(\tau)\|_{L^\infty}^{2\sigma} \right)$$

$$\leqslant C \left(\|w^\varepsilon(\tau)\|_{L^\infty}^{2\sigma} + \frac{1}{(\varepsilon + |\tau - 1|)^\sigma} \right).$$

Since w^ε is expected to be a relatively small error estimate, it should satisfy at least the same estimates as v_0^ε. From Proposition 1.26, $u^\varepsilon \in C(\mathbb{R}; \Sigma)$, hence $w^\varepsilon \in C(\mathbb{R}; \Sigma)$. Since $w^\varepsilon_{|t=0} = 0$, there exists $t^\varepsilon > 0$ such that

$$\|J^\varepsilon(\tau)w^\varepsilon(\tau)\|_{L^2} \leqslant 1 \tag{7.10}$$

for $\tau \in [0, t^\varepsilon]$. Recall that from the conservation of the energy for u^ε, Eq. (1.25),

$$\frac{d}{dt}\left(\frac{1}{2}\|\varepsilon\nabla u^\varepsilon(t)\|_{L^2}^2 + \frac{\varepsilon^\alpha}{\sigma+1}\|u^\varepsilon(t)\|_{L^{2\sigma+2}}^{2\sigma+2} \right) = 0.$$

Therefore, there exists C independent of ε such that

$$\|\varepsilon\nabla u^\varepsilon(t)\|_{L^2} \leqslant C, \quad \forall t \in \mathbb{R}.$$

So, there exists C' independent of ε, such that

$$\|\varepsilon\nabla w^\varepsilon(t)\|_{L^2} \leqslant C', \quad \forall t \in \mathbb{R}.$$

In view of (7.7), we infer, so long as (7.10) holds,

$$\|w^\varepsilon(\tau)\|_{L^\infty} \leqslant \frac{C}{(\varepsilon + |\tau - 1|)^{1/2}},$$

for some constant C independent of ε. We infer, so long as (7.10) holds:

$$\|w^\varepsilon\|_{L^\infty([0,t];L^2)} \leqslant C\varepsilon^{\alpha-1} \int_0^t \frac{d\tau}{(\varepsilon + |\tau - 1|)^\sigma}.$$

Fix $T > 1$. Distinguishing the regions $\{|\tau - 1| \geqslant \varepsilon\}$ and $\{|\tau - 1| > \varepsilon\}$, for $t \leqslant T$, the latest integral is controlled by:

$$\int_0^t \frac{d\tau}{(\varepsilon + |\tau - 1|)^\sigma} \leqslant \int_0^{1-\varepsilon} \frac{d\tau}{(1 - \tau)^\sigma} + \int_{1-\varepsilon}^{1+\varepsilon} \frac{d\tau}{\varepsilon^\sigma} + \int_{1-\varepsilon}^T \frac{d\tau}{(\tau - 1)^\sigma}$$

$$\leqslant C \left(\max \left(\varepsilon^{1-\sigma}, \log \frac{1}{\varepsilon}, 1 \right) + \varepsilon^{1-\sigma} \right),$$

where we have distinguished the three cases, $\sigma > 1$, $\sigma = 1$ and $0 < \sigma < 1$. Therefore, if (7.10) holds on $[0, T]$, we infer:

$$\|w^\varepsilon\|_{L^\infty([0,t];L^2)} \leqslant C \max \left(\varepsilon^{\alpha-\sigma}, \varepsilon^{\alpha-1} \log \frac{1}{\varepsilon} \right). \tag{7.11}$$

The strategy is to obtain similar estimates for $\varepsilon \nabla w^\varepsilon$ and $J^\varepsilon w^\varepsilon$. Applying the operator $\varepsilon \nabla$ to Eq. (7.9), we find:

$$\left(i\varepsilon \partial_t + \frac{\varepsilon^2}{2} \Delta \right) \varepsilon \nabla w^\varepsilon = \varepsilon^{1+\alpha} \nabla \left(|u^\varepsilon|^{2\sigma} u^\varepsilon \right) = (\sigma + 1)\varepsilon^\alpha |u^\varepsilon|^{2\sigma} \varepsilon \nabla u^\varepsilon$$

$$+ \sigma \varepsilon^\alpha (u^\varepsilon)^{\sigma+1} (\overline{u}^\varepsilon)^{\sigma-1} \varepsilon \nabla \overline{u}^\varepsilon,$$

along with the Cauchy data $\varepsilon \nabla w^\varepsilon_{|t=0} = 0$. From the conservation of the energy for u^ε, we can mimic the previous computations, and find, so long as (7.10) holds:

$$\|\varepsilon \nabla w^\varepsilon\|_{L^\infty([0,t];L^2)} \leqslant C \max \left(\varepsilon^{\alpha-\sigma}, \varepsilon^{\alpha-1} \log \frac{1}{\varepsilon} \right). \tag{7.12}$$

To complete the argument, apply the operator J^ε to Eq. (7.9). We have seen that J^ε behaves like the gradient: it commute with the linear Schrödinger operator, and acts on gauge invariant nonlinearities like a derivatives. Therefore, so long as (7.10) holds:

$$\|J^\varepsilon w^\varepsilon\|_{L^\infty([0,t];L^2)} \leqslant C \max \left(\varepsilon^{\alpha-\sigma}, \varepsilon^{\alpha-1} \log \frac{1}{\varepsilon} \right) \|J^\varepsilon u^\varepsilon\|_{L^\infty([0,t];L^2)}.$$

Since $u^\varepsilon = w^\varepsilon + v_0^\varepsilon$, we have:

$$\|J^\varepsilon u^\varepsilon\|_{L^\infty([0,t];L^2)} \leqslant \|J^\varepsilon w^\varepsilon\|_{L^\infty([0,t];L^2)} + \|J^\varepsilon v^\varepsilon\|_{L^\infty([0,t];L^2)} \leqslant 1 + \|\nabla a_0\|_{L^2},$$

so long as (7.10) holds. We infer:

$$\|J^\varepsilon w^\varepsilon\|_{L^\infty([0,t];L^2)} \leqslant C \max \left(\varepsilon^{\alpha-\sigma}, \varepsilon^{\alpha-1} \log \frac{1}{\varepsilon} \right). \tag{7.13}$$

Therefore, for every $T > 1$, there exists $\varepsilon(T) > 0$ such that (7.10) holds on $[0, T]$ for $0 < \varepsilon \leqslant \varepsilon(T)$. The proposition in the case $d = 1$ then follows from (7.11), (7.12) and (7.13).

The case $d \geqslant 2$. First, notice that since

$$\sigma > \frac{2}{d+2} \quad (\text{even for } d = 2),$$

we always have $d\sigma > 1$, so $\max(1, d\sigma) = d\sigma$.

Since we work at the level of Σ regularity, we cannot expect L^∞ estimates when $d \geqslant 2$. To overcome this issue, we do not use the mere energy estimate provided by Lemma 1.2, but rather Strichartz estimates, which we now recall.

Definition 7.4. A pair (q, r) is **admissible** if $2 \leqslant r < \frac{2d}{d-2}$ ($2 \leqslant r \leqslant \infty$ if $d = 1$, $2 \leqslant r < \infty$ if $d = 2$) and

$$\frac{2}{q} = \delta(r) := d\left(\frac{1}{2} - \frac{1}{r}\right).$$

Notation 7.5. For $f^\varepsilon = f^\varepsilon(t, x)$ and $t > 0$, we write

$$\|f^\varepsilon\|_{L_t^q(L^r)} := \|f^\varepsilon\|_{L^q(0,t;L^r(\mathbb{R}^d))} = \left(\int_0^t \left(\int_{\mathbb{R}^d} |f^\varepsilon(\tau, x)|^r \, dx\right)^{q/r} d\tau\right)^{1/q},$$

with the usual modification when q or r is infinite.

Strichartz estimates are classically given with $\varepsilon = 1$ (see [Ginibre and Velo (1985b); Kato (1987); Yajima (1987); Ginibre and Velo (1992); Keel and Tao (1998)]). Using the scaling

$$u^\varepsilon(t, x) = \frac{1}{\varepsilon^{d/2}} \psi^\varepsilon\left(\frac{t}{\varepsilon}, \frac{x}{\varepsilon}\right),$$

we get the following lemma.

Lemma 7.6 (Strichartz estimates). *Denote* $U_0^\varepsilon(t) = e^{i\varepsilon\frac{t}{2}\Delta}$.
(1) Homogeneous Strichartz estimate. *For any admissible pair* (q, r), *there exists* C_q *independent of* ε *such that*

$$\varepsilon^{1/q}\|U_0^\varepsilon\varphi\|_{L^q(\mathbb{R};L^r)} \leqslant C_q\|\varphi\|_{L^2}, \quad \forall\varphi \in L^2(\mathbb{R}^d).$$

(2) Inhomogeneous Strichartz estimate. *For a time interval* I, *denote*

$$D_I^\varepsilon(F)(t, x) = \int_{I \cap \{\tau \leqslant t\}} U_0^\varepsilon(t - \tau)F(\tau, x)d\tau.$$

For all admissible pairs (q_1, r_1) *and* (q_2, r_2), *and any interval* I, *there exists* $C = C_{r_1, r_2}$ *independent of* ε *and* I *such that*

$$\varepsilon^{1/q_1 + 1/q_2}\|D_I^\varepsilon(F)\|_{L^{q_1}(I;L^{r_1})} \leqslant C\|F\|_{L^{q_2'}(I;L^{r_2'})}, \quad (7.14)$$

for all $F \in L^{q_2'}(I;L^{r_2'})$.

The proof of Proposition 7.3 highly relies on the technical Proposition 7.8 below, which can be understood as an adaptation of the Gronwall lemma. Before stating and proving it, we need some preliminaries:

Lemma 7.7. *Let $d \geqslant 2$, and assume $\frac{2}{d+2} < \sigma < \frac{2}{d-2}$. There exists $\underline{q}$, $\underline{r}$, $\underline{s}$ and $\underline{k}$ satisfying*

$$\frac{1}{\underline{r}'} = \frac{1}{\underline{r}} + \frac{2\sigma}{\underline{s}} \quad ; \quad \frac{1}{\underline{q}'} = \frac{1}{\underline{q}} + \frac{2\sigma}{\underline{k}}, \tag{7.15}$$

and the additional conditions:

- *The pair $(\underline{q}, \underline{r})$ is admissible,*
- $0 < \frac{1}{\underline{k}} < \delta(\underline{s}) < 1$.

Proof. With $\delta(\underline{s}) = 1$, the first part of Eq. (7.15) becomes

$$\delta(\underline{r}) = \sigma\left(\frac{d}{2} - 1\right),$$

and this expression is less than 1 for $\sigma < \frac{2}{d-2}$. Still with $\delta(\underline{s}) = 1$, the second part of Eq. (7.15) yields

$$\frac{2}{\underline{k}} = 1 - \frac{d}{2} + \frac{1}{\sigma},$$

which lies in $]0, 2[$ for $\frac{2}{d+2} < \sigma < \frac{2}{d-2}$. By continuity, these conditions are still satisfied for $\delta(\underline{s})$ close to 1 and $\delta(\underline{s}) < 1$. □

Consider again $w^\varepsilon = u^\varepsilon - v_0^\varepsilon$. It solves Eq. (7.9). We now prove a general estimate for the integral equation,

$$\begin{aligned}
w^\varepsilon(t) = U_0^\varepsilon(t - t_0)w_0^\varepsilon &- i\varepsilon^{\alpha-1}\int_{t_0}^t U_0^\varepsilon(t - \tau)F^\varepsilon(w^\varepsilon(\tau))d\tau \\
&- i\varepsilon^{-1}\int_{t_0}^t U_0^\varepsilon(t - \tau)h^\varepsilon(\tau)d\tau.
\end{aligned} \tag{7.16}$$

Writing

$$|u^\varepsilon|^{2\sigma} u^\varepsilon = |u^\varepsilon|^{2\sigma} w^\varepsilon + |u^\varepsilon|^{2\sigma} v_0^\varepsilon,$$

the goal is to consider Eq. (7.16) with

$$F^\varepsilon(w^\varepsilon) = |u^\varepsilon|^{2\sigma} w^\varepsilon \quad ; \quad h^\varepsilon = \varepsilon^\alpha |u^\varepsilon|^{2\sigma} v_0^\varepsilon.$$

Proposition 7.8. *Let $t_1 > t_0$, with $|t_1 - t_0| \leqslant 2$. Assume that there exists a constant C independent of t and ε such that for $t_0 \leqslant t \leqslant t_1$,*

$$\|F^\varepsilon(w^\varepsilon)(t)\|_{L^{\underline{r}'}} \leqslant \frac{C}{(\varepsilon + |1 - t|)^{2\sigma\delta(\underline{s})}}\|w^\varepsilon(t)\|_{L^{\underline{r}}}, \tag{7.17}$$

and define

$$D^\varepsilon(t_0, t_1) := \left(\int_{t_0}^{t_1} \frac{dt}{(\varepsilon + |t - 1|)^{\underline{k}\delta(\underline{s})}} \right)^{2\sigma/\underline{k}}.$$

Then there exist C^* independent of ε, t_0 and t_1, such that for any admissible pair (q, r),

$$\|w^\varepsilon\|_{L^{\underline{q}}(t_0,t_1;L^{\underline{r}})} \leqslant C^* \varepsilon^{-1/\underline{q}} \|w_0^\varepsilon\|_{L^2} + C_{\underline{q},q} \varepsilon^{-1-1/\underline{q}-1/q} \|h^\varepsilon\|_{L^{q'}(t_0,t_1;L^{r'})}$$
$$+ C^* \varepsilon^{\alpha - d\sigma + 2\sigma(\delta(\underline{s}) - 1/\underline{k})} D^\varepsilon(t_0, t_1) \|w^\varepsilon\|_{L^{\underline{q}}(t_0,t_1;L^{\underline{r}})}.$$

Proof. Apply Strichartz inequalities to Eq. (7.16) with $q_1 = \underline{q}$, $r_1 = \underline{r}$, and $q_2 = \underline{q}$, $r_2 = \underline{r}$ for the term with $F^\varepsilon(w^\varepsilon)$, $q_2 = q$, $r_2 = r$ for the term with h^ε, it yields

$$\|w^\varepsilon\|_{L^{\underline{q}}(t_0,t_1;L^{\underline{r}})} \leqslant C\varepsilon^{-1/\underline{q}} \|w_0^\varepsilon\|_{L^2} + C_{\underline{q},q} \varepsilon^{-1-1/\underline{q}-1/q} \|h^\varepsilon\|_{L^{q'}(t_0,t_1;L^{r'})}$$
$$+ C\varepsilon^{\alpha-1-2/\underline{q}} \|F^\varepsilon(w^\varepsilon)\|_{L^{\underline{q}'}(t_0,t_1;L^{\underline{r}'})}.$$

Then estimate the space norm of the last term by (7.17) and apply Hölder inequality in time, thanks to Eq. (7.15). Using Eq. (7.15) and the fact that $(\underline{q}, \underline{r})$ is admissible, we compute:

$$-1 - \frac{2}{\underline{q}} = -\frac{2}{\underline{q}} - 1 + \frac{2\sigma}{\underline{k}} - \frac{2\sigma}{\underline{k}} = -\frac{4}{\underline{q}} - \frac{2\sigma}{\underline{k}}$$
$$= -2d\left(\frac{1}{2} - \frac{1}{\underline{r}}\right) - \frac{2\sigma}{\underline{k}}$$
$$= -d + d\left(1 - \frac{2\sigma}{\underline{s}}\right) - \frac{2\sigma}{\underline{k}}$$
$$= 2\sigma d\left(\frac{1}{2} - \frac{1}{\underline{s}}\right) - d\sigma - \frac{2\sigma}{\underline{k}}$$
$$= -d\sigma + 2\sigma\delta(\underline{s}) - \frac{2\sigma}{\underline{k}}.$$

The result follows. $\square$

We will rather use the following corollary:

Corollary 7.9. *Suppose the assumptions of Proposition 7.8 are satisfied. Assume moreover that $\alpha \geqslant d\sigma$ and $C^* \varepsilon^{2\sigma(\delta(\underline{s})-1/\underline{k})} D^\varepsilon(t_0, t_1) \leqslant 1/2$. Since $\underline{k}\delta(\underline{s}) > 1$, this holds in either of the two cases:*

- *$0 \leqslant t_0 \leqslant t_1 \leqslant 1 - \Lambda\varepsilon$ or $1 + \Lambda\varepsilon \leqslant t_0 \leqslant t_1$, with $\Lambda \geqslant \Lambda_0$ sufficiently large, or*

- $t_0, t_1 \in [1 - \Lambda\varepsilon, 1 + \Lambda\varepsilon]$, with $|t_1 - t_0|/\varepsilon \leqslant \eta$ sufficiently small.

Note that the parameters Λ_0 and η are independent of ε. Then for all admissible pair (q, r), we have:

$$\|w^\varepsilon\|_{L^\infty(t_0,t_1;L^2)} \leqslant C\|w_0^\varepsilon\|_{L^2} + C_{\underline{q},q}\varepsilon^{-1-1/q}\|h^\varepsilon\|_{L^{q'}(t_0,t_1;L^{r'})}. \qquad (7.18)$$

Proof. The additional assumption implies that the last term in the estimate of Proposition 7.8 can be "absorbed" by the left-hand side, up to doubling the constants,

$$\|w^\varepsilon\|_{L^{\underline{q}}(t_0,t_1;L^{\underline{r}})} \leqslant C\varepsilon^{-1/\underline{q}}\|w_0^\varepsilon\|_{L^2} + C\varepsilon^{-1-1/\underline{q}-1/q}\|h^\varepsilon\|_{L^{q'}(t_0,t_1;L^{r'})}. \qquad (7.19)$$

Now apply Strichartz inequalities to Eq. (7.16) again, but with $q_1 = \infty$, $r_1 = 2$, and $q_2 = \underline{q}$, $r_2 = \underline{r}$ for the term with $F^\varepsilon(w^\varepsilon)$, $q_2 = q$, $r_2 = r$ for the term with h^ε. It yields

$$\|w^\varepsilon\|_{L^\infty(t_0,t_1;L^2)} \leqslant C\|w_0^\varepsilon\|_{L^2} + C\varepsilon^{-1-1/q}\|h^\varepsilon\|_{L^{q'}(t_0,t_1;L^{r'})}$$
$$+ C\varepsilon^{\alpha-1-1/\underline{q}}\|F^\varepsilon(w^\varepsilon)\|_{L^{\underline{q}'}(t_0,t_1;L^{\underline{r}'})}.$$

Like before,

$$\varepsilon^{d\sigma-1}\|F^\varepsilon(w^\varepsilon)\|_{L^{\underline{q}'}(t_0,t_1;L^{\underline{r}'})} \lesssim \varepsilon^{2/\underline{q}+2\sigma(\delta(\underline{s})-1/\underline{k})}D^\varepsilon(t_0,t_1)\|w^\varepsilon\|_{L^{\underline{q}}(t_0,t_1;L^{\underline{r}})}$$
$$\lesssim \varepsilon^{2/\underline{q}}\|w^\varepsilon\|_{L^{\underline{q}}(t_0,t_1;L^{\underline{r}})},$$

and the corollary follows from (7.19), since $\alpha \geqslant d\sigma$. $\qquad \square$

We now essentially proceed like in the one-dimensional case. Inequality (7.7) is now replaced by (see (7.6))

$$\|u\|_{L^p} \leqslant \frac{C}{(\varepsilon + |t-1|)^{\delta(p)}} \|u\|_{L^2}^{1-\delta(p)} \left(\|\varepsilon\nabla u\|_{L^2} + \|J^\varepsilon(t)u\|_{L^2}\right)^{\delta(p)}, \qquad (7.20)$$

for all $p \in [2, 2/(d-2)[$. Note that we still have the *a priori* estimates:

$$\|u^\varepsilon(t)\|_{L^2} = \|v_0^\varepsilon(t)\|_{L^2} = \|a_0\|_{L^2},$$
$$\|\varepsilon\nabla u^\varepsilon(t)\|_{L^2} + \|\varepsilon\nabla v_0^\varepsilon(t)\|_{L^2} = \mathcal{O}(1),$$
$$\|J^\varepsilon(t)v_0^\varepsilon(t)\|_{L^2} = \|J^\varepsilon(0)v_0^\varepsilon(0)\|_{L^2} = \|\nabla a_0\|_{L^2}.$$

Since $w^\varepsilon \in C(\mathbb{R}; \Sigma)$, there exists $t^\varepsilon > 0$ such that

$$\|J^\varepsilon(\tau)w^\varepsilon(\tau)\|_{L^2} \leqslant 1 \qquad (7.21)$$

for $\tau \in [0, t^\varepsilon]$. So long as (7.21) holds, (7.20) yields, since $\underline{s} \in [2, 2/(d-2)[$,

$$\|u^\varepsilon(\tau)\|_{L^{\underline{s}}} \leqslant \frac{C}{(\varepsilon + |\tau-1|)^{\delta(\underline{s})}}. \qquad (7.22)$$

Let $T > 1$. Split the time interval

$$[1 - \Lambda_0 \varepsilon, 1 + \Lambda_0 \varepsilon]$$

provided by Corollary 7.9 into $\approx 2\Lambda_0/\eta$ intervals of length $\leqslant \eta$. Applying Corollary 7.9 $\approx 2 + 2\Lambda_0/\eta$ times yields, for $t \leqslant T$ and so long as Eq. (7.21) holds:

$$\|w^\varepsilon\|_{L^\infty(0,t;L^2)} \lesssim \varepsilon^{\alpha - 1 - 1/q} \left\| |u^\varepsilon|^{2\sigma} v_0^\varepsilon \right\|_{L^{q'}(0,t;L^{r'})},$$

for all admissible pair (q, r). Take $(q, r) = (\underline{q}, \underline{r})$. Hölder's inequality yields, in view of Lemma 7.7:

$$\left\| |u^\varepsilon|^{2\sigma} v_0^\varepsilon \right\|_{L^{\underline{q}'}(0,t;L^{\underline{r}'})} \leqslant \|u^\varepsilon\|^{2\sigma}_{L^{\underline{k}}(0,t;L^{\underline{s}})} \|v_0^\varepsilon\|_{L^{\underline{q}}(0,t;L^{\underline{r}})}$$

$$\leqslant C \left(\int_0^T \frac{d\tau}{(\varepsilon + |\tau - 1|)^{\underline{k}\delta(\underline{s})}} \right)^{2\sigma/\underline{k}} \varepsilon^{-1/\underline{q}} \|a_0\|_{L^2},$$

where we have used (7.22) and the homogeneous Strichartz estimate for v_0^ε. Distinguishing the regions $\{|\tau - 1| \geqslant \varepsilon\}$ and $\{|\tau - 1| < \varepsilon\}$, we infer, since $\underline{k}\delta(\underline{s}) > 1$,

$$\|w^\varepsilon\|_{L^\infty(0,t;L^2)} \leqslant C\varepsilon^{\alpha - 1 - 2/\underline{q} - 2\sigma\delta(\underline{s}) + 2\sigma/\underline{k}}.$$

Again, notice that

$$-1 - 2/\underline{q} - 2\sigma\delta(\underline{s}) + 2\sigma/\underline{k} = -d\sigma.$$

Therefore, so long as (7.21) holds:

$$\|w^\varepsilon\|_{L^\infty(0,t;L^2)} \leqslant C\varepsilon^{\alpha - d\sigma}. \tag{7.23}$$

For $\mathcal{B}^\varepsilon \in \{\varepsilon\nabla, J^\varepsilon\}$, apply $\mathcal{B}^\varepsilon$ to Eq. (7.9):

$$\left(i\varepsilon\partial_t + \frac{\varepsilon^2}{2}\Delta \right) \mathcal{B}^\varepsilon w^\varepsilon = \varepsilon^\alpha \mathcal{B}^\varepsilon \left(|u^\varepsilon|^{2\sigma} u^\varepsilon \right) ; \ \mathcal{B}^\varepsilon w^\varepsilon_{|t=0} = 0.$$

Since $\mathcal{B}^\varepsilon \left(|u^\varepsilon|^{2\sigma} u^\varepsilon \right)$ is a linear combination of terms of the form

$$(u^\varepsilon)^j (\overline{u}^\varepsilon)^{2\sigma - j} \mathcal{B}^\varepsilon u^\varepsilon \text{ and } (u^\varepsilon)^\ell (\overline{u}^\varepsilon)^{2\sigma - \ell} \overline{\mathcal{B}^\varepsilon u^\varepsilon},$$

we mimic the above approach. Write $\mathcal{B}^\varepsilon u^\varepsilon = \mathcal{B}^\varepsilon w^\varepsilon + \mathcal{B}^\varepsilon v_0^\varepsilon$. The function F^ε is now chosen in order to contain all the terms of the form

$$(u^\varepsilon)^j (\overline{u}^\varepsilon)^{2\sigma - j} \mathcal{B}^\varepsilon w^\varepsilon \text{ and } (u^\varepsilon)^\ell (\overline{u}^\varepsilon)^{2\sigma - \ell} \overline{\mathcal{B}^\varepsilon w^\varepsilon},$$

and h^ε contains all the terms of the form

$$\varepsilon^\alpha (u^\varepsilon)^j (\overline{u}^\varepsilon)^{2\sigma - j} \mathcal{B}^\varepsilon v_0^\varepsilon \text{ and } \varepsilon^\alpha (u^\varepsilon)^\ell (\overline{u}^\varepsilon)^{2\sigma - \ell} \overline{\mathcal{B}^\varepsilon v_0^\varepsilon}.$$

Since $\mathcal{B}^\varepsilon v_0^\varepsilon$ solves the free semi-classical Schrödinger equation, Strichartz inequalities yield:

$$\|\mathcal{B}^\varepsilon v_0^\varepsilon\|_{L^q(\mathbb{R};L^r)} \leqslant C_q \varepsilon^{-1/q} \|a_0\|_\Sigma,$$

where C_q is independent of ε. We can therefore follow the same lines as above, and conclude: so long as (7.21) holds,

$$\|\mathcal{B}^\varepsilon w^\varepsilon\|_{L^\infty(0,t;L^2)} \leqslant C\varepsilon^{\alpha-d\sigma}. \tag{7.24}$$

Like in the one-dimensional case, a continuity argument shows that for any $T > 1$, there exists $\varepsilon(T) > 0$ such that (7.21) holds on $[0,T]$ for all $\varepsilon \in]0, \varepsilon(T)]$. The proposition follows. Note that we have the more precise error estimate:

$$\|\mathcal{B}^\varepsilon w^\varepsilon\|_{L^\infty(0,T;L^2)} \leqslant C(T)\varepsilon^{\alpha-d\sigma}, \quad \forall \mathcal{B}^\varepsilon \in \{\mathrm{Id}, \varepsilon\nabla, J^\varepsilon\}.$$

7.3   Nonlinear propagation, linear caustic

In this paragraph, we assume $\alpha = 1 > d\sigma$. For technical reasons, we will treat the case $d = 1$ only, and we assume in addition $\sigma \geqslant 1/2$. Essentially, notice that Lemma 7.7 cannot be used when $d \geqslant 2$ and $\sigma < 1/d$. For results in the case $d \geqslant 2$ and $\alpha = 1 > d\sigma$, we invite the reader to consult [Carles (2000b)]. Here, we first explain how to derive suitable approximate solutions in the general case $d \geqslant 1$, and then we justify the asymptotics for $d = 1$.

Outside the focal point, we can use the approximate solution studied in Chap. 2. We first make the expressions given in Sec. 2.3 as explicit as possible. The rays of geometric optics are now given by

$$x(t,y) = (1-t)y.$$

Therefore, when $t < 1$, the inverse mapping is given by:

$$y(t,x) = \frac{x}{1-t}.$$

In the general case of the space dimension d, the Jacobi's determinant is given by

$$J_t(y) = \det \nabla_y x(t,y) = (1-t)^d.$$

We infer that for $t < 1$, the leading order approximate solution constructed in Chap. 2 is given by:

$$u_{\mathrm{app}}^\varepsilon(t,x) = \frac{1}{(1-t)^{d/2}} a_0 \left(\frac{x}{1-t}\right) e^{i\frac{|x|^2}{2\varepsilon(t-1)}} e^{iG(t,x)},$$

where G is given by:

$$G(t,x) = -\left|a_0\left(\frac{x}{1-t}\right)\right|^{2\sigma}\int_0^t \frac{d\tau}{|1-\tau|^{d\sigma}}.$$

Loosely speaking, since no nonlinear effect is expected near the focal point, a natural candidate for an approximate solution past the caustic consists in continuing $u_{\mathrm{app}}^\varepsilon$ for $t > 1$ by taking into account linear effects at a focal point. Since the linear effects at a focal point consist of a phase shift at leading order (Maslov index), define:

$$u_{\mathrm{app}}^\varepsilon(t,x) = \begin{cases} \dfrac{1}{(1-t)^{d/2}}a_0\left(\dfrac{x}{1-t}\right)e^{i\frac{|x|^2}{2\varepsilon(t-1)}}e^{iG(t,x)} & \text{if } t < 1, \\[2ex] \dfrac{e^{-id\pi/2}}{(t-1)^{d/2}}a_0\left(\dfrac{x}{1-t}\right)e^{i\frac{|x|^2}{2\varepsilon(t-1)}}e^{iG(t,x)} & \text{if } t > 1. \end{cases}$$

Note that the phase shift G is defined globally in time: the map

$$\tau \mapsto \frac{1}{|1-\tau|^{d\sigma}}$$

is locally integrable, since $d\sigma < 1$.

We now explain how to derive a global in time approximate solution from the Lagrangian integral point of view. Suppose that $A^\varepsilon(t,\xi)$ converges to some function $A(t,\xi)$ as $\varepsilon \to 0$, on some time interval $[0,T]$, $T > 0$. Formally, Eq. (7.3) yields

$$u^\varepsilon(t,x) \approx \frac{1}{(2\pi\varepsilon)^{d/2}}\int_{\mathbb{R}^d} e^{-i\frac{t-1}{2\varepsilon}|\xi|^2 + i\frac{x\cdot\xi}{\varepsilon}}A(t,\xi)d\xi.$$

For $t \neq 1$, we can apply stationary phase formula to the right-hand side, and find:

$$\frac{1}{(2\pi\varepsilon)^{d/2}}\int_{\mathbb{R}^d}e^{-i\frac{t-1}{2\varepsilon}|\xi|^2 + i\frac{x\cdot\xi}{\varepsilon}}A(t,\xi)d\xi \approx \frac{e^{id\pi/4\,\mathrm{sgn}(1-t)}}{|t-1|^{d/2}}A\left(t,\frac{x}{t-1}\right)e^{i\frac{|x|^2}{2\varepsilon(t-1)}}.$$

We infer

$$|u^\varepsilon|^{2\sigma}u^\varepsilon(t,x) \approx \frac{e^{id\pi/4\,\mathrm{sgn}(1-t)}}{|t-1|^{d/2+d\sigma}}|A|^{2\sigma}A\left(t,\frac{x}{t-1}\right)e^{i\frac{|x|^2}{2\varepsilon(t-1)}}.$$

Using stationary phase formula again, we obtain

$$\mathcal{F}\left(|u^\varepsilon|^{2\sigma}u^\varepsilon\right)\left(t,\frac{\xi}{\varepsilon}\right)$$

$$\approx \frac{1}{(2\pi)^{d/2}}\frac{e^{id\pi/4\,\mathrm{sgn}(1-t)}}{|t-1|^{d/2+d\sigma}}\int e^{i\frac{|x|^2}{2\varepsilon(t-1)} - i\frac{x\cdot\xi}{\varepsilon}}|A|^{2\sigma}A\left(t,\frac{x}{t-1}\right)dx$$

$$\approx \frac{\varepsilon^{d/2}}{|t-1|^{d\sigma}}|A|^{2\sigma}A(t,\xi)e^{-i\frac{t-1}{2\varepsilon}|\xi|^2}.$$

Since the time evolution of A^ε is given by Eq. (7.4), we expect the limiting equation ($\alpha = 1$):

$$i\partial_t A(t,\xi) = \frac{1}{|t-1|^{d\sigma}} |A(t,\xi)|^{2\sigma} A(t,\xi) \quad ; \quad A_{|t=0} = A_0, \qquad (7.25)$$

where we recall that A_0 is defined by

$$A_0(\xi) = e^{-id\pi/4} a_0(-\xi).$$

Notice that the modulus of A is independent of time (since $\partial_t |A|^2 = 0$), so

$$A(t,\xi) = A_0(\xi) e^{ig(t,\xi)},$$

where

$$g(t,\xi) = -|A_0(\xi)|^{2\sigma} \int_0^t \frac{d\tau}{|\tau-1|^{d\sigma}} = -|a_0(-\xi)|^{2\sigma} \int_0^t \frac{d\tau}{|\tau-1|^{d\sigma}}.$$

Note that up to a scaling in time, we recover the previous function G:

$$g(t,\xi) = G\left(t, (t-1)\xi\right).$$

Unlike G, g is defined for all t: the Lagrangian integral unfolds the singularity at $t = 1$.

Proposition 7.10. *Assume $d = 1$, and $\alpha = 1 > \sigma \geqslant 1/2$. Then*

$$A^\varepsilon \xrightarrow[\varepsilon \to 0]{} A_0 e^{ig} \quad in \ L^\infty_{\mathrm{loc}}(\mathbb{R}; \Sigma).$$

This implies, for all $\mathcal{B}^\varepsilon \in \{\mathrm{Id}, \varepsilon\partial_x, J^\varepsilon\}$, all $\beta \in [0,1[$ and all $T > 0$,

$$\left\| \mathcal{B}^\varepsilon \left(u^\varepsilon - v^\varepsilon_{1-\varepsilon^\beta} \right) \right\|_{L^\infty(1-\varepsilon^\beta, 1+\varepsilon^\beta; L^2)} \lesssim \varepsilon^{\beta-\sigma}$$

$$\sup_{|t-1| \geqslant \varepsilon^\beta, t \leqslant T} \left\| \mathcal{B}^\varepsilon(t) \left(u^\varepsilon(t) - u^\varepsilon_{\mathrm{app}}(t) \right) \right\|_{L^2} \xrightarrow[\varepsilon \to 0]{} 0.$$

Remark 7.11. The statement of the proposition suggests that more information is available in terms of the Lagrangian symbol than in terms of the wave function directly. This aspect is also present in the analysis of the case "nonlinear propagation, nonlinear caustic", see §7.5. Of course, by construction, the Lagrangian integral unfolds the singularity at the focal point, which makes it possible to have a uniform in time statement. Moreover, from a technical point of view, Lagrangian integrals make it easier to consider non-smooth functions (here, $z \mapsto |z|^\sigma$), since they come along with energy estimates which cost one derivative less than the error estimates outside the focal point presented in Chap. 2. This aspect is also crucial in §7.5.

Proof. First, we verify that the first point implies the second assertion. We use the following lemma.

Lemma 7.12. *Let $d \geqslant 1$ and $f \in C(\mathbb{R}; L^2)$. Denote*

$$\Lambda^\varepsilon(t, x) = \frac{1}{(2\pi\varepsilon)^{d/2}} \int_{\mathbb{R}^d} e^{-i\frac{t-1}{2\varepsilon}|\xi|^2 + i\frac{x \cdot \xi}{\varepsilon}} f(t, \xi) d\xi,$$

and

$$F^\varepsilon(t, x) = e^{i\frac{|x|^2}{2\varepsilon(t-1)}} \left(\frac{i}{1-t}\right)^{d/2} f\left(t, \frac{x}{t-1}\right).$$

Then there exists $h \in C(\mathbb{R}^2; \mathbb{R}_+)$ with $h(t, 0) = 0$ such that for all $t \neq 1$,

$$\|\Lambda^\varepsilon(t) - F^\varepsilon(t)\|_{L^2} = h\left(t, \frac{\varepsilon}{1-t}\right).$$

If, in addition, $f \in C(\mathbb{R}; H^2)$, then a little more can be said about h:

$$\exists C > 0, \quad h(t, \lambda) \leqslant C\lambda \|f(t, \cdot)\|_{H^2}.$$

Remark 7.13. We do not use the last point of this lemma in this section. It will be used in Sec. 7.5.

Proof. [Proof of Lemma 7.12] From Parseval's formula,

$$\Lambda^\varepsilon(t, x) = e^{i\frac{|x|^2}{2\varepsilon(t-1)}} \left(\frac{i}{1-t}\right)^{d/2} \int_{\mathbb{R}^d} e^{i\varepsilon|\eta|^2/(2(t-1))} e^{ix \cdot \eta/(1-t)} \mathcal{F}_{\xi \to \eta}^{-1} f(t, \eta) d\eta.$$

Define, for $\lambda \in \mathbb{R}$,

$$h(t, \lambda) = \left\|\left(e^{i\frac{\lambda}{2}|\cdot|^2} - 1\right) \mathcal{F}^{-1} f(t, \cdot)\right\|_{L^2}.$$

Since $f \in C(\mathbb{R}; L^2)$, $h \in C(\mathbb{R}^2; \mathbb{R})$. The property $h(t, 0) = 0$ then follows from the Dominated Convergence Theorem.

When $f \in C(\mathbb{R}; H^2)$, we use the general inequality $|e^{i\theta} - 1| \leqslant |\theta|$. $\square$

Introduce

$$\widetilde{u}_{\mathrm{app}}^\varepsilon(t, x) = \frac{1}{\sqrt{2\pi\varepsilon}} \int_{\mathbb{R}} e^{-i\frac{t-1}{2\varepsilon}|\xi|^2 + i\frac{x \cdot \xi}{\varepsilon}} A_0(\xi) e^{ig(t, \xi)} d\xi.$$

With $f(t, \xi) = A_0(\xi) e^{ig(t, \xi)}$, we check that $f \in C(\mathbb{R}; \Sigma)$. Lemma 7.12 shows that for any $\beta \in [0, 1[$ and any $T > 0$,

$$\sup_{|t-1| \geqslant \varepsilon^\beta, t \leqslant T} \left\|\widetilde{u}_{\mathrm{app}}^\varepsilon(t) - u_{\mathrm{app}}^\varepsilon(t)\right\|_{L^2} \xrightarrow[\varepsilon \to 0]{} 0.$$

We check easily that since $\partial_x f, xf \in C(\mathbb{R}; L^2)$, we have moreover

$$\sup_{|t-1| \geqslant \varepsilon^\beta, t \leqslant T} \left\| \mathcal{B}^\varepsilon(t) \left(\widetilde{u}^\varepsilon_{\mathrm{app}}(t) - u^\varepsilon_{\mathrm{app}}(t) \right) \right\|_{L^2} \xrightarrow[\varepsilon \to 0]{} 0, \quad \forall \mathcal{B}^\varepsilon \in \{\mathrm{Id}, \varepsilon\partial_x, J^\varepsilon\}.$$

Notice, on the other hand,

$$\left\| A^\varepsilon(t) - A_0 e^{ig(t)} \right\|_\Sigma = \sum_{\mathcal{B}^\varepsilon \in \{\mathrm{Id}, \varepsilon\partial_x, J^\varepsilon\}} \left\| \mathcal{B}^\varepsilon(t) \left(u^\varepsilon(t) - \widetilde{u}^\varepsilon_{\mathrm{app}}(t) \right) \right\|_{L^2}.$$

Therefore, the first part of the proposition implies: for any $\beta \in [0, 1[$ and any $T > 0$,

$$\sup_{|t-1| \geqslant \varepsilon^\beta, t \leqslant T} \sum_{\mathcal{B}^\varepsilon \in \{\mathrm{Id}, \varepsilon\partial_x, J^\varepsilon\}} \left\| \mathcal{B}^\varepsilon(t) \left(u^\varepsilon(t) - \widetilde{u}^\varepsilon_{\mathrm{app}}(t) \right) \right\|_{L^2} \xrightarrow[\varepsilon \to 0]{} 0.$$

For the region the region $\{|t - 1| \leqslant \varepsilon^\beta\}$, denote

$$w^\varepsilon_\beta = u^\varepsilon - v^\varepsilon_{1-\varepsilon^\beta}.$$

By definition of $v^\varepsilon_{1-\varepsilon^\beta}$, it solves:

$$i\varepsilon\partial_t w^\varepsilon_\beta + \frac{\varepsilon^2}{2}\partial_x^2 w^\varepsilon_\beta = \varepsilon |u^\varepsilon|^{2\sigma} u^\varepsilon \quad ; \quad w^\varepsilon_{\beta|t=1-\varepsilon^\beta} = 0.$$

For $\mathcal{B}^\varepsilon \in \{\mathrm{Id}, \varepsilon\partial_x, J^\varepsilon\}$, apply the operator $\mathcal{B}^\varepsilon$ to the above equation, and the energy estimate of Lemma 1.2:

$$\sup_{|t-1| \leqslant \varepsilon^\beta} \left\| \mathcal{B}^\varepsilon(t) w^\varepsilon_\beta(t) \right\|_{L^2} \lesssim \int_{|\tau - 1| \leqslant \varepsilon^\beta} \|u^\varepsilon(\tau)\|_{L^\infty}^{2\sigma} d\tau \sup_{|t-1| \leqslant \varepsilon^\beta} \left\| \mathcal{B}^\varepsilon(t) u^\varepsilon(t) \right\|_{L^2}.$$

From the conservations of mass and energy for u^ε, the L^2 norms of u^ε and $\varepsilon\partial_x u^\varepsilon$ are bounded independent of ε. Gagliardo-Nirenberg inequality yields

$$\sup_{t \in \mathbb{R}} \|u^\varepsilon(t)\|_{L^\infty} \leqslant C\varepsilon^{-1/2},$$

hence

$$\sup_{|t-1| \leqslant \varepsilon^\beta} \left\| \mathcal{B}^\varepsilon(t) w^\varepsilon_\beta(t) \right\|_{L^2} \lesssim \varepsilon^{\beta-\sigma} \sup_{|t-1| \leqslant \varepsilon^\beta} \left\| \mathcal{B}^\varepsilon(t) u^\varepsilon(t) \right\|_{L^2}.$$

Since

$$\sum_{\mathcal{B}^\varepsilon \in \{\mathrm{Id}, \varepsilon\partial_x, J^\varepsilon\}} \left\| \mathcal{B}^\varepsilon(t) u^\varepsilon(t) \right\|_{L^2} = \|A^\varepsilon(t)\|_\Sigma,$$

the first part of the proposition yields:

$$\sup_{|t-1| \leqslant \varepsilon^\beta} \left\| \mathcal{B}^\varepsilon(t) w^\varepsilon_\beta(t) \right\|_{L^2} \lesssim \varepsilon^{\beta-\sigma}.$$

Therefore, we just have to prove the first assertion of the proposition.

Fix $T > 0$ and $\beta \in\,]0, 1[$. Again, we distinguish the regions $\{t \leqslant 1 - \varepsilon^\beta\}$, $\{|t - 1| < \varepsilon^\beta\}$ and $1 + \varepsilon^\beta \leqslant t \leqslant T\}$. We check that by construction, $\widetilde{u}_{\text{app}}^\varepsilon$ satisfies

$$i\varepsilon \partial_t \widetilde{u}_{\text{app}}^\varepsilon + \frac{\varepsilon^2}{2} \partial_x^2 \widetilde{u}_{\text{app}}^\varepsilon = \varepsilon \left|\widetilde{u}_{\text{app}}^\varepsilon\right|^{2\sigma} \widetilde{u}_{\text{app}}^\varepsilon - \varepsilon r^\varepsilon,$$

where the error term r^ε is given by:

$$r^\varepsilon(t, x) = \left|\widetilde{u}_{\text{app}}^\varepsilon(t, x)\right|^{2\sigma} \widetilde{u}_{\text{app}}^\varepsilon(t, x)$$
$$- \frac{1}{\sqrt{2\pi\varepsilon}} \int_{\mathbb{R}} e^{-i\frac{t-1}{2\varepsilon}|\xi|^2 + i\frac{x \cdot \xi}{\varepsilon}} A_0(\xi) e^{ig(t,\xi)} \times (-\partial_t g(t,\xi))\, d\xi$$
$$= \left|\widetilde{u}_{\text{app}}^\varepsilon(t, x)\right|^{2\sigma} \widetilde{u}_{\text{app}}^\varepsilon(t, x)$$
$$- \frac{1}{\sqrt{2\pi\varepsilon}} \frac{1}{|1 - t|^\sigma} \int_{\mathbb{R}} e^{-i\frac{t-1}{2\varepsilon}|\xi|^2 + i\frac{x \cdot \xi}{\varepsilon}} |A_0(\xi)|^{2\sigma} A_0(\xi) e^{ig(t,\xi)}\, d\xi.$$

Note that the two terms involved in the definition of r^ε are of the form F^ε and Λ^ε respectively, as in Lemma 7.12, with

$$f(t, \xi) = |A_0(\xi)|^{2\sigma} A_0(\xi) e^{ig(t,\xi)}.$$

We check that $f \in C(\mathbb{R}; \Sigma)$. Therefore, since $\beta < 1$, Lemma 7.12 and Eq. (7.5) yield

$$\sup_{|t-1| \geqslant \varepsilon^\beta, t \leqslant T} \sum_{\mathcal{B}^\varepsilon \in \{\text{Id}, \varepsilon\partial_x, J^\varepsilon\}} \|\mathcal{B}^\varepsilon(t) r^\varepsilon(t)\|_{L^2} \xrightarrow[\varepsilon \to 0]{} 0. \tag{7.26}$$

Let $w^\varepsilon = u^\varepsilon - \widetilde{u}_{\text{app}}^\varepsilon$. It solves

$$i\varepsilon \partial_t w^\varepsilon + \frac{\varepsilon^2}{2} \partial_x^2 w^\varepsilon = \varepsilon \left(|u^\varepsilon|^{2\sigma} u^\varepsilon - \left|\widetilde{u}_{\text{app}}^\varepsilon\right|^{2\sigma} \widetilde{u}_{\text{app}}^\varepsilon\right) + \varepsilon r^\varepsilon. \tag{7.27}$$

From Lemma 7.1,

$$\sum_{\mathcal{B}^\varepsilon \in \{\text{Id}, \varepsilon\partial_x, J^\varepsilon\}} \|\mathcal{B}^\varepsilon(0) w^\varepsilon(0)\|_{L^2} \xrightarrow[\varepsilon \to 0]{} 0.$$

We then proceed as in Sec. 7.2: there exists $\varepsilon_0 > 0$ such that for every $\varepsilon \in\,]0, \varepsilon_0]$, there exists $t^\varepsilon > 0$ such that

$$\|J^\varepsilon(\tau) w^\varepsilon(\tau)\|_{L^2} \leqslant 1 \tag{7.28}$$

for $\tau \in [0, t^\varepsilon]$. So long as (7.28) holds, the weighted Gagliardo–Nirenberg inequality (7.7) yields, along with the conservations of mass and energy for u^ε:

$$\|w^\varepsilon(t)\|_{L^\infty} \leqslant \frac{C}{\left(\varepsilon + |t - 1|^{1/2}\right)},$$

for some C independent of ε. Since on the other hand $A_0 e^{ig} \in C(\mathbb{R}; \Sigma)$, Eq. (7.5) shows that there exists C independent of ε such that

$$\sum_{\mathcal{B}^\varepsilon \in \{\mathrm{Id}, \varepsilon\partial_x, J^\varepsilon\}} \left\| \mathcal{B}^\varepsilon(t)\widetilde{u}^\varepsilon_{\mathrm{app}}(t) \right\|_{L^2} \leqslant C, \quad \forall t \in \mathbb{R}.$$

Apply $\mathcal{B}^\varepsilon \in \{\mathrm{Id}, \varepsilon\partial_x, J^\varepsilon\}$ to Eq. (7.27), and write $u^\varepsilon = w^\varepsilon + \widetilde{u}^\varepsilon_{\mathrm{app}}$. The energy estimate of Lemma 1.2 shows that so long as (7.28) holds,

$$\begin{aligned}
\left\| \mathcal{B}^\varepsilon w^\varepsilon \right\|_{L^\infty(0,t;L^2)} &\lesssim \int_0^t \left\| |u^\varepsilon(\tau)|^{2\sigma} \mathcal{B}^\varepsilon(\tau)w^\varepsilon(\tau) \right\|_{L^2} d\tau \\
&\quad + \int_0^t \left\| \left(|u^\varepsilon(\tau)|^{2\sigma} - |\widetilde{u}^\varepsilon_{\mathrm{app}}(\tau)|^{2\sigma} \right) \mathcal{B}^\varepsilon(\tau)\widetilde{u}^\varepsilon_{\mathrm{app}}(\tau) \right\|_{L^2} d\tau \\
&\quad + \int_0^t \left\| \mathcal{B}^\varepsilon(\tau)r^\varepsilon(\tau) \right\|_{L^2} d\tau + o(1) \\
&\lesssim \int_0^t \left\| u^\varepsilon(\tau) \right\|_{L^\infty}^{2\sigma} \left\| \mathcal{B}^\varepsilon(\tau)w^\varepsilon(\tau) \right\|_{L^2} d\tau \\
&\quad + \int_0^t \left\| \left(|u^\varepsilon(\tau)|^{2\sigma-1} + |\widetilde{u}^\varepsilon_{\mathrm{app}}(\tau)|^{2\sigma-1} \right) |w^\varepsilon(\tau)| \right\|_{L^\infty} d\tau \\
&\quad + \int_0^t \left\| \mathcal{B}^\varepsilon(\tau)r^\varepsilon(\tau) \right\|_{L^2} d\tau + o(1),
\end{aligned}$$

where we have used the assumption $\sigma \geqslant 1/2$ and the uniform boundedness of $\mathcal{B}^\varepsilon \widetilde{u}^\varepsilon_{\mathrm{app}}$ in L^2. We infer

$$\begin{aligned}
\left\| \mathcal{B}^\varepsilon w^\varepsilon \right\|_{L^\infty(0,t;L^2)} &\lesssim \int_0^t \frac{1}{(\varepsilon + |\tau - 1|)^\sigma} \left\| \mathcal{B}^\varepsilon(\tau)w^\varepsilon(\tau) \right\|_{L^2} d\tau \\
&\quad + \int_0^t \frac{1}{(\varepsilon + |\tau - 1|)^{\sigma-1/2}} \left\| w^\varepsilon(\tau) \right\|_{L^\infty} d\tau \\
&\quad + \int_0^t \left\| \mathcal{B}^\varepsilon(\tau)r^\varepsilon(\tau) \right\|_{L^2} d\tau + o(1) \\
&\lesssim \int_0^t \frac{1}{(\varepsilon + |\tau - 1|)^\sigma} \left\| \mathcal{B}^\varepsilon(\tau)w^\varepsilon(\tau) \right\|_{L^2} d\tau \\
&\quad + \int_0^t \frac{1}{(\varepsilon + |\tau - 1|)^\sigma} \sum_{\mathcal{K}^\varepsilon \in \{\mathrm{Id}, \varepsilon\partial_x, J^\varepsilon\}} \left\| \mathcal{K}^\varepsilon(\tau)w^\varepsilon(\tau) \right\|_{L^2} d\tau \\
&\quad + \int_0^t \left\| \mathcal{B}^\varepsilon(\tau)r^\varepsilon(\tau) \right\|_{L^2} d\tau + o(1).
\end{aligned}$$

Summing over $\mathcal{B}^\varepsilon \in \{\mathrm{Id}, \varepsilon\partial_x, J^\varepsilon\}$, Gronwall lemma and (7.26) yield, so long as (7.28) holds:

$$\sum_{\mathcal{B}^\varepsilon \in \{\mathrm{Id}, \varepsilon\partial_x, J^\varepsilon\}} \left\| \mathcal{B}^\varepsilon w^\varepsilon \right\|_{L^\infty(0,t;L^2)} \xrightarrow[\varepsilon\to 0]{} 0,$$

provided that $t \leqslant 1 - \varepsilon^\beta$ and $\beta < 1$. Therefore, for every $\beta \in [0, 1[$, there exists $\varepsilon(\beta) > 0$ such that (7.28) holds on $[0, 1 - \varepsilon^\beta]$ for $\varepsilon \in]0, \varepsilon(\beta)]$.

For the region $\{|t-1| \leqslant \varepsilon^\beta\}$, assume moreover $\beta > \sigma$, which is consistent with the previous assumption $\beta \in [0, 1[$, since $\sigma < 1$. We go back to the evolution equation for A^ε, Eq. (7.4): since $\alpha = 1$, and in view of Eq. (7.5),

$$\partial_t \|A^\varepsilon(t)\|_\Sigma \leqslant \|\partial_t A^\varepsilon(t)\|_\Sigma \leqslant \sum_{\mathcal{B}^\varepsilon \in \{\mathrm{Id}, \varepsilon \partial_x, J^\varepsilon\}} \left\| \mathcal{B}^\varepsilon(t) \left(|u^\varepsilon(t)|^{2\sigma} u^\varepsilon(t) \right) \right\|_{L^2}$$

$$\lesssim \|u^\varepsilon(t)\|_{L^\infty}^{2\sigma} \sum_{\mathcal{B}^\varepsilon(t) \in \{\mathrm{Id}, \varepsilon \partial_x, J^\varepsilon\}} \|\mathcal{B}^\varepsilon(t) u^\varepsilon(t)\|_{L^2}$$

$$\lesssim \|u^\varepsilon(t)\|_{L^\infty}^{2\sigma} \|A^\varepsilon(t)\|_\Sigma \lesssim \varepsilon^{-\sigma} \|A^\varepsilon(t)\|_\Sigma,$$

where we have used the conservations of mass and energy for u^ε, and Gagliardo–Nirenberg inequality. Since we have seen that A^ε is uniformly bounded in Σ at time $t = 1 - \varepsilon^\beta$, Gronwall lemma yields

$$\sup_{|t-1| \leqslant \varepsilon^\beta} \|A^\varepsilon(t)\|_\Sigma \lesssim e^{C\varepsilon^{\beta-\sigma}} \lesssim 1,$$

since $\beta > \sigma$. On the other hand, we have

$$\left\| \partial_t \left(A_0 e^{ig(t)} \right) \right\|_\Sigma = \left\| A_0 e^{ig(t)} \partial_t g(t) \right\|_\Sigma$$

$$= \frac{1}{|t-1|^\sigma} \left\| |a_0|^{2\sigma} a_0 \right\|_\Sigma \leqslant \frac{C}{|t-1|^\sigma},$$

where we have used $\Sigma \hookrightarrow L^\infty(\mathbb{R})$. Write

$$\sup_{|t-1| \leqslant \varepsilon^\beta} \left\| A^\varepsilon(t) - A_0 e^{ig(t)} \right\|_\Sigma \leqslant \left\| A^\varepsilon \left(1 - \varepsilon^\beta \right) - A_0 e^{ig(1-\varepsilon^\beta)} \right\|_\Sigma$$

$$+ \int_{|\tau-1| \leqslant \varepsilon^\beta} \|\partial_t A^\varepsilon(\tau)\|_\Sigma \, d\tau + \int_{|\tau-1| \leqslant \varepsilon^\beta} \left\| \partial_t \left(A_0 e^{ig(\tau)} \right) \right\|_\Sigma d\tau$$

$$\leqslant o(1) + \int_{|\tau-1| \leqslant \varepsilon^\beta} \varepsilon^{-\sigma} d\tau + \int_{|\tau-1| \leqslant \varepsilon^\beta} \frac{d\tau}{|\tau-1|^\sigma}.$$

Since $t \mapsto |t-1|^{-\sigma}$ is locally integrable, we infer, for all $\beta \in]\sigma, 1[$,

$$\sup_{|t-1| \leqslant \varepsilon^\beta} \left\| A^\varepsilon(t) - A_0 e^{ig(t)} \right\|_\Sigma \xrightarrow[\varepsilon \to 0]{} 0.$$

This allows us to mimic the approach used for $t \leqslant 1 - \varepsilon^\beta$, in the region $\{1 + \varepsilon^\beta \leqslant t \leqslant T\}$, since in particular,

$$\left\| A^\varepsilon \left(1 + \varepsilon^\beta \right) - A_0 e^{ig(1+\varepsilon^\beta)} \right\|_\Sigma \xrightarrow[\varepsilon \to 0]{} 0.$$

This concludes the proof of the proposition. $\qquad\square$

7.4   Linear propagation, nonlinear caustic

We now resume the framework of arbitrary space dimension, $d \geqslant 1$, and consider, for $d\sigma > 1$,

$$i\varepsilon\partial_t u^\varepsilon + \frac{\varepsilon^2}{2}\Delta u^\varepsilon = \varepsilon^{d\sigma}\left|u^\varepsilon\right|^{2\sigma} u^\varepsilon \quad ; \quad u^\varepsilon(0,x) = a_0(x)e^{-i|x|^2/(2\varepsilon)}. \quad (7.29)$$

From the discussion in Sec. 6.3, the nonlinear effects are expected to be relevant at leading order in the limit $\varepsilon \to 0$ only near the focal point $(t,x) = (1,0)$. Moreover, in the linear case, the concentration phenomenon occurs at scale ε about the focal point. We blow up the variables at that scale about the focal point:

$$u^\varepsilon(t,x) = \frac{1}{\varepsilon^{d/2}}\psi^\varepsilon\left(\frac{t-1}{\varepsilon},\frac{x}{\varepsilon}\right).$$

The factor $\varepsilon^{-d/2}$ may be viewed as a normalization in $L^2(\mathbb{R}^d)$: for all t,

$$\left\|u^\varepsilon(t)\right\|_{L^2(\mathbb{R}^d)} = \left\|\psi^\varepsilon\left(\frac{t-1}{\varepsilon}\right)\right\|_{L^2(\mathbb{R}^d)}.$$

We first note that ψ^ε satisfies an equation where ε is absent:

$$i\partial_t\psi^\varepsilon + \frac{1}{2}\Delta\psi^\varepsilon = \left|\psi^\varepsilon\right|^{2\sigma}\psi^\varepsilon.$$

However, ψ^ε *does* depend on ε, through its Cauchy data:

$$\psi^\varepsilon\left(\frac{-1}{\varepsilon},\frac{x}{\varepsilon}\right) = \varepsilon^{d/2}a_0(x)e^{-i|x|^2/(2\varepsilon)},$$

hence

$$\psi^\varepsilon\left(\frac{-1}{\varepsilon},x\right) = \varepsilon^{d/2}a_0(\varepsilon x)e^{-i\varepsilon|x|^2/2}.$$

Two things must be noticed in the above expression. First, the data are prescribed at a time which is not fixed: it goes to $-\infty$ as $\varepsilon \to 0$. On the other hand, these data become flatter and flatter as $\varepsilon \to 0$, but their L^2 norm is fixed: the wave scatters. These two points of view are reminiscent of scattering theory for dispersive partial differential equations. Before going further into details in the semi-classical analysis of Eq. (7.29), we recall more results on the nonlinear Schrödinger equation.

7.4.1   *Elements of scattering theory for the nonlinear Schrödinger equation*

The first result we recall concerns the Cauchy problem for nonlinear Schrödinger equations with data prescribed near $t = -\infty$. The proof can be found in [Ginibre (1995, 1997); Ginibre *et al.* (1994)]. We recall that the notion of admissible pair was introduced in Definition 7.4.

Theorem 7.14 (Existence and continuity of wave operators). *Let* $d \geqslant 1$, $t_0 \in [-\infty, 0]$ *and* $\psi_- \in \Sigma$. *Denote* $U_0(t) = e^{i\frac{t}{2}\Delta}$, *and consider the Cauchy problem*

$$i\partial_t \psi + \frac{1}{2}\Delta\psi = |\psi|^{2\sigma}\psi \quad ; \quad U_0(-t)\psi(t)\big|_{t=t_0} = \psi_-. \tag{7.30}$$

If $\frac{2}{d+2} < \sigma < \frac{2}{d-2}$ *($\sigma > 1$ if $d = 1$), then Eq. (7.30) has a unique solution*

$$\psi \in Y := \Big\{ \varphi \in C(\mathbb{R}; \Sigma) \ ; \ \varphi, \nabla\varphi, (x + it\nabla)\varphi \in L^q(] - \infty, 0]; L^r)$$

for all admissible pair $(q, r)\Big\}.$

The solution ψ *is strongly continuous from* $(t_0, \psi_-) \in [-\infty, 0] \times \Sigma$ *to* Y, *and if we denote* $\widetilde{\psi}(t) = U_0(-t)\psi(t)$, *then* $\widetilde{\psi} \in C([-\infty, 0]; \Sigma)$. *If* $t_0 = -\infty$, *then*

$$\|U_0(-t)\psi(t) - \psi_-\|_\Sigma \underset{t \to -\infty}{\longrightarrow} 0,$$

and the map $W_- : \psi_- \mapsto \psi_{|t=0}$ *is called* wave operator.

Remark 7.15. The above result is *false* as soon as $\sigma \leqslant 1/d$: for instance, if $d = \sigma = 1$ and if $\psi \in C(\mathbb{R}; L^2)$ solves Eq. (7.30) with $t_0 = -\infty$, then necessarily $\psi_- = \psi = 0$. See [Barab (1984); Strauss (1974, 1981)] or [Ginibre (1997)].

The above result shows that it is possible to construct a solution to the nonlinear Schrödinger equation in prescribing an asymptotically free behavior as $t \to -\infty$. This is the first step in the nonlinear scattering theory: proving the existence of wave operators. Now that $\psi \in C(\mathbb{R}; \Sigma)$, the converse question is the following: does ψ behave asymptotically like a solution to the free Schrödinger equation as $t \to +\infty$? One can give a positive answer to this question, up to making an extra assumption on the power σ. See for instance [Cazenave and Weissler (1992); Nakanishi and Ozawa (2002)] or [Cazenave (2003)].

Theorem 7.16 (Asymptotic completeness in Σ). *Let $d \geqslant 1$, $\varphi \in \Sigma$. Consider the Cauchy problem*

$$i\partial_t \psi + \frac{1}{2}\Delta\psi = |\psi|^{2\sigma}\psi \quad ; \quad \psi\big|_{t=0} = \varphi. \tag{7.31}$$

Assume

$$\sigma \geqslant \sigma_0(d) := \frac{2 - d + \sqrt{d^2 + 12d + 4}}{4d},$$

and in addition, $\sigma < 2/(d-2)$ when $d \geqslant 3$. Then there exists a unique $\psi_+ \in \Sigma$ such that the solution $\psi \in C(\mathbb{R}; \Sigma)$ to Eq. (7.31) satisfies

$$\|U_0(-t)\psi(t) - \psi_+\|_\Sigma \xrightarrow[t \to +\infty]{} 0.$$

Moreover, the map $W_+^{-1} : \varphi \mapsto \psi_+$ is continuous from Σ to itself.

Remark 7.17. We check that $1/d < \sigma_0(d) < 2/d$, and $\sigma_0(d) > 2/(d+2)$ when $d \geqslant 2$.

Recall that the existence of such a solution $\psi \in C(\mathbb{R}; \Sigma)$ to Eq. (7.31) follows from Proposition 1.26.

Definition 7.18. The map $S : \psi_- \mapsto \psi_+$ given by Theorems 7.14 and 7.16 is the (nonlinear) *scattering operator* associated to Eq. (7.30).

The scattering operator can be understood as follows. Since the operator $U_0(t)$ is well-known, one first tries to construct a solution to the nonlinear Schrödinger equation that behaves like $U_0(t)\psi_-$ as $t \to -\infty$ for some prescribed ψ_-. This yields $\varphi = \psi(0)$. Conversely, can we neglect the nonlinearity for $t \to +\infty$ as well? If yes, then $\psi(t)$ behaves like $U_0(t)\psi_+$ for some function ψ_+. See Fig. 7.1. Note that the group $U_0(t)$ is unitary on $H^1(\mathbb{R}^d)$, but not on Σ:

$$x + it\nabla = U_0(t)xU_0(-t) \implies xU_0(t) = U_0(t)(x - it\nabla).$$

This explains why the error estimate is $\|U_0(-t)\psi(t) - \psi_\pm\|_\Sigma$, and not $\|\psi(t) - U_0(t)\psi_\pm\|_\Sigma$.

There is no reason to expect $S\psi_- = \psi_-$. However, besides the existence of the scattering operator S, very few of its properties are known. We can check, however, that at least for small data, it is not trivial; see Sec. 7.4.3.

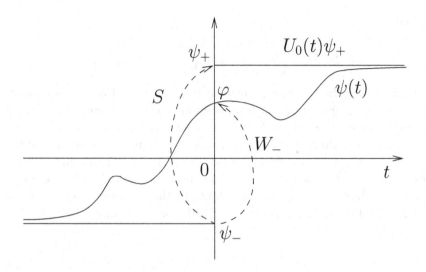

Fig. 7.1 Scattering operator.

7.4.2 Main result

Proposition 7.19. *Let* $d \geqslant 1$ *and* $a_0 \in \Sigma$. *In addition to the condition* $\sigma < 2/(d-2)$ *for* $d \geqslant 3$, *assume that* $\sigma \geqslant \sigma_0(d)$, *given in Theorem 7.16. Then the following limits hold:*

$$\limsup_{\varepsilon \to 0} \sup_{|t-1| \geqslant \Lambda \varepsilon} \left\| \mathcal{B}^{\varepsilon}(t) \left(u^{\varepsilon}(t) - u_{\mathrm{app}}^{\varepsilon}(t) \right) \right\|_{L^2} \xrightarrow[\Lambda \to +\infty]{} 0, \quad \forall \mathcal{B}^{\varepsilon} \in \{\mathrm{Id}, \varepsilon\nabla, J^{\varepsilon}\},$$

where the (discontinuous) function $u_{\mathrm{app}}^{\varepsilon}$ *is given by:*

$$u_{\mathrm{app}}^{\varepsilon}(t,x) = \begin{cases} \dfrac{1}{(1-t)^{d/2}} a_0 \left(\dfrac{x}{1-t} \right) e^{i\frac{|x|^2}{2\varepsilon(t-1)}} & \text{if } t < 1, \\[3ex] \dfrac{e^{-id\pi/4}}{\varepsilon^{d/2}} \left(W_- \circ \mathcal{F}a_0 \right) \left(\dfrac{x}{\varepsilon} \right) & \text{if } t = 1, \\[3ex] \dfrac{e^{-id\pi/2}}{(t-1)^{d/2}} \left(\mathcal{F}^{-1} \circ S \circ \mathcal{F}a_0 \right) \left(\dfrac{x}{1-t} \right) e^{i\frac{|x|^2}{2\varepsilon(t-1)}} & \text{if } t > 1, \end{cases}$$

where W_- *and* S *denote the wave and scattering operators respectively. For* $t = 1$, *we have*

$$\sum_{\mathcal{A}^{\varepsilon} \in \{\mathrm{Id}, \varepsilon\nabla, x/\varepsilon\}} \left\| \mathcal{A}^{\varepsilon} \left(u^{\varepsilon}(1) - u_{\mathrm{app}}^{\varepsilon}(1) \right) \right\|_{L^2} \xrightarrow[\varepsilon \to 0]{} 0.$$

This implies, in terms of the Lagrangian symbol:

$$\limsup_{\varepsilon \to 0} \sup_{|t-1| \geqslant \Lambda \varepsilon} \left\| A^{\varepsilon}(t) - \underline{A}(t) \right\|_{\Sigma} \xrightarrow[\Lambda \to +\infty]{} 0,$$

where

$$\underline{A}(t,\xi) = \begin{cases} A_0(\xi) = e^{-id\pi/4}a_0(-\xi) & \text{if } t < 1, \\ \left(\mathcal{F} \circ S \circ \mathcal{F}^{-1}A_0\right)(\xi) & \text{if } t > 1. \end{cases}$$

Before proving this result, a few comments are in order. First, the statement of the convergence outside the focal point may seem intricate. The meaning is that outside a boundary layer in time of order ε, nonlinear effects are negligible at leading order. On the other hand, since the definition of the approximate solution is different whether $t < 1$ or $t > 1$, nonlinear effects affect the wave function at leading order inside this boundary layer of order ε. These effects are measured, in average, by the scattering operator S; see Fig. 7.2. More precisely, compare with the linear asymptotics (6.2). For

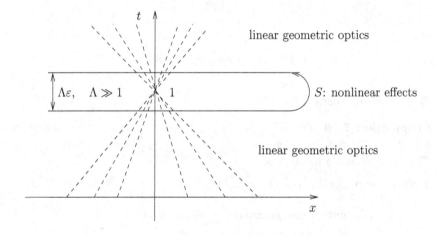

Fig. 7.2   Nonlinear caustic crossing.

$t < 1$, we recover the same asymptotic behavior. For $t = 1$, we still have the phase shift $e^{-id\pi/4}$ (which corresponds to half a Maslov index), but $\mathcal{F}a_0$ is replaced by $W_- \circ \mathcal{F}a_0$: nonlinear effects are not negligible any more. For $t > 1$, we retrieve the linear phenomenon measured by the Maslov index, plus a modification of the amplitude, in terms of S.

Concerning the Lagrangian symbol A^ε, we see that, unless $\mathcal{F}^{-1}A_0$ is a fixed point for S, its limit is discontinuous: this is a typical nonlinear effect, since by construction, the limit of A^ε is continuous in the linear case.

Such a description of caustic crossing in terms of a nonlinear scattering operator first appears in [Bahouri and Gérard (1999)], for the energy-critical wave equation

$$\left(\partial_t^2 - \Delta\right) u + u^5 = 0, \quad (t, x) \in \mathbb{R} \times \mathbb{R}^3. \tag{7.32}$$

For a sequence of initial data $(u^\varepsilon(0), \partial_t u^\varepsilon(0))_\varepsilon$ bounded in the energy space $\dot{H}^1(\mathbb{R}^3) \times L^2(\mathbb{R}^3)$, H. Bahouri and P. Gérard prove that if the weak limit of u^ε is zero, then the only leading order nonlinear effects correspond precisely to the existence of caustics reduced to one point (in general, an infinite number of focal points). In that case, the caustic crossing is described by the scattering operator associated to the nonlinear wave equation. See also [Gallagher and Gérard (2001)] for the case of a wave equation outside a convex obstacle. The proof uses the notion of *profile decomposition*, introduced in [Gérard (1998)]. See Sec. 7.6 for a presentation of this notion in the context of nonlinear Schrödinger equations.

In the case of the semi-linear wave equation (6.5), the case "linear propagation, nonlinear caustic" is studied in [Carles and Rauch (2004a)]. The main result asserts that the caustic crossing is described in terms of a nonlinear scattering operator, and most of the work in [Carles and Rauch (2004a)] is dedicated to constructing this nonlinear scattering operator.

Proof. Resume the change of unknown function introduced in the beginning of this paragraph:

$$u^\varepsilon(t, x) = \frac{1}{\varepsilon^{d/2}} \psi^\varepsilon \left(\frac{t-1}{\varepsilon}, \frac{x}{\varepsilon}\right).$$

Recall that ψ^ε satisfies

$$i\partial_t \psi^\varepsilon + \frac{1}{2}\Delta\psi^\varepsilon = |\psi^\varepsilon|^{2\sigma}\,\psi^\varepsilon \ ; \ \psi^\varepsilon\left(\frac{-1}{\varepsilon}, x\right) = \varepsilon^{d/2}a_0(\varepsilon x)e^{-i\varepsilon|x|^2/2}.$$

As suggested by the previous paragraph, we compute

$$U_0\left(\frac{1}{\varepsilon}\right)\psi^\varepsilon\left(\frac{-1}{\varepsilon}, x\right) = \left(\frac{\varepsilon}{2i\pi}\right)^{d/2} \int_{\mathbb{R}^d} e^{i\varepsilon|x-y|^2/2}\varepsilon^{d/2}a_0(\varepsilon y)e^{-i\varepsilon|y|^2/2}dy$$

$$= \frac{e^{-id\pi/4}}{(2\pi)^{d/2}} e^{i\varepsilon|x|^2/2} \int_{\mathbb{R}^d} e^{-ix\cdot y}a_0(y)dy$$

$$= e^{-id\pi/4}e^{i\varepsilon|x|^2/2}\widehat{a_0}(x).$$

By the Dominated Convergence Theorem, we infer:

$$U_0\left(\frac{1}{\varepsilon}\right)\psi^\varepsilon\left(\frac{-1}{\varepsilon}\right) \xrightarrow[\varepsilon\to0]{} e^{-id\pi/4}\widehat{a_0} \quad \text{in } \Sigma.$$

Define

$$\psi_- := e^{-id\pi/4}\widehat{a_0} \in \Sigma,$$

and introduce the function ψ, solving

$$i\partial_t \psi + \frac{1}{2}\Delta\psi = |\psi|^{2\sigma}\psi \quad ; \quad U_0(-t)\psi(t)\big|_{t=-\infty} = \psi_-.$$

Theorems 7.14 and 7.16 then imply:

$$\sup_{t\in\mathbb{R}}\|U_0(-t)\left(\psi^\varepsilon(t) - \psi(t)\right)\|_\Sigma \xrightarrow[\varepsilon\to 0]{} 0.$$

Back to the function u^ε, this yields

$$\sum_{\mathcal{B}^\varepsilon\in\{\mathrm{Id},\varepsilon\nabla,J^\varepsilon\}}\sup_{t\in\mathbb{R}}\left\|\mathcal{B}^\varepsilon(t)\left(u^\varepsilon(t,\cdot) - \frac{1}{\varepsilon^{d/2}}\psi\left(\frac{t-1}{\varepsilon},\frac{\cdot}{\varepsilon}\right)\right)\right\|_{L^2} \xrightarrow[\varepsilon\to 0]{} 0.$$

From the scattering for ψ, we infer, for all $\mathcal{B}^\varepsilon\in\{\mathrm{Id},\varepsilon\nabla,J^\varepsilon\}$:

$$\limsup_{\varepsilon\to 0}\sup_{t\leqslant 1-\Lambda\varepsilon}\left\|\mathcal{B}^\varepsilon(t)\left(u^\varepsilon(t,\cdot) - \frac{1}{\varepsilon^{d/2}}U_0\left(\frac{t-1}{\varepsilon}\right)\psi_-\left(\frac{\cdot}{\varepsilon}\right)\right)\right\|_{L^2} \xrightarrow[\Lambda\to+\infty]{} 0,$$

$$\limsup_{\varepsilon\to 0}\left\|\mathcal{B}^\varepsilon(1)\left(u^\varepsilon(1,\cdot) - \frac{1}{\varepsilon^{d/2}}W_-\psi_-\left(\frac{\cdot}{\varepsilon}\right)\right)\right\|_{L^2} = 0,$$

$$\limsup_{\varepsilon\to 0}\sup_{t\geqslant 1+\Lambda\varepsilon}\left\|\mathcal{B}^\varepsilon(t)\left(u^\varepsilon(t,\cdot) - \frac{1}{\varepsilon^{d/2}}U_0\left(\frac{t-1}{\varepsilon}\right)\psi_+\left(\frac{\cdot}{\varepsilon}\right)\right)\right\|_{L^2} \xrightarrow[\Lambda\to+\infty]{} 0,$$

where $\psi_+ = S\psi_-$. The proposition then stems from the following lemma:

Lemma 7.20. *Let $\varphi\in L^2$. Then the following asymptotics hold in $L^2(\mathbb{R}^d)$, as $t\to\pm\infty$:*

$$U_0(t)\varphi(x) = \frac{e^{-id\pi/4\,\mathrm{sgn}\,t}}{|t|^{d/2}}\widehat{\varphi}\left(\frac{x}{t}\right)e^{i|x|^2/(2t)} + o(1).$$

If in addition $\varphi\in\Sigma$, then for all $\mathcal{A}\in\{\mathrm{Id},\nabla,x+it\nabla\}$,

$$\left\|\mathcal{A}(t)\left(U_0(t)\varphi - \frac{e^{-id\pi/4\,\mathrm{sgn}\,t}}{|t|^{d/2}}\widehat{\varphi}\left(\frac{\cdot}{t}\right)e^{i|\cdot|^2/(2t)}\right)\right\|_{L^2} \xrightarrow[t\to\pm\infty]{} 0.$$

This lemma can be proved in the same way as Lemma 7.1, by writing

$$U_0(t)\varphi(x) = \frac{1}{(2i\pi t)^{d/2}}\int_{\mathbb{R}^d}e^{i|x-y|^2/(2t)}\varphi(y)dy$$

$$= \frac{1}{(2i\pi t)^{d/2}}e^{i|x|^2/(2t)}\int_{\mathbb{R}^d}e^{-ix\cdot y/t}e^{i|y|^2/(2t)}\varphi(y)dy$$

$$\approx \frac{1}{(2i\pi t)^{d/2}}e^{i|x|^2/(2t)}\int_{\mathbb{R}^d}e^{-ix\cdot y/t}\varphi(y)dy.$$

The proof is omitted.

With this lemma, the proposition follows easily. $\square$

7.4.3 On the propagation of Wigner measures

Since Proposition 7.19 provides a strong convergence of the wave function u^ε, we can infer the expression of the (unique) Wigner measure associated to u^ε. The definition of a Wigner measure was given in Sec. 3.4: it is the limit, up to a subsequence, of the Wigner transform of u^ε,

$$w^\varepsilon(t, x, \xi) = (2\pi)^{-d} \int_{\mathbb{R}^d} u^\varepsilon\left(t, x - \varepsilon\frac{\eta}{2}\right) \overline{u}^\varepsilon\left(t, x + \varepsilon\frac{\eta}{2}\right) e^{i\eta \cdot \xi} d\eta.$$

Proposition 7.19 implies:

$$w^\varepsilon(t, x, \xi) \xrightarrow[\varepsilon \to 0]{} \begin{cases} \mu_-(t, dx, d\xi) & \text{if } t < 1, \\ \mu_+(t, dx, d\xi) & \text{if } t > 1, \end{cases}$$

where

$$\mu_-(t, dx, d\xi) = \frac{1}{(1-t)^d} \left| a_0\left(\frac{x}{1-t}\right) \right|^2 dx \otimes \delta_{\xi = x/(t-1)},$$

$$\mu_+(t, dx, d\xi) = \frac{1}{(t-1)^d} \left| (\mathcal{F}^{-1} \circ S \circ \mathcal{F} a_0)\left(\frac{x}{1-t}\right) \right|^2 dx \otimes \delta_{\xi = x/(t-1)}.$$

In addition,

$$\lim_{t \to 1^-} \mu_-(t, dx, d\xi) = \delta_{x=0} \otimes |a_0(\xi)|^2 d\xi,$$

$$\lim_{t \to 1^+} \mu_+(t, dx, d\xi) = \delta_{x=0} \otimes \left| (\mathcal{F}^{-1} \circ S \circ \mathcal{F} a_0)(\xi) \right|^2 d\xi.$$

We see that, in general, the Wigner measure has a jump at $t = 1$. We can go even further in the analysis, and prove that the propagation of the Wigner measures past the focal point is an ill-posed Cauchy problem: we can find two initial amplitudes a_0 and b_0 such that the corresponding measures μ_- coincide, while the measures μ_+ are different. Consider $b_0(x) = a_0(x)e^{ih(x)}$, where h is a smooth, real-valued, function: the two measures μ_- coincide. We must then compare

$$\left| \mathcal{F}^{-1} \circ S \circ \mathcal{F} a_0 \right|^2 \quad \text{and} \quad \left| \mathcal{F}^{-1} \circ S \circ \mathcal{F}\left(a_0 e^{ih}\right) \right|^2.$$

Note that if h is constant, then the above two functions coincide. The same holds if h is linear in x, since the Fourier transform maps the multiplication by $e^{ix \cdot \xi_0}$ to a translation, and the nonlinear Schrödinger equation is invariant by translation. However, it seems very unlikely that

$$\left| \mathcal{F}^{-1} \circ S \circ \mathcal{F}\left(a_0 e^{ih}\right) \right|^2$$

does not depend on h, for any smooth, real-valued, function h. To prove this vague intuition, we can compute the first two terms of the asymptotic expansion of the scattering operator S near the origin. To simplify the computations, we consider the L^2-critical case $\sigma = 2/d$:

Proposition 7.21. *Let $d \geqslant 1$, $\sigma = 2/d$, and $\psi_- \in L^2(\mathbb{R}^d)$. Then for $\delta > 0$ sufficiently small, $S(\delta\psi_-)$ is well defined in $L^2(\mathbb{R}^d)$, and, as $\delta \to 0$:*

$$S(\delta\psi_-) = \delta\psi_- - i\delta^{1+4/d} \int_{-\infty}^{+\infty} U_0(-\tau) \left(|U_0(\tau)\psi_-|^{4/d} U_0(\tau)\psi_- \right) d\tau$$
$$+ \mathcal{O}_{L^2} \left(\delta^{1+8/d} \right).$$

Proof. The proof follows the same perturbative analysis as in [Gérard (1996)] (see also [Carles (2001b)] for the nonlinear Schrödinger equation). First, it follows from [Cazenave and Weissler (1989)] that $S(\delta\psi_-)$ is well defined in $L^2(\mathbb{R}^d)$ for $\delta > 0$ sufficiently small. We could also assume that $\psi_- \in \Sigma$, and invoke Theorems 7.14 and 7.16.

Let $\delta \in]0,1]$, and consider ψ^δ solving:

$$i\partial_t\psi^\delta + \frac{1}{2}\Delta\psi^\delta = \left|\psi^\delta\right|^{4/d}\psi^\delta \quad ; \quad U_0(-t)\psi^\delta(t)\big|_{t=-\infty} = \delta\psi_-.$$

Plugging an expansion of the form $\psi^\delta = \delta(\varphi_0 + \delta^{4/d}\varphi_1 + \delta^{4/d}r^\delta)$ into the above equation, and ordering in powers of δ, it is natural to impose the following conditions:

- Leading order: $\mathcal{O}(\delta)$.

$$i\partial_t\varphi_0 + \frac{1}{2}\Delta\varphi_0 = 0 \quad ; \quad U_0(-t)\varphi_0(t)\big|_{t=-\infty} = \psi_-.$$

- First corrector: $\mathcal{O}(\delta^{1+4/d})$.

$$i\partial_t\varphi_1 + \frac{1}{2}\Delta\varphi_1 = |\varphi_0|^{4/d}\varphi_0 \quad ; \quad U_0(-t)\varphi_1(t)\big|_{t=-\infty} = 0.$$

The first equation yields

$$\varphi_0(t) = U_0(t)\psi_-.$$

From the second equation, we have:

$$\varphi_1(t) = -i \int_{-\infty}^{t} U_0(t-\tau) \left(|\varphi_0(\tau)|^{4/d}\varphi_0(\tau) \right) d\tau.$$

Let $\gamma = 2 + 4/d$. Remark that the pair (γ, γ) is admissible (see Definition 7.4), and denote $L_{t,x}^r = L^r(]-\infty, -t] \times \mathbb{R}^d)$. Strichartz estimates ($\varepsilon = 1$ in Lemma 7.6) yield:

$$\|\varphi_0\|_{L^\gamma(\mathbb{R}\times\mathbb{R}^d)} \leqslant C \|\psi_-\|_{L^2},$$

$$\|\varphi_1\|_{L^\gamma(\mathbb{R}\times\mathbb{R}^d)} \leqslant C \left\||\varphi_0|^{4/d}\varphi_0\right\|_{L^{\gamma'}(\mathbb{R}\times\mathbb{R}^d)} \leqslant C \|\varphi_0\|_{L^\gamma(\mathbb{R}\times\mathbb{R}^d)}^{1+4/d}$$

$$\leqslant C \|\psi_-\|_{L^2}^{1+4/d},$$

where we have used Hölder's inequality, and the relation $1/\gamma' = (1+4/d)/\gamma$. We also have:

$$i\partial_t r^\delta + \frac{1}{2}\Delta r^\delta = g\left(\varphi_0 + \delta^{4/d}\varphi_1 + \delta^{4/d}r^\delta\right) - g(\varphi_0) \; ; \; U_0(-t)r^\delta(t)\big|_{t=-\infty} = 0,$$

where $g(z) = |z|^{4/d}z$. Strichartz and Hölder inequalities yield

$$\left\|r^\delta\right\|_{L_{t,x}^\gamma} \lesssim \left\|\left(|\varphi_0|^{4/d} + \left|\delta^{4/d}\varphi_1\right|^{4/d} + \left|\delta^{4/d}r^\delta\right|^{4/d}\right)\delta^{4/d}\left(|\varphi_1| + |r^\delta|\right)\right\|_{L_{t,x}^{\gamma'}}$$

$$\lesssim \left(\|\varphi_0\|_{L_{t,x}^\gamma}^{4/d} + \left\|\delta^{4/d}\varphi_1\right\|_{L_{t,x}^\gamma}^{4/d} + \left\|\delta^{4/d}r^\delta\right\|_{L_{t,x}^\gamma}^{4/d}\right)\delta^{4/d}\left(\|\varphi_1\|_{L_{t,x}^\gamma} + \|r^\delta\|_{L_{t,x}^\gamma}\right)$$

$$\lesssim \left(1 + \left\|\delta^{4/d}r^\delta\right\|_{L_{t,x}^\gamma}^{4/d}\right)\delta^{4/d}\left(1 + \|r^\delta\|_{L_{t,x}^\gamma}\right)$$

$$\lesssim \delta^{4/d} + \delta^{4/d}\left\|r^\delta\right\|_{L_{t,x}^\gamma} + \left\|\delta^{4/d}r^\delta\right\|_{L_{t,x}^\gamma}^{1+4/d}.$$

The second term on the right-hand side is absorbed by the left-hand side, provided that δ is sufficiently small (up to doubling the constants). For the last term, we use a standard result:

Lemma 7.22 (Bootstrap argument). *Let $M = M(t)$ be a nonnegative continuous function on $[0,T]$ such that, for every $t \in [0,T]$,*

$$M(t) \leqslant a + bM(t)^\theta,$$

where $a, b > 0$ and $\theta > 1$ are constants such that

$$a < \left(1 - \frac{1}{\theta}\right)\frac{1}{(\theta b)^{1/(\theta-1)}} \; , \quad M(0) \leqslant \frac{1}{(\theta b)^{1/(\theta-1)}} \; .$$

Then, for every $t \in [0,T]$, we have

$$M(t) \leqslant \frac{\theta}{\theta - 1}\, a.$$

This argument shows that for $0 < \delta \ll 1$, $r^\delta \in L^\gamma(\mathbb{R} \times \mathbb{R}^d)$, and

$$\|r^\delta\|_{L^\gamma(\mathbb{R}\times\mathbb{R}^d)} \lesssim \delta^{4/d}.$$

Using Strichartz estimates again, we infer:

$$\|r^\delta\|_{L^\infty(\mathbb{R};L^2(\mathbb{R}^d))} \lesssim \delta^{4/d} + \delta^{4/d}\left\|r^\delta\right\|_{L_{t,x}^\gamma} + \left\|\delta^{4/d}r^\delta\right\|_{L_{t,x}^\gamma}^{1+4/d} \lesssim \delta^{4/d}.$$

Therefore,

$$U_0(-t)\psi^\delta(t) = \delta U_0(-t)\varphi_0(t) + \delta^{1+4/d}U_0(-t)\varphi_1(t) + \delta^{1+4/d}U_0(-t)r^\delta(t)$$

$$= \delta\psi_- - i\delta^{1+4/d}\int_{-\infty}^t U_0(-\tau)\left(|\varphi_0(\tau)|^{4/d}\varphi_0(\tau)\right)d\tau$$

$$+ \mathcal{O}_{L^2}\left(\delta^{1+8/d}\right),$$

where the last term is uniform with respect to $t \in \mathbb{R}$. Letting $t \to +\infty$, the result follows. $\qquad\square$

We proceed as in [Carles (2001b)]. Denote

$$P(\psi_-) = -i \int_{-\infty}^{+\infty} U_0(-\tau) \left(|U_0(\tau)\psi_-|^{4/d} U_0(\tau)\psi_- \right) d\tau.$$

Obviously,

$$|\mathcal{F} \circ S\, (\delta\psi_-)|^2 = \delta^2 \left| \widehat{\psi_-} \right|^2 + 2\delta^{2+4/d} \operatorname{Re} \left(\widehat{\psi_-} \, \overline{\widehat{P\psi_-}} \right) + \mathcal{O} \left(\delta^{2+8/d} \right),$$

and we have to prove that we can find $\psi_- \in \Sigma$, and h smooth and real-valued, such that

$$\operatorname{Re} \left(\overline{\mathcal{F}\psi_-} \mathcal{F}\,(P\psi_-) \right) \neq \operatorname{Re} \left(\overline{\mathcal{F}\,(\psi_h)} \mathcal{F}\,(P\,(\psi_h)) \right) =: R(\psi_-, h),$$

where ψ_h is defined by

$$\widehat{\psi_h}(\xi) = e^{ih(\xi)} \widehat{\psi_-}(\xi).$$

If this was not true, then for every $\psi_- \in \Sigma$, the differential of the map $h \mapsto R(\psi_-, h)$ would be zero at every smooth, real-valued function h. An elementary but tedious computation shows that

$$D_h R(\psi_-, 0)(h) \not\equiv 0,$$

with $h(x) = |x|^2/2$ and $\psi_-(x) = e^{-|x|^2/2}$. Indeed with this choice, computations are explicit:

$$\psi_-(x) = \widehat{\psi}_-(x) = e^{-|x|^2/2}.$$

With $h_a(x) = a|x|^2/2$, $a \in \mathbb{R}$, we introduce ψ_a such that $\widehat{\psi}_a = e^{ih_a} \widehat{\psi}_-$. It is given by

$$\psi_a(x) = (1 + ia)^{-d/2} e^{-|x|^2/(2(1+ia))}.$$

The evolution of Gaussian functions under the action of the free Schrödinger group can be computed explicitly, and we find:

$$U_0(t)\psi_a(x) = (1 + i(a + t))^{-d/2} e^{-|x|^2/(2(1+i(a+t)))}.$$

Therefore,

$$R\,(\psi_-, h_a) = \operatorname{Im} \left(e^{-(1+ia)\frac{|x|^2}{2}} \times \right.$$

$$\left. \times \int_{-\infty}^{+\infty} \frac{(1 + i(a + t))^{-d/2}}{1 + (a + t)^2} \,(1 - it\zeta(a))^{-d/2} e^{-\frac{\zeta(a)}{1-it\zeta(a)} \frac{|x|^2}{2}} dt \right),$$

where $\quad \zeta(a) = \dfrac{4/d + 1 - i(a + t)}{1 + (a + t)^2}.$

To prove the above claim, we differentiate this quantity with respect to a, and assess the result at $a = 0$. Considering for simplicity $x = 0$, we check that $D_h R(\psi_-, 0)(h) \not\equiv 0$. This shows that the caustic crossing is an ill-posed Cauchy problem as far as Wigner measures are concerned:

Proposition 7.23. *Let $d \geqslant 1$ and $\sigma = 2/d$.*
(1) *There exists $a_0 \in \Sigma$ such that the Wigner measure associated to u^ε, the solution of Eq. (7.29), is discontinuous at $t = 1$:*

$$\lim_{t \to 1^-} \mu_-(t, dx, d\xi) \neq \lim_{t \to 1^+} \mu_+(t, dx, d\xi).$$

(2) *There exist a_0 and $\widetilde{a}_0$ in Σ, such that if u^ε and $\widetilde{u}^\varepsilon$ denote the solutions to Eq. (7.29) with these data, we have:*

$$\mu_-(t, dx, d\xi) = \widetilde{\mu}_-(t, dx, d\xi), \quad t < 1,$$
$$\mu_+(t, dx, d\xi) \neq \widetilde{\mu}_+(t, dx, d\xi), \quad t > 1,$$

where $\mu_\pm$ and $\widetilde{\mu}_\pm$ denote the Wigner measures associated to u^ε and $\widetilde{u}^\varepsilon$, respectively.

Remark 7.24. This result is not specific to the value $\sigma = 2/d$. It is stated in this case because we have studied the asymptotic expansion of the scattering operator near the origin for $\sigma = 2/d$: Proposition 7.21 could be extended to other values of σ, with a slightly longer proof, allowing the extension of Proposition 7.23 to other values of σ.

This reveals a difference between WKB régime and caustic crossing for the propagation of Wigner measures. We have seen in Chap. 2 that in the critical case of the WKB régime, the Wigner measures are propagated like in the linear case; the leading order nonlinear effect does not affect the Wigner measure. On the other hand, the above discussion shows that as soon as nonlinear effects affect the wave function u^ε at leading order ($\alpha = d\sigma > 1$), the propagation of the Wigner measure undergoes nonlinear phenomena.

7.5 Nonlinear propagation, nonlinear caustic

In the previous paragraph, we have seen that when $\alpha = d\sigma > 1$, the caustic crossing is described in terms of a nonlinear scattering operator. Suppose that this aspect remains when $\alpha = d\sigma = 1$. Then because $\alpha = 1$, we know that also outside the focal point, the nonlinearity cannot be neglected at leading order (see Chap. 2). This suggests that it is not possible to compare the dynamics of the nonlinear Schrödinger equation to such a simple dynamics as that of the free Schrödinger equation; see also Remark 7.15. To compare the nonlinear dynamics with a simpler one (but necessarily not "too" simple), we need the notion of *long range scattering*.

We suppose $d = 1$. Note that removing this assumption is everything but easy [Ginibre and Ozawa (1993)]. The existence of modified wave operators (wave operators adapted to the long range scattering framework) was first established in [Ozawa (1991)]. A notion of asymptotic completeness appears in [Hayashi and Naumkin (1998)]. The most advanced results (so far) on the long range scattering for the one-dimensional, cubic, nonlinear Schrödinger equation, can be found in [Hayashi and Naumkin (2006)].

We present the main result of [Carles (2001a)] without giving all the details. The main technical idea is to work with oscillatory integrals, like we did in Sec. 7.3. When $d = \sigma = 1$, we modify the initial data, and consider:

$$\begin{cases} i\varepsilon\partial_t u^\varepsilon + \dfrac{\varepsilon^2}{2}\partial_x^2 u^\varepsilon = \varepsilon \left|u^\varepsilon\right|^2 u^\varepsilon, \\ u^\varepsilon(0,x) = e^{-ix^2/(2\varepsilon)-i|a_0(x)|^2\log\varepsilon}a_0(x). \end{cases} \tag{7.33}$$

The new term in the phase is closely related to the fact that we need long range scattering to describe the caustic crossing. It is also suggested by a formal computation. Recall that the expected limiting equation for the Lagrangian amplitude A^ε is given by Eq. (7.25). In the current case $d = \sigma = 1$, this equation is:

$$i\partial_t A(t,\xi) = \frac{1}{|t-1|}\left|A(t,\xi)\right|^2 A(t,\xi).$$

For some general initial data $\widetilde{A}_0$ (not necessarily equal to A_0 derived in Lemma 7.1), we find, for $t < 1$:

$$A(t,\xi) = \widetilde{A}_0(\xi)e^{i|\widetilde{A}_0(\xi)|^2\log(1-t)}.$$

Applying stationary phase formula like in §7.3, we get, for $t < 1$, on a formal level:

$$u^\varepsilon(t,x) \approx \frac{e^{i\pi/4}}{\sqrt{1-t}}A\left(t,\frac{x}{t-1}\right)e^{i\frac{|x|^2}{2\varepsilon(t-1)}}$$

$$\approx \frac{e^{i\pi/4}}{\sqrt{1-t}}\widetilde{A}_0\left(\frac{x}{t-1}\right)e^{i|\widetilde{A}_0(x/(t-1))|^2\log(1-t)}e^{i\frac{|x|^2}{2\varepsilon(t-1)}}.$$

Motivated by the approach of Sec. 7.4, change the unknown function u^ε to ψ^ε, where

$$u^\varepsilon(t,x) = \frac{1}{\sqrt{\varepsilon}}\psi^\varepsilon\left(\frac{t-1}{\varepsilon}, \frac{x}{\varepsilon}\right).$$

In the present case, ψ^ε is given by:

$$\psi^\varepsilon(t,x) = \frac{e^{i\pi/4}}{\sqrt{|t|}}\widetilde{A}_0\left(\frac{x}{t}\right)e^{i|\widetilde{A}_0(x/t)|^2 \log(\varepsilon|t|)}e^{i\varepsilon|x|^2/(2t)}, \quad t < 0.$$

Passing to the strong limit in L^2, the last exponential becomes negligible as $\varepsilon \to 0$. On the other hand, the term $\log(\varepsilon|t|)$ causes a weak convergence to zero, if $\widetilde{A}_0$ is independent of ε. Since the L^2-norm of the wave functions u^ε and ψ^ε is independent of time, this convergence cannot be strong. If, instead of considering $\widetilde{A}_0$ independent of ε, we choose

$$\widetilde{A}_0(\xi) = A_0(\xi)e^{-i|A_0(\xi)|^2 \log \varepsilon},$$

where A_0 is given by Lemma 7.1, then we have:

$$\psi^\varepsilon(t,x) = \frac{e^{i\pi/4}}{\sqrt{|t|}}A_0\left(\frac{x}{t}\right)e^{i|A_0(x/t)|^2 \log|t|}e^{i\varepsilon|x|^2/(2t)}.$$

This function converges strongly in L^2, to

$$\psi(t,x) = \frac{e^{i\pi/4}}{\sqrt{|t|}}A_0\left(\frac{x}{t}\right)e^{i|A_0(x/t)|^2 \log|t|}.$$

This formal approach explains why it is convenient to modify the initial data, like we did in Eq. (7.33). Note that the above phase term in $\log|t|$ is exactly the modification which is needed in long range scattering; see also $S^\pm$ below. With this preliminary explanation, we can state our main result:

Proposition 7.25 ([Carles (2001a)]). *Assume $d = \alpha = \sigma = 1$.*
1. We can define a modified scattering operator for

$$i\partial_t\psi + \frac{1}{2}\partial_x^2\psi = |\psi|^2\psi, \tag{7.34}$$

and data in $\mathcal{F}(\mathcal{H})$, where:

$$\mathcal{H} = \{f \in H^3(\mathbb{R}); \ xf \in H^2(\mathbb{R})\}.$$

More precisely, there exists $\delta > 0$ such that if $\psi_- \in \mathcal{F}(\mathcal{H})$ with $\|\psi_-\|_\Sigma < \delta$, we can find unique $\psi \in C(\mathbb{R}_t, \Sigma)$ solving (7.34), and $\psi_+ \in L^2$, such that

$$\left\|\psi(t) - e^{iS^\pm(t)}U_0(t)\psi_\pm\right\|_{L^2} \xrightarrow[t\to\pm\infty]{} 0,$$

where $S^{\pm}$ are defined by:

$$S^{\pm}(t,x) := \left| \widehat{\psi_{\pm}} \left(\frac{x}{t} \right) \right|^2 \log |t|.$$

Denote $S : \psi_{-} \mapsto \psi_{+}$.

2. *Let $a_0 \in \mathcal{H}$, with $\|a_0\|_{\Sigma}$ sufficiently small. Let u^{ε} be the solution of (7.33) (which is in $C(\mathbb{R}; L^2)$). Define $\psi_{-} = e^{-i\pi/4}\widehat{a_0}$. The following asymptotics hold in L^2:*

- *If $t < 1$, then:*

$$u^{\varepsilon}(t,x) \underset{\varepsilon \to 0}{\sim} \frac{e^{i\pi/4}}{\sqrt{1-t}} e^{i\frac{x^2}{2\varepsilon(t-1)} + i \left| \widehat{\psi_{-}} \left(\frac{x}{t-1} \right) \right|^2 \log \frac{1-t}{\varepsilon}} \widehat{\psi_{-}} \left(\frac{x}{t-1} \right).$$

- *If $t > 1$, then:*

$$u^{\varepsilon}(t,x) \underset{\varepsilon \to 0}{\sim} \frac{e^{-i\pi/4}}{\sqrt{t-1}} e^{i\frac{x^2}{2\varepsilon(t-1)} + i \left| \widehat{\psi_{+}} \left(\frac{x}{t-1} \right) \right|^2 \log \frac{t-1}{\varepsilon}} \widehat{\psi_{+}} \left(\frac{x}{t-1} \right),$$

where $\psi_{+} = S\psi_{-}$.

The $-\pi/2$ phase shift between the two asymptotics (before and after focusing) is the Maslov index. The change in the amplitude, measured by a scattering operator, is like in Proposition 7.19. The new phenomenon is the phase shift

$$\left| \widehat{\psi_{+}} \left(\frac{x}{t-1} \right) \right|^2 \log \frac{t-1}{\varepsilon} - \left| \widehat{\psi_{-}} \left(\frac{x}{t-1} \right) \right|^2 \log \frac{t-1}{\varepsilon},$$

which appears when comparing the asymptotics for the wave function u^{ε} before and after the focal point. It is "highly nonlinear", and depends on ε. Following an idea due to Guy Métivier, we called it a "random" phase shift: it depends on the subsequence ε going to zero which is considered.

In [Carles (2001a)], this result is proved by revisiting the approach of [Ozawa (1991)] for the asymptotics on $t \in [0,1[$, so that we can use the result of [Hayashi and Naumkin (1998)] to describe the asymptotic behavior of u^{ε} on $]1, T]$ for any $T > 1$. In view of the improvement of [Hayashi and Naumkin (2006)] (the domain and range of the above operator S are improved), it should be possible to relax the assumption $a_0 \in \mathcal{H}$, and require less regularity for a_0. We shall not pursue this issue.

We shall merely outline the proof of Proposition 7.25. We leave out the proof of the first point, which is now a consequence of [Hayashi and Naumkin (2006)], and we focus our attention on the second point. We have already seen that a natural candidate as an approximate solution for $t < 1$ is given by:

$$u_{\mathrm{app}}^\varepsilon(t,x) = \frac{1}{\sqrt{2\pi\varepsilon}} \int_{\mathbb{R}} e^{-i\frac{t-1}{2\varepsilon}|\xi|^2 + i\frac{x\cdot\xi}{\varepsilon}} e^{i|A_0(\xi)|^2 \log\frac{1-t}{\varepsilon}} A_0(\xi) d\xi.$$

The last point of Lemma 7.12 shows that since $a_0 \in \mathcal{H}$,

$$i\varepsilon\partial_t u_{\mathrm{app}}^\varepsilon + \frac{\varepsilon^2}{2}\partial_x^2 u_{\mathrm{app}}^\varepsilon = \varepsilon \left|u_{\mathrm{app}}^\varepsilon\right|^2 u_{\mathrm{app}}^\varepsilon - \varepsilon r^\varepsilon,$$

with

$$\|r^\varepsilon(t)\|_{L^2} \lesssim \frac{\varepsilon}{(1-t)^2} \left(\log\frac{1-t}{\varepsilon}\right)^2,$$

$$\sum_{\mathcal{B}^\varepsilon \in \{\varepsilon\partial_x, J^\varepsilon\}} \|\mathcal{B}^\varepsilon(t)r^\varepsilon(t)\|_{L^2} \lesssim \frac{\varepsilon}{(1-t)^2} \left(\log\frac{1-t}{\varepsilon}\right)^3.$$

Introduce $w^\varepsilon = u^\varepsilon - u_{\mathrm{app}}^\varepsilon$. Lemma 7.12 also yields

$$\|w^\varepsilon(0)\|_{L^2} \lesssim \varepsilon \left(\log\frac{1}{\varepsilon}\right)^2,$$

$$\sum_{\mathcal{B}^\varepsilon \in \{\varepsilon\partial_x, J^\varepsilon\}} \|\mathcal{B}^\varepsilon(0)w^\varepsilon(0)\|_{L^2} \lesssim \varepsilon \left(\log\frac{1}{\varepsilon}\right)^3.$$

Let $\delta > 0$. Lemma 7.12 and weighted Gagliardo–Nirenberg inequality (7.6) show that there exists C_* depending on $\|a_0\|_{\mathcal{H}}$ and δ such that for $1 - t \geqslant C_*\varepsilon$,

$$\left\|u_{\mathrm{app}}^\varepsilon(t)\right\|_{L^\infty} \leqslant \frac{\|a_0\|_{L^\infty} + \delta}{\sqrt{1-t}},$$

$$\left\|\varepsilon\partial_x u_{\mathrm{app}}^\varepsilon(t)\right\|_{L^\infty} \leqslant \frac{C}{\sqrt{1-t}},$$

$$\left\|J^\varepsilon(t)u_{\mathrm{app}}^\varepsilon(t)\right\|_{L^\infty} \leqslant \frac{C}{\sqrt{1-t}} \log\frac{1-t}{\varepsilon},$$

where C depends on $\|a_0\|_{\mathcal{H}}$. The error term w^ε solves:

$$i\varepsilon\partial_t w^\varepsilon + \frac{\varepsilon^2}{2}\partial_x^2 w^\varepsilon = \varepsilon \left|u^\varepsilon\right|^2 u^\varepsilon - \varepsilon \left|u_{\mathrm{app}}^\varepsilon\right|^2 u_{\mathrm{app}}^\varepsilon + \varepsilon r^\varepsilon$$

$$= \varepsilon \left(|u^\varepsilon|^2 w^\varepsilon + \left(\left|w^\varepsilon + u_{\mathrm{app}}^\varepsilon\right|^2 - \left|u_{\mathrm{app}}^\varepsilon\right|^2\right) u_{\mathrm{app}}^\varepsilon\right) + \varepsilon r^\varepsilon.$$

From the above estimates and (7.6), there exists $\varepsilon_0 > 0$ such that for $\varepsilon \in]0, \varepsilon_0]$,

$$\|w^\varepsilon(0)\|_{L^\infty} \leqslant \frac{\delta}{2}.$$

By continuity, there exists $t^\varepsilon > 0$ such that

$$\|w^\varepsilon(\tau)\|_{L^\infty} \leqslant \frac{\delta}{\sqrt{1-\tau}} \qquad (7.35)$$

for $t \in [0, t^\varepsilon]$. So long as (7.35) holds, Lemma 1.2 yields:

$$\|w^\varepsilon(t)\|_{L^2} \leqslant \|w^\varepsilon(0)\|_{L^2} + \int_0^t \|r^\varepsilon(\tau)\|_{L^2}\, d\tau$$

$$+ \int_0^t \left(2\left\|u_{\mathrm{app}}^\varepsilon(\tau)\right\|_{L^\infty}^2 + \left\|u_{\mathrm{app}}^\varepsilon(\tau)\right\|_{L^\infty} \|w^\varepsilon(\tau)\|_{L^\infty} \right) \|w^\varepsilon(\tau)\|_{L^2}\, d\tau$$

$$\leqslant \|w^\varepsilon(0)\|_{L^2} + C \int_0^t \frac{\varepsilon}{(1-\tau)^2} \left(\log \frac{1-\tau}{\varepsilon} \right)^2 d\tau$$

$$+ C_0 \int_0^t \|w^\varepsilon(\tau)\|_{L^2} \frac{d\tau}{1-\tau},$$

where

$$C_0 = 2\left(\|a_0\|_{L^\infty} + \delta\right)^2 + \delta\left(\|a_0\|_{L^\infty} + \delta\right).$$

Gronwall lemma yields, so long as (7.35) holds:

$$\|w^\varepsilon(t)\|_{L^2} \leqslant \frac{\|w^\varepsilon(0)\|_{L^2}}{(1-t)^{C_0}} + C \int_0^t \frac{\varepsilon}{(1-\tau)^2} \left(\log \frac{1-\tau}{\varepsilon} \right)^2 \left(\frac{1-\tau}{1-t} \right)^{C_0} d\tau.$$

Rewrite the last term as

$$\int_0^t \frac{\varepsilon}{(1-\tau)^2} \left(\log \frac{1-\tau}{\varepsilon} \right)^2 \left(\frac{1-\tau}{1-t} \right)^{C_0} d\tau = \left(\frac{\varepsilon}{1-t} \right)^{C_0} \int_{(1-t)/\varepsilon}^{1/\varepsilon} \frac{(\log \tau)^2}{\tau^{2-C_0}}\, d\tau.$$

For $C_0 < 1$, an integration by parts shows that

$$\int_a^b \frac{\log \tau}{\tau^{2-C_0}}\, d\tau = \mathcal{O}\left(\frac{(\log a)^2}{a^{1-C_0}} \right) \qquad \text{as } b \geqslant a \to +\infty.$$

Therefore, if $C_0 < 1$ and $1 - t \geqslant C_* \varepsilon$ with C_* sufficiently large,

$$\|w^\varepsilon(t)\|_{L^2} \leqslant \frac{C}{(1-t)^{C_0}} \varepsilon \left(\log \frac{1}{\varepsilon} \right)^2 + C \frac{\varepsilon}{1-t} \left(\log \left(\frac{1-t}{\varepsilon} \right) \right)^2$$

$$\lesssim \frac{\varepsilon}{1-t} \left(\log \left(\frac{1-t}{\varepsilon} \right) \right)^2.$$

With this first estimate, we can infer

$$\sum_{\mathcal{B}^\varepsilon \in \{\varepsilon\partial_x, J^\varepsilon\}} \|\mathcal{B}^\varepsilon(t) w^\varepsilon(t)\|_{L^2} \leqslant C \frac{\varepsilon}{1-t} \left(\log \left(\frac{1-t}{\varepsilon} \right) \right)^3.$$

Using the weighted Gagliardo–Nirenberg inequality (7.6), we deduce

$$\|w^\varepsilon(t)\|_{L^2} \leqslant \frac{C}{\sqrt{1-t}} \frac{\varepsilon}{1-t} \left(\log\left(\frac{1-t}{\varepsilon} \right) \right)^{5/2} .$$

Therefore, if we choose C_* sufficiently large, there exists $\varepsilon_* > 0$ such that for all $\varepsilon \in]0, \varepsilon_*]$, (7.35) holds on $[0, 1 - C_*\varepsilon]$. Applying Lemma 7.12 to $u_{\mathrm{app}}^\varepsilon$ yields the asymptotic behavior of u^ε for $t < 1$ in Proposition 7.25. Note that the smallness condition that we have used so far is $C_0 < 1$. Since $\delta > 0$ is arbitrarily small, this assumption boils down to

$$\|a_0\|_{L^\infty} < \frac{1}{\sqrt{2}}.$$

For the asymptotics when $t > 1$, the smallness condition ceases to be explicit, since we have to use the results of [Hayashi and Naumkin (1998)] or [Hayashi and Naumkin (2006)]. Introduce ψ^ε given by

$$u^\varepsilon(t,x) = \frac{1}{\varepsilon^{d/2}} \psi^\varepsilon \left(\frac{t-1}{\varepsilon}, \frac{x}{\varepsilon} \right).$$

It is easy to deduce from the above analysis that

$$\sup_{t \leqslant -C_*} \|U_0(-t)\left(\psi^\varepsilon(t) - \psi(t)\right)\|_\Sigma \xrightarrow[\varepsilon \to 0]{} 0,$$

where ψ is the unique solution to Eq. (7.34) with

$$\left\| \psi(t) - e^{iS^-(t)} U_0(t) \psi_- \right\|_{L^2} \xrightarrow[t \to -\infty]{} 0.$$

The local well-posedness for Eq. (7.34) shows that for all $T > 0$,

$$\sup_{t \leqslant T} \|U_0(-t)\left(\psi^\varepsilon(t) - \psi(t)\right)\|_\Sigma \xrightarrow[\varepsilon \to 0]{} 0.$$

Theorem 7.26 ([Hayashi and Naumkin (1998)]). *Let $\varphi \in \Sigma$, with $\|\varphi\|_\Sigma = \delta' \leqslant \delta$, where δ is sufficiently small. Let $\psi \in C(\mathbb{R}_t, \Sigma)$ be the solution of the initial value problem*

$$i\partial_t \psi + \frac{1}{2}\partial_x^2 \psi = |\psi|^2 \psi \quad ; \quad \psi_{|t=0} = \varphi.$$

There exists a unique pair $(W, \phi) \in \left(L^2 \cap L^\infty\right) \times L^\infty$ such that for $t \geqslant 1$,

$$\left\| \mathcal{F}\left(U_0(-t)\psi\right)(t) \exp\left(-i \int_1^t |\widehat{\psi}(\tau)|^2 \frac{d\tau}{\tau} \right) - W \right\|_{L^2 \cap L^\infty} \leqslant C\delta' t^{-\alpha + C(\delta')^2},$$

$$\left\| \int_1^t |\widehat{\psi}(\tau)|^2 \frac{d\tau}{\tau} - |W|^2 \log t - \phi \right\|_{L^\infty} \leqslant C\delta' t^{-\alpha + C(\delta')^2},$$

where $C\delta' < \alpha < 1/4$, and ϕ is a real valued function. Furthermore we have the asymptotic formula for large time,

$$\psi(t,x) = \frac{1}{(it)^{1/2}} W\left(\frac{x}{t}\right) \exp\left(i\frac{x^2}{2t} + i\left|W\left(\frac{x}{t}\right)\right|^2 \log t + i\phi\left(\frac{x}{t}\right)\right)$$
$$+ \mathcal{O}_{L^2}\left(\delta' t^{-1/2-\alpha+C(\delta')^2}\right),$$

and the estimate

$$\left\|\mathcal{F}\left(U_0(-t)\psi(t)\right) - W\exp(i|W|^2\log t + i\phi)\right\|_{L^2\cap L^\infty} \leqslant C\delta' t^{-\alpha+C(\delta')^2}.$$

Finally, the map $\varphi \mapsto (W,\phi)$ is continuous on the above spaces.

First, note that in view of Lemma 7.20, the above result yields the first part of Proposition 7.25 with

$$\psi_+ = \mathcal{F}^{-1}\left(We^{i\phi}\right).$$

We now briefly explain how to conclude the proof of Proposition 7.25. Inspired by the linear long range scattering theory (see e.g. [Dereziński and Gérard (1997)]), introduce

$$\phi^\varepsilon(t,\xi) = \int_{-1/\varepsilon}^{(t-1)/\varepsilon} |\psi(\tau,\tau\xi)|^2 d\tau + |a_0(-\xi)|^2 \log\frac{1}{\varepsilon}.$$

Write

$$A^\varepsilon(t,\xi) = e^{i\phi^\varepsilon(t,\xi)} B^\varepsilon(t,\xi),$$

so that

$$u^\varepsilon(t,x) = \frac{1}{\sqrt{2\pi\varepsilon}} \int_{\mathbb{R}} e^{-i\frac{t-1}{2\varepsilon}|\xi|^2 + i\frac{x\cdot\xi}{\varepsilon}} e^{i\phi^\varepsilon(t,\xi)} B^\varepsilon(t,\xi) d\xi.$$

The asymptotics for $t < 1$ and Lemma 7.20 imply, in $L^\infty_{\mathrm{loc}}(]0,1[; L^\infty(\mathbb{R}))$,

$$\phi^\varepsilon(t,\xi) = |a_0(-\xi)|^2 \log\frac{1-t}{\varepsilon} + o(1),$$

hence

$$\|B^\varepsilon(t,\cdot) - A_0\|_{L^2} \xrightarrow[\varepsilon\to0]{} 0,$$

with $A_0(\xi) = e^{-i\pi/4}a_0(-\xi)$. For $t > 1$, Theorem 7.26 yields $W \in L^2 \cap L^\infty$ and $H \in L^\infty$ such that, in $L^\infty_{\mathrm{loc}}(]1,+\infty[; L^\infty(\mathbb{R}))$,

$$\phi^\varepsilon(t,\xi) = |W(\xi)|^2 \log\frac{t-1}{\varepsilon} + H(\xi) + o(1).$$

By construction,

$$B^\varepsilon(t,\xi) = e^{-i\phi^\varepsilon(t,\xi)} \mathcal{F}\left(U_0\left(\frac{1-t}{\varepsilon}\right)\psi^\varepsilon\left(\frac{t-1}{\varepsilon}\right)\right).$$

Since the map $\varphi \mapsto (W,\phi)$ of Theorem 7.26 is continuous, we conclude:

$$B^\varepsilon(t,\xi) \xrightarrow[\varepsilon\to0]{} e^{-iH(\xi)+i\phi(\xi)} W(\xi) = e^{-iH(\xi)}\widehat{\psi_+}(\xi) \quad \text{in } L^\infty_{\mathrm{loc}}(]1,+\infty[; L^2(\mathbb{R})).$$

Using Lemma 7.12, we infer Proposition 7.25. Note that the function H is not present in the limit, due to some cancellations.

7.6   Why initial quadratic oscillations?

In this chapter, except in the case "nonlinear propagation, nonlinear caustic", we have considered initial data with quadratic oscillations, exactly of the form

$$u^\varepsilon(0, x) = a_0(x) e^{-i|x|^2/(2\varepsilon)}. \tag{7.36}$$

We have seen that in order to observe nonlinear effects at leading order near the focal point, we have to impose $\alpha \leqslant d\sigma$. In this section, we consider the critical case

$$i\varepsilon \partial_t u^\varepsilon + \frac{\varepsilon^2}{2} \Delta u^\varepsilon = \varepsilon^{d\sigma} |u^\varepsilon|^{2\sigma} u^\varepsilon, \tag{7.37}$$

with $d\sigma > 1$ (linear propagation). The formal computations of Chap. 6 suggest that since a focal point is expected to concentrate the maximum of energy when a caustic is formed, any other caustic should be "linear". All in all, this means that we expect nonlinear effects to be visible at leading order only when a focal point is present. In this paragraph, we show that this intuition can be made rigorous at least when $\sigma \geqslant 2/d$. The complete proofs appear in [Carles *et al.* (2003)] for the case $\sigma > 2/d$, and in [Carles and Keraani (2007)] for the case $\sigma = 2/d$. Since the papers are rather technical, we only give a flavor of their content, and invite the reader to consult the articles for details. Throughout this section, we assume that if $d \geqslant 3$, $\sigma < 2/(d-2)$ (the nonlinearity is H^1-subcritical).

7.6.1   *Notion of linearizability*

The notion of linearizability that we shall use is the analogue for nonlinear Schrödinger equations of the concept introduced by P. Gérard [Gérard (1996)] in the case of the semi-linear wave equation (7.32). Consider the linear evolution of the initial data for u^ε:

$$i\varepsilon \partial_t v^\varepsilon + \frac{\varepsilon^2}{2} \Delta v^\varepsilon = 0 \quad ; \quad v^\varepsilon_{|t=0} = u^\varepsilon_{|t=0}. \tag{7.38}$$

Roughly speaking, the nonlinearity in Eq. (7.37) is relevant at leading order if and only if the relation $u^\varepsilon(t) - v^\varepsilon(t) = o(v^\varepsilon(t))$ ceases to hold for some $t > 0$ (we consider only forward in time propagation here, since backward propagation is similar). To make this vague statement precise, we clarify our assumptions on the initial data.

Assumption 7.27. The initial data $u^\varepsilon_{|t=0} = u^\varepsilon_0$ belong to $H^1(\mathbb{R}^d)$, uniformly in the following sense: if we denote

$$\|f^\varepsilon\|_{H^1_\varepsilon} = \|f^\varepsilon\|_{L^2} + \|\varepsilon \nabla f^\varepsilon\|_{L^2},$$

then

$$\sup_{0 < \varepsilon \leqslant 1} \|u_0^\varepsilon\|_{H_\varepsilon^1} < \infty.$$

This assumption is satisfied for data of the form (7.36), provided that we assume $a_0 \in \Sigma$. More generally, if u_0^ε is of the form considered in WKB régime,

$$u_0^\varepsilon(x) = a_0(x) e^{i\phi_0(x)/\varepsilon},$$

where ϕ_0 is smooth and subquadratic, and $a_0 \in \Sigma$, then Assumption 7.27 is satisfied.

Definition 7.28 (Linearizability). *Let I^ε be an interval of $\mathbb{R}$, possibly depending on ε, containing the origin; u^ε is linearizable on I^ε if*

$$\limsup_{\varepsilon \to 0} \sup_{t \in I^\varepsilon} \|u^\varepsilon(t) - v^\varepsilon(t)\|_{H_\varepsilon^1} = 0.$$

Recall the conservations of mass and energy in this case:

$$\frac{d}{dt}\|u^\varepsilon(t)\|_{L^2}^2 = \frac{d}{dt}\|v^\varepsilon(t)\|_{L^2}^2 = 0,$$

$$\frac{d}{dt}\|\varepsilon \nabla v^\varepsilon(t)\|_{L^2}^2 = 0, \tag{7.39}$$

$$\frac{d}{dt}\left(\frac{1}{2}\|\varepsilon \nabla u^\varepsilon(t)\|_{L^2}^2 + \frac{\varepsilon^{d\sigma}}{\sigma + 1}\|u^\varepsilon(t)\|_{L^{2\sigma+2}}^{2\sigma+2}\right) = 0.$$

We have the first result:

Lemma 7.29. *Let u_0^ε satisfying Assumption 7.27. Assume in addition that initially, the potential energy goes to zero:*

$$\varepsilon^{d\sigma}\|u_0^\varepsilon\|_{L^{2\sigma+2}}^{2\sigma+2} \xrightarrow[\varepsilon \to 0]{} 0.$$

Let $T > 0$. If u^ε is linearizable on $[0, T]$, then

$$\limsup_{\varepsilon \to 0} \sup_{0 \leqslant t \leqslant T} \varepsilon^{d\sigma}\|v^\varepsilon(t)\|_{L^{2\sigma+2}}^{2\sigma+2} = 0.$$

Proof. The proof follows the same lines as for the energy-critical wave equation[Gérard (1996)]. Let

$$R := \limsup_{\varepsilon \to 0} \sup_{0 \leqslant t \leqslant T} \left| \frac{1}{2}\|\varepsilon \nabla u^\varepsilon(t)\|_{L^2}^2 + \frac{\varepsilon^{d\sigma}}{\sigma + 1}\|u^\varepsilon(t)\|_{L^{2\sigma+2}}^{2\sigma+2} \right.$$

$$\left. - \frac{1}{2}\|\varepsilon \nabla v^\varepsilon(t)\|_{L^2}^2 - \frac{\varepsilon^{d\sigma}}{\sigma + 1}\|v^\varepsilon(t)\|_{L^{2\sigma+2}}^{2\sigma+2} \right|.$$

On the one hand, linearizability implies that

$$R \leqslant \limsup_{\varepsilon \to 0} \sup_{0 \leqslant t \leqslant T} \frac{\varepsilon^{d\sigma}}{\sigma + 1} \int_{\mathbb{R}^d} \left| |u^\varepsilon(t,x)|^{2\sigma+2} - |v^\varepsilon(t,x)|^{2\sigma+2} \right| dx.$$

Writing

$$\left| |u^\varepsilon(t,x)|^{2\sigma+2} - |v^\varepsilon(t,x)|^{2\sigma+2} \right|$$
$$\lesssim \left(|u^\varepsilon(t,x)|^{2\sigma+1} + |v^\varepsilon(t,x)|^{2\sigma+1} \right) |u^\varepsilon(t,x) - v^\varepsilon(t,x)|,$$

Hölder's inequality yields

$$R \lesssim \limsup_{\varepsilon \to 0} \sup_{0 \leqslant t \leqslant T} \varepsilon^{d\sigma} \|u^\varepsilon(t) - v^\varepsilon(t)\|_{L^{2\sigma+2}} \left(\|u^\varepsilon(t)\|_{L^{2\sigma+2}} + \|v^\varepsilon(t)\|_{L^{2\sigma+2}} \right)^{2\sigma+1}.$$

Assumption 7.27, and the conservation of linear and nonlinear energy for v^ε and u^ε respectively, yield, along with Gagliardo–Nirenberg inequality (for v^ε):

$$\|u^\varepsilon(t)\|_{L^{2\sigma+2}}^{2\sigma+2} + \|v^\varepsilon(t)\|_{L^{2\sigma+2}}^{2\sigma+2} \lesssim \varepsilon^{-d\sigma}.$$

Using Gagliardo–Nirenberg inequality, this implies

$$R \lesssim \limsup_{\varepsilon \to 0} \sup_{0 \leqslant t \leqslant T} \varepsilon^{d\sigma} \|u^\varepsilon(t) - v^\varepsilon(t)\|_{L^{2\sigma+2}} \varepsilon^{-d\sigma \frac{2\sigma+1}{2\sigma+2}}$$
$$\lesssim \limsup_{\varepsilon \to 0} \sup_{0 \leqslant t \leqslant T} \varepsilon^{\delta(2\sigma+2)} \|u^\varepsilon(t) - v^\varepsilon(t)\|_{L^{2\sigma+2}}$$
$$\lesssim \limsup_{\varepsilon \to 0} \sup_{0 \leqslant t \leqslant T} \|u^\varepsilon(t) - v^\varepsilon(t)\|_{L^2}^{1-\delta(2\sigma+2)} \|\varepsilon\nabla u^\varepsilon(t) - \varepsilon\nabla v^\varepsilon(t)\|_{L^2}^{\delta(2\sigma+2)}.$$

We conclude from the linearizability assumption that $R = 0$. On the other hand, the conservation of energy (7.39) yields

$$R = \limsup_{\varepsilon \to 0} \sup_{0 \leqslant t \leqslant T} \frac{\varepsilon^{d\sigma}}{\sigma + 1} \left| \|u_0^\varepsilon\|_{L^{2\sigma+2}}^{2\sigma+2} - \|v^\varepsilon(t)\|_{L^{2\sigma+2}}^{2\sigma+2} \right|.$$

By assumption, the first term of the right-hand side goes to zero, and the lemma follows. $\qquad\square$

The above lemma announces an important idea in the linearizability criterion, first introduced in [Gérard (1996)]: to assess the nonlinear effects affecting u^ε, we consider the evolution of a quantity involving the solution v^ε to the companion *linear* equation only.

It turns out that this necessary condition for linearizability is sufficient when $\sigma > 2/d$. It is not so when $\sigma = 2/d$. We present here a general sufficient condition for linearizability, which yields the result when $\sigma > 2/d$, and which is also a necessary condition when $\sigma = 2/d$. This condition, and its proof, may be viewed as a simplification of the approach presented in [Carles *et al.* (2003)]. Instead of the usual inhomogeneous Strichartz estimates (7.14), we use similar estimates for pairs (q_j, r_j) not necessarily admissible. Such estimates were derived in [Cazenave and Weissler (1992)], and generalized in [Foschi (2005)]:

Lemma 7.30. *Let (q, r) be an admissible pair with $r > 2$. Let $k > q/2$, and define $\widetilde{k}$ by*

$$\frac{1}{\widetilde{k}} + \frac{1}{k} = \frac{2}{q}.$$

There exists C depending only on d, r and k such that for all interval I,

$$\varepsilon^{2/q} \left\| \int_{I \cap \{\tau \leqslant t\}} U_0^\varepsilon(t - \tau) F(\tau) d\tau \right\|_{L^k(I;L^r)} \leqslant C \left\| F \right\|_{L^{\widetilde{k}'}(I;L^{r'})},$$

for all $F \in L^{\widetilde{k}'}(I; L^{r'})$.

Proof. In view of the dispersive estimate

$$\left\| U_0^\varepsilon(t) \right\|_{L^{r'} \to L^r} \lesssim |\varepsilon t|^{-\delta(r)},$$

we have

$$\left\| \int_{I \cap \{\tau \leqslant t\}} U_0^\varepsilon(t - \tau) F(\tau) d\tau \right\|_{L^r} \lesssim \int_0^t \varepsilon^{-\delta(r)} (t - \tau)^{-\delta(r)} \left\| F(\tau) \right\|_{L^{r'}} d\tau.$$

By assumption, $\delta(r) = 2/q$. The lemma then follows from Riesz potential inequalities (see e.g. [Stein (1993)]). $\square$

Proposition 7.31. *Let Assumption 7.27 be satisfied. Assume that*

$$\sigma > \sigma_0(d) = \frac{2 - d + \sqrt{d^2 + 12d + 4}}{4d},$$

where $\sigma_0(d)$ appeared in Theorem 7.16. Let

$$r = 2\sigma + 2 \quad ; \quad q = \frac{4\sigma + 4}{d\sigma} \quad ; \quad k = \frac{4\sigma(\sigma + 1)}{2 - \sigma(d - 2)}.$$

If

$$\varepsilon^{2/q - 1/k} \left\| v^\varepsilon \right\|_{L^k(I^\varepsilon;L^r)} \xrightarrow[\varepsilon \to 0]{} 0, \tag{7.40}$$

then u^ε is linearizable on I^ε.

Proof. We first note that (q, r) is admissible, and that $k > q/2$ if and only if $\sigma > \sigma_0(d)$. Define $w^\varepsilon = u^\varepsilon - v^\varepsilon$. It solves

$$i\varepsilon\partial_t w^\varepsilon + \frac{\varepsilon^2}{2}\Delta w^\varepsilon = \varepsilon^{d\sigma} |u^\varepsilon|^{2\sigma} u^\varepsilon \quad ; \quad w^\varepsilon_{|t=0} = 0.$$

Writing $|u^\varepsilon|^{2\sigma} u^\varepsilon = |u^\varepsilon|^{2\sigma} u^\varepsilon - |v^\varepsilon|^{2\sigma} v^\varepsilon + |v^\varepsilon|^{2\sigma} v^\varepsilon$, Lemma 7.30 yields:

$$\|w^\varepsilon\|_{L^k L^r} \lesssim \varepsilon^{d\sigma - 1 - 2/q} \left\| |u^\varepsilon|^{2\sigma} u^\varepsilon - |v^\varepsilon|^{2\sigma} v^\varepsilon \right\|_{L^{\tilde{k}'} L^{r'}}$$

$$+ \varepsilon^{d\sigma - 1 - 2/q} \left\| |v^\varepsilon|^{2\sigma} v^\varepsilon \right\|_{L^{\tilde{k}'} L^{r'}},$$

where $L^j L^s$ stands for $L^j(I^\varepsilon; L^s)$. Note that

$$\frac{1}{r'} = \frac{2\sigma + 1}{r} \quad ; \quad \frac{1}{\tilde{k}'} = \frac{2\sigma + 1}{k}.$$

From Taylor formula, Hölder's inequality and the relation $u^\varepsilon = w^\varepsilon + v^\varepsilon$, we infer:

$$\|w^\varepsilon\|_{L^k L^r} \lesssim \varepsilon^{d\sigma - 1 - 2/q} \left(\|w^\varepsilon\|^{2\sigma}_{L^k L^r} + \|v^\varepsilon\|^{2\sigma}_{L^k L^r} \right) \|w^\varepsilon\|_{L^k L^r}$$

$$+ \varepsilon^{d\sigma - 1 - 2/q} \|v^\varepsilon\|^{2\sigma + 1}_{L^k L^r}.$$

Again, note that

$$d\sigma - 1 - \frac{1}{k} = (2\sigma + 1)\left(\frac{2}{q} - \frac{1}{k} \right),$$

to deduce:

$$\varepsilon^{2/q - 1/k} \|w^\varepsilon\|_{L^k L^r} \lesssim \left(\varepsilon^{2/q - 1/k} \|w^\varepsilon\|_{L^k L^r} \right)^{2\sigma} \varepsilon^{2/q - 1/k} \|w^\varepsilon\|_{L^k L^r}$$

$$+ \left(\varepsilon^{2/q - 1/k} \|v^\varepsilon\|_{L^k L^r} \right)^{2\sigma} \varepsilon^{2/q - 1/k} \|w^\varepsilon\|_{L^k L^r}$$

$$+ \left(\varepsilon^{2/q - 1/k} \|v^\varepsilon\|_{L^k L^r} \right)^{2\sigma + 1}.$$

By assumption, the second term on the right-hand side can be absorbed by the left-hand side, provided that ε is sufficiently small:

$$\varepsilon^{2/q - 1/k} \|w^\varepsilon\|_{L^k L^r} \lesssim \left(\varepsilon^{2/q - 1/k} \|w^\varepsilon\|_{L^k L^r} \right)^{2\sigma} \varepsilon^{2/q - 1/k} \|w^\varepsilon\|_{L^k L^r}$$

$$+ \left(\varepsilon^{2/q - 1/k} \|v^\varepsilon\|_{L^k L^r} \right)^{2\sigma + 1}.$$

Using the bootstrap argument of Lemma 7.22, we infer, for ε sufficiently small:

$$\varepsilon^{2/q - 1/k} \|w^\varepsilon\|_{L^k L^r} \lesssim \left(\varepsilon^{2/q - 1/k} \|v^\varepsilon\|_{L^k L^r} \right)^{2\sigma + 1} \ll 1.$$

Note that this implies, along with (7.40),

$$\varepsilon^{2/q-1/k} \left\| u^\varepsilon \right\|_{L^k L^r} \ll 1.$$

This estimate allows us to conclude, by applying Strichartz estimates. Indeed, we first find

$$\left\| w^\varepsilon \right\|_{L^q L^r} \lesssim \varepsilon^{d\sigma-1-2/q} \left\| |u^\varepsilon|^{2\sigma} u^\varepsilon - |v^\varepsilon|^{2\sigma} v^\varepsilon \right\|_{L^{q'} L^{r'}}$$

$$+ \varepsilon^{d\sigma-1-2/q} \left\| |v^\varepsilon|^{2\sigma} v^\varepsilon \right\|_{L^{q'} L^{r'}}.$$

Noticing that

$$\frac{1}{r'} = \frac{2\sigma + 1}{r} \quad ; \quad \frac{1}{q'} = \frac{1}{q} + \frac{2\sigma}{k},$$

Taylor formula and Hölder's inequality yield

$$\left\| w^\varepsilon \right\|_{L^q L^r} \lesssim \varepsilon^{d\sigma-1-2/q} \left(\left\| w^\varepsilon \right\|_{L^k L^r}^{2\sigma} + \left\| v^\varepsilon \right\|_{L^k L^r}^{2\sigma} \right) \left\| w^\varepsilon \right\|_{L^q L^r}$$

$$+ \varepsilon^{d\sigma-1-2/q} \left\| v^\varepsilon \right\|_{L^k L^r}^{2\sigma} \left\| v^\varepsilon \right\|_{L^q L^r}.$$

Again, since

$$d\sigma - 1 - \frac{1}{q} = 2\sigma \left(\frac{2}{q} - \frac{1}{k} \right) + \frac{1}{q},$$

we infer

$$\varepsilon^{1/q} \left\| w^\varepsilon \right\|_{L^q L^r} \lesssim \left(\varepsilon^{2/q-1/k} \left\| w^\varepsilon \right\|_{L^k L^r} \right)^{2\sigma} \varepsilon^{1/q} \left\| w^\varepsilon \right\|_{L^q L^r}$$

$$+ \left(\varepsilon^{2/q-1/k} \left\| v^\varepsilon \right\|_{L^k L^r} \right)^{2\sigma} \varepsilon^{1/q} \left\| w^\varepsilon \right\|_{L^q L^r}$$

$$+ \left(\varepsilon^{2/q-1/k} \left\| v^\varepsilon \right\|_{L^k L^r} \right)^{2\sigma} \varepsilon^{1/q} \left\| v^\varepsilon \right\|_{L^q L^r}$$

$$\lesssim \left(\varepsilon^{2/q-1/k} \left\| w^\varepsilon \right\|_{L^k L^r} \right)^{2\sigma} \varepsilon^{1/q} \left\| w^\varepsilon \right\|_{L^q L^r}$$

$$+ \left(\varepsilon^{2/q-1/k} \left\| v^\varepsilon \right\|_{L^k L^r} \right)^{2\sigma} \varepsilon^{1/q} \left\| w^\varepsilon \right\|_{L^q L^r}$$

$$+ \left(\varepsilon^{2/q-1/k} \left\| v^\varepsilon \right\|_{L^k L^r} \right)^{2\sigma} \left\| u_0^\varepsilon \right\|_{L^2},$$

where we have used the homogeneous Strichartz inequality for v^ε. Therefore, the first two terms of the right-hand side can be absorbed by the left-hand side for ε sufficiently small, and

$$\varepsilon^{1/q} \left\| w^\varepsilon \right\|_{L^q L^r} \ll 1.$$

Applying Strichartz inequality again, we have

$$\|w^\varepsilon\|_{L^\infty L^2} \lesssim \varepsilon^{d\sigma-1-1/q} \left\| |u^\varepsilon|^{2\sigma} u^\varepsilon - |v^\varepsilon|^{2\sigma} v^\varepsilon \right\|_{L^{q'} L^{r'}}$$
$$+ \varepsilon^{d\sigma-1-2/q} \left\| |v^\varepsilon|^{2\sigma} v^\varepsilon \right\|_{L^{q'} L^{r'}}$$
$$\lesssim \left(\varepsilon^{2/q-1/k} \|w^\varepsilon\|_{L^k L^r} \right)^{2\sigma} \varepsilon^{1/q} \|w^\varepsilon\|_{L^q L^r}$$
$$+ \left(\varepsilon^{2/q-1/k} \|v^\varepsilon\|_{L^k L^r} \right)^{2\sigma} \varepsilon^{1/q} \|w^\varepsilon\|_{L^q L^r}$$
$$+ \left(\varepsilon^{2/q-1/k} \|v^\varepsilon\|_{L^k L^r} \right)^{2\sigma} \|u_0^\varepsilon\|_{L^2},$$

so

$$\|w^\varepsilon\|_{L^\infty L^2} \ll 1,$$

which is exactly the first part of the definition of linearizability. Differentiating the equation for w^ε and applying Strichartz inequalities, we find

$$\varepsilon^{1/q} \|\varepsilon\nabla w^\varepsilon\|_{L^q L^r} \lesssim \varepsilon^{d\sigma-1-2/q}\varepsilon^{1/q} \left\| |u^\varepsilon|^{2\sigma} \varepsilon\nabla u^\varepsilon \right\|_{L^{q'} L^{r'}}$$
$$\lesssim \varepsilon^{d\sigma-1-2/q} \|u^\varepsilon\|_{L^k L^r}^{2\sigma} \varepsilon^{1/q} \|\varepsilon\nabla u^\varepsilon\|_{L^q L^r}$$
$$\lesssim \varepsilon^{d\sigma-1-2/q} \|u^\varepsilon\|_{L^k L^r}^{2\sigma} \varepsilon^{1/q} \|\varepsilon\nabla w^\varepsilon\|_{L^q L^r}$$
$$+ \varepsilon^{d\sigma-1-2/q} \|u^\varepsilon\|_{L^k L^r}^{2\sigma} \varepsilon^{1/q} \|\varepsilon\nabla v^\varepsilon\|_{L^q L^r}$$
$$\lesssim \left(\varepsilon^{2/q-1/k} \|u^\varepsilon\|_{L^k L^r} \right)^{2\sigma} \varepsilon^{1/q} \|\varepsilon\nabla w^\varepsilon\|_{L^q L^r}$$
$$+ \left(\varepsilon^{2/q-1/k} \|u^\varepsilon\|_{L^k L^r} \right)^{2\sigma} \|\varepsilon\nabla u_0^\varepsilon\|_{L^2},$$

where we have used the homogeneous Strichartz estimate for $\varepsilon\nabla v^\varepsilon$. Thus,

$$\varepsilon^{1/q} \|\varepsilon\nabla w^\varepsilon\|_{L^q L^r} \ll 1.$$

Using Strichartz inequality again, we conclude

$$\|\varepsilon\nabla w^\varepsilon\|_{L^\infty L^2} \ll 1,$$

which completes the proof of the proposition. $\qquad\square$

We now distinguish the cases $\sigma > 2/d$ and $\sigma = 2/d$.

7.6.2 The L^2-supercritical case: $\sigma > 2/d$

From Lemma 7.29 and Proposition 7.31, we infer:

Corollary 7.32. *Let $\sigma > 2/d$ and u_0^ε satisfying Assumption 7.27. Assume in addition that initially, the potential energy goes to zero:*

$$\varepsilon^{d\sigma} \|u_0^\varepsilon\|_{L^{2\sigma+2}}^{2\sigma+2} \xrightarrow[\varepsilon\to 0]{} 0. \tag{7.41}$$

Let $T > 0$. Then u^ε is linearizable on $[0, T]$, if and only if

$$\limsup_{\varepsilon \to 0} \sup_{0 \leqslant t \leqslant T} \varepsilon^{d\sigma} \|v^\varepsilon(t)\|_{L^{2\sigma+2}}^{2\sigma+2} = 0. \tag{7.42}$$

Proof. In view of Lemma 7.29, we need only prove that the above condition is sufficient for linearizability. Since $\sigma > 2/d$, we have $k > q$. Using Strichartz inequality, we obtain

$$\varepsilon^{2/q-1/k} \|v^\varepsilon\|_{L^k L^r} \leqslant \left(\varepsilon^{2/q} \|v^\varepsilon\|_{L^\infty L^r} \right)^{1-q/k} \left(\varepsilon^{1/q} \|v^\varepsilon\|_{L^q L^r} \right)^{q/k}$$

$$\lesssim \left(\varepsilon^{2/q} \|v^\varepsilon\|_{L^\infty L^r} \right)^{1-q/k} \|u_0^\varepsilon\|_{L^2}^{q/k}$$

$$\lesssim \left(\varepsilon^{2/q} \|v^\varepsilon\|_{L^\infty L^r} \right)^{1-q/k}.$$

Recalling that $r = 2\sigma + 2$ and $2/q = d\sigma/(2\sigma + 2)$, (7.42) implies

$$\varepsilon^{2/q-1/k} \|v^\varepsilon\|_{L^k L^r} \ll 1,$$

which in turn yields linearizability, from Proposition 7.31. $\qquad\square$

The next step in the analysis consists in answering the following question: when is (7.42) violated? As pointed out before, it must be noticed that Corollary 7.32 turns the analysis of a nonlinear problem into the analysis of the behavior of v^ε, solving a *linear* Schrödinger equation.

To study the negation of (7.42), our approach relies on the notion of *profile decomposition*, introduced by P. Gérard [Gérard (1998)] to measure the lack of compactness of critical Sobolev embeddings, inspired by the approach of [Métivier and Schochet (1998)]. For the linear Schrödinger equation, we use more precisely the decomposition in the homogeneous Sobolev space $\dot{H}^1$, due to S. Keraani in the case $d = 3$ [Keraani (2001)]. It can easily be generalized to any spatial dimension. For simplicity, we assume $d = 3$.

Theorem 7.33 ([Keraani (2001)], Theorem 1.6). *Assume $d = 3$. Let $V^\varepsilon = V^\varepsilon(s, y)$ solve*

$$i\partial_s V^\varepsilon + \frac{1}{2}\Delta V^\varepsilon = 0, \tag{7.43}$$

with $V^\varepsilon_{|t=0} = V_0^\varepsilon$, where the family $(V_0^\varepsilon)_{0 < \varepsilon \leqslant 1}$ is bounded in $\dot{H}^1(\mathbb{R}^3)$. Up to extracting a subsequence, we have:

$$V^\varepsilon(s, y) = \sum_{j=0}^{\ell} \frac{1}{\sqrt{\eta_j^\varepsilon}} V_j \left(\frac{s - s_j^\varepsilon}{(\eta_j^\varepsilon)^2}, \frac{y - y_j^\varepsilon}{\eta_j^\varepsilon} \right) + W_\ell^\varepsilon(s, y), \tag{7.44}$$

where $\eta_j^\varepsilon \in \mathbb{R}_+ \setminus \{0\}$ are the scales of concentration, satisfying the following orthogonality condition:

$$\forall j \neq k, \quad either \quad \limsup_{\varepsilon \to 0} \frac{\eta_j^\varepsilon}{\eta_k^\varepsilon} + \frac{\eta_k^\varepsilon}{\eta_j^\varepsilon} = +\infty,$$

$$or \quad \eta_j^\varepsilon = \eta_k^\varepsilon \quad and \quad \limsup_{\varepsilon \to 0} \frac{|s_j^\varepsilon - s_k^\varepsilon| + |y_j^\varepsilon - y_k^\varepsilon|}{\eta_j^\varepsilon} = +\infty.$$

The remainder W_ℓ^ε satisfies

$$\limsup_{\varepsilon \to 0} \|W_\ell^\varepsilon\|_{L^q(\mathbb{R}, L^r)} \underset{\ell \to \infty}{\longrightarrow} 0, \tag{7.45}$$

for $\dfrac{2}{q} + \dfrac{3}{r} = \dfrac{1}{2}$, with $r < +\infty$. Such (q, r) are said to be $\dot{H}^1$-admissible (as opposed to the L^2-admissible pairs of Definition 7.4).
Finally the V_j's and W_ℓ^ε are solutions to Eq. (7.43) in $L^\infty(\mathbb{R}, \dot{H}^1)$.

We define the rescaled function (recall that $d = 3$)

$$V^\varepsilon(s, y) := \varepsilon^{3/2} v^\varepsilon(\varepsilon s, \varepsilon y),$$

which satisfies the linear equation (7.43) with data

$$V_0^\varepsilon(y) := \varepsilon^{3/2} u_0^\varepsilon(\varepsilon y).$$

Clearly, $(V_0^\varepsilon)_{0 < \varepsilon \leqslant 1}$ is bounded in $\dot{H}^1(\mathbb{R}^3)$. Applying Theorem 7.33, we prove that up to a subsequence, the scales η_j^ε all have a non-zero, finite limit. For that, we use the notion of ε_n–oscillatory sequences. For more details on the subject, we refer to [Bahouri and Gérard (1999); Gérard *et al.* (1997)].

Definition 7.34. Let $(\varepsilon_n)_{n \in \mathbb{N}}$ be a given sequence in $\mathbb{R}_+ \setminus \{0\}$, and let (V^n) be a bounded sequence in $\dot{H}^1$. The sequence (V^n) is ε_n–oscillatory if the following property holds:

$$\limsup_{n \to \infty} \int_{\varepsilon_n |\xi| \leqslant R^{-1}} |\xi|^2 |\mathcal{F}(V^n)(\xi)|^2 \, d\xi + \int_{\varepsilon_n |\xi| \geqslant R} |\xi|^2 |\mathcal{F}(V^n)(\xi)|^2 \, d\xi \underset{R \to +\infty}{\longrightarrow} 0.$$

Remark 7.35. For a time-dependent sequence (V^n), uniformly bounded in $L^\infty(\mathbb{R}_+, \dot{H}^1)$, the definition holds taking the limit uniformly in time.

It is easy to see (see [Gérard (1998)] or [Keraani (2001)]) that V^ε is η_j^ε-oscillatory for every sequence η_j^ε appearing in the decomposition (7.44).

Lemma 7.36. *Suppose the sequence (V^ε) is η^ε-oscillatory for some sequence η^ε. Then up to a subsequence, $\eta^\varepsilon = \lambda$ for some $\lambda > 0$.*

Proof.   We can write, uniformly in time,

$$\|\nabla V^\varepsilon(s)\|_{L^2}^2 \lesssim \int_{R^{-1}\leqslant \eta^\varepsilon|\xi|\leqslant R} |\xi|^2 |\mathcal{F}(V^\varepsilon)|^2 \, d\xi + \delta(\varepsilon, R)$$

$$\lesssim \left(\frac{R}{\eta^\varepsilon}\right)^2 \|V^\varepsilon(s)\|_{L^2}^2 + \delta(\varepsilon, R),$$

where $\limsup_{\varepsilon\to 0} \delta(\varepsilon, R) \to 0$ as $R \to \infty$. The conservation of the energy yields

$$\|\nabla V^\varepsilon(s)\|_{L^2}^2 = \|\nabla V^\varepsilon(0)\|_{L^2}^2 = \|\varepsilon \nabla u_0^\varepsilon\|_{L^2}^2.$$

Up to a subsequence, we can suppose that this quantity is bounded from below by some $c > 0$ independent of $\varepsilon \in]0,1]$ (otherwise, Condition (7.42) would not be violated, from Gagliardo–Nirenberg inequality). Fixing R such that

$$\limsup_{\varepsilon\to 0} \delta(\varepsilon, R) \leqslant \frac{c}{2},$$

yields, up to an extraction,

$$\infty > \limsup_{\varepsilon\to 0} \eta^\varepsilon = \lambda \geqslant 0.$$

Now suppose that $\lambda = 0$. Write, for all time,

$$V^\varepsilon = V_R^\varepsilon + W_R^\varepsilon, \quad \text{with} \quad \mathcal{F}V_R^\varepsilon(t,\xi) := \mathbf{1}_{R^{-1}\leqslant \eta^\varepsilon|\xi|\leqslant R}\mathcal{F}V^\varepsilon(t,\xi),$$

and for all $\delta > 0$, if R is large enough uniformly in ε and η^ε, we have

$$\|W_R^\varepsilon\|_{L^\infty(\mathbb{R},H^1)} \leqslant \delta.$$

Gagliardo–Nirenberg inequality implies that $\|W_R^\varepsilon\|_{L^\infty(\mathbb{R},L^{2\sigma+2})}$ can be chosen arbitrarily small if R is large enough, uniformly in ε and η^ε. Thus,

$$\|V^\varepsilon\|_{L^\infty([0,T],L^{2\sigma+2})}^{2\sigma+2} \lesssim \|V_R^\varepsilon\|_{L^\infty([0,T],L^{2\sigma+2})}^{2\sigma+2} + o(1) \tag{7.46}$$

$$\lesssim \|V_R^\varepsilon\|_{L^\infty([0,T],L^2)}^{(2\sigma+2)(1-\delta(2\sigma+2))} + o(1),$$

where the second inequality is due again to Gagliardo–Nirenberg inequality and the boundedness of V^ε in H^1.

Frequency localization implies that for all $s \in [0,T]$,

$$\|V_R^\varepsilon(s)\|_{L^2}^2 \lesssim (\eta^\varepsilon)^2 \int_{\frac{1}{R}\leqslant \eta^\varepsilon\xi\leqslant R} |\xi|^2 \left|\mathcal{F}V_R^\varepsilon(s,\xi)\right|^2 d\xi.$$

So the result follows, since by assumption the left-hand side in (7.46) does not go to zero and $(2\sigma + 2)(1 - \delta(2\sigma + 2)) = 2 - \sigma > 0$ when $d = 3$. □

Lemma 7.36 has several important consequences. First, the V_j's and W_ℓ^ε are bounded in $L^\infty(\mathbb{R}, H^1)$. Indeed, by the orthogonality properties, one has

$$V^\varepsilon(s + s_j^\varepsilon, y + y_j^\varepsilon) \rightharpoonup V_j(s, y) \quad \text{in} \quad \mathcal{D}'(\mathbb{R} \times \mathbb{R}^3).$$

But V^ε is bounded in $L^\infty(\mathbb{R}, L^2)$, so it follows that for all $j \in \mathbb{N}$, the profiles V_j are bounded in $L^\infty(\mathbb{R}, L^2)$. This implies also that

$$(W_\ell^\varepsilon)_{0 < \varepsilon \leqslant 1} \quad \text{is bounded in} \quad L^\infty(\mathbb{R}, L^2), \quad \text{uniformly in} \quad \ell \in \mathbb{N}. \quad (7.47)$$

Second,

$$\limsup_{\varepsilon \to 0} \|W_\ell^\varepsilon\|_{L^\infty(\mathbb{R}; L^{2\sigma+2})} \xrightarrow[\ell \to \infty]{} 0.$$

This follows from (7.45) with $q = +\infty$, (7.47) and Hölder's inequality.

Note also that the family $(V_j)_{j \in \mathbb{N}}$ is not trivial. Indeed,

$$\|V^\varepsilon\|_{L^\infty([0,T], L^{2\sigma+2})} \leqslant \sum_{j=1}^\ell \|V_j\|_{L^\infty([0,T], L^{2\sigma+2})} + \|W_\ell^\varepsilon\|_{L^\infty([0,T], L^{2\sigma+2})},$$

so since (7.42) is not satisfied, the above limit implies that all of the V_j's cannot be zero.

Back to the definition of v^ε, it follows that one can write

$$u_0^\varepsilon(x) = \sum_{j=0}^\ell \frac{1}{\varepsilon^{3/2}} V_j \left(-\frac{t_j^\varepsilon}{\varepsilon}, \frac{x - x_j^\varepsilon}{\varepsilon} \right) + w_\ell^\varepsilon(x),$$

where we have set

$$t_j^\varepsilon := \varepsilon s_j^\varepsilon, \quad x_j^\varepsilon := \varepsilon y_j^\varepsilon \quad \text{and} \quad w_\ell^\varepsilon := \frac{1}{\varepsilon^{3/2}} W_\ell^\varepsilon \left(0, \frac{\cdot}{\varepsilon} \right).$$

Note that w_ℓ^ε is such that

$$\limsup_{\varepsilon \to 0} \varepsilon^{d\sigma} \|w_\ell^\varepsilon\|_{L^{2\sigma+2}}^{2\sigma+2} \xrightarrow[\ell \to \infty]{} 0.$$

Using the assumption (7.41), one can prove

$$\limsup_{\varepsilon \to 0} \frac{t_j^\varepsilon}{\varepsilon} = +\infty.$$

Finally, quadratic oscillations appear like in Lemma 7.20, up to a rescaling, and we obtain (see [Carles *et al.* (2003)] for details):

Theorem 7.37. *Let $\sigma > 2/d$. Let Assumption 7.27 be satisfied, and assume that (7.41) holds. Assume that (7.42) is not satisfied for some $T > 0$.*

Then up to the extraction of a subsequence, there exist an orthogonal family $(t_j^\varepsilon, x_j^\varepsilon)_{j \in \mathbb{N}}$ in $\mathbb{R}_+ \times \mathbb{R}^d$, that is

$$\limsup_{\varepsilon \to 0} \left(\frac{|t_j^\varepsilon - t_k^\varepsilon|}{\varepsilon} + \frac{|x_j^\varepsilon - x_k^\varepsilon|}{\varepsilon} \right) = \infty \quad \forall j \neq k,$$

a family $(\Psi_\ell^\varepsilon)_{\ell \in \mathbb{N}}$, bounded in $H_\varepsilon^1(\mathbb{R}^d)$, and a (non-trivial) family $(\varphi_j)_{j \in \mathbb{N}}$, bounded in $\mathcal{F}(H^1)$, such that:

$$u_0^\varepsilon(x) = \Psi_\ell^\varepsilon(x) + r_\ell^\varepsilon(x), \quad \text{with } \limsup_{\varepsilon \to 0} \varepsilon^{d\sigma} \|U_0^\varepsilon(t) r_\ell^\varepsilon\|_{L^\infty(\mathbb{R}_+, L^{2\sigma+2})}^{2\sigma+2} \xrightarrow[\ell \to \infty]{} 0,$$

and for every $\ell \in \mathbb{N}$, the following asymptotics holds in $L^2(\mathbb{R}^d)$, as $\varepsilon \to 0$,

$$\Psi_\ell^\varepsilon(x) = \sum_{j=0}^\ell \frac{1}{(t_j^\varepsilon)^{d/2}} \varphi_j \left(\frac{x - x_j^\varepsilon}{t_j^\varepsilon} \right) e^{-i|x - x_j^\varepsilon|^2/(2\varepsilon t_j^\varepsilon)} + o(1).$$

Moreover, we have

$$\limsup_{\varepsilon \to 0} \frac{t_j^\varepsilon}{\varepsilon} = +\infty \text{ and } \limsup_{\varepsilon \to 0} t_j^\varepsilon \in [0, T], \quad \forall j \in \mathbb{N}.$$

In view of Corollary 7.32, the condition on r_ℓ^ε means that in the limit $\ell \to \infty$, the evolution of r_ℓ^ε is essentially linear. Therefore, the only obstruction to linearizability stems from Ψ_ℓ^ε. The asymptotic behavior of Ψ_ℓ^ε then shows that linearizability fails to hold because of the presence of quadratic oscillations in the initial data.

7.6.3   The L^2-critical case: $\sigma = 2/d$

When $\sigma = 2/d$, it is easy to see that not only initial data of Theorem 7.37 have a truly nonlinear evolution. Let U solve

$$i\partial_t U + \frac{1}{2}\Delta U = |U|^{4/d}U, \tag{7.48}$$

with $U_{|t=0} = \phi$. If $\phi \in \Sigma$, then U is defined globally in time, $U \in C(\mathbb{R}_t; \Sigma)$. Let $(t_0, x_0) \in \mathbb{R} \times \mathbb{R}^d$. Then the function

$$u^\varepsilon(t, x) = \frac{1}{\varepsilon^{n/4}} U\left(t - t_0, \frac{x - x_0}{\sqrt{\varepsilon}} \right) \tag{7.49}$$

solves

$$i\varepsilon\partial_t u^\varepsilon + \frac{\varepsilon^2}{2}\Delta u^\varepsilon = \varepsilon^2 |u^\varepsilon|^{4/d} u^\varepsilon. \tag{7.50}$$

Moreover, $u^\varepsilon(0, \cdot)$ and $\varepsilon\nabla u^\varepsilon(0, \cdot)$ are bounded in $L^2(\mathbb{R}^d)$. This peculiar solution is such that the nonlinearity in (7.50) has a leading order influence for any (finite) time, near $x = x_0$.

We check that the solutions (7.49) are deduced from those of Theorem 7.37 by the scaling

$$\widetilde{U}(t,x) = \lambda^{d/2} U\left(\lambda^2 t, \lambda x\right),$$

with $\lambda = \sqrt{\varepsilon}$. This scaling leaves Eq. (7.48) invariant (for any positive λ). We saw above that one of the key steps in the proof of Theorem 7.37 consists in singling out the scale $\eta_j^\varepsilon = 1$ by showing that scales going to zero or to infinity yield a linearizable evolution. This step must be modified when $\sigma = 2/d$.

As a matter of fact, we have to resume the study from the very start. It is easy to check that in the above example, (7.42) is verified, even though the solution u^ε is obviously not linearizable. The conclusion of Corollary 7.32 is no longer true in the case $\sigma = 2/d$, and the limitation $\sigma > 2/d$ in the proof of Corollary 7.32 was not a technical artifice. We have the following new linearizability criterion:

Theorem 7.38 ([Carles and Keraani (2007)]). *Suppose* $\sigma = 2/d$, *with* $d = 1$ *or* 2, *and that Assumption 7.27 is satisfied. Let* I^ε *be a time interval containing the origin. Then* u^ε *is linearizable on* I^ε *if and only if*

$$\limsup_{\varepsilon \to 0} \varepsilon \|v^\varepsilon\|_{L^{2+4/d}(I^\varepsilon \times \mathbb{R}^d)}^{2+4/d} = 0. \tag{7.51}$$

The proof that (7.51) implies linearizability is a direct consequence of Proposition 7.31, since when $\sigma = 2/d$, we have

$$r = q = k = \frac{4}{d} + 2.$$

It is in the proof of the converse that the assumption $d = 1$ or 2 appears in [Carles and Keraani (2007)]. This assumption could be removed, in view of the results in [Bégout and Vargas (2007)]. This part of the proof relies on a profile decomposition, in $L^2(\mathbb{R}^d)$, as opposed to the profile decomposition in $\dot{H}^1(\mathbb{R}^d)$ which was used in the case $\sigma > 2/d$. Moreover, we apply this technique not only to solutions to the linear Schrödinger equation, but also to solutions of the nonlinear equation, like in [Bahouri and Gérard (1999); Keraani (2001)]. This part therefore relies on the existence of these two decompositions, which are established after improved Strichartz estimates. In the case $d = 2$, such estimates were proved after the work of J. Bourgain [Bourgain (1995)], in [Bourgain (1998); Moyua *et al.* (1999)], and profile decomposition were given by F. Merle and L. Vega [Merle and Vega (1998)]. The case $d = 1$ was established by S. Keraani in his PhD thesis, and appears in [Carles and Keraani (2007)]. For $d \geqslant 3$, improved Strichartz estimates are proved in [Bégout and Vargas (2007)]; as noted in [Carles and Keraani (2007)], once such estimates are available, a general technique to prove profile decompositions yields the result.

Since the proof of the two profile decompositions and the proof of Theorem 7.38 are rather technical, we leave them out here, and invite the interested reader to consult directly [Carles and Keraani (2007)] and [Bégout and Vargas (2007)]. To make the comparison with the L^2 supercritical case complete, we conclude this paragraph by stating the analogue of Theorem 7.37 in the L^2-critical case.

Definition 7.39. If $(h_j^\varepsilon, t_j^\varepsilon, x_j^\varepsilon, \xi_j^\varepsilon)_{j \in \mathbb{N}}$ is a family of sequences in $\mathbb{R}_+ \setminus \{0\} \times \mathbb{R} \times \mathbb{R}^d \times \mathbb{R}^d$, then we say that $(h_j^\varepsilon, t_j^\varepsilon, x_j^\varepsilon, \xi_j^\varepsilon)_{j \in \mathbb{N}}$ is an orthogonal family if

$$\limsup_{\varepsilon \to 0} \left(\frac{h_j^\varepsilon}{h_k^\varepsilon} + \frac{h_k^\varepsilon}{h_j^\varepsilon} + \frac{|t_j^\varepsilon - t_k^\varepsilon|}{(h_j^\varepsilon)^2} + \left| \frac{x_j^\varepsilon - x_k^\varepsilon}{h_j^\varepsilon} + \frac{t_j^\varepsilon \xi_j^\varepsilon - t_k^\varepsilon \xi_k^\varepsilon}{h_j^\varepsilon} \right| \right) = \infty, \quad \forall j \neq k.$$

Theorem 7.40. *Assume $d = 1$ or 2, and let Assumption 7.27 be satisfied. Let $T > 0$ and assume that (7.51) is not satisfied with $I^\varepsilon = [0, T]$. Then up to the extraction of a subsequence, there exist an orthogonal family $(h_j^\varepsilon, t_j^\varepsilon, x_j^\varepsilon, \xi_j^\varepsilon)_{j \in \mathbb{N}}$, a family $(\phi_j)_{j \in \mathbb{N}}$, bounded in $L^2(\mathbb{R}^d)$, such that:*

$$u_0^\varepsilon(x) = \sum_{j=1}^{\ell} \widetilde{H}_j^\varepsilon(\phi_j)(x) + w_\ell^\varepsilon(x),$$

where $\widetilde{H}_j^\varepsilon(\phi_j)(x) = e^{ix \cdot \xi_j^\varepsilon / \sqrt{\varepsilon}} e^{-i\varepsilon \frac{t_j^\varepsilon}{2} \Delta} \left(\frac{1}{(h_j^\varepsilon \sqrt{\varepsilon})^{d/2}} \phi_j \left(\frac{x - x_j^\varepsilon}{h_j^\varepsilon \sqrt{\varepsilon}} \right) \right),$

and $\displaystyle \limsup_{\varepsilon \to 0} \varepsilon \, \|U_0^\varepsilon(t) w_\ell^\varepsilon\|_{L^{2+4/d}(\mathbb{R} \times \mathbb{R}^d)}^{2+4/d} \xrightarrow{\ell \to +\infty} 0.$

We have $\liminf t_j^\varepsilon / (h_j^\varepsilon)^2 \neq -\infty$, $\liminf (T - t_j^\varepsilon) / (h_j^\varepsilon)^2 \neq -\infty$ (as $\varepsilon \to 0$), and $\sqrt{\varepsilon} \leqslant h_j^\varepsilon \leqslant 1$ for every $j \in \mathbb{N}$.
If $t_j^\varepsilon / (h_j^\varepsilon)^2 \to +\infty$ as $\varepsilon \to 0$, then we also have, in $L^2(\mathbb{R}^d)$:

$$\widetilde{H}_j^\varepsilon(\phi_j)(x) = e^{ix \cdot \xi_j^\varepsilon / \sqrt{\varepsilon} + id\pi/4 - i|x - x_j^\varepsilon|^2 / (2\varepsilon t_j^\varepsilon)}$$

$$\times \left(\frac{h_j^\varepsilon}{t_j^\varepsilon \sqrt{\varepsilon}} \right)^{d/2} \widehat{\phi_j} \left(-\frac{h_j^\varepsilon}{t_j^\varepsilon \sqrt{\varepsilon}} (x - x_j^\varepsilon) \right) + o(1) \text{ as } \varepsilon \to 0.$$

A few comments are in order. First, w_ℓ^ε plays the same role as r_ℓ^ε in Theorem 7.37: for large ℓ, it does not see nonlinear effects at leading order, since it satisfies the linearizability condition (now Eq. (7.51)) with $I^\varepsilon = \mathbb{R}$. In the last case of the theorem, we recover the quadratic oscillations. However, there are other obstructions to linearizability, since we may have $t_j^\varepsilon = T/2$ and $h_j^\varepsilon = 1$: this is the case in the example given in the beginning of this paragraph, Eq. (7.49). Roughly speaking, the above result shows that this is the other borderline case: the scales of concentration, $h_j^\varepsilon \sqrt{\varepsilon}$, lie between ε and $\sqrt{\varepsilon}$. Therefore, (7.49) and (7.36) are essentially the two extreme cases when $\sigma = 2/d$. Finally, almost all the conclusions of the above theorem remain true, if instead of Assumption 7.27, we simply suppose that $(u_0^\varepsilon)_{0<\varepsilon\leqslant 1}$ is bounded in $L^2(\mathbb{R}^d)$. The only difference in the conclusions is that we lose the lower bound for the scales h_j^ε: we cannot say more than $0 < h_j^\varepsilon \leqslant 1$.

7.6.4 *Nonlinear superposition*

When Theorems 7.37 and 7.40 are available, one can ask: how does an initial sum of data with quadratic oscillations evolve under the nonlinear dynamics of Eq. (7.37)? The answer, both when $\sigma > 2/d$ and when $\sigma = 2/d$, is that each part of the initial data evolves independently of the others at leading order, when $\varepsilon \to 0$. In other words, there exists a superposition principle, even though we consider nonlinear equations.

Heuristically, the explanation is rather simple. We know that in the linear case, each part of the data with the form

$$a_j \left(x - x_j \right) e^{-i|x-x_j|^2/(2\varepsilon t_j)}$$

focuses at the point $(t, x) = (t_j, x_j)$. It is of order $\mathcal{O}(1)$ away from the focus, and of order $\varepsilon^{-d/2}$ at the focal point, in a neighborhood of order ε. In the nonlinear case, since $d\sigma > 1$, the nonlinearity is negligible in WKB régime. Therefore, outside the focal points, the nonlinear superposition principle is simply the usual linear superposition principle. Near the focal points, nonlinear effects become relevant at leading order. The decoupling of nonlinear interactions is due to the orthogonality of the cores, denoted $(t_j^\varepsilon, x_j^\varepsilon)$. Roughly speaking, concentration at focal points occurs in balls which do not intersect. From this point of view, the nonlinear superposition is a consequence of precise geometric properties.

In the L^2-supercritical case, the rigorous justification of this statement relies on a precise use of the linearizability criterion (7.42). In the L^2-critical case, the nonlinear superposition principle is a direct consequence of the nonlinear profile decomposition, which is at the heart of the proof of Theorem 7.38.

7.7   Focusing on a line

To conclude this long chapter, we mention a result where focusing at one point is replaced by focusing on a line. Consider the Cauchy problem, in space dimension $d = 2$,

$$i\varepsilon\partial_t u^\varepsilon + \frac{\varepsilon^2}{2}\Delta u^\varepsilon = \varepsilon^\alpha \left|u^\varepsilon\right|^{2\sigma} u^\varepsilon \quad ; \quad u^\varepsilon(0, x) = a_0(x)e^{-x_1^2/(2\varepsilon)}, \qquad (7.52)$$

with $x = (x_1, x_2) \in \mathbb{R}^2$. The rays of geometric optics are given by

$$\dot{x} = \xi \quad ; \quad \dot{\xi} = 0 \quad ; \quad x_{|t=0} = y \quad ; \quad \xi_{1|t=0} = -y_1 \quad ; \quad \xi_{2|t=0} = 0.$$

Therefore, we have

$$x_1(t) = y_1(1 - t) \quad ; \quad x_2(t) = y_2.$$

Rays meet on the line $\{x_1 = 0\}$ at time $t = 1$, see Fig. 7.3. Heuristically, x_2

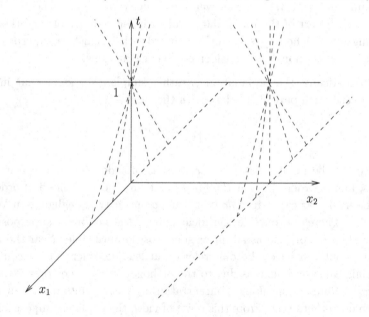

Fig. 7.3   Focusing on a line in $\mathbb{R}^2$.

plays the role of a parameter, and the geometric information is carried in the x_1 variable. Roughly speaking, there is focusing at a point ($x_1 = 0$) along a continuous range of a parameter ($x_2 \in \mathbb{R}$). This suggests the following distinctions for Eq. (7.52):

	$\alpha > 1$	$\alpha = 1$
$\alpha > \sigma$	linear propagation linear caustic	nonlinear propagation linear caustic
$\alpha = \sigma$	linear propagation, nonlinear caustic	nonlinear propagation nonlinear caustic

This is exactly the general table considered in the beginning of this chapter, in the case $d = 1$. This table is justified in [Carles (2000a)], where the case "nonlinear propagation, nonlinear caustic", is not studied. In particular, in the case "linear propagation, nonlinear caustic", the caustic crossing is described by a nonlinear scattering operator, associated to the equation:

$$i\partial_t \psi + \frac{1}{2}\partial_{x_1}^2 \psi = |\psi|^{2\sigma}\psi \quad ; \quad e^{-i\frac{t}{2}\partial_{x_1}^2}\psi(t, x_1, x_2)\big|_{t=\infty} = \psi_-(x_1, x_2).$$

By working in suitable spaces, the usual scattering theory for the nonlinear Schrödinger equation, recalled in Theorems 7.14 and 7.16, is adapted for $\sigma > \sigma_0(1)$, the critical power for scattering in space dimension one. Like in the case of a single focal point, the crossing of the line caustic is described in terms of this operator, in the limit $\varepsilon \to 0$. We invite the interested reader to consult [Carles (2000a)] for technical issues.

Chapter 8

Focal Point in the Presence of an External Potential

In this chapter, we continue the analysis of the previous one, in the presence of a non-trivial external potential:

$$i\varepsilon\partial_t u^\varepsilon + \frac{\varepsilon^2}{2}\Delta u^\varepsilon = V u^\varepsilon + \varepsilon^\alpha \left|u^\varepsilon\right|^{2\sigma} u^\varepsilon.$$

We consider two cases:

- When the initial data are independent of ε, and the external potential is harmonic.
- When the initial data are concentrated at scale ε, and the external potential is more general.

We will see that in the first case, it is easy to include rapid plane oscillations in the initial data (Remark 8.3). In the second case, we present a rather complete picture when the external potential is exactly a polynomial of degree at most two. On the other hand, for general subquadratic potentials, we give partial results only.

8.1 Isotropic harmonic potential

First we consider the case of an isotropic harmonic potential, with ε-independent initial data:

$$i\varepsilon\partial_t u^\varepsilon + \frac{\varepsilon^2}{2}\Delta u^\varepsilon = \frac{|x|^2}{2}u^\varepsilon + \varepsilon^\alpha \left|u^\varepsilon\right|^{2\sigma} u^\varepsilon \quad ; \quad u^\varepsilon(0,x) = a_0(x). \quad (8.1)$$

Rays of geometric optics are given by the Hamiltonian system

$$\dot{x} = \xi \quad ; \quad \dot{\xi} = -x \quad ; \quad x(0,y) = y \quad ; \quad \xi(0,y) = 0. \quad (8.2)$$

We find: $\ddot{x} + x = 0$, along with the initial conditions $x(0,y) = y$ and $\dot{x}(0,y) = 0$. Therefore, $x(t,y) = y\cos t$: rays are sinusoids, which meet at

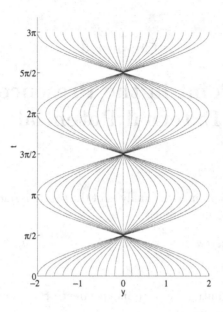

Fig. 8.1   Rays of geometric optics: sinusoids.

the origin for $t \in \pi/2 + \pi\mathbb{Z}$, see Fig. 8.1. Geometrically, this is quite the same thing as in the previous paragraph (no external potential, but initial quadratic oscillations), repeated indefinitely many times. The parallel is even analytical. Consider the linear analogue of Eq. (8.1):

$$i\varepsilon\partial_t v^\varepsilon + \frac{\varepsilon^2}{2}\Delta v^\varepsilon = \frac{|x|^2}{2}v^\varepsilon \quad ; \quad v^\varepsilon(0,x) = a_0(x). \tag{8.3}$$

Because the potential is exactly quadratic, this solution is known explicitly in terms of an oscillatory integral, given by the *Mehler's formula* (see e.g. [Feynman and Hibbs (1965); Hörmander (1995)]): for $|t| < \pi$,

$$v^\varepsilon(t,x) = \frac{1}{(2i\pi\varepsilon\sin t)^{d/2}}\int_{\mathbb{R}^d} e^{i\left(\frac{|x|^2+|y|^2}{2}\cos t - x\cdot y\right)/(\varepsilon\sin t)}a_0(y)dy. \tag{8.4}$$

For $t \in \pi + 2\pi\mathbb{Z}$, the fundamental solution is singular, and some extra phase shifts must be included in the above formula when $|t| > \pi$; see e.g. [Yajima (1996)] and references therein. For $0 < t < \pi/2$, we can apply stationary phase formula in (8.4). We find:

$$v^\varepsilon(t,x) \underset{\varepsilon\to 0}{\sim} \frac{1}{(\cos t)^{d/2}}a_0\left(\frac{x}{\cos t}\right)e^{-i|x|^2\tan t/(2\varepsilon)}. \tag{8.5}$$

Note that we retrieve the solutions to the eikonal equation and to the transport equation of WKB analysis, derived in Examples 1.12 and 1.18 respectively. For $t = \pi/2$, we find directly:

$$v^\varepsilon \left(\frac{\pi}{2}, x\right) = \frac{1}{(2i\pi\varepsilon)^{d/2}} \int_{\mathbb{R}^d} e^{-ix \cdot y/\varepsilon} a_0(y) dy = \frac{e^{-in\pi/4}}{\varepsilon^{d/2}} \widehat{a}_0 \left(\frac{x}{\varepsilon}\right). \tag{8.6}$$

The general picture we obtain for the solution of the linear equation is therefore closely akin to what we got in the case of a single focal point, with no external potential, see (6.2). Thus, the parallel is not only geometrical, it is also analytical. As a consequence, the same table as in Chap. 7 is expected. We shall consider only the critical case "linear propagation, nonlinear caustic", that is

$$i\varepsilon\partial_t u^\varepsilon + \frac{\varepsilon^2}{2}\Delta u^\varepsilon = \frac{|x|^2}{2} u^\varepsilon + \varepsilon^{d\sigma} |u^\varepsilon|^{2\sigma} u^\varepsilon \quad ; \quad u^\varepsilon(0, x) = a_0(x), \tag{8.7}$$

with $d\sigma > 1$. This case was studied in [Carles (2003b)]. As mentioned there, the proof in the case $\alpha = d\sigma > 1$ would make it possible to show that when $\alpha > \max(d\sigma, 1)$, the nonlinear term is indeed negligible on any (finite) time interval, in the limit $\varepsilon \to 0$. Since we have given a rather detailed proof of this fact in the case of a focal point without external potential, we do not pursue this issue here.

To introduce two useful operators, resume one of the points of view that led us to the introduction of the operator $x/\varepsilon + i(t-1)\nabla$ in the previous chapter. Using stationary phase argument on Mehler's formula (8.4), or solving directly the eikonal equation, we see that outside the focal point, the rapid oscillations of the linear solution are given by

$$\phi_{\text{eik}}(t, x) = -\frac{|x|^2}{2} \tan t.$$

To recover the fact that the L^p norm of the approximation of v^ε is independent of ε by using Gagliardo–Nirenberg inequalities, we replace the usual operator ∇ by

$$e^{i\phi_{\text{eik}}(t,x)/\varepsilon}\nabla \left(e^{-i\phi_{\text{eik}}(t,x)/\varepsilon}.\right) = e^{-i|x|^2 \tan t/(2\varepsilon)}\nabla \left(e^{i|x|^2 \tan t/(2\varepsilon)}.\right).$$

Also, to compensate the scaling factor $\cos t$ in the approximation of v^ε, multiply the above operator by $\cos t$. Up to an irrelevant factor i, we therefore consider

$$J^\varepsilon(t) = \frac{1}{i} \cos t\, e^{-i|x|^2 \tan t/(2\varepsilon)}\nabla \left(e^{i|x|^2 \tan t/(2\varepsilon)}.\right) = \frac{x}{\varepsilon} \sin t - i \cos t \nabla. \tag{8.8}$$

Like in the case of Chap. 7 with $x + it\nabla$, we note that this operator has been known for quite a long time (*Heisenberg derivative*, see e.g. [Robert (1987); Thirring (1981)]). Denote

$$U^\varepsilon(t) = \exp\left(-i\frac{t}{2\varepsilon}\left(-\varepsilon^2\Delta + |x|^2\right)\right)$$

the propagator associated to (8.3). Then we also have

$$J^\varepsilon(t) = U^\varepsilon(t)\left(\frac{1}{i}\nabla\right)U^\varepsilon(-t). \tag{8.9}$$

Therefore, it commutes with the linear part of the equation:

$$\left[i\varepsilon\partial_t + \frac{\varepsilon^2}{2}\Delta - \frac{|x|^2}{2}, J^\varepsilon(t)\right] = 0.$$

Similarly, another important Heisenberg derivative can be computed:

$$
\begin{aligned}
H^\varepsilon(t) &= U^\varepsilon(t)xU^\varepsilon(-t) = x\cos t + i\varepsilon\sin t\nabla \\
&= i\varepsilon\sin t\, e^{i|x|^2/(2\varepsilon\tan t)}\nabla\left(e^{-i|x|^2/(2\varepsilon\tan t)}\cdot\right).
\end{aligned}
\tag{8.10}
$$

By the first relation, H^ε also commutes with the linear part of the equation. The phase $\phi(t,x) = |x|^2/(2\tan t)$ solves the eikonal equation

$$\partial_t\phi + \frac{1}{2}|\nabla\phi|^2 + \frac{|x|^2}{2} = 0 \quad;\quad \phi\left(\frac{\pi}{2}, x\right) = 0.$$

Note that, as Heisenberg derivatives or as linear combinations of x and ∇ with time dependent coefficients, J^ε and H^ε are well-defined for all time. On the other hand, the factorization with a phase solving the eikonal equation is valid for almost all time only. Finally, note that the operators J^ε and H^ε make it possible to rewrite the (conserved) energy associated to Eq. (8.3) (which is the kinetic part of the energy associated to Eq. (8.1)):

$$
\begin{aligned}
E^\varepsilon_{\lin}(0) = E^\varepsilon_{\lin}(t) &= \frac{1}{2}\|\varepsilon\nabla v^\varepsilon(t)\|^2_{L^2} + \frac{1}{2}\int_{\mathbb{R}^d}|x|^2\,|v^\varepsilon(t,x)|^2\,dx \\
&= \frac{1}{2}\|\varepsilon J^\varepsilon(t)v^\varepsilon(t)\|^2_{L^2} + \frac{1}{2}\|H^\varepsilon(t)v^\varepsilon(t)\|^2_{L^2}.
\end{aligned}
$$

In view of the nonlinear analysis, we list the most interesting properties of these operators below.

Lemma 8.1. *The operators J^ε and H^ε defined by (8.8) and (8.10) respectively satisfy the following properties.*
(1) *They commute with the linear part of the equation, since*

$$J^\varepsilon(t) = U^\varepsilon(t)\left(\frac{1}{i}\nabla\right)U^\varepsilon(-t) \;;\; H^\varepsilon(t) = U^\varepsilon(t)xU^\varepsilon(-t).$$

(2) *There are two real-valued functions* $\phi_1(t,x)$ *and* $\phi_2(t,x)$ *such that we can write*

$$J^\varepsilon(t) = -i\cos t\, e^{i\phi_1(t,x)/\varepsilon} \nabla\left(e^{-i\phi_1(t,x)/\varepsilon}.\right), \qquad t \notin \frac{\pi}{2} + \pi\mathbb{Z},$$

$$H^\varepsilon(t) = i\varepsilon \sin t\, e^{i\phi_2(t,x)/\varepsilon} \nabla\left(e^{-i\phi_2(t,x)/\varepsilon}.\right), \qquad t \notin \pi\mathbb{Z}.$$

(3) *Weighted Gagliardo–Nirenberg estimates are available: for* $0 \leqslant \delta(p) < 1$, *there exists* C_p *such that for all* $u \in \Sigma$,

$$\|u\|_{L^p} \leqslant \frac{C_p}{|\cos t|^{\delta(p)}} \|u\|_{L^2}^{1-\delta(p)} \|J^\varepsilon(t)u\|_{L^2}^{\delta(p)}, \qquad t \notin \frac{\pi}{2} + \pi\mathbb{Z}, \qquad (8.11)$$

$$\|u\|_{L^p} \leqslant \frac{C_p}{|\varepsilon \sin t|^{\delta(p)}} \|u\|_{L^2}^{1-\delta(p)} \|H^\varepsilon(t)u\|_{L^2}^{\delta(p)}, \qquad t \notin \pi\mathbb{Z}. \qquad (8.12)$$

(4) *They act on gauge invariant nonlinearities like derivatives. If* $G(z) = F\left(|z|^2\right)z$ *is* C^1, *then*

$$J^\varepsilon(t)G(u) = \partial_z G(u)J^\varepsilon(t)u - \partial_{\bar z}G(u)\overline{J^\varepsilon(t)u},$$

$$H^\varepsilon(t)F(u) = \partial_z G(u)H^\varepsilon(t)u - \partial_{\bar z}G(u)\overline{H^\varepsilon(t)u}.$$

The last two points are direct consequences of the second. Remark that the third point implies that J^ε yields good L^p estimates away from focuses, while H^ε is better suited near focal points. This lemma shows that these operators enjoy similar properties to that of Killing vector-fields, whose use has proven efficient in the study of the nonlinear wave equation; see e.g. [Klainerman (1985)]. Before stating our main result, we point out the following formal approximation, which turns out to be not so formal during the proof:

$$J^\varepsilon(t) \underset{t\to\pi/2}{\sim} \frac{x}{\varepsilon} + i\left(t - \frac{\pi}{2}\right)\nabla \quad ; \quad H^\varepsilon(t) \underset{t\to\pi/2}{\sim} i\varepsilon\nabla.$$

Up to replacing the focusing time $t = 1$ by $t = \pi/2$, we retrieve the two operators used in Chap. 7. The geometrical interpretation is that near the focuses, sinusoids can be approximated by straight lines. The analytical interpretation is that near the focuses, the harmonic potential becomes negligible.

Theorem 8.2. *Let* $1 \leqslant d \leqslant 5$, $a_0 \in \Sigma$ *and* $\sigma > 1/2$ *with* $\sigma_0(d) \leqslant \sigma < \frac{2}{d-2}$, *where* $\sigma_0(d)$ *is defined in Theorem 7.16. Let* $k \in \mathbb{N}$. *Then the following asymptotics holds for* u^ε *when* $\pi/2 + (k-1)\pi < a \leqslant b < \pi/2 + k\pi$: *for all* $\mathcal{B}^\varepsilon \in \{\mathrm{Id}, J^\varepsilon, H^\varepsilon\}$,

$$\sup_{a\leqslant t\leqslant b} \left\|\mathcal{B}^\varepsilon(t)\left(u^\varepsilon(t) - \right.\right.$$

$$\left.\left. -\frac{e^{-idk\pi/2}}{|\cos t|^{d/2}}\left(\mathcal{F}\circ S^k \circ \mathcal{F}^{-1}\right)a_0\left(\frac{\cdot}{\cos t}\right)e^{-i|\cdot|^2\tan t/(2\varepsilon)}\right)\right\|_{L^2} \underset{\varepsilon\to 0}{\longrightarrow} 0,$$

where S^k stands for the kth iterate of the scattering operator S associated to the nonlinear Schrödinger equation

$$i\partial_t \psi + \frac{1}{2}\Delta\psi = |\psi|^{2\sigma}\psi. \tag{8.13}$$

At focal points, for all $\mathcal{A}^\varepsilon \in \{\mathrm{Id}, \frac{x}{\varepsilon}, \varepsilon\nabla_x\}$,

$$\left\| \mathcal{A}^\varepsilon \left(u^\varepsilon \left(\frac{\pi}{2} + k\pi\right) - \frac{e^{-id\pi/4 - idk\pi/2}}{\varepsilon^{d/2}} \left(W_- \circ S^k \circ \mathcal{F}^{-1}\right) a_0 \left(\frac{\cdot}{\varepsilon}\right) \right) \right\|_{L^2} \xrightarrow[\varepsilon \to 0]{} 0,$$

where W_- is the wave operator associated to (8.13) (see Theorem 7.14).

Remark 8.3. We can infer a similar result for $\widetilde{u}^\varepsilon$ solving

$$i\varepsilon\partial_t \widetilde{u}^\varepsilon + \frac{\varepsilon^2}{2}\Delta\widetilde{u}^\varepsilon = \frac{|x|^2}{2}\widetilde{u}^\varepsilon + \varepsilon^{d\sigma}|\widetilde{u}^\varepsilon|^{2\sigma}\widetilde{u}^\varepsilon \quad ; \quad \widetilde{u}^\varepsilon(0,x) = a_0(x)e^{ix\cdot\xi_0/\varepsilon},$$

for $\xi_0 \in \mathbb{R}^d$. Indeed, we check that

$$u^\varepsilon(t,x) = \widetilde{u}^\varepsilon\left(t, x + \xi_0 \sin t\right)e^{-i\left(x + \frac{\xi_0}{2}\sin t\right)\cdot\xi_0 \cos t/\varepsilon}$$

solves Eq. (8.7).

We now comment on Theorem 8.2. First, we assume $\sigma > 1/2$ so that the nonlinearity $z \mapsto |z|^{2\sigma}z$ is twice differentiable. Since on the other hand, we have to assume $\sigma < 2/(d-2)$, we suppose $d \leqslant 5$. If we consider the case $k = 0$ in Theorem 8.2, we see that the asymptotic behavior of u^ε is described by the right-hand side of (8.5), which is exactly the approximate solution provided by WKB analysis for the linear equation (8.3). This shows that for $|t| < \pi/2$, the nonlinearity is negligible at leading order in Eq. (8.7), and the geometry of the propagation is dictated by the harmonic oscillator. The same is true for any $k \in \mathbb{N}$: outside the focal points, nonlinear effects are negligible. On the other hand, nonlinear effects are relevant at leading order at every focus. Each caustic crossing is described by the Maslov index, plus a change in the amplitude, measured by the scattering operator associated to Eq. (8.13). From this respect, the result is quite similar to Proposition 7.19. It should be noticed though, that the harmonic oscillator is absent from the description of the caustic crossing. The explanation is the following: because of the influence of the harmonic potential, u^ε focuses at the origin as $t \to \pi/2$. For $t \approx \pi/2$, u^ε is concentrated at scale ε, in the same fashion as in Eq. (8.6) (but with a different profile). Therefore, the "right" space variable is x/ε, and not x. Writing

$$|x|^2 u^\varepsilon = \varepsilon^2 \left|\frac{x}{\varepsilon}\right|^2 u^\varepsilon,$$

this explains why the harmonic potential becomes negligible near the focus. Note finally that under our assumptions on a_0 and σ, $\left(\mathcal{F} \circ S^k \circ \mathcal{F}^{-1}\right) a_0$ is well defined as a function of Σ, for all $k \in \mathbb{N}$. Therefore, two dynamics dominate alternatingly in the behavior of u^ε: the linear dynamics of the harmonic oscillator outside the focal points, and the nonlinear dynamics of Eq. (8.13) near the focal points. It is remarkable that these dynamics act in a rather decoupled way.

Remark 8.4. A similar result was established by S. Ibrahim in the case of a nonlinear wave equation [Ibrahim (2004)]. In this case, the geometry of the propagation is not dictated by an external potential but by the fact that the space variable lies on a sphere. This causes several focusing phenomena, like in the present case, and for suitable scalings, each caustic crossing is described by a scattering operator.

In view of the next section, we restate Theorem 8.2 by considering $t = \pi/2$ as the initial time: after the change of variable $t \mapsto t - \pi/2$, we have

Corollary 8.5. *Suppose that $1 \leqslant d \leqslant 5$ and $\sigma > 1/2$. Let $\varphi \in \Sigma$, and $\sigma_0(d) \leqslant \sigma < \frac{2}{d-2}$. Assume that u^ε solves*

$$\begin{cases} i\varepsilon\partial_t u^\varepsilon + \dfrac{\varepsilon^2}{2}\Delta u^\varepsilon = \dfrac{|x|^2}{2}u^\varepsilon + \varepsilon^{d\sigma}\left|u^\varepsilon\right|^{2\sigma}u^\varepsilon, \\ u^\varepsilon(0,x) = \dfrac{1}{\varepsilon^{d/2}}\varphi\left(\dfrac{x}{\varepsilon}\right) + \dfrac{1}{\varepsilon^{d/2}}r^\varepsilon\left(\dfrac{x}{\varepsilon}\right), \end{cases} \tag{8.14}$$

with $\|r^\varepsilon\|_\Sigma \to 0$ as $\varepsilon \to 0$. Denote $\psi_\pm = W_\pm^{-1}\varphi$, where $W_\pm$ are the wave operators associated to Eq. (8.13). Then if $0 \leqslant \delta(r) < 1$, the following asymptotics hold in $L^2 \cap L^r$:

- *For $-\pi < t < 0$,* $\quad u^\varepsilon(t,x) \underset{\varepsilon \to 0}{\sim} \dfrac{e^{id\pi/4}}{|\sin t|^{d/2}} \widehat{\psi_-}\left(\dfrac{x}{\sin t}\right) e^{i|x|^2/(2\varepsilon \tan t)}.$

- *For $0 < t < \pi$,* $\quad u^\varepsilon(t,x) \underset{\varepsilon \to 0}{\sim} \dfrac{e^{-id\pi/4}}{(\sin t)^{d/2}} \widehat{\psi_+}\left(\dfrac{x}{\sin t}\right) e^{i|x|^2/(2\varepsilon \tan t)}.$

Remark 8.6. In [Nier (1996)], the author considers equations which can be compared to Eq. (8.7), that is

$$i\varepsilon\partial_t v^\varepsilon + \dfrac{\varepsilon^2}{2}\Delta v^\varepsilon = V(x)v^\varepsilon + U\left(\dfrac{x}{\varepsilon}\right)v^\varepsilon \ ; \ v^\varepsilon(0,x) = \dfrac{1}{\varepsilon^{d/2}}\varphi\left(\dfrac{x}{\varepsilon}\right), \tag{8.15}$$

where U is a short range potential. The potential V in that case cannot be the harmonic potential, for it has to be bounded as well as all its derivatives. In that paper, the author proves that under suitable assumptions, the influence of U occurs near $t = 0$ and is localized near the origin, while only the value $V(0)$ of V at the origin is relevant in this régime. For times $\varepsilon \ll |t| < T_*$, the situation is different: the potential U becomes negligible, while V dictates the propagation. As in our paper, the transition between these two régimes is measured by the scattering operator associated to U.

Our assumption $\sigma \geqslant \sigma_0(d) > 1/d$ makes the nonlinear term short range. With our scaling for the nonlinearity, this perturbation is relevant only near the focus, where the harmonic potential is negligible, while the opposite occurs for $\varepsilon \ll |t| < \pi$.

Remark 8.7. Since near the focal point, the description of the wave function u^ε does not involve the external potential at leading order, the phenomenon is the same as in Sec. 7.4. In particular, the discussion of Sec. 7.4.3 can be repeated: the conclusions of Proposition 7.23 remain valid in the present case, up to replacing the focusing time. So, the Cauchy problem for the propagation of Wigner measures is ill-posed.

To simplify the presentation, we sketch the argument of the proof of Theorem 8.2 at a formal level only. To justify these computations, the key point in [Carles (2003b)] consists in introducing the analogues of Proposition 7.8 and Corollary 7.9 (in space dimension $d \geqslant 1$), in two cases:

- With the same operator $U_0^\varepsilon(t) = e^{i\varepsilon \frac{t}{2} \Delta}$. This is because near the focal points, the harmonic potential can be neglected in Eq. (8.7).

- When U_0^ε is replaced by U^ε. Indeed, Eq. (8.4) shows that U^ε enjoys the same dispersive properties as U_0^ε, *locally in time* (and only locally in time, because $-\varepsilon^2 \Delta + |x|^2$ has eigenvalues). Therefore, Strichartz inequalities are available for U^ε. The only difference is that in the homogeneous Strichartz estimate, the time interval $\mathbb{R}$ must be replaced by a *finite* time interval, and in this case as well as in the inhomogeneous estimate (7.14), the constants C depend on the *finite* time interval I.

The organization of the proof of Theorem 8.2 is the following:

- Justify WKB analysis until t is as close as possible to $\pi/2$: this allows $t \leqslant \pi/2 - \Lambda \varepsilon$, in the limit $\Lambda \to +\infty$.
- Show that there exists a transition régime for $t = \pi/2 - \Lambda \varepsilon$, when $\Lambda \to +\infty$: the harmonic potential becomes negligible. We then approximate u^ε by essentially the same approximate solution as the one studied in the proof of Proposition 7.19.

- For $|t - \pi/2| \leqslant \Lambda\varepsilon$, we proceed as in the proof of Proposition 7.19. However, some extra source terms appear because of the harmonic potential. This forces us to consider smoother initial data and use a density argument relying on the global well-posedness for Eq. (8.13). This is where we need to assume that $z \mapsto |z|^{2\sigma}z$ is twice differentiable.
- For $t \geqslant \pi/2 + \Lambda\varepsilon$, we can repeat the analysis of the second, then of the first point.

Before the first focus

For $t \in [0, \pi/2[$, our natural candidate as an approximate solution is v^ε, solution of Eq. (8.3). We can also approximate v^ε, by

$$v^\varepsilon_{\text{app}}(t, x) = \frac{1}{(\cos t)^{d/2}} a_0 \left(\frac{x}{\cos t}\right) e^{-i|x|^2 \tan t/(2\varepsilon)}.$$

The error term $w^\varepsilon_{\text{lin}} = v^\varepsilon - v^\varepsilon_{\text{app}}$ satisfies

$$i\varepsilon\partial_t w^\varepsilon_{\text{lin}} + \frac{\varepsilon^2}{2}\Delta w^\varepsilon_{\text{lin}} = \frac{|x|^2}{2} w^\varepsilon_{\text{lin}} + \left(\frac{\varepsilon}{\cos t}\right)^2 \frac{e^{-i|x|^2 \tan t/(2\varepsilon)}}{2(\cos t)^{d/2}} (\Delta a_0) \left(\frac{x}{\cos t}\right),$$

along with the initial condition $w^\varepsilon_{\text{lin}}(0, \cdot) = 0$. Assume that a_0 is relatively smooth:

$$a_0 \in \mathcal{H} = \{f \in H^3(\mathbb{R}^d); \quad xf \in H^2(\mathbb{R}^d)\}.$$

The basic energy estimate of Lemma 1.2 then yields:

$$\|w^\varepsilon_{\text{lin}}(t)\|_{L^2} \leqslant \frac{1}{2} \int_0^t \frac{\varepsilon}{(\cos \tau)^2} \|\Delta a_0\|_{L^2} \, d\tau.$$

Apply J^ε to the equation satisfied by $w^\varepsilon_{\text{lin}}$:

$$i\varepsilon\partial_t J^\varepsilon w^\varepsilon_{\text{lin}} + \frac{\varepsilon^2}{2}\Delta J^\varepsilon w^\varepsilon_{\text{lin}} = \frac{|x|^2}{2} J^\varepsilon w^\varepsilon_{\text{lin}}$$
$$- i\cos t \left(\frac{\varepsilon}{\cos t}\right)^2 \frac{e^{-i|x|^2 \tan t/(2\varepsilon)}}{2(\cos t)^{d/2}} \nabla (\Delta a_0) \left(\frac{x}{\cos t}\right),$$

by the definition of J^ε given in Eq. (8.8). We infer

$$\|J^\varepsilon(t)w^\varepsilon_{\text{lin}}(t)\|_{L^2} \leqslant \frac{1}{2} \int_0^t \frac{\varepsilon}{(\cos \tau)^2} \|a_0\|_{H^3} \, d\tau.$$

Finally, apply H^ε to the equation satisfied by $w_{\text{lin}}^\varepsilon$:

$$i\varepsilon \partial_t H^\varepsilon w_{\text{lin}}^\varepsilon + \frac{\varepsilon^2}{2}\Delta H^\varepsilon w_{\text{lin}}^\varepsilon = \frac{|x|^2}{2}H^\varepsilon w_{\text{lin}}^\varepsilon$$

$$+x\cos t\left(\frac{\varepsilon}{\cos t}\right)^2 \frac{e^{-i|x|^2 \tan t/(2\varepsilon)}}{2(\cos t)^{d/2}}(\Delta a_0)\left(\frac{x}{\cos t}\right)$$

$$+i\varepsilon\sin t\left(\frac{\varepsilon}{\cos t}\right)^2 \nabla\left(\frac{e^{-i|x|^2 \tan t/(2\varepsilon)}}{2(\cos t)^{d/2}}(\Delta a_0)\left(\frac{x}{\cos t}\right)\right).$$

We deduce:

$$\|H^\varepsilon(t)w_{\text{lin}}^\varepsilon(t)\|_{L^2} \lesssim \int_0^t \frac{\varepsilon}{(\cos\tau)^2}\|xa_0\|_{H^2}\,d\tau + \int_0^t \frac{\varepsilon^2\sin\tau}{(\cos\tau)^3}\|a_0\|_{H^3}\,d\tau.$$

Therefore, if $a_0 \in \mathcal{H}$, we have:

$$\sum_{\mathcal{B}^\varepsilon \in \{\text{Id}, J^\varepsilon, H^\varepsilon\}} \|\mathcal{B}^\varepsilon(t)w_{\text{lin}}^\varepsilon(t)\|_{L^2} \lesssim \int_0^t \frac{\varepsilon}{(\cos\tau)^2}\,d\tau + \left(\frac{\varepsilon}{\cos t}\right)^2$$

$$\lesssim \frac{\varepsilon}{\pi/2 - t} + \left(\frac{\varepsilon}{\pi/2 - t}\right)^2.$$

A density argument yields:

Lemma 8.8. *Let $a_0 \in \Sigma$. We have*

$$\sum_{\mathcal{B}^\varepsilon \in \{\text{Id}, J^\varepsilon, H^\varepsilon\}} \limsup_{\varepsilon \to 0}\ \sup_{0 \leqslant t \leqslant \pi/2 - \Lambda\varepsilon} \left\|\mathcal{B}^\varepsilon(t)\left(v^\varepsilon(t) - v_{\text{app}}^\varepsilon(t)\right)\right\|_{L^2} \underset{\Lambda \to +\infty}{\longrightarrow} 0.$$

We now have to compare u^ε and v^ε. Let $w^\varepsilon = u^\varepsilon - v^\varepsilon$. It solves

$$i\varepsilon\partial_t w^\varepsilon + \frac{\varepsilon^2}{2}\Delta w^\varepsilon = \frac{|x|^2}{2}w^\varepsilon + \varepsilon^{d\sigma}|u^\varepsilon|^{2\sigma}u^\varepsilon.$$

The initial datum for w^ε is zero. In view of the analysis for $t > \pi/2$, notice that it suffices to assume

$$\sum_{\mathcal{B}^\varepsilon \in \{\text{Id}, J^\varepsilon, H^\varepsilon\}} \|\mathcal{B}^\varepsilon(0)w_{\text{lin}}^\varepsilon(0)\|_{L^2} \underset{\varepsilon \to 0}{\longrightarrow} 0.$$

It is this assumption which is needed when we iterate the argument, from $t \in [0, \pi]$ to $t \in [\pi, 2\pi]$, and so on (finitely many times). Proceeding the same way as in Sec. 7.4, thanks to the operator J^ε which provides sharp L^p estimates before the focus (sharp in term of the dependence upon t and ε), we can prove:

Lemma 8.9. *Let $a_0 \in \Sigma$, and $\sigma > \max(1/d, 2/(d+2))$, with $\sigma < 2/(d-2)$ if $d \geqslant 3$. We have*

$$\sum_{\mathcal{B}^\varepsilon \in \{\text{Id}, J^\varepsilon, H^\varepsilon\}} \limsup_{\varepsilon \to 0}\ \sup_{0 \leqslant t \leqslant \pi/2 - \Lambda\varepsilon} \left\|\mathcal{B}^\varepsilon(t)\left(u^\varepsilon(t) - v^\varepsilon(t)\right)\right\|_{L^2} \underset{\Lambda \to +\infty}{\longrightarrow} 0.$$

Matching linear and nonlinear régimes

Lemmas 8.8 and 8.9 show that up to $t = \pi/2 - \Lambda\varepsilon$, u^ε can be approximated by $v^\varepsilon_{\mathrm{app}}$ as $\varepsilon \to 0$, in the limit $\Lambda \to \infty$. In this régime, we have:

$$v^\varepsilon_{\mathrm{app}}(t,x) = \frac{1}{(\cos t)^{d/2}} a_0 \left(\frac{x}{\cos t}\right) e^{-i|x|^2 \tan t/(2\varepsilon)}$$

$$\approx \frac{1}{(\pi/2 - t)^{d/2}} a_0 \left(\frac{x}{\pi/2 - t}\right) e^{-i|x|^2/(2\varepsilon(\pi/2 - t))}.$$

Up to replacing $\pi/2$ by 1, we retrieve the same approximation as in Sec. 7.4, before the focus: this is a hint that the harmonic potential is becoming negligible at the approach of the focal point. Resume the notation

$$\psi_- = e^{-id\pi/4} \widehat{a_0}.$$

We can then prove:

Proposition 8.10. *Let $a_0 \in \Sigma$, and σ as in Lemma 8.9. We have*

$$\limsup_{\varepsilon \to 0} \left\| \mathcal{A}^\varepsilon_\Lambda \left(u^\varepsilon \left(\frac{\pi}{2} - \Lambda\varepsilon\right) - \frac{1}{\varepsilon^{d/2}} (U_0(-\Lambda)\psi_-) \left(\frac{\cdot}{\varepsilon}\right) \right) \right\|_{L^2} \xrightarrow[\Lambda \to +\infty]{} 0,$$

for all $\mathcal{A}^\varepsilon_\Lambda \in \{\mathrm{Id}, \varepsilon\nabla, \frac{x}{\varepsilon} - i\Lambda\varepsilon\nabla\}$.

Two things must be said about this proposition. First, we match the evolution of u^ε with a large time asymptotics for the free Schrödinger equation (without potential), after some rescaling, as in Sec. 7.4. This indicates that the harmonic potential loses its influence at leading order. In addition, we do not measure the error in terms of the operators J^ε and H^ε, but in terms of their counterparts used in Sec. 7.4. The two measurements are actually equivalent at leading order, and computations show that we can approximate the rays of geometric optics by lines, near $t = \pi/2$. See Fig. 8.2.

Inside the boundary layer

Inside the boundary layer in time about $t = \pi/2$, of width $\Lambda\varepsilon$, we neglect the harmonic potential. As in Sec. 7.4, introduce ψ solution to

$$i\partial_t \psi + \frac{1}{2}\Delta\psi = |\psi|^{2\sigma}\psi \quad ; \quad U_0(-t)\psi(t)\big|_{t=-\infty} = \psi_- := e^{-id\pi/4}\widehat{a_0}.$$

It scatters like $U_0(t)\psi_+$ as $t \to +\infty$, for some $\psi_+ \in \Sigma$, provided that $\sigma \geqslant \sigma_0(d)$ (Theorems 7.14 and 7.16). Rescale this function as follows:

$$u^\varepsilon_{\mathrm{app}}(t,x) = \frac{1}{\varepsilon^{d/2}} \psi\left(\frac{t - \pi/2}{\varepsilon}, \frac{x}{\varepsilon}\right).$$

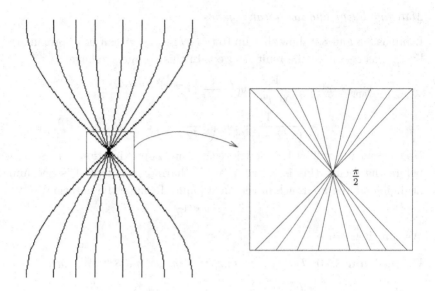

Fig. 8.2   Rays of geometric optics are straightened near $t = \frac{\pi}{2}$.

In [Carles (2003b)], the proof proceeds in two steps: first, it is shown that the exact solution u^ε can be truncated in a neighborhood of the origin of size ε^γ, for any $0 < \gamma < 1$. Then, this truncated solution is shown to be close to $u^\varepsilon_{\mathrm{app}}$. More precisely, let $\chi \in C_0^\infty(\mathbb{R}^d; \mathbb{R}_+)$, with $\chi = 1$ on $B(0,1)$ and $\mathrm{supp}\, \chi \subset B(0,2)$. For $\lambda \in]0, 1[$, let

$$u^\varepsilon_\lambda(t,x) = \chi\left(\frac{x}{\varepsilon^\lambda}\right) u^\varepsilon(t,x).$$

Introduce the error $w^\varepsilon_\lambda = u^\varepsilon - u^\varepsilon_\lambda$. It has small initial data at time $\pi/2 - \Lambda\varepsilon$, in the sense of Proposition 8.10, and solves

$$i\varepsilon \partial_t w^\varepsilon_\lambda + \frac{\varepsilon^2}{2}\Delta w^\varepsilon_\lambda = \frac{|x|^2}{2} w^\varepsilon_\lambda + \varepsilon^{d\sigma} |u^\varepsilon|^{2\sigma} w^\varepsilon_\lambda - \frac{\varepsilon^{2-\lambda}}{2}\nabla\chi\left(\frac{x}{\varepsilon^\lambda}\right) \cdot \nabla u^\varepsilon$$
$$- \varepsilon^{2-2\lambda}\Delta\chi\left(\frac{x}{\varepsilon^\lambda}\right) u^\varepsilon.$$

If $a_0 \in \mathcal{H}$, we can establish extra bounds on u^ε that make it possible to show that w^ε_λ, $J^\varepsilon w^\varepsilon_\lambda$ and $H^\varepsilon w^\varepsilon_\lambda$ are small in L^2. It is in proving that such a regularity is well propagated that we need to assume that the nonlinearity $z \mapsto |z|^{2\sigma}z$ is C^2, hence $\sigma > 1/2$. The last two terms of the right-hand side are then treated like small source terms. Here, we use Strichartz estimates associated to the group U^ε.

The second step consists in considering $\widetilde{w}_\lambda^\varepsilon = u_\lambda^\varepsilon - u_{\mathrm{app}}^\varepsilon$. In view of the identity

$$\chi(x) = \chi\left(\frac{x}{2}\right)\chi(x),$$

it solves:

$$i\varepsilon\partial_t \widetilde{w}_\lambda^\varepsilon + \frac{\varepsilon^2}{2}\Delta\widetilde{w}_\lambda^\varepsilon = \chi\left(\frac{x}{2\varepsilon^\lambda}\right)\frac{|x|^2}{2}u_\lambda^\varepsilon + \varepsilon^{d\sigma}\left(|u^\varepsilon|^{2\sigma}u_\lambda^\varepsilon - |u_{\mathrm{app}}^\varepsilon|^{2\sigma}u_{\mathrm{app}}^\varepsilon\right)$$

$$+ \frac{\varepsilon^{2-\lambda}}{2}\nabla\chi\left(\frac{x}{\varepsilon^\lambda}\right)\cdot\nabla u^\varepsilon + \varepsilon^{2-2\lambda}\Delta\chi\left(\frac{x}{\varepsilon^\lambda}\right)u^\varepsilon.$$

Here, we use Strichartz estimates associated to U_0^ε, and view the first term of the right-hand side as a small source term:

$$\chi\left(\frac{x}{2\varepsilon^\lambda}\right)\frac{|x|^2}{2} = \frac{\varepsilon^{2\lambda}}{2}\chi\left(\frac{x}{2\varepsilon^\lambda}\right)\left|\frac{x}{\varepsilon^\lambda}\right|^2 = \varepsilon^{2\lambda}\widetilde{\chi}\left(\frac{x}{2\varepsilon^\lambda}\right), \quad \text{where } \widetilde{\chi} \in C_0^\infty(\mathbb{R}^d; \mathbb{R}).$$

Then it is possible to factor out $\widetilde{w}_\lambda^\varepsilon$ in the difference of the two nonlinear terms, up to an extra, small, source term, which appears when replacing u^ε with u_λ^ε. The approximation $a_0 \in \mathcal{H}$ is removed by a density argument for $\sigma \geqslant \sigma_0(d)$, since (8.13) is globally well-posed in Σ. We can prove finally:

Proposition 8.11. *Under the assumptions of Theorem 8.2,*

$$\sum_{\mathcal{B}^\varepsilon \in \{\mathrm{Id}, J^\varepsilon, H^\varepsilon\}} \limsup_{\varepsilon \to 0} \sup_{|\pi/2 - t| \leqslant \Lambda\varepsilon} \left\|\mathcal{B}^\varepsilon(t)\left(u^\varepsilon(t) - u_{\mathrm{app}}^\varepsilon(t)\right)\right\|_{L^2} \xrightarrow[\Lambda \to +\infty]{} 0.$$

The proof of this result, barely sketched above, is fairly technical, and is not reproduced here for the sake of readability.

Past the first boundary layer

For $t = \pi/2 + \Lambda\varepsilon$, we can then prove the analogue of Proposition 8.10, with $-\Lambda$ replaced by $+\Lambda$, and ψ_- replaced by ψ_+. Then, we can mimic the proofs and results of Lemmas 8.8 and 8.9, by using the remark that u^ε and the solution of the linear equation don't have to match exactly at the initial time, but only up to a small error in L^2, when any of the operators Id, J^ε or H^ε acts on the difference. Using this remark again, we see that on $[\pi, 2\pi]$, we can repeat what was done on $[0, \pi]$. Hence Theorem 8.2.

8.2   General quadratic potentials

When the external potential $V = V(x)$ is time-independent and is exactly a polynomial of degree at most two, we can extend Corollary 8.5. We consider

$$\begin{cases} i\varepsilon\partial_t u^\varepsilon + \dfrac{\varepsilon^2}{2}\Delta u^\varepsilon = V u^\varepsilon + \varepsilon^{d\sigma}\,|u^\varepsilon|^{2\sigma}\,u^\varepsilon, \\[2mm] u^\varepsilon(0,x) = \dfrac{1}{\varepsilon^{d/2}}\varphi\left(\dfrac{x - x_0}{\varepsilon}\right)e^{ix\cdot\xi_0/\varepsilon}, \end{cases} \tag{8.16}$$

for some $x_0, \xi_0 \in \mathbb{R}^d$. In this paragraph, we make the following assumptions:

Assumption 8.12. We suppose that $\sigma_0(d) \leqslant \sigma < 2/(d-2)$, $\sigma > 1/2$, and therefore $1 \leqslant d \leqslant 5$.
The initial profile φ is in Σ.
The potential V is time-independent, and is exactly a polynomial of degree at most two:

$$\partial^3_{jk\ell}V \equiv 0, \quad \forall j,k,\ell \in \{1,\dots,d\}.$$

Introduce the Hamiltonian system with initial data (x_0, ξ_0):

$$\dot{x}(t) = \xi(t) \quad ; \quad \dot{\xi}(t) = -\nabla V(x(t)) \quad ; \quad x(0) = x_0 \quad ; \quad \xi(0) = \xi_0. \tag{8.17}$$

Note that under our assumption on V, $x(t)$ and $\xi(t)$ can be computed exactly (see Eq. (8.20) below). Introduce the solution ψ to

$$i\partial_t \psi + \frac{1}{2}\Delta\psi = |\psi|^{2\sigma}\psi \quad ; \quad \psi_{|t=0} = \varphi. \tag{8.18}$$

Assumption 8.12 and Theorem 7.16 show that there exist $\psi_\pm \in \Sigma$ such that

$$\|U_0(-t)\psi(t) - \psi_\pm\|_\Sigma \underset{t\to\pm\infty}{\longrightarrow} 0.$$

To state our main result, introduce the quantity

$$\Phi(t,x) = x\cdot\xi(t) - \frac{1}{2}\left(x(t)\cdot\xi(t) - x_0\cdot\xi_0\right).$$

Proposition 8.13. *Under Assumption 8.12, suppose in addition that V is of the form*

$$V(x) = \frac{1}{2}\sum_{j=1}^{d}\delta_j\omega_j^2 x_j^2 + \sum_{j=1}^{d}b_j x_j,$$

for some real constants b_j, where $\omega_j > 0$, $\delta_j \in \{-1,0,+1\}$, and $\delta_j b_j = 0$ for all j. There exists $T > 0$ independent of $\varepsilon \in\,]0,1]$ and $\varepsilon(T) > 0$ such that for $0 < \varepsilon \leqslant \varepsilon(T)$, Eq. (8.16) has a unique solution $u^\varepsilon \in C([-T,T];\Sigma)$.

In addition, its behavior on $[-T,T]$ *is given by the following régimes:*
(1) For any $\Lambda > 0$,

$$\limsup_{\varepsilon \to 0} \sup_{|t| \leqslant \Lambda \varepsilon} \Big(\big\| u^\varepsilon(t) - u^\varepsilon_{\text{app}}(t) \big\|_{L^2} + \big\| \varepsilon \nabla u^\varepsilon(t) - \varepsilon \nabla u^\varepsilon_{\text{app}}(t) \big\|_{L^2}$$

$$+ \Big\| \frac{x - x(t)}{\varepsilon} \big(u^\varepsilon(t) - u^\varepsilon_{\text{app}}(t) \big) \Big\|_{L^2} \Big) = 0,$$

where $u^\varepsilon_{\text{app}}(t,x) = \dfrac{1}{\varepsilon^{d/2}} \psi \left(\dfrac{t}{\varepsilon}, \dfrac{x}{\varepsilon} \right) e^{i\Phi(t,x)/\varepsilon}$, *and* ψ *solves Eq. (8.18).*
(2) Beyond this boundary layer, we have:

$$\limsup_{\varepsilon \to 0} \sup_{\Lambda \varepsilon \leqslant \pm t \leqslant T} \Big(\big\| u^\varepsilon(t) - v^\varepsilon_\pm(t) \big\|_{L^2} + \big\| \varepsilon \nabla u^\varepsilon(t) - \varepsilon \nabla v^\varepsilon_\pm(t) \big\|_{L^2}$$

$$+ \big\| (x - x(t)) \big(u^\varepsilon(t) - v^\varepsilon_\pm(t) \big) \big\|_{L^2} \Big) \xrightarrow[\Lambda \to +\infty]{} 0,$$

where $v^\varepsilon_\pm$ *solve the linear equations*

$$i\varepsilon \partial_t v^\varepsilon_\pm + \frac{\varepsilon^2}{2} \Delta v^\varepsilon_\pm = V v^\varepsilon_\pm \quad ; \quad v^\varepsilon_\pm(0,x) = \frac{1}{\varepsilon^{d/2}} \psi_\pm \left(\frac{x - x_0}{\varepsilon} \right) e^{ix \cdot \xi_0 / \varepsilon}.$$

Like in the case of the isotropic harmonic potential, we obtain a nonlinear extension of the result of [Nier (1996)] (see Remark 8.6). We compare this result with yet another problem at the end of this section, when the nonlinearity is focusing instead of defocusing. Proposition 8.13 is proved in [Carles and Miller (2004)]. We present the main steps of the proof only. Before doing so, we point out that in some cases, more can be said on the time T in Proposition 8.13. Essentially, if there is no refocusing at one point, then T can be taken arbitrarily large: the nonlinearity remains negligible off $\{t = 0\}$. On the other hand, if the potential V causes at least one of the solutions $v^\varepsilon_\pm$ to refocus at one point for $\pm t > 0$, then so does u^ε. Nonlinear effects then affect u^ε at leading order again, like in Sec. 8.1. Note that the refocusing phenomenon has to occur at one point: for any other caustic, the nonlinearity is subcritical. Only a focal point can ignite the nonlinearity $\varepsilon^{d\sigma} |u^\varepsilon|^{2\sigma} u^\varepsilon$ at leading order. More precisely, such refocusing can occur only if $\delta_j = +1$ for all j and the ω_j's are pairwise rationally dependent; see below.

We now explain why the assumption on the form of the potential made in Proposition 8.13 is not really one. Up to an orthonormal change of basis in $\mathbb{R}^d$ (which leaves the Laplacian invariant), we can assume that V is of the form

$$V(x) = \frac{1}{2} \sum_{j=1}^{d} \delta_j \omega_j^2 x_j^2 + \sum_{j=1}^{d} b_j x_j + c, \tag{8.19}$$

for some real constants b_j and c, where $\omega_j > 0$, and $\delta_j \in \{-1, 0, +1\}$. This form is obtained by diagonalizing the quadratic part of V. Up to completing the square and changing the origin, we may even assume $\delta_j b_j = 0$ for all j. By changing the origin, we may modify the form of the initial data, by a multiplicative factor $e^{ia/\varepsilon}$ for some constant $a \in \mathbb{R}$. Finally, we may assume that $a = c = 0$ by considering $u^\varepsilon(t, x)e^{i(ct+a)/\varepsilon}$ instead of $u^\varepsilon(t, x)$.

Note that the real numbers $\delta_j \omega_j^2/2$ are the eigenvalues of the quadratic part of V. Then refocusing at one point can happen only if $\delta_j = +1$ for all j and the ω_j's are pairwise rationally dependent:

$$\frac{\omega_j}{\omega_k} \in \mathbb{Q} \quad \forall j, k \in \{1, \ldots, d\}.$$

This case is very similar to that described in Sec. 8.1. Note that before refocusing at one point, there may be focusing on an affine space of dimension at least one.

Example 8.14. Assume $d = 2$ and $x_0 = \xi_0 = 0$. If $\omega_1 = 2\omega_2$ for instance, then at time $t = \pi/\omega_1$, u^ε focuses on the line $\{x_1 = 0\}$, and at time $t = 2\pi/\omega_1 = \pi/\omega_2$, u^ε refocuses at the origin. We have seen in Sec. 7.7 that the critical indexes for focusing on a line correspond to focusing at one point in space dimension one, $\alpha = \sigma$. Since we consider the case $\alpha = 2\sigma$, the nonlinearity is negligible when the wave focuses on the line.

When u^ε focuses on a set which is not reduced to a point, the only effect at leading order is linear, and is measured by the Maslov index. When u^ε focuses at one point, nonlinear effects are described in terms of scattering operator.

The first step in the proof of Proposition 8.13 consists in reducing the analysis to the case $x_0 = \xi_0 = 0$. This can be achieved because V is a polynomial of degree at most two, and time-independent. Indeed, solve the above Hamiltonian system when V has the form (8.19). Introduce

$$g_j(t) = \begin{cases} \dfrac{\sin(\omega_j t)}{\omega_j}, & \text{if } \delta_j = 1, \\ t, & \text{if } \delta_j = 0, \\ \dfrac{\sinh(\omega_j t)}{\omega_j}, & \text{if } \delta_j = -1. \end{cases} \quad ; \quad h_j(t) = \begin{cases} \cos(\omega_j t), & \text{if } \delta_j = 1, \\ 1, & \text{if } \delta_j = 0, \\ \cosh(\omega_j t), & \text{if } \delta_j = -1. \end{cases}$$

Then the solution of Eq. (8.17) is given by

$$x_j(t) = h_j(t)x_{0j} + g_j(t)\xi_{0j} - \frac{1}{2}b_j t^2,$$

$$\xi_j(t) = h_j(t)\xi_{0j} - \delta_j \omega_j^2 g_j(t)x_{0j} - b_j t. \tag{8.20}$$

Assume that u^ε solves Eq. (8.16). Introduce

$$\widetilde{u}^\varepsilon(t,x) = u^\varepsilon(t, x + x(t)) e^{-i\Phi(t, x + x(t))/\varepsilon}.$$

We check that $\widetilde{u}^\varepsilon$ solves Eq. (8.16) with $x_0 = \xi_0 = 0$. Without loss of generality, we therefore assume that u^ε solves Eq. (8.16) with $x_0 = \xi_0 = 0$.

A second consequence of the fact that V is polynomial is the existence of "nice" operators. In the case of a focal point at $(t, x) = (1, 0)$, we have used the operators $x/\varepsilon + i(t-1)\nabla$ and $\varepsilon\nabla$. In the case of the isotropic harmonic potential, we have used the operators J^ε and H^ε, defined in Eq. (8.8) and Eq. (8.10) respectively. The definition of "nice" operators is somehow summarized in Lemma 8.1. Essentially, they commute with the linear part of the equation, they act on gauge invariant nonlinearities like derivatives, and they provide sharp (for the limit $\varepsilon \to 0$) weighted Gagliardo–Nirenberg inequalities. We recall that the second point of Lemma 8.1 is just stated in order to infer the last two points without computations. In the present case, we introduce:

$$J^\varepsilon(t) := U^\varepsilon(t)\frac{x}{\varepsilon}U^\varepsilon(-t) \quad ; \quad H^\varepsilon(t) := U^\varepsilon(t)i\varepsilon\nabla_x U^\varepsilon(-t),$$

where we denote

$$U^\varepsilon(t) = \exp\left(-i\frac{t}{\varepsilon}\left(-\frac{\varepsilon^2}{2}\Delta + V\right)\right).$$

By computing commutators, we check:

$$\partial_t J^\varepsilon(t) = U^\varepsilon(t)i\nabla U^\varepsilon(-t) = \frac{1}{\varepsilon}H^\varepsilon(t) \quad ; \quad \partial_t H^\varepsilon(t) = -U^\varepsilon(t)\nabla V U^\varepsilon(-t).$$

Therefore,

$$\partial_t^2 J_j^\varepsilon(t) = -\frac{1}{\varepsilon}U^\varepsilon(t)\partial_j V U^\varepsilon(-t)$$

$$= -\delta_j \omega_j^2 U^\varepsilon(t)\frac{x_j}{\varepsilon}U^\varepsilon(-t) - \frac{b_j}{\varepsilon} = -\delta_j \omega_j^2 J_j^\varepsilon(t) - \frac{b_j}{\varepsilon}.$$

We thus have explicitly,

$$J_j^\varepsilon(t) = \frac{x_j}{\varepsilon}h_j(t) + ig_j(t)\partial_j - \frac{b_j}{2\varepsilon}t^2,$$

$$H_j^\varepsilon(t) = -\delta_j\omega_j^2 x_j g_j(t) + ih_j(t)\varepsilon\partial_j - b_j t. \tag{8.21}$$

Remark 8.15. In view of the definition *via* the group U^ε, the notations J^ε and H^ε may not seem consistent with Eqs. (8.9) and (8.10) (first expression) in the case where V is an isotropic harmonic potential. However, it is easily checked that Eq. (8.21) agrees with the expressions (8.8) and (8.10) (second expression). The explanation lies in the fact that we have changed the origin of time, from 0 to $\pi/2$, if we compare with Sec. 8.1. More precisely, we check we following identities:

$$U^\varepsilon_{\text{i.h.}}(t)\left(\frac{1}{i}\nabla\right)U^\varepsilon_{\text{i.h.}}(-t) = U^\varepsilon_{\text{i.h.}}\left(t-\frac{\pi}{2}\right)\frac{x}{\varepsilon}U^\varepsilon_{\text{i.h.}}\left(\frac{\pi}{2}-t\right),$$

$$U^\varepsilon_{\text{i.h.}}(t)xU^\varepsilon_{\text{i.h.}}(-t) = U^\varepsilon_{\text{i.h.}}\left(t-\frac{\pi}{2}\right)(i\varepsilon\nabla)U^\varepsilon_{\text{i.h.}}\left(\frac{\pi}{2}-t\right),$$

where $U^\varepsilon_{\text{i.h.}}$ is the group associated to the semi-classical isotropic harmonic potential

$$U^\varepsilon_{\text{i.h.}}(t) = \exp\left(-i\frac{t}{2\varepsilon}\left(-\varepsilon^2\Delta + |x|^2\right)\right).$$

These identities are consequences of the fact that the harmonic oscillator rotates the phase space at angular velocity one (see Eq. (8.2)), so after $\pi/2$ time units, x has become $-\xi$, and ξ has become x.

These operators inherit interesting properties which we list below.

Lemma 8.16. *The operators J^ε and H^ε satisfy the following properties.*
(1) *They commute with the linear part of (8.16).*
(2) *Denote*

$$\phi_1(t,x) := \frac{1}{2}\sum_{k=1}^d\left(\frac{h_k(t)}{g_k(t)}x_k^2 - b_k tx_k - \frac{t^3}{12}b_k^2\right),$$

$$\phi_2(t,x) := -\frac{1}{2}\sum_{k=1}^d\left(\delta_k\omega_k^2\frac{g_k(t)}{h_k(t)}x_k^2 + 2b_k tx_k + \frac{t^3}{3}b_k^2\right).$$

Then ϕ_1 and ϕ_2 are well-defined and smooth for almost all t, and

$$J_j^\varepsilon(t) = ig_j(t)e^{i\phi_1(t,x)/\varepsilon}\partial_j\left(e^{-i\phi_1(t,x)/\varepsilon}\,\cdot\,\right),$$
$$H_j^\varepsilon(t) = i\varepsilon h_j(t)e^{i\phi_2(t,x)/\varepsilon}\partial_j\left(e^{-i\phi_2(t,x)/\varepsilon}\,\cdot\,\right).$$

(8.22)

(3) *Let $r \geqslant 2$ such that $\delta(r) < 1$. Define $P^\varepsilon(t)$ by*

$$P^\varepsilon(t) := \prod_{j=1}^d\left(|g_j(t)| + \varepsilon|h_j(t)|\right)^{1/d}.$$

There exists C_r such that, for any $u \in \Sigma$,

$$\|u\|_{L^r} \le \frac{C_r}{P^\varepsilon(t)^{\delta(r)}} \|u\|_{L^2}^{1-\delta(r)} \left(\|J^\varepsilon(t)u\|_{L^2} + \|H^\varepsilon(t)u\|_{L^2} \right)^{\delta(r)}. \qquad (8.23)$$

(4) *The operators J^ε and H^ε act on gauge invariant nonlinearities like derivatives (see the last point of Lemma 8.1).*

The first point stems from the definition of the operators J^ε and H^ε. The second point can easily be checked, and the last two points are direct consequences of the second.

Remark 8.17 (Avron–Herbst formula). *In the form* (8.19), *we could have assumed $b_j = 0$ for all j. As noticed in [Carles and Nakamura (2004)], since the nonlinearity is gauge invariant, Avron–Herbst formula reduces the case $\delta_j b_j = 0$ to the case $b_j = 0$. For $b = (b_1, \ldots, b_n)$, we check that $\widetilde{u}^\varepsilon$ solves Eq.* (8.16) *with $b_j = 0$ for all j in Eq.* (8.19), *where $\widetilde{u}^\varepsilon$ is given by:*

$$\widetilde{u}^\varepsilon(t,x) = u^\varepsilon\left(t, x - \frac{t^2}{2}b\right) \exp\left(i\left(tb \cdot x - \frac{t^3}{3}|b|^2\right)/\varepsilon\right).$$

See also [Carles (2011)] for a similar formula when b depends on time.

We can now explain why, unless all the ω_j's are pairwise rationally dependent, there is no refocusing at one point off $\{t = 0\}$. Since $d\sigma > 1$, the geometry of the propagation is the same as in the linear case: nonlinear effects become relevant only near focal points. We can then use the estimate provided by Eq. (8.23). Let v^ε solve

$$i\varepsilon\partial_t v^\varepsilon + \frac{\varepsilon^2}{2}\Delta v^\varepsilon = V v^\varepsilon \quad ; \quad v^\varepsilon(0, x) = \frac{1}{\varepsilon^{d/2}}\psi_0\left(\frac{x}{\varepsilon}\right).$$

From the first point of Lemma 8.16, we infer that not only the L^2 norm of v^ε is time independent, but also the L^2 norms of $J^\varepsilon v^\varepsilon$ and $H^\varepsilon v^\varepsilon$. Therefore, if $0 \le \delta(r) < 1$, the following estimate holds uniformly in time:

$$\|v^\varepsilon(t)\|_{L^r} \lesssim P^\varepsilon(t)^{-\delta(r)}.$$

The case of a focal point corresponds to the situation where

$$\|v^\varepsilon(t)\|_{L^r} \approx \varepsilon^{-\delta(r)}.$$

In view of the definition of the weight P^ε, this shows that there is focusing at one point $(t, x) = (t_0, x_0)$ if and only if $g_j(t_0) = 0$ for all j. Notice that whenever $g_j(t) = 0$, $h_j(t) = 1$. Of course, Eq. (8.20) shows that $g_j(t_0) = 0$ for all j and $t_0 \ne 0$ is possible only if $\delta_j = +1$ for all j: the quadratic part of V is positive definite. Finally, $g_j(t_0) = 0$ for all j and $t_0 \ne 0$ if and only if $\omega_j t_0 \in \pi\mathbb{Z}$ for all j, which in turn is equivalent to the property announced above.

After these preliminary reductions and properties, the proof of Proposition 8.13 becomes very similar to the proof of Theorem 8.2. We first show that for $|t| \leqslant \Lambda \varepsilon$, we can neglect the external potential (recall that up to a phase shift of the form $e^{ict/\varepsilon}$ in u^ε, we have assumed $V(0) = 0$). To do this, we can assume that φ is in the Schwartz class, rather than just $\varphi \in \Sigma$ (this is where we assume that $z \mapsto |z|^{2\sigma}z$ is C^2). Then, the global well-posedness for Eq. (8.18) makes it possible to use a density argument. This yields

$$\limsup_{\varepsilon \to 0} \sup_{|t| \leqslant \Lambda\varepsilon} \Big(\left\| u^\varepsilon(t) - u_{\mathrm{app}}^\varepsilon(t) \right\|_{L^2} + \left\| J^\varepsilon(t) \left(u^\varepsilon(t) - u_{\mathrm{app}}^\varepsilon(t) \right) \right\|_{L^2}$$
$$+ \left\| H^\varepsilon(t) \left(u^\varepsilon(t) - u_{\mathrm{app}}^\varepsilon(t) \right) \right\|_{L^2} \Big) = 0.$$

Note also that it is equivalent to use the operators J^ε and H^ε or the operators $x/\varepsilon + it\nabla$ and $\varepsilon\nabla$, as we have seen in the case of the isotropic harmonic potential; see Fig. 8.2. To see that this implies the first point of Proposition 8.13, recall that we have assumed $x(t) \equiv 0$, and that if we also suppose $b = 0$, then

$$\begin{pmatrix} J_j^\varepsilon \\ H_j^\varepsilon \end{pmatrix} = \begin{pmatrix} h_j & g_j/\varepsilon \\ -\varepsilon\delta_j\omega_j^2 g_j & h_j \end{pmatrix} \begin{pmatrix} x_j/\varepsilon \\ i\varepsilon\partial_j \end{pmatrix}.$$

The determinant of the above matrix is $h_j^2 + \delta_j\omega_j^2 g_j^2 \equiv 1$, and we have

$$\frac{x_j}{\varepsilon} = h_j(t)J_j^\varepsilon(t) - \frac{g_j(t)}{\varepsilon}H_j^\varepsilon(t) \quad ; \quad i\varepsilon\partial_j = \varepsilon\delta_j\omega_j^2 g_j(t)J_j^\varepsilon(t) + h_j(t)H_j^\varepsilon(t).$$

Since $g_j(t) = \mathcal{O}(t)$ as t goes to zero, the first point of Proposition 8.13 follows.

Suppose $t > 0$, since the case $t < 0$ is similar. For the matching region $\{t = \Lambda\varepsilon\}$, we can show that v_+^ε can be approximated by $\widetilde{v}_+^\varepsilon$, solutions to the free equation (recall that we now assume $x_0 = \xi_0 = 0$):

$$i\varepsilon\partial_t \widetilde{v}_+^\varepsilon + \frac{\varepsilon^2}{2}\Delta\widetilde{v}_+^\varepsilon = 0 \quad ; \quad \widetilde{v}_+^\varepsilon(0, x) = \frac{1}{\varepsilon^{d/2}}\psi_+\left(\frac{x}{\varepsilon}\right).$$

We can prove that for all $\Lambda > 0$,

$$\sum_{\mathcal{B}^\varepsilon \in \{\mathrm{Id}, J^\varepsilon, H^\varepsilon\}} \limsup_{\varepsilon \to 0} \sup_{0 \leqslant t \leqslant \Lambda\varepsilon} \left\| \mathcal{B}^\varepsilon(t) \left(v_+^\varepsilon(t) - \widetilde{v}_+^\varepsilon(t) \right) \right\|_{L^2} = 0.$$

This means that on the linear level, V is negligible for $0 \leqslant t \leqslant \Lambda\varepsilon$. Lemma 7.20 shows that $u_{\mathrm{app}}^\varepsilon$ and $\widetilde{v}_+^\varepsilon$ match at time $t = \Lambda\varepsilon$ in the limit $\varepsilon \to 0$ for large Λ:

$$\sum_{\mathcal{B}^\varepsilon \in \{\mathrm{Id}, J^\varepsilon, H^\varepsilon\}} \limsup_{\varepsilon \to 0} \left\| \mathcal{B}^\varepsilon(\Lambda\varepsilon) \left(u_{\mathrm{app}}^\varepsilon(\Lambda\varepsilon) - \widetilde{v}_+^\varepsilon(\Lambda\varepsilon) \right) \right\|_{L^2} \xrightarrow[\Lambda \to +\infty]{} 0.$$

Gathering all the informations together, we infer

$$\sum_{\mathcal{B}^\varepsilon \in \{\mathrm{Id}, J^\varepsilon, H^\varepsilon\}} \limsup_{\varepsilon \to 0} \left\| \mathcal{B}^\varepsilon(\Lambda\varepsilon) \left(u^\varepsilon(\Lambda\varepsilon) - v_+^\varepsilon(\Lambda\varepsilon) \right) \right\|_{L^2} \xrightarrow[\Lambda \to +\infty]{} 0.$$

Past this boundary layer, we can show that the nonlinearity remains negligible in the limit $\varepsilon \to 0$, essentially thanks to the operator J^ε, like in the case of the isotropic harmonic potential. The details can be found in [Carles and Miller (2004)].

To conclude this paragraph, and to prepare the next one, we compare the results of Proposition 8.13 with some others, concerning the case of a focusing nonlinearity:

$$\begin{cases} i\varepsilon\partial_t u^\varepsilon + \dfrac{\varepsilon^2}{2}\Delta u^\varepsilon = V u^\varepsilon - \varepsilon^{d\sigma}|u^\varepsilon|^{2\sigma} u^\varepsilon, \\[2mm] u^\varepsilon(0, x) = \dfrac{1}{\varepsilon^{d/2}} Q\left(\dfrac{x - x_0}{\varepsilon} \right) e^{ix\cdot\xi_0/\varepsilon}. \end{cases} \tag{8.24}$$

Now, $V = V(x)$ is not supposed to be exactly a polynomial. Here, Q is the unique positive, radially symmetric ([Kwong (1989)]), solution of:

$$-\frac{1}{2}\Delta Q + Q = |Q|^{2\sigma} Q.$$

The problem (8.24) was introduced in [Bronski and Jerrard (2000)]. This first result was then refined in [Keraani (2002, 2006)]. The focusing nonlinearity is an obstruction to dispersive phenomena, which were measured by the presence of the scattering operator in the previous defocusing case. The solution u^ε is expected to keep the ground state Q as a leading order profile. Nevertheless, the point where it is centered in the phase space, initially (x_0, ξ_0), should evolve according to the Hamiltonian flow. In the absence of external potential, $V \equiv 0$, we have explicitly:

$$u^\varepsilon(t, x) = \frac{1}{\varepsilon^{d/2}} Q\left(\frac{x - x(t)}{\varepsilon} \right) e^{ix\cdot\xi(t)/\varepsilon + i\theta(t)/\varepsilon},$$

where $(x(t), \xi(t)) = (x_0 + t\xi_0, \xi_0)$ solves the Hamiltonian system with initial data (x_0, ξ_0), and $\theta(t) = t - t|\xi_0|^2/2$. When V is not trivial, seek u^ε of the form of a rescaled WKB expansion:

$$u^\varepsilon(t, x) \underset{\varepsilon \to 0}{\sim} \frac{1}{\varepsilon^{d/2}} \left(\sum_{j \geqslant 0} \varepsilon^j U_j \left(\frac{t}{\varepsilon}, \frac{x - x(t)}{\varepsilon} \right) \right) e^{i\phi(t,x)/\varepsilon}.$$

Plugging this expansion into Eq. (8.24) and canceling the $\mathcal{O}(\varepsilon^0)$ term, we get:

$$i\partial_t U_0 + \frac{1}{2}\Delta U_0 + U_0 \left(-\partial_t \phi - \frac{1}{2}|\nabla\phi|^2 - V + |U_0|^{2\sigma} \right) - i\left(\dot{x}(t) - \nabla\phi \right)\cdot\nabla U_0 = 0.$$

Impose the leading order profile to be the standing wave given by

$$U_0(t, x) = e^{it} Q(x).$$

Then the above equation becomes:

$$U_0 \left(-\partial_t \phi - \frac{1}{2} |\nabla \phi|^2 - V \right) - i \left(\dot{x}(t) - \nabla \phi \right) \cdot \nabla U_0 = 0.$$

Since $U_0 e^{-it}$ is real-valued, and since we seek a real-valued phase ϕ, this yields:

$$\partial_t \phi + \frac{1}{2} |\nabla \phi|^2 + V = 0 \quad ; \quad \phi(0, x) = x \cdot \xi_0.$$

$$\dot{x}(t) = \nabla \phi(t, x).$$

The first equation is the eikonal equation. We infer that we have exactly

$$\nabla \phi \left(t, x(t) \right) = \xi(t).$$

See Lemma 1.5. The form of U_0 and the exponential decay of Q show that we can formally assume that $x = x(t) + \mathcal{O}(\varepsilon)$. In this case,

$$\nabla \phi(t, x) = \nabla \phi \left(t, x(t) \right) + \mathcal{O}(\varepsilon) = \xi(t) + \mathcal{O}(\varepsilon) = \dot{x}(t) + \mathcal{O}(\varepsilon).$$

Thus, we have canceled the $\mathcal{O}(\varepsilon^0)$ term, up to adding extra terms of order ε, that would be considered in the next step of the analysis, which we stop here. Back to u^ε, this formal computation yields

$$u^\varepsilon(t, x) \sim \frac{1}{\varepsilon^{d/2}} Q \left(\frac{x - x(t)}{\varepsilon} \right) e^{i\phi(t,x)} \sim \frac{1}{\varepsilon^{d/2}} Q \left(\frac{x - x(t)}{\varepsilon} \right) e^{ix \cdot \xi(t)/\varepsilon + i\theta(t)/\varepsilon},$$

where $\theta(t) = t \left(1 - |\xi_0|^2/2 - V(x_0) \right) + \int_0^t x(s) \cdot \nabla V(x(s)) ds$.

To give the above formal analysis a rigorous justification, the following assumptions are made in [Keraani (2006)]:

Assumption 8.18. The nonlinearity is L^2-subcritical: $\sigma < 2/d$. The potential $V = V(x)$ is real-valued, and can be written as $V = V_1 + V_2$, where

- $V_1 \in W^{3,\infty}(\mathbb{R}^d)$.
- $\partial^\alpha V_2 \in W^{2,\infty}(\mathbb{R}^d)$ for every multi-index α with $|\alpha| = 2$.

For instance, V can be a polynomial of degree at most two ($V_1 = 0$).

Theorem 8.19 ([Keraani (2006)]). *Let* $x_0, \xi_0 \in \mathbb{R}^d$. *Under Assumption 8.18, the solution* u^ε *to Eq. (8.24) can be approximated as follows:*

$$u^\varepsilon(t,x) = \frac{1}{\varepsilon^{d/2}} Q\left(\frac{x - x(t)}{\varepsilon}\right) e^{ix \cdot \xi(t)/\varepsilon + i\theta^\varepsilon(t)/\varepsilon} + \mathcal{O}(\varepsilon) \quad in \ L^\infty_{\text{loc}}(\mathbb{R}_t; H^1_\varepsilon),$$

where $(x(t), \xi(t))$ *is given by the Hamiltonian flow, the real-valued function* θ^ε *depends on* t *only, and* H^1_ε *is defined in Assumption 7.27.*

Remark 8.20. The assumption $\sigma < 2/d$ is crucial for the above result to hold. Indeed, if $\sigma = 2/d$, $V(x) = |x|^2/2$ is an isotropic harmonic potential, and $x_0 = \xi_0 = 0$, then (see [Carles (2002); Keraani (2006)]):

$$u^\varepsilon(t,x) = \frac{1}{(\cos t)^{d/2}} Q\left(\frac{x}{\varepsilon \cos t}\right) e^{i\frac{\tan t}{\varepsilon} - i\frac{|x|^2}{2\varepsilon}\tan t}, \quad 0 \leqslant t < \frac{\pi}{2}.$$

The proof of the above result heavily relies on the orbital stability of the ground state, which holds when $\sigma < 2/d$. For $v \in H^1(\mathbb{R}^d)$, denote

$$\mathcal{E}(v) = \frac{1}{2}\|\nabla v\|^2_{L^2} - \frac{1}{\sigma+1}\|v\|^{2\sigma+2}_{L^{2\sigma+2}}.$$

The ground state Q is the unique solution, up to translation and rotation, to the minimization problem:

$$\mathcal{E}(Q) = \inf\{\mathcal{E}(v) \ ; \ v \in H^1(\mathbb{R}^d) \text{ and } \|v\|_{L^2} = \|Q\|_{L^2}\}.$$

The orbital stability is given by the following result:

Proposition 8.21 ([Weinstein (1985)]). *Let* $\sigma < 2/d$. *There exist* $C, h > 0$ *such that if* $\phi \in H^1(\mathbb{R}^d)$ *is such that* $\|\phi\|_{L^2} = \|Q\|_{L^2}$ *and* $\mathcal{E}(\phi) - \mathcal{E}(Q) < h$, *then:*

$$\inf_{y \in \mathbb{R}^d, \theta \in \mathbb{T}} \left\|\phi - e^{i\theta}Q(\cdot - y)\right\|^2_{H^1} \leqslant C\left(\mathcal{E}(\phi) - \mathcal{E}(Q)\right).$$

The strategy in [Keraani (2006)] consists in applying the above result to the function

$$v^\varepsilon(t,x) = u^\varepsilon(t, \varepsilon x + x(t)) e^{-i(\varepsilon x + x(t)) \cdot \xi(t)/\varepsilon}.$$

The proof eventually relies on Gronwall lemma and a continuity argument. In order to invoke these arguments, S. Keraani uses Proposition 8.21 and the scheme of the proof of [Bronski and Jerrard (2000)], based on duality arguments and estimates on measures. This yields the result on a time interval $[-T_0, T_0]$. Since this T_0 depends only on constants of the motion, the argument can be repeated, to get the L^∞_{loc} estimate of Theorem 7.33.

In the particular case where the external potential V is an harmonic potential (isotropic or anisotropic), the proof can be simplified. We invite the reader to pay attention to the short note [Keraani (2005)], where this simplification is available.

There are several differences between Proposition 8.13 and Theorem 8.19. First, the profile is very particular in Theorem 8.19: it has to be the ground state. Moreover, in order for the orbital stability property to hold, the assumption $\sigma < 2/d$ is necessary. The assumption is not inconsistent with the requirement $\sigma_0(d) \leqslant \sigma < 2/(d-2)$ in Proposition 8.13, since $\sigma_0(d) < 2/d$. On the other hand, the external potential V does not have to be exactly a polynomial in Theorem 8.19. In the next paragraph, we discuss the validity of Proposition 8.13 when V is a subquadratic potential.

8.3   About general subquadratic potentials

Consider the case of a more general external potential, with initial data concentrated at the origin in the phase space:

$$i\varepsilon\partial_t u^\varepsilon + \frac{\varepsilon^2}{2}\Delta u^\varepsilon = V u^\varepsilon + \varepsilon^{d\sigma}|u^\varepsilon|^{2\sigma}u^\varepsilon \; ; \; u^\varepsilon(0,x) = \frac{1}{\varepsilon^{d/2}}\varphi\left(\frac{x}{\varepsilon}\right), \quad (8.25)$$

where $V = V(t,x)$ may depend on time. We have seen that for several issues, it is convenient to work with subquadratic potentials. Such a property was more than useful to solve the eikonal equation globally in space (see Proposition 1.9). Moreover, this assumption is fairly natural to study the nonlinear Cauchy problem (fix $\varepsilon > 0$ in Eq. (8.25)), to construct strong solutions (see Sec. 1.4.2), as well as to construct mild solutions thanks to local in time Strichartz estimates, in view of the results in [Fujiwara (1979)] (see Sec. 1.4.3). Suppose that V is smooth and subquadratic:

$$V \in C^\infty(\mathbb{R} \times \mathbb{R}^d), \quad \partial_x^\alpha V \in C(\mathbb{R}; L^\infty(\mathbb{R}^d)), \; \forall|\alpha| \geqslant 2.$$

The first point in Proposition 8.13 can be adapted to this more general framework. Up to considering $u^\varepsilon(t,x)e^{i\int_0^t V(\tau,0)d\tau/\varepsilon}$ instead of $u^\varepsilon(t,x)$, we may assume that $V(t,0) = 0$. Since for $|t| \leqslant \Lambda\varepsilon$, the curvature of rays of geometric optics is negligible, we can prove that for all $\Lambda > 0$,

$$\sum_{\mathcal{A}^\varepsilon \in \{\mathrm{Id}, x/\varepsilon + it\nabla, \varepsilon\nabla\}} \limsup_{\varepsilon \to 0} \sup_{0 \leqslant t \leqslant \Lambda\varepsilon} \left\|\mathcal{A}^\varepsilon(t)\left(u^\varepsilon(t) - u^\varepsilon_{\mathrm{app}}(t)\right)\right\|_{L^2} = 0,$$

for the same $u^\varepsilon_{\mathrm{app}}$ as in Proposition 8.13 (with $\Phi \equiv 0$). By assuming first that $\varphi \in \mathcal{S}(\mathbb{R}^d)$, we conclude as in Proposition 8.13. Moreover, the transition as $\Lambda \to \infty$ occurs in the same way.

Problems arise when trying to prove the analogue of the second régime in Proposition 8.13. Assume for simplicity that V is time-independent, $V = V(x)$. In all the previous cases, the proof that the nonlinearity is negligible for $\varepsilon \ll |t| \leqslant T$ relies on the Heisenberg derivative

$$J^\varepsilon(t) = U^\varepsilon(t)\frac{x}{\varepsilon}U^\varepsilon(-t), \quad \text{where } U^\varepsilon(t) = \exp\left(-i\frac{t}{\varepsilon}\left(-\frac{\varepsilon^2}{2}\Delta + V\right)\right).$$

We start with the good news:

Lemma 8.22. *Suppose that V is time-independent, and subquadratic. The operator J^ε defined above satisfies the following properties:*
(1) It commutes with the linear part of Eq. (8.25).
(2) Weighted Gagliardo–Nirenberg inequalities are available: there exists $\delta > 0$ independent of ε, such that the following holds. If $0 \leqslant \delta(r) < 1$, then there exists C_r such that for all $u \in \Sigma$,

$$\|u\|_{L^r} \leqslant \frac{C_r}{|t|^{\delta(r)}}\|u\|_{L^2}^{1-\delta(r)}\|J^\varepsilon(t)u\|_{L^2}^{\delta(r)}, \quad |t| \leqslant \delta.$$

Proof. The first point stems from the definition of J^ε. For the second point, we use the local dispersive estimate established in [Fujiwara (1979)]: there exist C and $\delta > 0$ independent of ε such that as soon as $|t| \leqslant \delta$,

$$\|U^\varepsilon(t)\|_{L^1 \to L^\infty} \leqslant \frac{C}{(\varepsilon|t|)^{d/2}}.$$

Since $U^\varepsilon(t)$ is unitary on L^2, interpolation yields, if $0 \leqslant \delta(r) < 1$:

$$\|U^\varepsilon(t)f\|_{L^r} \lesssim |\varepsilon t|^{-\delta(r)}\|f\|_{L^{r'}}, \quad |t| \leqslant \delta, \ \forall f \in L^{r'}.$$

Let $g^\varepsilon(t,x) = U^\varepsilon(-t)u(x)$. We have

$$\|U^\varepsilon(t)g^\varepsilon(t)\|_{L^r} \lesssim |\varepsilon t|^{-\delta(r)}\|g^\varepsilon(t)\|_{L^{r'}}.$$

Let $\lambda > 0$, and write

$$\|g^\varepsilon(t)\|_{L^{r'}}^{r'} = \int_{|x| \leqslant \lambda} |g^\varepsilon(t,x)|^{r'}dx + \int_{|x| > \lambda} |g^\varepsilon(t,x)|^{r'}dx.$$

Estimate the first term by Hölder's inequality, by writing $|g^\varepsilon|^{r'} = 1 \times |g^\varepsilon|^{r'}$,

$$\int_{|x| \leqslant \lambda} |g^\varepsilon(t,x)|^{r'}dx \lesssim \lambda^{d/p'}\left(\int_{|x| \leqslant \lambda} |g^\varepsilon(t,x)|^{r'p}dx\right)^{1/p},$$

and choose $p = 2/r'(\geqslant 1)$. Estimate the second term by the same Hölder's inequality, after inserting the factor x as follows,

$$\int_{|x|>\lambda} |g^\varepsilon(t,x)|^{r'} dx = \int_{|x|>\lambda} |x|^{-r'} |x|^{r'} |g^\varepsilon(t,x)|^{r'} dx$$

$$\leqslant \left(\int_{|x|>\lambda} |x|^{-r'p'} dx \right)^{1/p'} \left(\int_{|x|>\lambda} |xg^\varepsilon(t,x)|^2 dx \right)^{1/p}$$

$$\lesssim \lambda^{d/p'-r'} \|xg^\varepsilon(t,x)\|_{L^2}^{2/p}.$$

In summary, we have the following estimate, for any $\lambda > 0$:

$$\|g^\varepsilon(t)\|_{L^{r'}} \lesssim \lambda^{d/(p'r')} \|g^\varepsilon(t)\|_{L^2} + \lambda^{d/(p'r')-1} \|xg^\varepsilon(t,x)\|_{L^2}.$$

Notice that $d/(p'r') = \delta(r)$, and equalize both terms of the right-hand side:

$$\lambda = \frac{\|xg^\varepsilon(t,x)\|_{L^2}}{\|g^\varepsilon(t)\|_{L^2}}.$$

This yields $\|g^\varepsilon(t)\|_{L^{r'}} \lesssim \|g^\varepsilon(t)\|_{L^2}^{1-\delta(r)} \|xg^\varepsilon(t,x)\|_{L^2}^{\delta(r)}$. Therefore,

$$\|U^\varepsilon(t)g^\varepsilon(t)\|_{L^r} \lesssim |\varepsilon t|^{-\delta(r)} \|g^\varepsilon(t)\|_{L^2}^{1-\delta(r)} \|xg^\varepsilon(t,x)\|_{L^2}^{\delta(r)}$$

$$\lesssim |t|^{-\delta(r)} \|g^\varepsilon(t)\|_{L^2}^{1-\delta(r)} \left\| \frac{x}{\varepsilon} g^\varepsilon(t,x) \right\|_{L^2}^{\delta(r)}.$$

Back to u, write $g^\varepsilon(t) = U^\varepsilon(-t)u$. This completes the proof of the lemma, since $U^\varepsilon(t)$ is unitary on L^2.   □

Recall that in applying Lemma 8.16, we have invoked the first point and the last two points only. The second point was used to infer the last two points. From the above lemma, the operator J^ε that we now consider also satisfies the first point and an analogue of the third point of Lemma 8.16. To be able to prove the analogue of Proposition 8.13 when the external potential is a general subquadratic potential, we would need to know how J^ε acts on gauge invariant nonlinearities. Unfortunately, this question seems to be open. Note that we may not need exactly the fourth point of Lemma 8.16: J^ε may act on gauge invariant nonlinearities "approximately" like a derivative. The goal would be to find what an acceptable definition of "approximately" could be, and to show that it is satisfied by J^ε. What can be proved, at least, is that the first two points in Lemma 8.16 hold *only* when the external potential is exactly a polynomial:

Proposition 8.23. *Let $V \in C^\infty(\mathbb{R}^d; \mathbb{R})$ be a smooth, subquadratic potential. Let $\phi \in C^4(]0, T] \times \mathbb{R}^d; \mathbb{R})$ and $f \in C^1(]0, T])$ for some $T > 0$. Assume that f has no zero on the interval $]0, T]$. Define the operator A^ε by:*

$$A^\varepsilon(t) = if(t)e^{i\phi(t,x)/\varepsilon} \nabla \left(e^{-i\phi(t,x)/\varepsilon} \cdot \right) = \frac{f(t)}{\varepsilon} \nabla \phi(t,x) + if(t)\nabla.$$

Then A^ε commutes with the linear part of Eq. (8.25) if and only if V is exactly a polynomial of degree at most two.

Remark 8.24. In Lemma 8.16, the phases ϕ_1 and ϕ_2 solve the eikonal equation

$$\partial_t \phi + \frac{1}{2} |\nabla \phi|^2 + V = 0. \tag{8.26}$$

In the above statement, we do not assume that ϕ solves the eikonal equation. However, we will see in the proof that it is essentially necessary (up to a phase shift which depends only on time, and thereby does not affect the definition of A^ε).

Proof. In view of Lemma 8.16, we only have to prove the "only if" part. Computations yield

$$\left[i\varepsilon \partial_t + \frac{\varepsilon^2}{2} \Delta - V, A_j^\varepsilon(t) \right] = f'(t) \partial_j \phi + f(t) \partial_{jt}^2 \phi + f(t) \partial_j V$$
$$+ \varepsilon \Big(-f'(t) \partial_j + f(t) \nabla(\partial_j \phi) \cdot \nabla + \frac{1}{2} f(t) \Delta(\partial_j \phi) \Big). \tag{8.27}$$

This bracket is zero if and only if the terms in ε^0 and ε^1 are zero. The term in ε is the sum of an operator of order one and of an operator of order zero. It is zero if and only if both operators are zero. The operator of order one is zero if and only if

$$f(t) \partial_{jj}^2 \phi = f'(t), \quad \partial_{jk}^2 \phi \equiv 0 \ \text{ if } j \neq k.$$

In particular, $\partial_{jj}^2 \phi$ is a function of time only, independent of x, and we have

$$\frac{1}{2} f(t) \Delta(\partial_j \phi) \equiv 0.$$

From the above computations, the first two terms in ε^0 also write

$$f'(t) \partial_j \phi + f(t) \partial_{jt}^2 \phi = \sum_{k=1}^{d} f(t) \partial_k \phi \partial_{jk}^2 \phi + f(t) \partial_{jt}^2 \phi = f(t) \partial_j \left(\partial_t \phi + \frac{1}{2} |\nabla \phi|^2 \right).$$

Canceling the term in ε^0 in (8.27) therefore yields, since f is never zero on $]0, T]$,

$$\partial_j \left(\partial_t \phi + \frac{1}{2} |\nabla \phi|^2 + V \right) = 0. \tag{8.28}$$

Differentiating the above equation with respect to x_k and x_ℓ, all the terms with ϕ vanish, since we noticed that the derivatives of order at least three of ϕ are zero. We deduce that for any triplet (j, k, ℓ), $\partial_{jk\ell}^3 V \equiv 0$, that is, V is a polynomial of degree at most two.

Notice that since (8.28) holds for any $j \in \{1, \ldots, d\}$, there exists a function Ξ of time only such that

$$\partial_t \phi + \frac{1}{2}|\nabla \phi|^2 + V = \Xi(t).$$

This means that ϕ is almost a solution to the eikonal equation (8.26). Replacing ϕ by $\widetilde{\phi}(t, x) := \phi(t, x) - \int_0^t \Xi(\tau)d\tau$ does not affect the definition of A^ε, and $\widetilde{\phi}$ solves (8.26). $\qquad \square$

Remark 8.25. In view of Eq. (8.27), if not zero, the commutator could be at best of order $\mathcal{O}(\varepsilon)$. Unfortunately, a term of this size cannot be considered as a small source term, because of the factor ε in front of the time derivative of the semi-classical Schrödinger operator; see Lemma 1.2.

We leave out the discussion on general subquadratic potentials at this stage. Recall that for instance, we have no complete picture when V is as in Assumption 8.18 and Theorem 8.19. Finding tools to study this more general case would be of course the key to extend Proposition 8.13. It might also open new possibilities for similar problems.

In a similar spirit, we invite the reader to consult the very interesting reference [Sacchetti (2005)]. There, a global in time analysis is obtained for a problem which is slightly different from Eq. (8.25). The external potential V is particular (double well), and the initial data are concentrated on the two eigenfunctions associated to the semi-classical Hamiltonian (this implies that the concentration of the initial data is not at scale ε as in Eq. (8.25), but at scale $\sqrt{\varepsilon}$). Also, the order of magnitude of the coupling constant is much weaker than in Eq. (8.25) (it is exponentially decreasing as $\varepsilon \to 0$). However, it is critical as far as leading order nonlinear effects are concerned. A global in time analysis is then established (in the same fashion as in Theorem 8.2), which yields a precise description of the interaction between the linear effects due to the double well, and the nonlinear effects. The main typically nonlinear effect is the following: if the initial data are concentrated on a single well, and if the coupling factor is sufficiently large, then the solution remains localized on the same well, while the linear solution would oscillate between the two wells (beating motion). See also the more recent work [Bambusi and Sacchetti (2007)] on this subject.

Chapter 9

Some Ideas for Supercritical Cases

To conclude these notes, we present partial results for a focal point in the supercritical case $\alpha < d\sigma$, without external potential:

$$i\varepsilon\partial_t u^\varepsilon + \frac{\varepsilon^2}{2}\Delta u^\varepsilon = \varepsilon^\alpha \left|u^\varepsilon\right|^{2\sigma} u^\varepsilon \quad ; \quad u^\varepsilon(0,x) = a_0(x)e^{-i|x|^2/(2\varepsilon)}.$$

Using a semi-classical lens transform (which amounts to adapting Eq. (9.6) in view of the approach of [Carles (2002)], see Remark 9.5), we could also consider the case with an isotropic harmonic potential. To observe non-linear effects which are due to focusing phenomenon, we assume $\alpha > 1$: the nonlinearity is negligible in a WKB régime. Also, we suppose that the nonlinearity is exactly cubic: $\sigma = 1$. All in all, we assume $1 < \alpha < d$. Note that in particular, this assumption excludes the one-dimensional case. This point appears several times in the proof. We therefore consider:

$$i\varepsilon\partial_t u^\varepsilon + \frac{\varepsilon^2}{2}\Delta u^\varepsilon = \varepsilon^\alpha \left|u^\varepsilon\right|^2 u^\varepsilon \quad ; \quad u^\varepsilon(0,x) = a_0(x)e^{-i|x|^2/(2\varepsilon)}. \tag{9.1}$$

In [Carles (2007a)], preliminary results are shown concerning the asymptotic behavior of u^ε. Besides these results, which we state and prove below, the computations in [Carles (2007a)] yield an interesting example. Indeed, a formal computation yields a function that solves Eq. (9.1), up to a source term which can be taken very small as $\varepsilon \to 0$. However, the usual stability analysis does not make it possible to conclude that it is close to the exact solution, since the exponential factor in the Gronwall lemma counterbalances the smallness of the source term. This problem is similar to the one encountered in the supercritical régime for WKB analysis (Chap. 4). Moreover, we can prove that this function is not a good approximation of the exact solution, when Gronwall lemma ceases to be interesting. The reason is more subtle than a spectral instability: it is related to the notion of good unknown functions, in order to observe the right oscillations.

First, qualitative arguments yield relevant scalings for Eq. (9.1). Recall that the mass and energy associated to Eq. (9.1) do not depend on time:

$$\|u^\varepsilon(t)\|_{L^2} = \|a_0\|_{L^2},$$

$$\|\varepsilon\nabla u^\varepsilon(t)\|_{L^2}^2 + \varepsilon^\alpha\|u^\varepsilon(t)\|_{L^4}^4 = \text{const.} = \mathcal{O}(1) \underset{\varepsilon\to0}{\sim} \|xa_0\|_{L^2}^2.$$

Intuitively, the nonlinear term is relevant at a focal point if the potential energy is of order $\mathcal{O}(1)$ exactly, while the modulus of u^ε is described by a concentrating profile,

$$|u^\varepsilon(t,x)| \underset{\varepsilon\to0}{\sim} \frac{1}{\varepsilon^{d\gamma/2}}\varphi\left(\frac{x}{\varepsilon^\gamma}\right), \tag{9.2}$$

for some $\gamma > 0$. The power of ε in front of φ is to ensure the L^2-norm conservation. We consider only the modulus of u^ε, because the forthcoming discussion will show that phenomena affecting the phase are crucial, and not completely understood. We then compute:

$$\varepsilon^\alpha\|u^\varepsilon(t)\|_{L^4}^4 \underset{\varepsilon\to0}{\sim} \varepsilon^{\alpha-d\gamma}\|\varphi\|_{L^4}^4.$$

We check that the "linear" value $\gamma = 1$ is forbidden, because in that case, the potential energy is unbounded ($\alpha < d$). Note also that the value $\gamma = 1$ was the one encountered in the critical case $\alpha = d\sigma$ (see Proposition 7.19). For the potential energy to be of order $\mathcal{O}(1)$ exactly, we have to choose:

$$\gamma = \frac{\alpha}{d} < 1. \tag{9.3}$$

We will not prove that the above argument is correct, but we will show that the scale ε^γ is an important feature of this problem. We show that there is a time at which u^ε behaves like in Eq. (9.2), but our analysis stops before $t = 1$. Notice also that the above argument suggests that the amplification of the solution u^ε as time goes to 1 is less important than in the linear case; supercritical phenomena may occur in the phase, and also affect the amplitude. We also point out that estimates agreeing with these heuristic arguments were established by S. Masaki [Masaki (2007)] in the analogous case, where the cubic nonlinearity is replaced by a Hartree type nonlinearity.

Definition 9.1. If $T > 0$, $(k_j)_{j\geqslant1}$ is an increasing sequence of real numbers, $(\phi_j)_{j\geqslant1}$ is a sequence in $H^\infty(\mathbb{R}^d)$, and $\phi \in C([0,T];H^\infty)$, the asymptotic relation

$$\phi(t,x) \sim \sum_{j\geqslant1} t^{k_j}\phi_j(x) \quad \text{as } t \to 0$$

means that for every integer $J \geqslant 1$ and every $s > 0$,

$$\left\| \phi(t, \cdot) - \sum_{j=1}^{J} t^{k_j} \phi_j \right\|_{H^s(\mathbb{R}^d)} = o\left(t^{k_J}\right) \quad \text{as } t \to 0.$$

Theorem 9.2. *Let* $a_0 \in \mathcal{S}(\mathbb{R}^d)$. *Assume* $d > \alpha > 1$. *Then there exist* $T > 0$ *independent of* $\varepsilon \in]0, 1]$, *a sequence* $(\phi_j)_{j \geqslant 1}$ *in* H^∞, *and* $\phi \in C([0, T]; H^\infty)$, *such that:*

1. $\phi(t, x) \sim \sum_{j \geqslant 1} t^{jd-1} \phi_j(x)$ *as* $t \to 0$.
2. *For* $1 - t \gg \varepsilon^\gamma$ *(*$\gamma = \alpha/d < 1$*), the asymptotic behavior of* u^ε *is given by:*

$$\limsup_{\varepsilon \to 0} \sup_{0 \leqslant t \leqslant 1 - \Lambda \varepsilon^\gamma} \| u^\varepsilon(t) - v^\varepsilon(t) \|_{L^2(\mathbb{R}^d)} \xrightarrow[\Lambda \to +\infty]{} 0,$$

where $\displaystyle v^\varepsilon(t, x) = \frac{e^{i \frac{|x|^2}{2\varepsilon(t-1)}}}{(1-t)^{d/2}} a_0 \left(\frac{x}{1-t} \right) \exp\left(i \varepsilon^{\gamma - 1} \phi \left(\frac{\varepsilon^\gamma}{1-t}, \frac{x}{1-t} \right) \right).$

Some comments are in order. In the linear case as well as in the critical case $\alpha = d > 1$, the above result holds with $\gamma = 1$ and $\phi \equiv 0$. The case $\alpha < n$ is supercritical as far as nonlinear effects near $t = 1$ are concerned. We emphasize two important features in the above result: the analysis stops sooner than $1 - t \gg \varepsilon$, and nonlinear effects cause the presence of the (non-trivial) phase ϕ. For $1 - t \gg \varepsilon^\gamma$, we have

$$\varepsilon^{\gamma - 1} \phi \left(\frac{\varepsilon^\gamma}{1-t}, \frac{x}{1-t} \right) \sim \sum_{j \geqslant 1} \frac{\varepsilon^{j\alpha - 1}}{(1-t)^{jd-1}} \phi_j \left(\frac{x}{1-t} \right).$$

The above phase shift starts being relevant for $1 - t \approx \varepsilon^{\frac{\alpha-1}{d-1}}$ (recall that $d > \alpha > 1$); this is the first boundary layer where nonlinear effects appear at leading order, measured by ϕ_1. We will check that this phase shift is relevant: ϕ_1 is not zero (unless $u^\varepsilon \equiv 0$, see (9.11) below). We then have a countable number of boundary layers in time, of size

$$1 - t \approx \varepsilon^{\frac{j\alpha - 1}{jd - 1}},$$

which reach the layer $1 - t \approx \varepsilon^\gamma$ in the limit $j \to +\infty$. At each new boundary layer, a new phase ϕ_j becomes relevant at leading order. In general, none of the ϕ_j's is zero: see e.g. (9.13) for ϕ_2. The result of a cascade of phases can be compared to the one discovered by C. Cheverry [Cheverry (2006)] in the case of fluid dynamics, although the phenomenon seems to be different. Yet, our result shares another property with [Cheverry (2006)], which does not appear in the above statement. Theorem 9.2 shows perturbations of the phase (the ϕ_j's), but not of the amplitude: the main profile is the same as in the linear case, that is, a rescaling of a_0. However, to compute the first N phase shifts, $(\phi_j)_{1 \leqslant j \leqslant N}$, one has to compute $d - 1$ corrector terms of the main profile a_0. This is due to the fact that the above result can be connected rather explicitly to the supercritical WKB analysis of Sec. 4.2.1, *via* a semi-classical conformal transform.

Each phase shift oscillates at a rate between $\mathcal{O}(1)$ (when it starts being relevant) and $\mathcal{O}(\varepsilon^{\gamma-1})$ (when it reaches the layer of size ε^γ). Since $\gamma > 0$, this means that each phase shift is rapidly oscillating at the scale of the amplitude, but oscillating strictly more slowly than the geometric phase $\frac{|x|^2}{2\varepsilon(t-1)}$, for $1 - t \gg \varepsilon^\gamma$. We will see that for $1 - t = \mathcal{O}(\varepsilon^\gamma)$, all the terms in ϕ, plus the geometric phase, have the same order: all these phases become comparable, see (9.16).

If we transpose the results of [Masaki (2007)] to the case of the cubic nonlinearity, we should have the uniform estimate:

$$\|J^\varepsilon(t)u^\varepsilon(t)\|_{L^2} \lesssim \varepsilon^{\alpha/d-1} = \varepsilon^{\gamma-1}.$$

Recalling that

$$J^\varepsilon(t) = i(t-1)e^{i|x|^2/(2\varepsilon(t-1))}\nabla\left(e^{-i|x|^2/(2\varepsilon(t-1))}\cdot\right),$$

the above estimate suggests that besides the oscillations carried by the solution of the eikonal equation, u^ε oscillates at scale $\varepsilon^{1-\gamma}$, which is between ε and 1. This estimate is in perfect agreement with the previous discussions, and with Theorem 9.2. One of the aspects in [Masaki (2007)] is that the above estimate is proved to be valid for all time, and not only for $t < 1$. An interesting question would be to know if this estimate is sharp (as $\varepsilon \to 0$) for $t > 1$. In other words, do intermediary scales of oscillations have appeared near $t = 1$, persisting for $t > 1$?

9.1   Cascade of phase shifts

In this paragraph, we prove Theorem 9.2. We start by some formal computations, whose conclusion is in good agreement with the statement of Theorem 9.2. Then, we prove Theorem 9.2, and notice that the two approaches disagree, past the first boundary layer. This aspect is explained in Sec. 9.1.3.

9.1.1 A formal computation

For simplicity, we assume $a_0 \in \mathcal{S}(\mathbb{R}^d)$. Resume the approach *via* Lagrangian integral (see Sec. 7.1),

$$u^\varepsilon(t,x) = \frac{1}{(2\pi\varepsilon)^{d/2}} \int_{\mathbb{R}^d} e^{-i\frac{t-1}{2\varepsilon}|\xi|^2 + i\frac{x\cdot\xi}{\varepsilon}} A^\varepsilon(t,\xi)d\xi.$$

From Lemma 7.1, we know that the Lagrangian symbol A^ε converges at time $t = 0$, as $\varepsilon \to 0$. For $t \neq 1$, we apply stationary phase formula at a very formal level, as in Sec. 7.3 (with now $\sigma = 1$):

$$\mathcal{F}\left(|u^\varepsilon|^2 u^\varepsilon\right)\left(t, \frac{\xi}{\varepsilon}\right) \approx \frac{\varepsilon^{d/2}}{|t-1|^n}|A^\varepsilon|^2 A^\varepsilon(t,\xi)e^{-i\frac{t-1}{2\varepsilon}|\xi|^2}.$$

Using the second part of Lemma 7.1, we infer that the evolution of A^ε should be described by

$$i\partial_t A^\varepsilon(t,\xi) = \frac{\varepsilon^{\alpha-1}}{|t-1|^d}|A^\varepsilon|^2 A^\varepsilon(t,\xi).$$

To be consistent, we should say that the right-hand side is negligible, since $\alpha > 1$. We keep it though. Like several times before, we notice that the modulus of A^ε is independent of time, and we compute explicitly:

$$A^\varepsilon(t,\xi) = A^\varepsilon(0,\xi)\exp\left(-i\varepsilon^{\alpha-1}|A^\varepsilon(0,\xi)|^2\int_0^t \frac{d\tau}{(1-\tau)^d}\right)$$

$$\approx A_0(\xi)\exp\left(-i\varepsilon^{\alpha-1}|A_0(\xi)|^2\int_0^t \frac{d\tau}{(1-\tau)^d}\right),$$

where $A_0(\xi) = e^{-id\pi/4}a_0(-\xi)$. Back to u^ε, this yields, for $t < 1$:

$$u^\varepsilon(t,x) \approx \frac{e^{id\pi/4}}{(1-t)^{d/2}}A^\varepsilon\left(t,\frac{x}{t-1}\right)e^{i\frac{|x|^2}{2\varepsilon(t-1)}}$$

$$\approx \frac{1}{(1-t)^{d/2}}a_0\left(\frac{x}{1-t}\right)e^{i\frac{|x|^2}{2\varepsilon(t-1)}}e^{-i\left|a_0\left(\frac{x}{1-t}\right)\right|^2\varepsilon^{\alpha-1}\int_0^t\frac{d\tau}{(1-\tau)^d}}.$$

Since $d \geqslant 2$,

$$\int_0^t \frac{d\tau}{(1-\tau)^d} \underset{t\to 1}{\sim} \frac{1}{(d-1)(1-t)^{d-1}},$$

we see that the last phase in the approximation of u^ε becomes relevant for

$$\varepsilon^{\alpha-1} \approx (1-t)^{d-1} : 1-t \approx \varepsilon^{\frac{\alpha-1}{d-1}}.$$

This is exactly the first boundary layer appearing in Theorem 9.2. To go further into this formal analysis, we adopt another point of view, which may appear as more explicit. Using a generalized WKB expansion, seek

$$u^\varepsilon(t,x) \underset{\varepsilon\to 0}{\sim} v_1^\varepsilon(t,x) = b^\varepsilon(t,x)e^{i\phi(t,x)/\varepsilon},$$

and change the usual hierarchy to force the contribution of the nonlinear term to appear in the transport equation:

$$\begin{cases} \partial_t \phi + \dfrac{1}{2}|\nabla \phi|^2 = 0 & ; \quad \phi(0,x) = -\dfrac{|x|^2}{2}. \\ \partial_t b^\varepsilon + \nabla \phi \cdot \nabla b^\varepsilon + \dfrac{1}{2} b^\varepsilon \Delta \phi = -i\varepsilon^{\alpha-1}|b^\varepsilon|^2 b^\varepsilon & ; \quad b^\varepsilon(0,x) = a_0(x). \end{cases}$$

The eikonal equation is the same as the linear case, as well as its solution. The transport equation is an ordinary differential equation along the rays of geometric optics $\frac{x}{1-t} = \text{const.}$, of the form

$$\dot{y} = -i\varepsilon^{\alpha-1}|y|^2 y.$$

The modulus of b^ε is constant along rays, and

$$b^\varepsilon(t,x) = \frac{1}{(1-t)^{d/2}} a_0\left(\frac{x}{1-t}\right) e^{-i\varepsilon^{\alpha-1}\left|a_0\left(\frac{x}{1-t}\right)\right|^2 \int_0^t \frac{d\tau}{(1-\tau)^d}}.$$

We retrieve the same approximation as with the Lagrangian integral approach. By construction,

$$i\varepsilon \partial_t v_1^\varepsilon + \frac{\varepsilon^2}{2}\Delta v_1^\varepsilon = \varepsilon^\alpha |v_1^\varepsilon|^2 v_1^\varepsilon + r_1^\varepsilon \quad ; \quad v_1^\varepsilon(0,x) = a_0(x)e^{-i|x|^2/(2\varepsilon)},$$

where

$$r_1^\varepsilon(t,x) = \frac{\varepsilon^2}{2}e^{i|x|^2/(2\varepsilon(t-1))}\Delta b^\varepsilon(t,x).$$

Denote $g_1^\varepsilon(t,x) = -\varepsilon^{\alpha-1}\left|a_0\left(\frac{x}{1-t}\right)\right|^2 \int_0^t \frac{d\tau}{(1-\tau)^d}$. We compute:

$$\Delta b^\varepsilon(t,x) = \frac{e^{ig_1^\varepsilon(t,x)}}{(1-t)^{2+d/2}}\Delta a_0\left(\frac{x}{1-t}\right)$$

$$- \frac{2ie^{ig_1^\varepsilon(t,x)}}{(1-t)^{2+d/2}}\nabla a_0\left(\frac{x}{1-t}\right)\cdot \nabla |a_0|^2\left(\frac{x}{1-t}\right)\varepsilon^{\alpha-1}\int_0^t \frac{d\tau}{(1-\tau)^d}$$

$$- \frac{ie^{ig_1^\varepsilon(t,x)}}{(1-t)^{2+d/2}}a_0\left(\frac{x}{1-t}\right)\Delta |a_0|^2\left(\frac{x}{1-t}\right)\varepsilon^{\alpha-1}\int_0^t \frac{d\tau}{(1-\tau)^d}$$

$$- \frac{e^{ig_1^\varepsilon(t,x)}}{(1-t)^{2+d/2}}a_0\left(\frac{x}{1-t}\right)\left|\nabla |a_0|^2\right|^2\left(\frac{x}{1-t}\right)\left(\varepsilon^{\alpha-1}\int_0^t \frac{d\tau}{(1-\tau)^d}\right)^2.$$

We infer

$$\|r_1^\varepsilon(t)\|_{L^2} \lesssim \frac{\varepsilon^2}{(1-t)^2} + \frac{\varepsilon^{\alpha+1}}{(1-t)^{d+1}} + \frac{\varepsilon^{\alpha+1}}{(1-t)^{d+1}} + \frac{\varepsilon^{2\alpha}}{(1-t)^{2d}}.$$

Therefore,

$$\frac{1}{\varepsilon}\int_0^t \|r_1^\varepsilon(\tau)\|_{L^2}d\tau \lesssim \frac{\varepsilon}{1-t} + \frac{\varepsilon^\alpha}{(1-t)^d} + \frac{\varepsilon^\alpha}{(1-t)^d} + \frac{\varepsilon^{2\alpha-1}}{(1-t)^{2d-1}}.$$

Following the energy estimates of Lemma 1.2, this quantity might be the one that dictates the size of the error $u^\varepsilon - v_1^\varepsilon$. The first term is the "linear" one: it is small for $1 - t \gg \varepsilon$. The second and third ones (which are the same, but which we keep to keep track of the corresponding terms in Δb^ε) are small for $1 - t \gg \varepsilon^{\alpha/d}$, and the last one is small for $1 - t \gg \varepsilon^{(2\alpha-1)/(2d-1)}$. Since

$$\frac{2\alpha - 1}{2d - 1} < \frac{\alpha}{d} = \gamma < 1,$$

the last term in Δb^ε is the first to cease to be negligible. We note that in the equation for v_1^ε, the corresponding term has the same argument as v_1^ε. This suggests that the last term in r_1^ε might be eliminated by adding an extra phase term in the approximate solution, in view of the identity

$$i\varepsilon\partial_t \left(a e^{i\theta^\varepsilon} \right) = -\partial_t \theta^\varepsilon \times a e^{i\theta^\varepsilon} + e^{i\theta^\varepsilon} i\varepsilon\partial_t a.$$

Seek an approximate solution of the form:

$$v^\varepsilon(t, x) = \frac{1}{(1-t)^{d/2}} a_0 \left(\frac{x}{1-t} \right) e^{i\phi^\varepsilon(t,x)}, \quad \phi^\varepsilon(t, x) = \frac{|x|^2}{2\varepsilon(t-1)} + g^\varepsilon(t, x).$$

We find

$$i\varepsilon\partial_t v^\varepsilon + \frac{\varepsilon^2}{2}\Delta v^\varepsilon = \left(i\frac{\varepsilon^2}{2}\Delta g^\varepsilon - \varepsilon\partial_t g^\varepsilon - \frac{\varepsilon^2}{2}|\nabla g^\varepsilon|^2 + \frac{\varepsilon}{1-t} x \cdot \nabla g^\varepsilon \right) v^\varepsilon$$

$$+ i\frac{\varepsilon^2}{(1-t)^{1+d/2}}\nabla g^\varepsilon \cdot \nabla a_0 \left(\frac{x}{1-t} \right) e^{i\phi^\varepsilon}$$

$$+ \frac{1}{2} \left(\frac{\varepsilon}{1-t} \right)^2 \frac{e^{i\phi^\varepsilon}}{(1-t)^{d/2}}\Delta a_0 \left(\frac{x}{1-t} \right).$$

As suggested by the previous computations, write

$$g^\varepsilon(t, x) = \frac{1}{\varepsilon}\int_0^t h\left(\frac{\varepsilon^\alpha}{(1-\tau)^d}, \frac{x}{1-t} \right) d\tau, \quad \text{with } h(z, \xi) \sim \sum_{j \geqslant 1} z^j g_j(\xi). \quad (9.4)$$

In the equation verified by v^ε, the last term is the "same" as in the linear case: it becomes relevant only in a boundary layer of size ε. Since our approach will lead us to the boundary layer of size ε^γ ($\gamma < 1$), we ignore that term. The remaining terms with a factor i are of order, in L^2,

$$\varepsilon^2\|\Delta g^\varepsilon(t)\|_{L^\infty} + \frac{\varepsilon^2}{1-t}\|\nabla g^\varepsilon(t)\|_{L^\infty} \lesssim \frac{\varepsilon}{(1-t)^2}\int_0^t \frac{\varepsilon^\alpha}{(1-\tau)^d} d\tau \lesssim \frac{\varepsilon^{\alpha+1}}{(1-t)^{d+1}},$$

and their contribution is also left out in this computation, since they become of order one only for $1 - t \approx \varepsilon^\gamma$.

Now we require that v^ε be an approximate solution to (9.1):

$$\left(\partial_t - \frac{x}{1-t} \cdot \nabla\right) g^\varepsilon + \frac{\varepsilon}{2}|\nabla g^\varepsilon|^2 = -\varepsilon^{\alpha-1}|v^\varepsilon(t,x)|^2$$

$$= -\frac{\varepsilon^{\alpha-1}}{(1-t)^d}\left|a_0\left(\frac{x}{1-t}\right)\right|^2.$$

Using (9.4), we get:

$$g_1(\xi) = -|a_0(\xi)|^2,$$

$$\text{for } j \geqslant 2, \quad g_j(\xi) = -\frac{1}{2}\sum_{p+q=j}\frac{1}{(pd-1)(qd-1)}\nabla g_p \cdot \nabla g_q, \quad (9.5)$$

with the convention $g_0 \equiv 0$. This algorithm produces smooth solutions, since $a_0 \in \mathcal{S}(\mathbb{R}^d)$. Define

$$g_N^\varepsilon(t,x) = \frac{1}{\varepsilon}\sum_{j=1}^d \int_0^t \left(\frac{\varepsilon^\gamma}{1-\tau}\right)^{dj} d\tau \times g_j\left(\frac{x}{1-t}\right),$$

$$v_N^\varepsilon(t,x) = \frac{1}{(1-t)^{d/2}}a_0\left(\frac{x}{1-t}\right)e^{i\frac{|x|^2}{2\varepsilon(t-1)}+g_N^\varepsilon(t,x)}.$$

Proposition 9.3 (Formal approximation to (9.1)). *Let $d > \alpha > 1$, $a_0 \in \mathcal{S}(\mathbb{R}^d)$, and fix $N \in \mathbb{N}^*$. The function v_N^ε solves*

$$i\varepsilon\partial_t v_N^\varepsilon + \frac{\varepsilon^2}{2}\Delta v_N^\varepsilon = \varepsilon^\alpha|v_N^\varepsilon|^2 v_N^\varepsilon + r_N^\varepsilon \quad ; \quad v_d^\varepsilon(0,x) = a_0(x)e^{-i|x|^2/(2\varepsilon)}.$$

For $1 - t \geqslant \varepsilon^\gamma = \varepsilon^{\alpha/d}$, the source term satisfies:

$$\frac{1}{\varepsilon}\int_0^t \|r_N^\varepsilon(\tau)\|_{L^2}d\tau \lesssim \frac{\varepsilon^{(N+1)\alpha-1}}{(1-t)^{(N+1)d-1}} + \frac{\varepsilon^\alpha}{(1-t)^d}.$$

For $1 \leqslant j \leqslant N$, the jth term of the series defining g_N^ε becomes relevant in a boundary layer of size $\varepsilon^{\frac{j\alpha-1}{jd-1}}$, since

$$\frac{1}{\varepsilon}\int_0^t \left(\frac{\varepsilon^\gamma}{1-\tau}\right)^{dj}d\tau \underset{t\to 1}{\sim} \frac{1}{jd-1}\frac{\varepsilon^{j\alpha-1}}{(1-t)^{jd-1}}.$$

In the limit $N \to +\infty$, a countable family of boundary layers appear. Qualitatively, this agrees with Theorem 9.2.

Remark 9.4. In the critical case $\alpha = d > 1$, we have $\beta = \gamma = 1$: the above boundary layers "collapse" one on another. There are no such phase shifts as above.

The sole estimate of the source term does not prove much. In a standard (semi-linear) stability argument, the nonlinearity $|u^\varepsilon|^2 u^\varepsilon - |v_N^\varepsilon|^2 v_N^\varepsilon$ is usually treated by a Gronwall type argument. If the nonlinearity is "too strong", then the above estimate, which is completely relevant in the linear case, does not necessarily account for the size of the error. Since we are in a supercritical case, it is not surprising that Proposition 9.3 is only a formal result. To see what information could be hoped from Proposition 9.3, set $w_N^\varepsilon = u^\varepsilon - v_N^\varepsilon$. It solves

$$i\varepsilon \partial_t w_N^\varepsilon + \frac{\varepsilon^2}{2} \Delta w_N^\varepsilon = \varepsilon^\alpha \left(|u^\varepsilon|^2 u^\varepsilon - |v_N^\varepsilon|^2 v_N^\varepsilon \right) - r_N^\varepsilon \quad ; \quad w_{N|t=0}^\varepsilon = 0.$$

Applying Lemma 1.2, we find, for $t > 0$:

$$\|w_N^\varepsilon(t)\|_{L^2} \lesssim \varepsilon^{\alpha-1} \int_0^t \left\| |u^\varepsilon|^2 u^\varepsilon - |v_N^\varepsilon|^2 v_N^\varepsilon \right\|_{L^2} d\tau + \frac{1}{\varepsilon} \int_0^t \|r_N^\varepsilon(\tau)\|_{L^2} d\tau$$

$$\lesssim \varepsilon^{\alpha-1} \int_0^t \left(\|u^\varepsilon(\tau)\|_{L^\infty}^2 + \|v_N^\varepsilon(\tau)\|_{L^\infty}^2 \right) \|w_N^\varepsilon(\tau)\|_{L^2} d\tau$$

$$+ \frac{\varepsilon^{(N+1)\alpha-1}}{(1-t)^{(N+1)d-1}}.$$

Assume that for $1 - t \gg \varepsilon^\gamma$, u^ε is of the same order of magnitude as v_N^ε, that is

$$\|u^\varepsilon(t)\|_{L^\infty} \lesssim \frac{1}{(1-t)^{d/2}}.$$

(This estimate will actually be proved in Sec. 9.1.2, see Remark 9.8.) Gronwall lemma then yields, for $t < 1$:

$$\|w_N^\varepsilon(t)\|_{L^2} \lesssim \frac{\varepsilon^{(N+1)\alpha-1}}{(1-t)^{(N+1)d-1}} e^{C\varepsilon^{\alpha-1}/(1-t)^{d-1}}.$$

Formally, take $N = \infty$: in L^2, the error w^ε is controlled by

$$\frac{\varepsilon^\alpha}{(1-t)^d} e^{C\varepsilon^{\alpha-1}/(1-t)^{d-1}}.$$

For $1 - t \gtrsim \varepsilon^{(\alpha-1)/(d-1)}$, the exponential is bounded. Moreover, if

$$\frac{\varepsilon^{\alpha-1}}{(1-t)^{d-1}} = c \log \frac{1}{\varepsilon},$$

we still infer

$$\|w_N^\varepsilon(t)\|_{L^2} \ll 1,$$

provided that c (independent of ε) is sufficiently small. This shows that the first phase shift, also derived with the Lagrangian integral approach, is the right one, since the approximation is good past the first boundary layer. On the other hand, the second boundary layer corresponds to

$$1 - t \approx \varepsilon^{\frac{2\alpha-1}{2d-1}}.$$

We check that

$$\frac{\varepsilon^\alpha}{(1-t)^d} e^{C\varepsilon^{\alpha-1}/(1-t)^{d-1}}\Big|_{t=1-\varepsilon^{(2\alpha-1)/(2d-1)}} \gg 1.$$

Therefore, we cannot conclude anything by this approach, as soon as the second boundary layer is reached (and even before). We will see in Sec. 9.1.3 that this approximation ceases indeed to be good between the first two boundary layers. We will also explain why.

9.1.2   *A rigorous computation*

Introduce the new unknown function ψ given by:

$$u^\varepsilon(t,x) = \frac{1}{(1-t)^{d/2}} \psi^\varepsilon \left(\frac{\varepsilon^\gamma}{1-t}, \frac{x}{1-t} \right) e^{i\frac{|x|^2}{2\varepsilon(t-1)}}. \qquad (9.6)$$

Recalling that $\gamma = \alpha/d < 1$, denote

$$h = \varepsilon^{1-\gamma} \underset{\varepsilon \to 0}{\longrightarrow} 0. \qquad (9.7)$$

Changing the notation $\psi^\varepsilon(\tau,\xi)$ into $\psi^h(t,x)$, we check that for $t < 1$, Eq. (9.1) is equivalent to:

$$ih\partial_t \psi^h + \frac{h^2}{2}\Delta\psi^h = t^{d-2}|\psi^h|^2\psi^h \quad ; \quad \psi^h\left(h^{\frac{\gamma}{1-\gamma}}, x\right) = a_0(x). \qquad (9.8)$$

Except for two aspects, this equation is the same as in Sec. 4.2.1:

- There is a factor t^{d-2} in front of the nonlinearity.
- The data are prescribed at $t = h^{\frac{\gamma}{1-\gamma}}$, instead of $t = 0$.

Note that the factor t^{d-2} is not present if $d = 2$, and would be singular at $t = 0$ if we wanted to treat the case $d = 1$.

Remark 9.5. It is easy to adapt this analysis to the case of a supercritical focal point, in an isotropic harmonic potential:

$$i\varepsilon\partial_t u^\varepsilon + \frac{\varepsilon^2}{2}\Delta u^\varepsilon = \frac{|x|^2}{2}u^\varepsilon + \varepsilon^\alpha |u^\varepsilon|^2 u^\varepsilon \quad ; \quad u^\varepsilon(0,x) = a_0(x).$$

Indeed, resuming the reduction of the end of Sec. 5.3, we can set

$$\psi^h(t,x) = U^\varepsilon\left(\frac{t}{\varepsilon^\gamma} - 1, x \right),$$

where U^ε is given by

$$U^\varepsilon(t,x) = \frac{1}{(1+t^2)^{d/4}} e^{i\frac{t}{1+t^2}\frac{|x|^2}{2\varepsilon}} u^\varepsilon\left(\arctan t, \frac{x}{\sqrt{1+t^2}}\right).$$

We find:

$$\begin{cases} ih\partial_t \psi^h + \dfrac{h^2}{2}\Delta\psi^h = \left(\left(t_0^h\right)^2 + \left(t - t_0^h\right)^2\right)^{d/2-1} \left|\psi^h\right|^2 \psi^h, \\ \psi^h\left(t_0^h, x\right) = a_0(x), \end{cases}$$

with $t_0^h = h^{\gamma/(1-\gamma)}$. This reduced problem is closely akin to Eq. (9.8). In particular, it is easy to adapt Theorem 9.2 to this case.

To study Eq. (9.8), we naturally adapt the approach of [Grenier (1998)], presented in Sec. 4.2.1. We want to write

$$\psi^h(t,x) = a^h(t,x)e^{i\phi^h(t,x)/h},$$

where

$$\begin{cases} \partial_t\phi^h + \dfrac{1}{2}\left|\nabla\phi^h\right|^2 + t^{d-2}\left|a^h\right|^2 = 0 & ;\ \phi^h_{|t=t_0^h} = 0, \\ \partial_t a^h + \nabla\phi^h\cdot\nabla a^h + \dfrac{1}{2}a^h\Delta\phi^h = i\dfrac{h}{2}\Delta a^h & ;\ a^h_{|t=t_0^h} = a_0, \end{cases} \tag{9.9}$$

and $t_0^h = h^{\gamma/(1-\gamma)}$. We introduce the velocity $v^h = \nabla\phi^h$, and force the initial time to be zero by a shift in time:

$$\widetilde{v}^h(t,x) = v^h\left(t + t_0^h, x\right) \quad;\quad \widetilde{a}^h(t,x) = a^h\left(t + t_0^h, x\right).$$

We now have to study a quasi-linear equation:

$$\partial_t \mathbf{u}^h + \sum_{j=1}^d A_j(\mathbf{u}^h)\partial_j\mathbf{u}^h = \frac{h}{2}L\mathbf{u}^h,$$

with $\mathbf{u}^h = \begin{pmatrix} \operatorname{Re}\widetilde{a}^h \\ \operatorname{Im}\widetilde{a}^h \\ \widetilde{v}_1^h \\ \vdots \\ \widetilde{v}_d^h \end{pmatrix} = \begin{pmatrix} \widetilde{a}_1^h \\ \widetilde{a}_2^h \\ \widetilde{v}_1^h \\ \vdots \\ \widetilde{v}_d^h \end{pmatrix}$, $L = \begin{pmatrix} 0 & -\Delta & 0 & \dots & 0 \\ \Delta & 0 & 0 & \dots & 0 \\ 0 & 0 & & 0_{d\times d} & \end{pmatrix}$, and

$$A(\mathbf{u},\xi) = \sum_{j=1}^d A_j(\mathbf{u})\xi_j = \begin{pmatrix} \widetilde{v}\cdot\xi & 0 & \frac{\widetilde{a}_1}{2}{}^t\xi \\ 0 & \widetilde{v}\cdot\xi & \frac{\widetilde{a}_2}{2}{}^t\xi \\ 2\left(t+t_0^h\right)^{d-2}\widetilde{a}_1\,\xi & 2\left(t+t_0^h\right)^{d-2}\widetilde{a}_2\,\xi & \widetilde{v}\cdot\xi I_d \end{pmatrix}.$$

The matrix $A(\mathbf{u}, \xi)$ can be symmetrized by

$$S^h = \begin{pmatrix} I_2 & 0 \\ 0 & \frac{1}{4\left(t+t_0^h\right)^{d-2}} I_d \end{pmatrix},$$

which depends only on t and h, since the nonlinearity that we consider is exactly cubic. In estimating $\partial_t S$, the assumption $d \geqslant 2$ is again helpful. Indeed, we can mimic the computations of Sec. 4.2.1. For $s > d/2 + 2$, we bound

$$\left\langle S^h \Lambda^s \mathbf{u}^h, \Lambda^s \mathbf{u}^h \right\rangle,$$

(scalar product in $L^2(\mathbb{R}^{d+2})$) by computing its time derivative:

$$\frac{d}{dt} \left\langle S^h \Lambda^s \mathbf{u}^h, \Lambda^s \mathbf{u}^h \right\rangle = \left\langle \partial_t S^h \Lambda^s \mathbf{u}^h, \Lambda^s \mathbf{u}^h \right\rangle + 2 \left\langle S^h \partial_t \Lambda^s \mathbf{u}^h, \Lambda^s \mathbf{u}^h \right\rangle,$$

since S^h is symmetric. Because $d \geqslant 2$,

$$\left\langle \partial_t S^h \Lambda^s \mathbf{u}^h, \Lambda^s \mathbf{u}^h \right\rangle \leqslant 0.$$

Therefore, we can easily infer the analogue of Theorem 4.1:

Proposition 9.6. *Let $d > \alpha > 1$ and $a_0 \in \mathcal{S}(\mathbb{R}^d)$. There exist $T^* > 0$ independent of $h \in]0,1]$ and a unique pair $(\phi^h, a^h) \in C([t_0^h, T^* + t_0^h]; H^\infty)^2$, solution to (9.9). Moreover, a^h and ϕ^h are bounded in $L^\infty([t_0^h, T^* + t_0^h]; H^s)$, uniformly in $h \in]0,1]$, for all s.*

Remark 9.7. In [Carles (2007a)], the homogeneous nonlinearity $\varepsilon^\alpha |u^\varepsilon|^2 u^\varepsilon$ is replaced with $f\left(\varepsilon^\alpha |u^\varepsilon|^2\right) u^\varepsilon$, where $f' > 0$. Computations are not more difficult, just a little lengthier to write. We point out that in estimating the time derivative of the symmetrizer S, the "new" term (compared to Sec. 4.2.1) is non-positive, as above, and therefore can be left out in the energy estimates leading to the analogue of the above proposition.

Remark 9.8. We infer that ψ^h is bounded in $L^\infty([t_0^h, T^* + t_0^h] \times \mathbb{R}^d)$. In view of Eq. (9.6), this shows that for $1 - t \geqslant \varepsilon^\gamma / T^*$,

$$\|u^\varepsilon(t)\|_{L^\infty} \lesssim \frac{1}{(1-t)^{d/2}}.$$

We have a similar result for the expected limit of (ϕ^h, a^h):

Proposition 9.9. *Let $d \geqslant 2$ and $a_0 \in \mathcal{S}(\mathbb{R}^d)$. There exists $T > 0$ such that the system*

$$\begin{cases} \partial_t \phi + \dfrac{1}{2} |\nabla \phi|^2 + t^{d-2} |a|^2 = 0 & ; \ \phi_{|t=0} = 0, \\[2mm] \partial_t a + \nabla \phi \cdot \nabla a + \dfrac{1}{2} a \Delta \phi = 0 & ; \ a_{|t=0} = a_0 \end{cases} \tag{9.10}$$

has a unique the solution $(a, \phi) \in C([0, T]; H^\infty)$. In addition, there exist sequences $(\phi_j)_{j \geqslant 1}$ and $(a_j)_{j \geqslant 1}$ in $H^\infty(\mathbb{R}^d)$, such that

$$\phi(t, x) \sim \sum_{j \geqslant 1} t^{jd-1} \phi_j(x), \quad \text{and} \quad a(t, x) \sim \sum_{j \geqslant 0} t^{jd} a_j(x) \quad \text{as } t \to 0.$$

The last part of the proposition is easily verified, by induction: plugging such asymptotic series into Eq. (9.10), a formal computation yields a source term which is $\mathcal{O}(t^\infty)$ as $t \to 0$. We can now measure the error:

Proposition 9.10. *Let $s \in \mathbb{N}$. We have $T^* \geqslant T$, and there exists C_s independent of h such that for every $t_0^h \leqslant t \leqslant T$,*

$$\|a^h(t) - a(t)\|_{H^s} + \|\phi^h(t) - \phi(t)\|_{H^s} \leqslant C_s \left(ht + h^{\frac{\gamma(d-1)}{1-\gamma}} \right).$$

Proof. We keep the same notations as above. Define $(\tilde{a}, \tilde{v})$ from (a, v) by the same shift in time. Denote by $\mathbf{u}$ the analogue of $\mathbf{u}^h$ corresponding to $(\tilde{a}, \tilde{v})$. We have

$$\partial_t \left(\mathbf{u}^h - \mathbf{u} \right) + \sum_{j=1}^d A_j(\mathbf{u}^h) \partial_j \left(\mathbf{u}^h - \mathbf{u} \right) + \sum_{j=1}^d \left(A_j(\mathbf{u}^h) - A_j(\mathbf{u}) \right) \partial_j \mathbf{u} = \frac{h}{2} L \mathbf{u}^h.$$

We know that $\mathbf{u}^h$ and $\mathbf{u}$ are bounded in $L^\infty([0, \min(T^*, T) - t_0^h]; H^s)$. Denoting $\mathbf{w}^h = \mathbf{u}^h - \mathbf{u}$, we get, for $s > 2 + d/2$:

$$\frac{d}{dt} \left(S^h \Lambda^s \mathbf{w}^h, \Lambda^s \mathbf{w}^h \right) \leqslant C \left(\|\mathbf{u}\|_{H^{s+2}}, \|\mathbf{u}^h\|_{H^s} \right) \left(S^h \Lambda^s \mathbf{w}^h, \Lambda^s \mathbf{w}^h \right)$$
$$+ C h \|\mathbf{u}\|_{H^{s+2}} \|\mathbf{w}^h\|_{H^s}$$
$$\lesssim \left(S^h \Lambda^s \mathbf{w}^h, \Lambda^s \mathbf{w}^h \right) + h^2,$$

where S^h is the previous symmetrizer, which depends only on t and h. Using Gronwall lemma, we infer:

$$\|\mathbf{w}^h(t)\|_{H^s} \lesssim ht + \left\| (a, v)\big|_{t=t_0^h} - (a, v)\big|_{t=0} \right\|_{H^s}$$
$$\lesssim ht + \left(t_0^h \right)^{d-1} \|(\partial_t a, \partial_t v)\|_{L^\infty([0,T]; H^s)} \quad \lesssim ht + \left(t_0^h \right)^{d-1},$$

where we have used Proposition 9.9. Along with a continuity argument, this proves that we also have $T^* \geqslant T$, since T^* does not depend on h. This completes the proof of Proposition 9.10. $\qquad \square$

Note that since $\alpha > 1$, $\frac{\gamma(d-1)}{1-\gamma} > 1$. Back to ψ^h, we infer:

$$
\begin{aligned}
\left\| \psi^h - a_0 e^{i\phi/h} \right\|_{L^\infty([t_0^h,\tau];L^2)} &\lesssim \left\| a^h e^{i\phi^h/h} - a e^{i\phi/h} \right\|_{L^\infty([t_0^h,\tau];L^2)} \\
&\quad + \left\| a e^{i\phi/h} - a_0 e^{i\phi/h} \right\|_{L^\infty([t_0^h,\tau];L^2)} \\
&\lesssim \left\| a^h - a \right\|_{L^\infty([t_0^h,T];L^2)} + \frac{1}{h} \left\| \phi^h - \phi \right\|_{L^\infty([t_0^h,\tau]\times\mathbb{R}^d)} \\
&\quad + \left\| a - a_0 \right\|_{L^\infty([t_0^h,\tau];L^2)} \\
&\lesssim h + \tau + h^{\frac{\alpha-1}{1-\gamma}} + \tau^d.
\end{aligned}
$$

Letting $\tau \to 0$, Theorem 9.2 follows, by using Eq. (9.6), and the last point of Proposition 9.9. Note that the cascade of phase shifts proceeds along the same spirit as in Sec. 5.3, since it stems from the Taylor expansion, as time goes to zero, of the rapidly oscillatory phase.

Remark 9.11. The cascade of phase shifts can be understood as the creation of a new phase, appearing discretely in time. With the transform (9.6) in mind, the asymptotic expansion of the phase shift ϕ stems from the last part of Proposition 9.9. The coupling in (9.10) shows that even if $\phi_{|t=0} = 0$, $\phi(\delta, x)$ is not identically zero, for $\delta > 0$ arbitrarily small. The phase of ψ^h is given asymptotically by

$$
\frac{\phi(t,x)}{h} \sim \sum_{j \geqslant 1} \frac{t^{jd-1}}{h} \phi_j(x).
$$

With the same line of reasoning as above, a phase shift appears for t of order $h^{1/(d-1)}$, then a second for t of order $h^{1/(2d-1)}$, and so on. The superposition of these phase shifts, which are oscillating faster and faster, finally leads to a continuous phase, corresponding to an oscillation associated to the wavelength h.

9.1.3   Why do the results disagree?

The construction of Sec. 9.1.1 and the results of the previous paragraph do not agree. To see this, we come back to Proposition 9.9. Plugging the

Taylor expansions in time for ϕ and a into Eq. (9.10), we find:

$$\mathcal{O}\left(t^{d-2}\right): \quad (d-1)\phi_1 + |a_0|^2 = 0, \tag{9.11}$$

$$\mathcal{O}\left(t^{d-1}\right): \quad da_1 + \nabla\phi_1 \cdot \nabla a_0 + \frac{1}{2}a_0\Delta\phi_1 = 0, \tag{9.12}$$

$$\mathcal{O}\left(t^{2d-2}\right): \quad (2d-1)\phi_2 + \frac{1}{2}|\nabla\phi_1|^2 + 2\operatorname{Re}(\overline{a_0}a_1) = 0. \tag{9.13}$$

The function ϕ_1 is the same as the one obtained by the approach of Sec. 9.1.1: the two approximate solutions are close to each other up to the first boundary layer, when the first phase shift appears. This also stems from the computations at the end of Sec. 9.1.1, in view of Remark 9.8. On the other hand, we see that to get ϕ_2, the modulation of the amplitude (a_1) must be taken into account (note that we consider approximations as $t \to 0$ here, not as $\varepsilon \to 0$: a_1 does not denote the same quantity as in Chap. 4); in Eq. (9.5), g_2 is computed without evaluating Δa_0, unlike ϕ_2, so $g_2 \neq \phi_2$ in general. This means in particular that the two approximate solutions diverge when reaching the second boundary layer: the approach of Sec. 9.1.1 is only formal, and does not lead to a good approximation. And yet, the source term in Proposition 9.3 is small. We will see below that this divergence is not due to a spectral instability, but to the fact that the approach followed to construct the formal approximation was too crude.

We apply the transform (9.6) to the intermediary approximate solution v_N^ε. We show that the formal approximation stops being a good approximation between the first and the second boundary layer exactly, as suspected at the end of Sec. 9.1.1. Like for the exact solution, write

$$v_N^\varepsilon(t,x) = \frac{1}{(1-t)^{d/2}}\psi_N^\varepsilon\left(\frac{\varepsilon^\gamma}{1-t},\frac{x}{1-t}\right)e^{i\frac{|x|^2}{2\varepsilon(t-1)}}.$$

We check that ψ_N^h is given by

$$\psi_N^h(\tau,\xi) = a_0(\xi)e^{i\varphi_N^h(\tau,\xi)}, \quad \text{where } g_N^\varepsilon(t,x) = \varphi_N^h\left(\frac{\varepsilon^\gamma}{1-t},\frac{x}{1-t}\right).$$

We compute

$$\varphi_N^h(t,x) = \frac{1}{ht}\sum_{j=1}^{d}\frac{1}{jd-1}\left(t^{jd} - \left(t_0^h\right)^{jd}\right)g_j(x).$$

Therefore, ψ_N^h satisfies

$$ih\partial_t\psi_N^h + \frac{h^2}{2}\Delta\psi_N^h = t^{d-2}|\psi_N^h|^2\psi_N^h + \theta_N^h(t,x) \quad ; \quad \psi_N^h\left(t_0^h,x\right) = a_0(x),$$

where:

$$\theta_N^h(t,x) = \left(t^{(N+1)d-2} K_0(x) + ihK_1(t,x) + \frac{\left(t_0^h\right)^d}{t^2} \Xi_0^h(x) \right) \psi_N^h(t,x)$$

$$+ ihK_2(t,x) + h^2 K_3(t,x) + ih\frac{\left(t_0^h\right)^d}{t^2} \Xi_1^h(x),$$

where the functions K_j are smooth and independent of h, and Ξ_j^h are bounded in all Sobolev spaces, uniformly in h. Note that the factor in front of Ξ_j^h is not singular, since we assume $t \geqslant t_0^h$, and $d \geqslant 2$. Now write $\psi_N^h(t,x) = a_N^h(t,x) e^{i\phi_N^h(t,x)/h}$, with

$$\partial_t \mathbf{v}^h + \sum_{j=1}^d A_j(\mathbf{v}^h)\partial_j \mathbf{v}^h = \frac{h}{2} L\mathbf{v}^h + \mathsf{S}^h(t,x), \text{ with } \mathbf{v}^h(t,x) = \begin{pmatrix} \mathrm{Re}\, a_N^h \\ \mathrm{Im}\, a_N^h \\ \partial_1 \phi_N^h \\ \vdots \\ \partial_d \phi_N^h \end{pmatrix},$$

$$\mathsf{S}^h(t,x) = \begin{pmatrix} \mathrm{Re}\left(K_1 a_N^h\right) + \mathrm{Re}\left(\left(K_2 - ihK_3 + \frac{\left(t_0^h\right)^d}{(t+t_0^h)^2}\Xi_1^h\right)e^{-i\phi_N^h/h}\right) \\ \mathrm{Im}\left(K_1 a_N^h\right) + \mathrm{Im}\left(\left(K_2 - ihK_3 + \frac{\left(t_0^h\right)^d}{(t+t_0^h)^2}\Xi_1^h\right)e^{-i\phi_N^h/h}\right) \\ -(t+t_0^h)^{(N+1)d-2}\partial_1 K_0 - \frac{\left(t_0^h\right)^d}{(t+t_0^h)^2}\partial_1\Xi_0^h \\ \vdots \\ -(t+t_0^h)^{(N+1)d-2}\partial_d K_0 - \frac{\left(t_0^h\right)^d}{(t+t_0^h)^2}\partial_n\Xi_0^h \end{pmatrix},$$

where the matrices A_j are the same as in Sec. 9.1.2 and the functions in the definitions of $\mathbf{v}^h$ and S^h are evaluated at $(t+t_0^h,x)$. We can proceed like in Sec. 9.1.2: the new term is the source S^h. Unlike for the exact solution, the oscillatory aspect of the problem has not disappeared: the first two components of S^h contain a highly oscillatory factor. Therefore, we cannot expect h-independent energy estimates here. To measure the effect of this oscillatory term, forget the shift in time, and take $t_0^h = 0$. Then assuming that for small times, $\partial_x^\beta \phi_N^h(t,x) = \mathcal{O}(t^{d-1})$ for any multi-index β (like for the exact solution), the H^s norms of the first two components of S^h are controlled by

$$\mathcal{O}\left(\frac{t^{1+s(d-1)}}{h^s}\right).$$

Back to the initial variables, this yields a control by

$$\left(\frac{\varepsilon^\gamma}{1-t}\right)^{1+s(d-1)} \varepsilon^{-s(1-\gamma)} = \frac{\varepsilon^{\gamma+s\alpha-s}}{(1-t)^{1+s(d-1)}}.$$

This is small for $1 - t \gg \varepsilon^\omega$, with
$$\omega = \frac{\gamma + s\alpha - s}{1 + s(d-1)}.$$
We check that for $d > \alpha > 1$, we have
$$\frac{\alpha - 1}{d - 1} < \omega = \frac{\gamma + s(\alpha - 1)}{1 + s(d-1)} < \frac{2\alpha - 1}{2d - 1}, \text{ for all } s \geqslant 0.$$
The first inequality means that we can expect the formal approximation to be a good approximation of the exact solution beyond the first boundary layer (which holds true). The second one explains why the approximation ceases to be relevant before the second boundary layer.

A possible way to understand the above computation is that the choice of the variables is crucial: working with the "usual" unknown v^ε (as in Sec. 9.1.1) is not very efficient. On the other hand, with the variables introduced by E. Grenier for his generalized WKB methods, a precise and rigorous analysis is possible, *via* the transform (9.6).

Remark 9.12. Even though there is stability in a reasonable sense for the limiting hyperbolic system (9.10), small perturbations of the initial amplitude a_0 may drastically alter the asymptotic behavior of u^ε; see Sec. 5.3.

9.2 And beyond?

Using the analysis of Sec. 4.2.1, we could not only describe u^ε for $t \leqslant 1 - \Lambda\varepsilon^\gamma$ in the limit $\Lambda \to +\infty$, but also for $t \leqslant 1 - \varepsilon^\gamma/T$, where T is given by Proposition 9.9. The main differences with the approximate solution of Theorem 9.2 is that the amplitude a can no longer be approximated by its initial value a_0, and a phase modulation must be inserted (which was denoted $\Phi^{(1)}$ in Sec. 4.2.1). This shows that for $t = 1 - \varepsilon^\gamma/T$, the amplitude of u^ε is of order $\varepsilon^{-d\gamma/2} = \varepsilon^{-\alpha/2}$ in L^∞. Moreover, the potential term in the nonlinear energy is of order $\mathcal{O}(1)$ exactly:
$$\varepsilon^\alpha \left\| u^\varepsilon \left(1 - \varepsilon^\gamma/T\right) \right\|_{L^4}^4 = \varepsilon^\alpha \int_{\mathbb{R}^d} \left(\frac{T}{\varepsilon^\gamma}\right)^{2d} \left| \psi^h \left(T, \frac{Tx}{\varepsilon^\gamma}\right) \right|^4 dx$$
$$= \varepsilon^\alpha \left(\frac{T}{\varepsilon^\gamma}\right)^d \left\| \psi^h(T) \right\|_{L^4}^4 = T^d \left\| \psi^h(T) \right\|_{L^4}^4 \approx 1.$$
Since we know that the potential energy is bounded for all time, from the conservation of the energy, this suggests that u^ε might have reached its maximal order of magnitude at time $t = 1 - \varepsilon^\gamma/T$, as guessed in the introduction of this chapter.

To know how u^ε evolves past this time, it seems reasonable to introduce the (L^2 unitary) scaling transform,

$$u^\varepsilon(t,x) = \frac{1}{\varepsilon^{d\gamma/2}}\varphi^\varepsilon\left(\frac{t-1}{\varepsilon^\gamma},\frac{x}{\varepsilon^\gamma}\right) = \frac{1}{\varepsilon^{\alpha/2}}\varphi^\varepsilon\left(\frac{t-1}{\varepsilon^\gamma},\frac{x}{\varepsilon^\gamma}\right). \qquad (9.14)$$

With the same change of notation as for ψ, we have

$$ih\partial_t\varphi^h + \frac{h^2}{2}\Delta\varphi^h = \left|\varphi^h\right|^2\varphi^h. \qquad (9.15)$$

For $-1/t_0^h \leqslant t \leqslant -1/(T+t_0^h)$, we have:

$$\varphi^h(t,x) = \left(\frac{-1}{t}\right)^{d/2}\psi^h\left(\frac{-1}{t},\frac{-x}{t}\right)e^{i|x|^2/(2ht)}$$

$$= \left(\frac{-1}{t}\right)^{d/2}a^h\left(\frac{-1}{t},\frac{-x}{t}\right)\exp\left(\frac{i}{h}\left(\frac{|x|^2}{2t}+\phi^h\left(\frac{-1}{t},\frac{-x}{t}\right)\right)\right)$$

$$=: \mathbf{a}^h(t,x)e^{i\Phi^h(t,x)/h}. \qquad (9.16)$$

The phase Φ^h is no longer in Sobolev spaces, because of the quadratic term. However, it enters the class studied in Sec. 4.2.2. So to go further into the analysis, we meet again a question which was natural in Chap. 4: what happens when singularities appear in the limiting Euler equation? What does it mean for u^ε?

We have seen that because we study a focal point in a supercritical régime, new frequencies have appeared. There are potentially other possible effects. For instance, is there a (different) notion of caustic for φ^h, that is, in a supercritical WKB régime? Indeed, we know that we have to expect the solution of the limiting Euler system to develop singularities in finite time ([Chemin (1990); Makino *et al.* (1986); Xin (1998)]), but this tells us nothing about φ^h. Typically, we do not expect the L^∞ norm of φ^h to be unbounded, since both its L^2 norm and its L^4 norm are bounded for all time and all h. Of course, this does not imply that the L^∞ norm of φ^h is bounded, but this is a rather appealing property. As suggested in Chap. 6, this might mean that two notions of caustic could be distinguished in supercritical régimes: a geometrical notion (the rays along which the amplitude is carried cease to form a diffeomorphism of the whole space), and an analytical notion (the L^∞ norm of the wave function becomes unbounded as the semi-classical parameter goes to zero). These two notions might be disconnected in supercritical cases, and other analytical mechanisms may be involved.

In the linear, subcritical and critical cases, we have seen that past a single focal point, one phase suffices to describe the rapid oscillations of the wave function, unlike what happens for a cusped caustic in the linear case, for instance. In the present supercritical case, it is not clear even how many phases are necessary to describe the wave function for $t > 1$. As a matter of fact, the geometry of singularities is not understood either: it is not clear that for $t > 1$, u^ε is of order $\mathcal{O}(1)$ again. It might for instance keep the form it has reached at time $t = 1 - \varepsilon^\gamma / T$.

This informal discussion shows that many questions remain open, both in a WKB régime and in a caustic régime. Moreover, these questions are more connected than it may seem at first glance.

PART 3

Coherent States

Chapter 10

The Linear Case

The fact that the center, in phase space, of wavepackets associated to (linear) Schrödinger equations evolves according to classical mechanics was noted first by Ehrenfest [Ehrenfest (1927)]. If ψ^ε solves

$$i\varepsilon\partial_t\psi^\varepsilon + \frac{\varepsilon^2}{2}\Delta\psi^\varepsilon = V(x)\psi^\varepsilon, \quad x \in \mathbb{R}^d,$$

then the center of mass

$$Q^\varepsilon(t) := \int_{\mathbb{R}^d} x|\psi^\varepsilon(t,x)|^2 dx \in \mathbb{R}^d,$$

and the momentum

$$P^\varepsilon(t) = \varepsilon \operatorname{Im} \int_{\mathbb{R}^d} \overline{\psi^\varepsilon}(t,x)\nabla\psi^\varepsilon(t,x)dx \in \mathbb{R}^d,$$

solve the system of ordinary differential equations

$$\dot{Q}^\varepsilon(t) = P^\varepsilon(t) \quad ; \quad \dot{P}^\varepsilon(t) = -\int_{\mathbb{R}^d} \nabla V(x)|\psi^\varepsilon(t,x)|^2 dx.$$

In the case where V is the harmonic potential, $V(x) = |x|^2/2$, $\dot{P}^\varepsilon = -Q^\varepsilon$, and we recover the classical Hamiltonian system (1.11).

This principle was resumed in chemistry: the use of Gaussian wavepackets for solving time-dependent Schrödinger equations seems to go back to Heller [Heller (1975)]. At about the same time as Heller, mathematicians have also considered states which are different from WKB states as in Part 1, namely *coherent states*. Contrary to WKB states, coherent states are localized in space and frequency or, to adopt the terminology from quantum mechanics, in position and momentum. The mathematical analysis of coherent states seems to go back to the seminal work of Hepp [Hepp (1974)], followed in this direction by a series of articles by Hagedorn, including [Hagedorn (1980, 1981)]. The term coherent states now covers several fields in mathematics, e.g. groups and algebras in quantum theory, complex geometry, Lie groups, Lie algebras, dynamical systems, analysis of partial differential equations (including many-body theory). For an approach guided by the analysis point of view (with also results concerning Lie groups), the book by Combescure and Robert, [Combescure and Robert (2012)], is certainly the most comprehensive reference on the subject. Several shorter reviews are very interesting too, like e.g. [Combescure and Robert (2006); Paul (1997)].

To be consistent with the standard notations in quantum mechanics, in this part, the position will be denoted by q, and the momentum by p. Correspondingly, the classical trajectories (solutions of (1.11)) will be denoted likewise. The notion of coherent states that we adopt consists in considering initial data of the form

$$\psi^\varepsilon(0,x) = \frac{1}{\varepsilon^{d/4}} a\left(\frac{x-q_0}{\sqrt{\varepsilon}}\right) e^{ip_0\cdot(x-q_0)/\varepsilon}, \quad (q_0,p_0) \in \mathbb{R}^{2d}, \qquad (10.1)$$

where, in agreement with several references considering the evolution of coherent states, and unlike the convention used so far in this book, the main unknown function is denoted by ψ^ε (as opposed to u^ε so far). This initial coherent state in centered at $(q_0,p_0) \in \mathbb{R}^{2d}$ in phase space. We note that the L^2-norm of $\psi^\varepsilon(0,\cdot)$ is independent of ε,

$$\|\psi^\varepsilon(0,\cdot)\|_{L^2(\mathbb{R}^d)} = \|a\|_{L^2(\mathbb{R}^d)}.$$

The specific scaling in terms of ε is motivated by the uncertainty principle, as explained below.

10.1   On the uncertainty principle

In quantum mechanics, the uncertainty principle, due to Heisenberg, asserts that it is impossible to measure precisely both the position and the momentum. From a mathematical point of view, a way to quantify this is to say that a function cannot be compactly supported with a compactly supported Fourier transform. We refer to [Fefferman (1983)] for important mathematical ideas refining this principle. A very loose and simple way to state the Heisenberg uncertainty principle is the following estimate:

Lemma 10.1. *For all* $u \in \mathcal{S}(\mathbb{R}^d)$,

$$\|u\|_{L^2}^2 \leqslant \frac{2}{d} \|xu\|_{L^2} \|\nabla u\|_{L^2}.$$

Proof. This inequality can easily be established thanks to an integration by parts, considering $\int \bar{u}x \cdot \nabla u$ (see e.g. [Rauch (1991)], §3.2). Alternatively, we may invoke the fact that the spectrum of the harmonic oscillator is well known (see e.g. [Landau and Lifschitz (1967)],

$$\sigma_p(H) = \left\{ \frac{d}{2} + k; \quad k \in \mathbb{N} \right\}, \quad H = -\frac{1}{2}\Delta + \frac{|x|^2}{2}.$$

This implies that for all $u \in \mathcal{S}(\mathbb{R}^d)$,

$$\frac{d}{2}\|u\|_{L^2}^2 \leqslant \langle u, Hu \rangle = \frac{1}{2}\|\nabla u\|_{L^2}^2 + \frac{1}{2}\|xu\|_{L^2}^2.$$

For $\lambda > 0$, replacing u with u_λ, $u_\lambda(x) = \lambda^{d/2}u(\lambda x)$, the above inequality yields

$$\frac{d}{2}\|u\|_{L^2}^2 \leqslant \frac{\lambda^2}{2}\|\nabla u\|_{L^2}^2 + \frac{1}{2\lambda^2}\|xu\|_{L^2}^2.$$

Optimize in λ by setting

$$\lambda^2 = \frac{\|xu\|_{L^2}}{\|\nabla u\|_{L^2}},$$

and the result follows. $\square$

Whether we proceed with an integration by parts and use Cauchy–Schwarz inequality, or we invoke the spectrum of the harmonic oscillator, the proof provides an important piece of information: the case of equality corresponds to the Gaussian

$$u(x) = \kappa e^{-|x|^2/2},$$

which is the ground state of the harmonic oscillator (the eigenstate corresponding to the lowest eigenvalue), up to the value of $\kappa \in \mathbb{R} \setminus \{0\}$ which is irrelevant for the above inequality to be an equality, by homogeneity.

In view of the fact that the Fourier transform exchanges differentiation and multiplication by a polynomial, Plancherel formula yields

$$\|u\|_{L^2}^2 \leqslant \frac{2}{d}\|xu\|_{L^2}\|\xi\hat{u}\|_{L^2}.$$

Let $(q, p) \in \mathbb{R}^{2d}$. Changing the origin in phase space, and using the same rescaling as in the above proof, we infer

$$\|u\|_{L^2}^2 \leqslant \frac{2}{d}\left\|\frac{x-q}{\sqrt{\varepsilon}}u\right\|_{L^2}\left\|(\sqrt{\varepsilon}\nabla - p)u\right\|_{L^2}.$$

The notion of coherent state (10.1) saturates the above inequality, *in terms of* ε, in the sense that the three terms in the above inequality are of order $\mathcal{O}(1)$ in the limit $\varepsilon \to 0$. On the other hand, the envelope a is not necessarily a Gaussian, that is, we do not impose equality in the uncertainty principle. The main reason is that the propagation of a Gaussian under a *nonlinear* Schrödinger flow is no longer a Gaussian (in general — a remarkable exception is the logarithmic Schrödinger equation, see [Bialynicki-Birula and Mycielski (1976); Carles and Nouri (2017)]), so even if a is a Gaussian, the envelope of the propagated coherent state below must not be expected to be a Gaussian.

Remark 10.2 (Wigner measures of coherent states). *Another way to realize that the scaling* (10.1) *saturates the uncertainty principle, as far as scaling is concerned, is to compute the Wigner measure associated to the right-hand side of* (10.1) *(the notion of Wigner measure was recalled in Sec. 3.4). We find that there is only one Wigner measure, given by (see e.g. [Lions and Paul (1993)])*

$$\mu(dx, d\xi) = \|a\|^2_{L^2(\mathbb{R}^d)} \delta_{x=q_0} \otimes \delta_{\xi=p_0}.$$

In the linear case, if the Hamiltonian is a polynomial of order at most two in (x, ξ) (e.g. harmonic oscillator), possibly time-dependent, then Gaussian initial data propagate as time-dependent Gaussian ([Hagedorn (1980); Combescure and Robert (2006)]). The coherent state approximation consists precisely of a quadratic approximation of the initial Hamiltonian, as we now explain.

10.2   Propagation of coherent states

In this chapter, we consider the linear Schrödinger equation

$$i\varepsilon \partial_t \psi^\varepsilon + \frac{\varepsilon^2}{2} \Delta \psi^\varepsilon = V(x)\psi^\varepsilon, \tag{10.2}$$

where V satisfies Assumption 1.7 (stating that V is smooth and at most quadratic), and the initial datum is given by (10.1). Even though Assumption 1.7 concerns time-dependent potentials, we choose to make them time independent in this part, mostly for simplicity. It is an easy exercise to check that under Assumption 1.7 for a time-dependent potential, all the results stated below *on fixed time intervals* remain valid. However, when large time is addressed (typically, Ehrenfest time), extra assumptions regarding the growth of V and its derivatives with respect to time would be needed. Guessing that initial coherent states propagate as coherent states, with a center $(q(t), p(t))$ in phase space, we seek an approximate solution of the form

$$\psi_{app}^{\varepsilon}(t, x) = \frac{1}{\varepsilon^{d/4}} u\left(t, \frac{x - q(t)}{\sqrt{\varepsilon}}\right) e^{i(S(t) + p(t) \cdot (x - q(t)))/\varepsilon}. \tag{10.3}$$

We proceed as in Sec. 1.2: we plug this approximate solution into (10.2), and cancel the lowest powers of ε. An important novelty is that now, the relevant space variable is no longer x, but

$$y = \frac{x - q(t)}{\sqrt{\varepsilon}},$$

and the envelope u is a function of (t, y). In particular, $x = q(t) + y\sqrt{\varepsilon}$, and expecting that wave packets are carried by localized functions (typically, Gaussians), it makes sense to perform Taylor expansions about the point $q(t)$, every time the variable x appears:

$$V(x) = V(q(t)) + \sqrt{\varepsilon} y \cdot \nabla V(q(t)) + \frac{\varepsilon}{2} \langle y, \nabla^2 V(q(t)) y \rangle + \mathcal{O}\left(\varepsilon^{3/2}\right),$$

where $\nabla^2 V$ is the Hessian matrix of V, and we will be more precise about the remainder term later. We emphasize however that if V is a polynomial of order at most two (which simply means under Assumption 1.7, that V is a polynomial), then the above Taylor expansion is exact, and the remainder $\mathcal{O}\left(\varepsilon^{3/2}\right)$ is actually zero. We then write

$$i\varepsilon \partial_t \psi_{app}^{\varepsilon} + \frac{\varepsilon^2}{2} \Delta \psi_{app}^{\varepsilon} - V(x) \psi_{app}^{\varepsilon} = \frac{1}{\varepsilon^{d/4}} \left(b_0 + \sqrt{\varepsilon} b_1 + \varepsilon b_2 + \mathcal{O}\left(\varepsilon^{3/2}\right)\right) e^{i\phi/\varepsilon},$$

where

$$\phi(t, x) = S(t) + p(t) \cdot (x - q(t)),$$

$$b_0 = -u\left(\dot{S}(t) - p(t) \cdot \dot{q}(t) + \frac{|p(t)|^2}{2} + V(q(t))\right),$$

$$b_1 = -i(\dot{q}(t) - p(t)) \cdot \nabla u - y \cdot (\dot{p}(t) + \nabla V(q(t))) u,$$

$$b_2 = i\partial_t u + \frac{1}{2}\Delta u - \frac{1}{2}\langle y, \nabla^2 V(q(t)) y \rangle u.$$

The approximate solution (at leading order) is characterized by setting $b_0 = b_1 = b_2 = 0$. Unlike what we did in the case of WKB expansion, we start by canceling b_1. This term is a time dependent linear combination of ∇u and yu, and we cancel each coefficient:

$$\dot{q}(t) = p(t), \quad \dot{p}(t) = -\nabla V(q(t)).$$

To agree with the initial coherent state, we impose $q(0) = q_0$ and $p(0) = p_0$, so that (q, p) is given by the classical Hamiltonian flow,

$$\dot{q} = p, \quad \dot{p} = -\nabla V(q); \quad q(0) = q_0, \quad p(0) = p_0. \tag{10.4}$$

Since we will be interested in the large time régime, we refine Lemma 1.3, under Assumption 1.7:

Lemma 10.3. *Let* $(q_0, p_0) \in \mathbb{R}^{2d}$. *For* V *at most quadratic like in Assumption 1.7, (10.4) has a unique global, smooth solution* $(q, p) \in C^{\infty}(\mathbb{R}; \mathbb{R}^d)^2$. *It grows at most exponentially:*

$$\exists C_0 > 0, \quad |q(t)| + |p(t)| \lesssim e^{C_0 |t|}, \quad \forall t \in \mathbb{R}.$$

We note that the lemma is sharp in the case $V(x) = -|x|^2$. We will return to this case at the end of the chapter.

Proof. The exponential control stems for instance from the fact that q solves

$$\ddot{q}(t) + \nabla V(q(t)) = 0,$$

which implies, after multiplication by $\dot{q}(t)$ and integration,

$$\frac{d}{dt} \left(\frac{1}{2} (\dot{q}(t))^2 + V(q(t)) \right) = 0. \tag{10.5}$$

Since V is at most quadratic, $|V(y)| \lesssim 1 + |y|^2$, and Gronwall lemma yields $|q(t)| \lesssim e^{C|t|}$. From (10.5), we also have $|\dot{q}(t)| \lesssim e^{C|t|}$, hence the lemma. $\square$

Now we choose to cancel b_0, and find

$$\dot{S}(t) = p(t) \cdot \dot{q}(t) - \frac{|p(t)|^2}{2} - V(q(t)) = \frac{|p(t)|^2}{2} - V(q(t)),$$

where the second equality is due to (10.4) (hence the interest of canceling b_1 first). Since $S(0) = 0$, we find

$$S(t) = \int_0^t \left(\frac{|p(s)|^2}{2} - V((q(s))) \right) ds, \tag{10.6}$$

which is the *classical action* (the action of classical mechanics). Finally, setting $b_2 = 0$, we get

$$i\partial_t u + \frac{1}{2} \Delta u = \frac{1}{2} \langle y, \nabla^2 V(q(t)) y \rangle u \quad ; \quad u_{|t=0} = a. \tag{10.7}$$

A remarkable feature of this Schrödinger evolution equation is that the potential is a polynomial in y (the actual space variable for u), of order at most two, a *time dependent* harmonic potential if the Hessian of V at

$q(t)$ is not zero. The approximate coherent state envelope thus consists of a quadratic approximation of the initial Hamiltonian along the classical trajectory. Note also that in the multidimensional case $d \geqslant 2$, there is no reason in general for the matrix $\nabla^2 V(q(t))$ (the Hessian of V at $q(t)$) to be diagonal, or at least diagonalizable in a time independent basis. We will return to more precise properties of the evolution operator associated to Eq. (10.7) later. At this stage, we simply note that $u(t) = U(t,0)a$, where the notion of propagator was recalled in Sec. 1.4.2 (and $\varepsilon = 1$ in the case of Eq. (10.7)).

The notion of approximate coherent state that we address in this book consists precisely in canceling b_0, b_1 and b_2. One could of course construct higher order approximate solutions, an aspect that we leave out here, and refer to [Combescure and Robert (1997)] instead. On the other hand, we recall that is V is a polynomial (of degree at most two), then we actually made no approximation, $\psi^\varepsilon_{\text{app}} = \psi^\varepsilon$.

10.3 The Gaussian case

As noted in (at least) [Hepp (1974); Heller (1975)] and [Hagedorn (1980, 1981)], considering quadratic Hamiltonians like in (10.7) provides a large family of explicit solutions: if the initial datum a is a Gaussian, then so is $u(t, \cdot)$, at all time $t \in \mathbb{R}$. In other words, solving the partial differential equation (10.7) turns out to simply solving ordinary differential equations, involving the time dependent coefficients of the Gaussian $u(t, \cdot)$. We simply outline the computation in the one-dimensional case, and refer to [Hagedorn (1980)] (see also [Lasser and Lubich (2020); Lubich (2008)]) for the general multidimensional case.

Suppose $d = 1$, and that a is of the form

$$a(y) = \beta_0 e^{-\gamma_0(y-y_0)^2/2}, \quad y_0 \in \mathbb{R}, \ \beta_0, \gamma_0 \in \mathbb{C}, \ \mathrm{Re}\,\gamma_0 > 0.$$

For $d = 1$, (10.7) simply reads

$$i\partial_t u + \frac{1}{2}\partial_y^2 u = \frac{1}{2}V''(q(t))\, y^2 u \quad ; \quad u_{|t=0} = a. \tag{10.8}$$

The solution is sought under the form $u(t, y) = \beta(t)e^{-\gamma(t)(y-y(t))^2/2+iy\eta(t)}$,

$y(t), \eta(t) \in \mathbb{R}$, and we compute

$$i\partial_t u + \frac{1}{2}\partial_y^2 u - \frac{1}{2}V''\left(q(t)\right)y^2 = e^{-\gamma(t)(y-y(t))^2/2}\left(c_0(t) + c_1(t)y + c_2(t)y^2\right),$$

where $c_0 = i\dot{\beta} - i\beta\dot{\gamma}\dfrac{y(t)^2}{2} - i\beta\gamma y(t)\dot{y} - \dfrac{\beta\gamma}{2} + \beta\gamma^2\dfrac{y(t)^2}{2} - \dfrac{\eta^2\beta}{2} + i\gamma\eta\beta y(t),$

$\quad c_1 = \beta(t)\left(i\dot{\gamma}y(t) + i\gamma\dot{y}(t) - \gamma^2 y(t) - \dot{\eta} - i\gamma\eta\right),$

$\quad c_2 = \dfrac{\beta(t)}{2}\left(-i\dot{\gamma}(t) + \gamma(t)^2 - V''\left(q(t)\right)\right).$

Canceling c_0, c_1 and c_2 yields, since $y(t)$ and $\eta(t)$ are real-valued:

$$i\dot{\gamma}(t) = \gamma(t)^2 - V''\left(q(t)\right),$$

$$\dot{y}(t) = \eta(t), \quad \ddot{y}(t) + V''\left(q(t)\right)y(t) = 0,$$

$$i\dot{\beta}(t) = \frac{1}{2}\left(\gamma(t) - V''\left(q(t)\right)y(t)^2 + \eta(t)^2\right)\beta(t).$$

Separating the real and imaginary parts of γ, $\gamma = \gamma_r + i\gamma_i$, the first equation is equivalent to

$$\dot{\gamma}_r = 2\gamma_i\gamma_r \quad ; \quad \dot{\gamma}_i = \gamma_i^2 - \gamma_r^2 + V''\left(q(t)\right).$$

It is then obvious that γ_r remains positive for all time, since it is at initial time. The center $y(t)$ of the Gaussian moves according to a classical oscillator, and the amplitude β is given explicitly by simple integration in time.

We note that for $d \geqslant 2$, if $\nabla^2 V$ is diagonal, that is if we consider an equation of the form

$$i\partial_t u + \frac{1}{2}\Delta u = \frac{1}{2}\sum_{j=1}^{d}\Omega_j(t)y_j^2 u,$$

then a tensorization formula is available: if

$$u(0, y) = \beta_0 \exp\left(-\sum_{j=1}^{d}\gamma_{0j}(y_j - y_{0j})^2/2\right),$$

then $u(t, y) = \beta(t)u_1(t, y_1)u_2(t, y_2)\ldots u_d(t, y_d)$, where

$$i\partial_t u_j + \frac{1}{2}\partial_y^2 u_j = \frac{1}{2}\Omega_j(t)y^2 u_j \quad ; \quad u_j(0, y) = e^{-\gamma_{0j}(y-y_{0j})^2/2}$$

is given by the one-dimensional case, and β is given by integration in time in a similar fashion as above.

To conclude this section, note that linear combinations of Gaussians provide L^2-bases, so by linear superposition, the propagation of Gaussian coherent states provides a precise, explicit description of the evolution of solutions to (10.2): decompose the initial datum $\psi^\varepsilon(0, \cdot)$ on a Gaussian wave packets basis, and consider the propagation of each coherent states. This idea is the starting point of several numerical methods, inspired by [Heller (1975); Hagedorn (1980)]; see e.g. [Jin *et al.* (2011); Lasser and Lubich (2020); Lubich (2008)] and references therein.

10.4 Error estimate and Ehrenfest time

To measure the error between ψ^ε and $\psi^\varepsilon_{\text{app}}$, we proceed in a way very similar to the approach followed in Sec. 1.3.2, and change the unknown function ψ^ε to u^ε, through the relation

$$\psi^\varepsilon(t, x) = \frac{1}{\varepsilon^{d/4}} u^\varepsilon\left(t, \frac{x - q(t)}{\sqrt{\varepsilon}}\right) e^{i(S(t) + p(t) \cdot (x - q(t)))/\varepsilon}, \tag{10.9}$$

where $(q(t), p(t))$ is given by (10.4), S is given by (10.6), so it is equivalent to consider ψ^ε or u^ε. In addition,

$$\|\psi^\varepsilon(t) - \psi^\varepsilon_{\text{app}}(t)\|_{L^2(\mathbb{R}^d)} = \|u^\varepsilon(t) - u(t)\|_{L^2(\mathbb{R}^d)}, \quad \forall t \in \mathbb{R}.$$

The error $w^\varepsilon = u^\varepsilon - u$ satisfies

$$i\partial_t w^\varepsilon + \frac{1}{2}\Delta w^\varepsilon = V^\varepsilon(t, y)w^\varepsilon + W^\varepsilon(t, y)u \quad ; \quad w^\varepsilon_{|t=0} = 0,$$

where

$$V^\varepsilon(t, y) = \frac{1}{\varepsilon}\left(V\left(q(t) + y\sqrt{\varepsilon}\right) - V\left(q(t)\right) - \sqrt{\varepsilon}y \cdot \nabla V\left(q(t)\right)\right),$$

and the potential W^ε is the remainder term in the previous Taylor expansion,

$$W^\varepsilon(t, y) = V^\varepsilon(t, y) - \frac{1}{2}\left\langle y, \nabla^2 V\left(q(t)\right) y\right\rangle.$$

Recall that if V is a polynomial (of degree at most two), $W^\varepsilon \equiv 0$, and $\psi^\varepsilon = \psi^\varepsilon_{\text{app}}$. If not, under Assumption 1.7, Taylor's formula yields

$$|W^\varepsilon(t, y)| \leqslant C\sqrt{\varepsilon}\left(1 + |y|^3\right),$$

for some constant C independent of $t \in \mathbb{R}$, $y \in \mathbb{R}^d$, and $\varepsilon \in [0, 1]$. The basic energy estimate (Lemma 1.2 with $\varepsilon = 1$ there) yields, along with this inequality,

$$\sup_{0 \leqslant t \leqslant T} \|w^\varepsilon(t)\|_{L^2(\mathbb{R}^d)} \leqslant \int_0^T \|W^\varepsilon(s)u(s)\|_{L^2(\mathbb{R}^d)} ds$$

$$\lesssim \sqrt{\varepsilon} \int_0^T \|\langle y\rangle^3 u(s)\|_{L^2(\mathbb{R}^d)} ds.$$

We readily infer that the error is of order at most $\sqrt{\varepsilon}$ in L^2, on any fixed time interval. But contrary to what we faced with WKB expansions, all the objects at stake are defined globally in time, and we have no restriction such as the formation of a caustic. We can then wonder about large time error estimates: up to what time T^ε (possibly depending on ε) can we ensure

$$\sup_{0\leqslant t\leqslant T^\varepsilon} \|w^\varepsilon(t)\|_{L^2(\mathbb{R}^d)} = o(1) \text{ as } \varepsilon \to 0?$$

The previous estimate reduces this question to the control of the growth of the third momentum of u in L^2 at time goes to infinity.

Lemma 10.4. *Let V satisfying Assumption 1.7, and $a \in \mathcal{S}(\mathbb{R}^d)$. For all $k \in \mathbb{N}$, there exists C_k such that the solution to* (10.7) *satisfies*

$$\|u(t)\|_{\Sigma^k} := \|u(t)\|_{H^k(\mathbb{R}^d)} + \||y|^k u(t)\|_{L^2(\mathbb{R}^d)} \lesssim e^{C_k t}, \quad \forall t \geqslant 0.$$

Proof. For $k = 0$, the result is straightforward, with $C_0 = 0$ (conservation of the L^2-norm). For $k = 1$, we apply the operators ∇ and y to (10.7):

$$\left(i\partial_t + \frac{1}{2}\Delta\right)\nabla u = \frac{1}{2}\left\langle y, \nabla^2 V\left(q(t)\right)y\right\rangle \nabla u + \nabla^2 V\left(q(t)\right)yu,$$

$$\left(i\partial_t + \frac{1}{2}\Delta\right)yu = \frac{1}{2}\left\langle y, \nabla^2 V\left(q(t)\right)y\right\rangle \nabla u + \nabla u.$$

The basic energy estimate yields, since V is at most quadratic,

$$\|\nabla u(t)\|_{L^2} \leqslant \|\nabla a\|_{L^2} + \|\nabla^2 V\|_{L^\infty} \int_0^t \|yu(s)\|_{L^2}ds,$$

$$\|yu(t)\|_{L^2} \leqslant \|ya\|_{L^2} + \int_0^t \|\nabla u(s)\|_{L^2}ds.$$

Gronwall lemma, applied to $\|\nabla u(t)\|_{L^2} + \|yu(t)\|_{L^2}$, yields the lemma for $k = 1$.

For $k \geqslant 2$, we use the commutator formulas, for $1 \leqslant j \leqslant d$,

$$[y_j^k, \Delta] = -ky_j^{k-1}\partial_j - k(k-1)y_j^{k-2},$$

$$[\partial_j^k, \langle y, \nabla^2 V\left(q(t)\right)y\rangle] = \sum_m \partial_{jm}^2 V\left(q(t)\right)y_m\partial_j^{k-1} + 2\partial_{jj}^2 V\left(q(t)\right)\partial_j^{k-2}.$$

Proceeding as in the case $k = 1$, we infer

$$\|u(t)\|_{H^k} + \||y|^k u(t)\|_{L^2} \leqslant \|a\|_{H^k} + \||y|^k a\|_{L^2}$$

$$+ C\int_0^t \left(\|u(s)\|_{H^k} + \||y|^k u(s)\|_{L^2}\right) ds$$

$$+ C\sum_{j=1}^d \int_0^t \left(\||y|^{k-1}\partial_j u(s)\|_{L^2} + \||y|\partial_j^{k-1} u(s)\|_{L^2}\right) ds.$$

We conclude by invoking the estimate

$$\left\| |y|^{k-1} \partial_j u(s) \right\|_{L^2} + \left\| |y| \partial_j^{k-1} u(s) \right\|_{L^2} \lesssim \| u(s) \|_{H^k} + \left\| |y|^k u(s) \right\|_{L^2},$$

which looks fairly natural, but whose complete proof can be cumbersome (see [Ben Abdallah *et al.* (2008); Helffer (1984)]). $\qquad\square$

Remark 10.5. The statement of the lemma relates the Sobolev norm $\| u(t) \|_{H^k}$ with the L^2-norm of the momenta $\left\| |y|^k u(t) \right\|_{L^2}$. The proof suggests some coupling between the two: indeed, in the case of the standard harmonic oscillator ($\nabla^2 V = \mathrm{Id}$), the action of the flow map is to rotate the phase space, and the two norms are definitely linked.

In view of Lemma 10.4 (with $k = 3$), the error estimate becomes

$$\sup_{0 \leqslant t \leqslant T} \| w^\varepsilon(t) \|_{L^2(\mathbb{R}^d)} \lesssim \sqrt{\varepsilon} e^{CT},$$

for some $C > 0$. To summarize, we have proved:

Proposition 10.6. *Let $d \geqslant 1$, V satisfying Assumption 1.7, $(q_0, p_0) \in \mathbb{R}^{2d}$, and $a \in \mathcal{S}(\mathbb{R}^d)$. Let $\psi^\varepsilon \in C(\mathbb{R}; L^2(\mathbb{R}^d))$ be the solution to (10.1)–(10.2), and let $\psi_{\mathrm{app}}^\varepsilon \in C(\mathbb{R}; L^2(\mathbb{R}^d))$ be defined by (10.3), (10.4), (10.6) and (10.7). There exist C_0, C_1 independent of $\varepsilon \in [0,1]$ and $t \geqslant 0$ such that*

$$\| \psi^\varepsilon(t) - \psi_{\mathrm{app}}^\varepsilon(t) \|_{L^2(\mathbb{R}^d)} \leqslant C_0 \sqrt{\varepsilon} e^{C_1 t}, \quad \forall t \geqslant 0.$$

In particular, there exists c independent of $\varepsilon \in [0,1]$ such that

$$\sup_{0 \leqslant t \leqslant c \log \frac{1}{\varepsilon}} \| \psi^\varepsilon(t) - \psi_{\mathrm{app}}^\varepsilon(t) \|_{L^2(\mathbb{R}^d)} \xrightarrow[\varepsilon \to 0]{} 0.$$

The time $c \log 1/\varepsilon$ (with an optimized c) is usually referred to as *Ehrenfest time*: see e.g. [Bambusi *et al.* (1999); Bouzouina and Robert (2002); Schubert *et al.* (2012)] for a more precise analysis, regarding also the sharpness of the order of magnitude $\log 1/\varepsilon$, according to the potential V.

10.5 Optimality of Lemma 10.4

With no extra assumption on the potential than Assumption 1.7, Lemma 10.4 may or may not be sharp, as we show on three cases.

In the case without potential, $V \equiv 0$, we have obviously

$$\left\| e^{i \frac{t}{2} \Delta} a \right\|_{H^k} = \| a \|_{H^k}, \quad \forall t \in \mathbb{R}.$$

On the other hand, the momenta grow algebraically in time, due to dispersion. To measure this phenomenon, resume the operator presented in Proposition 1.26,

$$J(t) = y + it\nabla = ite^{-i|y|^2/(2t)}\nabla\left(e^{-i|y|^2/(2t)} \cdot\right) = e^{i\frac{t}{2}\Delta}\, y\, e^{-i\frac{t}{2}\Delta},$$

where the above identities have been discussed in Sec. 7.1 and Sec. 8.2. The last identity implies that $J(t)e^{i\frac{t}{2}\Delta}a = e^{i\frac{t}{2}\Delta}(ya)$, and so

$$\left\|ye^{i\frac{t}{2}\Delta}a\right\|_{L^2} \leqslant \left\|J(t)e^{i\frac{t}{2}\Delta}a\right\|_{L^2} + t\left\|\nabla e^{i\frac{t}{2}\Delta}a\right\|_{L^2} \lesssim 1 + t, \quad t \geqslant 0.$$

Iterating the operator J, we can prove $\||y|^k u(t)\|_{L^2} \lesssim 1 + t^k$. This algebraic control is sharp, as can be seen, in the case $k = 1$, from the virial identity

$$\frac{d^2}{dt^2}\|yu(t)\|_{L^2}^2 = 2\|\nabla u\|_{L^2}^2 = 2\|\nabla a\|_{L^2}^2,$$

which is a consequence of the conservation of the L^2-norms of ∇u and $J(t)u$.

In the case of the harmonic oscillator, $V(y) = |y|^2/2$, the Hermite functions form an L^2 eigenbasis of $H = -\frac{1}{2}\Delta + V$:

$$a(y) = \sum_{\ell \in N} \alpha_\ell \psi_\ell(y),$$

where $(\psi_\ell)_{\ell \in \mathbb{N}}$ are Hermite functions. We infer

$$u(t, y) = \sum_{\ell \in N} \alpha_\ell e^{-it\lambda_\ell}\psi_\ell(y),$$

where λ_ℓ is the eigenvalue associated to ψ_ℓ, $H\psi_\ell = \lambda_\ell\psi_\ell$. But as we have recalled in Sec. 10.1, $\lambda_\ell \in d/2 + \mathbb{N}$, so $e^{-it\lambda_\ell}$ is 4π-periodic for all $\ell \in \mathbb{N}$. Therefore, u is 4π-periodic in time, and one can take $C_k = 0$ in the conclusion of Lemma 10.4: the Sobolev norms and the L^2-norms of the momenta of u are bounded in time.

The exponential bound is sharp in the case of the repulsive, or inverted, harmonic potential, $V(y) = -|y|^2/2$. A quick way to verify that the momentum of u grows exponentially in time in this case is to resume the vector-fields from [Carles (2003a)], presented in Sec. 8.2,

$$J(t) = y\sinh t + i\cosh t\nabla \quad ; \quad H(t) = y\cosh t + i\sinh t\nabla.$$

Since they commute with the equation,

$$\|J(t)u(t)\|_{L^2} = \|\nabla a\|_{L^2}, \quad \|H(t)u(t)\|_{L^2} = \|ya\|_{L^2}, \quad \forall t \in \mathbb{R}.$$

On the other hand,

$$\begin{pmatrix} y \\ i\nabla \end{pmatrix} = \begin{pmatrix} -\sinh t & \cosh t \\ \cosh t & -\sinh t \end{pmatrix}\begin{pmatrix} J(t) \\ H(t) \end{pmatrix},$$

and we infer the virial identity

$$\frac{d^2}{dt^2}\|yu(t)\|_{L^2}^2 = 4\|yu(t)\|_{L^2}^2 + 4E,$$

where E is the conserved energy,

$$E := \frac{1}{2}\|\nabla u(t)\|_{L^2}^2 - \frac{1}{2}\|yu(t)\|_{L^2}^2 \equiv \frac{1}{2}\|\nabla a\|_{L^2}^2 - \frac{1}{2}\|ya\|_{L^2}^2.$$

Therefore,

$$\|yu(t)\|_{L^2}^2 = \|ya\|_{L^2}^2 \cosh(2t) + \sinh(2t)\,\mathrm{Im}\int \bar{a}(y)y\cdot\nabla a(y)dy$$
$$+ (\cosh(2t) - 1)\,E.$$

We conclude that the L^2-norm of the first momentum of u grows exponentially in time (in general), and by conservation of the energy, so does the H^1-norm of u.

10.6 Wave packet transform

For $a \in L^2(\mathbb{R}^d)$ and $z = (q,p) \in \mathbb{R}^{2d}$, denote

$$a_z(x) = \frac{1}{\varepsilon^{d/4}}a\left(\frac{x-q}{\sqrt{\varepsilon}}\right)e^{ip\cdot(x-q)/\varepsilon}.$$

The wave packet transform of a function $\psi : \mathbb{R}^d \to \mathbb{C}$ is

$$W_a^\varepsilon\psi(z) = \frac{1}{(2\pi\varepsilon)^{d/2}}\langle\psi, a_z\rangle = \frac{1}{(2\pi\varepsilon)^{d/2}}\int_{\mathbb{R}^d}\psi(x)\bar{a}_z(x)dx.$$

Like for the Fourier transform, we have an inversion formula, and a Plancherel type formula:

Lemma 10.7. *Let $a \in L^2(\mathbb{R}^d)$, with $\|a\|_{L^2(\mathbb{R}^d)} = 1$. For all $\psi \in L^2(\mathbb{R}^d)$,*

$$\psi(x) = \frac{1}{(2\pi\varepsilon)^d}\int_{\mathbb{R}^{2d}}\langle\psi, a_z\rangle\, a_z(x)dz, \quad a.e.\ x \in \mathbb{R}^d,$$

$$\|\psi\|_{L^2(\mathbb{R}^d)}^2 = \frac{1}{(2\pi\varepsilon)^d}\int_{\mathbb{R}^{2d}}|\langle\psi, a_z\rangle|^2\,dz.$$

Proof. We prove the above formulas for a and ψ in the Schwartz class, so all the computations make sense directly. Expanding the inner product, we have

$$\frac{1}{(2\pi\varepsilon)^d}\int_{\mathbb{R}^{2d}}\langle\psi, a_z\rangle\, a_z(x)dz = \frac{1}{(2\pi\varepsilon)^d}\int_{\mathbb{R}^{3d}}\psi(y)\bar{a}_z(y)a_z(x)dydz.$$

Using Fubini Theorem, we focus on the integral in z:

$$\int_{\mathbb{R}^{2d}} \overline{a}_z(y) a_z(x) dz = \frac{1}{\varepsilon^{d/2}} \int_{\mathbb{R}^{2d}} \overline{a}\left(\frac{y-q}{\sqrt{\varepsilon}}\right) a\left(\frac{x-q}{\sqrt{\varepsilon}}\right) e^{ip\cdot(x-y)/\varepsilon} dq dp$$

$$= \frac{(2\pi\varepsilon)^d}{\varepsilon^{d/2}} \int_{\mathbb{R}^d} \overline{a}\left(\frac{x-q}{\sqrt{\varepsilon}}\right) a\left(\frac{x-q}{\sqrt{\varepsilon}}\right) dq,$$

where we have used Fourier inversion formula. Integrating in q,

$$\frac{1}{\varepsilon^{d/2}} \int_{\mathbb{R}^d} \overline{a}\left(\frac{x-q}{\sqrt{\varepsilon}}\right) a\left(\frac{x-q}{\sqrt{\varepsilon}}\right) dq = \|a\|^2_{L^2(\mathbb{R}^d)},$$

hence the first formula.

For the second formula, consider the inner product of the first formula with ψ (in x),

$$\|\psi\|^2_{L^2(\mathbb{R}^d)} = \frac{1}{(2\pi\varepsilon)^d} \left\langle \int_{\mathbb{R}^{2d}} \langle \psi, a_z \rangle\, a_z dz, \psi \right\rangle$$

$$= \frac{1}{(2\pi\varepsilon)^d} \int_{\mathbb{R}^{2d}} \langle \psi, a_z \rangle \langle a_z, \psi \rangle\, dz = \frac{1}{(2\pi\varepsilon)^d} \int_{\mathbb{R}^{2d}} |\langle \psi, a_z \rangle|^2 dz,$$

and the lemma follows. $\square$

The advantage of the inversion formula is that when a propagator is considered, $e^{-itH/\varepsilon}$, then we can write

$$e^{-itH/\varepsilon}\psi = \frac{1}{(2\pi\varepsilon)^d} \int_{\mathbb{R}^{2d}} \langle \psi, a_z \rangle\, e^{-itH/\varepsilon} a_z dz, \qquad (10.10)$$

so the problem boils down to the propagation of coherent states.

We do not examine the wave packet transform into more details, and conclude by noticing that it goes under different names: in [Combescure and Robert (2012)], the wave-packet transform is called Fourier–Bargmann transform. In the case where a is a Gaussian, the wave-packet transform is usually referred to as FBI transform (for Fourier–Bros–Iagolnitzer), see e.g. [Delort (1992); Martinez (2002); Zworski (2012)]. Such transformations have many applications in microlocal analysis (linear or nonlinear), and we simply evoke one, in the context of semi-classical analysis related to quantum chemistry, namely the Herman–Kluk formula, which exploits Eq. (10.10) to obtain a simplified expression for $e^{-itH/\varepsilon}$, up to some small error in the semi-classical limit; see [Robert (2010); Swart and Rousse (2009)].

Chapter 11

Nonlinear Coherent States: Main Tools

11.1 Notion of criticality

Mimicking the approach followed in the first part of the book, we consider a nonlinear counterpart to (10.2), keeping the initial data (10.1) fixed, and introducing a dependence upon ε in the coupling constant in front of the nonlinearity:

$$i\varepsilon\partial_t\psi^\varepsilon + \frac{\varepsilon^2}{2}\Delta\psi^\varepsilon = V(x)\psi^\varepsilon + \lambda\varepsilon^\alpha|\psi^\varepsilon|^{2\sigma}\psi^\varepsilon, \tag{11.1}$$

with $\sigma > 0$ and $\lambda \in \mathbb{R}$. Throughout all the third part of this book, V satisfies Assumption 1.7. Because striking results are available then, we also consider nonlocal nonlinearities here,

$$i\varepsilon\partial_t\psi^\varepsilon + \frac{\varepsilon^2}{2}\Delta\psi^\varepsilon = V(x)\psi^\varepsilon + \varepsilon^\alpha\left(K * |\psi^\varepsilon|^2\right)\psi^\varepsilon. \tag{11.2}$$

We begin by identifying critical values for α in terms of the influence of the nonlinearity, according to the type of nonlinearity. To do so, we simply resume the computation presented in Sec. 10.2, and consider the presence not only of the term b_0, $\sqrt{\varepsilon}b_1$ and εb_2, but also of a term $b_{\mathrm{nl}}^\varepsilon$, accounting for nonlinear effects at leading order.

Power-like nonlinearity

In the case of (11.1), the new term $b_{\mathrm{nl}}^\varepsilon$ is given by

$$b_{\mathrm{nl}}^\varepsilon = \lambda\varepsilon^\alpha|\psi^\varepsilon_{\mathrm{app}}|^{2\sigma}\psi^\varepsilon_{\mathrm{app}}e^{-i\phi/\varepsilon} = \lambda\varepsilon^\alpha \times \left(\frac{1}{\varepsilon^{d/4}}\right)^{2\sigma}|u|^{2\sigma}u.$$

Like in the case of WKB analysis, we consider the value α_c critical when the nonlinear term alters the definition of the approximate solution $\psi^\varepsilon_{\mathrm{app}}$, while for $\alpha > \alpha_c$, the ansatz is the same as in the linear case. In the present case, we find

$$\alpha_c = 1 + \frac{d\sigma}{2}.$$

For $\alpha = \alpha_c$, the envelope equation is different, and (10.7) becomes

$$i\partial_t u + \frac{1}{2}\Delta u = \frac{1}{2}\langle y, \nabla^2 V\left(q(t)\right) y\rangle u + \lambda|u|^{2\sigma}u \quad ; \quad u_{|t=0} = a. \qquad (11.3)$$

Hartree-type nonlinearity: homogeneous kernel

In the case of (11.2) with $K(x) = \lambda/|x|^\gamma$, $\lambda \in \mathbb{R}$ and $\gamma > 0$, we compute

$$b_{\text{nl}}^\varepsilon = \lambda\varepsilon^\alpha \left(\int_{\mathbb{R}^d} |z|^{-\gamma}|\psi_{\text{app}}^\varepsilon(t, x - z)|^2 dz\right) \psi_{\text{app}}^\varepsilon(t, x)e^{-i\phi/\varepsilon}$$

$$= \lambda\varepsilon^\alpha \left(\frac{1}{\varepsilon^{d/2}}\int_{\mathbb{R}^d} |z|^{-\gamma}\left|u\left(t, \frac{x - z - q(t)}{\sqrt{\varepsilon}}\right)\right|^2 dz\right) u(t, y),$$

where we recall the identity $x = q(t) + y\sqrt{\varepsilon}$. Using this identity to simplify the convolution, we find

$$b_{\text{nl}}^\varepsilon = \lambda\varepsilon^\alpha \left(\frac{1}{\varepsilon^{d/2}}\int_{\mathbb{R}^d} |z|^{-\gamma}\left|u\left(t, y - \frac{z}{\sqrt{\varepsilon}}\right)\right|^2 dz\right) u(t, y)$$

$$= \lambda\varepsilon^{\alpha-\gamma/2} \left(\int_{\mathbb{R}^d} |z|^{-\gamma}\left|u\left(t, y - z\right)\right|^2 dz\right) u(t, y),$$

where we have changed the variable $z \to z\sqrt{\varepsilon}$ and used the homogeneity of the kernel. Therefore,

$$\alpha_c = 1 + \frac{\gamma}{2},$$

and the envelope equation in the critical case becomes

$$i\partial_t u + \frac{1}{2}\Delta u = \frac{1}{2}\langle y, \nabla^2 V\left(q(t)\right) y\rangle u + \lambda\left(|y|^{-\gamma} * |u|^2\right) u \quad ; \quad u_{|t=0} = a.$$

Hartree-type nonlinearity: smooth kernel

In the case of (11.2) with K smooth, real-valued, bounded as well as its derivatives (for instance $K \in \mathcal{S}(\mathbb{R}^d; \mathbb{R})$), we have now

$$b_{\text{nl}}^\varepsilon = \varepsilon^\alpha \left(\frac{1}{\varepsilon^{d/2}}\int_{\mathbb{R}^d} K(z)\left|u\left(t, y - \frac{z}{\sqrt{\varepsilon}}\right)\right|^2 dz\right) u(t, y)$$

$$= \varepsilon^\alpha \left(\int_{\mathbb{R}^d} K(z\sqrt{\varepsilon})\left|u\left(t, y - z\right)\right|^2 dz\right) u(t, y)$$

$$= \varepsilon^\alpha \left(K(0)\|u(t)\|_{L^2(\mathbb{R}^d)}^2 u(t, y) + \mathcal{O}\left(\sqrt{\varepsilon}\right)\right),$$

where we have performed a Taylor expansion for the term $K\left(z\sqrt{\varepsilon}\right)$. Therefore, we find, like in the case of WKB analysis,

$$\alpha_c = 1,$$

and the envelope equation in the critical case becomes

$$i\partial_t u + \frac{1}{2}\Delta u = \frac{1}{2}\left\langle y, \nabla^2 V\left(q(t)\right)y\right\rangle u + K(0)\|u(t)\|_{L^2(\mathbb{R}^d)}^2 u \quad ; \quad u_{|t=0} = a.$$

It turns out that the effect of the nonlinearity is very weak there: the L^2-norm of u is independent of time (K is real-valued), and $u(t,y)e^{itK(0)\|a\|_{L^2}^2}$ solves (10.7).

11.2 NLS with a time-dependent potential

In this section, we focus our attention on the nonlinear equation (11.3). We shall not address the counterpart of this equation with a Hartree-type non-linearity and a homogeneous kernel, and refer to [Cao and Carles (2011)] for nonlinear results in that case. Discarding the origin of the time dependence of the potential in (11.3), we consider

$$i\partial_t u + \frac{1}{2}\Delta u = \frac{1}{2}\left\langle y, M(t)y\right\rangle u + \lambda|u|^{2\sigma}u \quad ; \quad u_{|t=0} = a, \tag{11.4}$$

where $M(t)$ is a symmetric, real-valued matrix, bounded (as a function of time) as well as all its derivatives.

11.2.1 *Some algebraic miracles*

The linear case $\lambda = 0$, with M constant and diagonal in (11.4), corresponds to the standard harmonic oscillator (8.3) (with $\varepsilon = 1$). The fundamental solution is explicit (see e.g. [Feynman and Hibbs (1965)]), the formula is known as *Mehler's formula*, see (8.4). In the case where M is symmetric and depends on time, a generalized Mehler's formula is available, as studied in details in [Hörmander (1995)]. The potential is not necessarily homogeneous of degree two: it has to be a polynomial in y, of degree at most two. Recall that in the more general case where the external potential is at most quadratic, in the sense of Assumption 1.7, we know from [Fujiwara (1979, 1980)] that there exists $\eta > 0$ such that for $|t| < \eta$, the solution to

$$i\partial_t u_{\mathrm{lin}} + \frac{1}{2}\Delta u_{\mathrm{lin}} = V(t,y)u_{\mathrm{lin}} \quad ; \quad u_{\mathrm{lin}|t=0} = a,$$

can be expressed as

$$u_{\text{lin}}(t,y) = \frac{1}{(2i\pi t)^{d/2}} \int_{\mathbb{R}^d} e^{i\varphi(t,y,z)} A(t,y,z) a(z) dz,$$

where $A(0,y,z) = 1$, $\partial_y^\alpha \partial_z^\beta A \in L^\infty(]-\eta,\eta[\times\mathbb{R}^d \times \mathbb{R}^d)$ for all $\alpha, \beta \in \mathbb{N}^d$, and

$$\varphi(t,x,y) = \frac{|x-y|^2}{2t} + t\xi(t,x,y),$$

with $\partial_x^\alpha \partial_y^\beta \xi \in L^\infty(]-\eta,\eta[\times\mathbb{R}^d \times \mathbb{R}^d)$ as soon as $|\alpha + \beta| \geqslant 2$. This general formula can be simplified, and made more or less explicit, in the case where V is polynomial, as we illustrate in the case $d = 1$: the solution to

$$i\partial_t u_{\text{lin}} + \frac{1}{2}\partial_y^2 u_{\text{lin}} = \frac{1}{2}\Omega(t)y^2 u_{\text{lin}} \quad ; \quad u_{\text{lin}|t=0} = a,$$

is given formally by

$$u_{\text{lin}}(t,y) = \frac{1}{\sqrt{2i\pi\mu(t)}} \int_{\mathbb{R}} e^{i(\alpha(t)y^2 + 2\beta(t)yz + \gamma(t)z^2)/2} a(z) dz,$$

where

$$\ddot{\mu} + \Omega(t)\mu = 0 \quad ; \quad \mu(0) = 0, \quad \dot{\mu}(0) = 1,$$

$$\alpha = \frac{\dot{\mu}}{\mu} \quad ; \quad \beta = -\frac{1}{\mu} \quad ; \quad \gamma(t) = \frac{1}{\mu(t)\dot{\mu}(t)} - \int_0^t \frac{\Omega(\tau)}{(\dot{\mu}(\tau))^2} d\tau.$$

If Ω is smooth, then the formula for u_{lin} is valid for $|t| < T$ and $T > 0$ sufficiently small. In general, T is finite, as shown by the case $\Omega \equiv 1$ (see (8.4)), where we recover the standard Mehler's formula

$$\mu(t) = \sin t, \quad \alpha(t) = \gamma(t) = \cot t, \quad \beta(t) = -\frac{1}{\sin t}.$$

In the multidimensional case, in the case where $M(t)$ (from (11.4)) is diagonal,

$$M(t) = \begin{pmatrix} \Omega_1(t) & 0 & \dots & 0 \\ 0 & \Omega_2(t) & \dots & 0 \\ \vdots & & \ddots & \\ 0 & \dots & 0 & \Omega_d(t) \end{pmatrix},$$

we can simply tensorize the one-dimensional formula,

$$u_{\text{lin}}(t,y) = u_1(t,y_1) u_2(t,y_2) \dots u_d(t,y_d),$$

where u_j stands for the one-dimensional solution associated with Ω_j like in the above formula. On the other hand, as already pointed out, for a general symmetric matrix $M(t)$, the diagonalization of $M(t)$ may involve a change of basis matrix depending on time, so the Schrödinger equation written in the moving frame may contain extra terms, ruining the expected algebraic simplification.

In the very specific case where $M(t)$ is proportional to the identity matrix (which is always the case if $d = 1$),

$$M(t) = \Omega(t)\mathrm{Id}, \quad \Omega(t) \in \mathbb{R},$$

then we have an extra algebraic miracle, known as *generalized lens transform*. The lens transform refers to the case where $\Omega(t) \equiv 1$ (harmonic oscillator). Consider μ as above, and ν its companion, that is

$$\ddot{\nu} + \Omega(t)\nu = 0 \quad ; \quad \nu(0) = 0, \quad \dot{\nu}(0) = 1.$$

Since no damping term is present in the above differential equation, the Wronskian of μ and ν is constant,

$$\dot{\mu}\nu - \mu\dot{\nu} \equiv 1.$$

If v solves

$$i\partial_t v + \frac{1}{2}\Delta v = 0 \quad ; \quad v_{|t=0} = a,$$

then

$$u(t,y) = \frac{1}{\nu(t)^{d/2}} v\left(\frac{\mu(t)}{\nu(t)}, \frac{y}{\nu(t)}\right) e^{i\dot{\nu}(t)|y|^2/(2\nu(t))} \tag{11.5}$$

solves formally

$$i\partial_t u + \frac{1}{2}\Delta u = \frac{1}{2}\Omega(t)|y|^2 u \quad ; \quad u_{|t=0} = a.$$

In the case $\Omega \equiv 1$, $\mu(t) = \sin t$, $\nu(t) = \cos t$, and so $\mu(t)/\nu(t) = \tan t$: the map $v \mapsto u$ compactifies time, as when t varies from $-\pi/2$ to $+\pi/2$, $\tan t$ varies from $-\infty$ to $+\infty$, so while the time variable for v ranges the whole line, the time variable for u remains bounded. This is where the difference between formal computation and rigorous formula lies: (11.5) makes sense as long as $t \mapsto \mu(t)/\nu(t)$ is invertible. This is always the case at least locally in time, since

$$\frac{d}{dt}\left(\frac{\mu(t)}{\nu(t)}\right) = \frac{1}{\nu(t)^2}, \quad \text{hence} \quad \frac{d}{dt}\left(\frac{\mu(t)}{\nu(t)}\right)\Big|_{t=0} = 1,$$

where we have used the conservation of the Wronskian for the first identity.

It turns out that the (generalized) lens transform has applications in the nonlinear case: we readily check that if v solves

$$i\partial_t v + \frac{1}{2}\Delta v = H(t)|v|^{2\sigma}v \quad ; \quad v_{|t=0} = a,$$

then u, given by (11.5) (as long as this makes sense), solves

$$i\partial_t u + \frac{1}{2}\Delta u = \frac{1}{2}\Omega(t)|y|^2 u + h(t)|u|^{2\sigma}u \quad ; \quad u_{|t=0} = a,$$

$$\text{where} \quad h(t) = \nu(t)^{d\sigma-2}H\left(\frac{\mu(t)}{\nu(t)}\right).$$

We refer to [Carles (2011)] for the derivation of the above formulas, to [Carles (2002); Tao (2009); Carles (2009); Duyckaerts *et al.* (2011)] for some previous results or applications involving the "standard" lens transform, to [Carles and Drumond Silva (2015)] for more discussions around the generalized lens transform, and to [Hari (2016)] for an application of the generalized lens transform in the context of propagation of nonlinear coherent states.

11.2.2   Strichartz estimates

As recalled in Sec. 1.4.3, Strichartz estimates are available for the linear part of (11.4), as stated in Lemma 7.6, with the difference that they are in general only local in time, as pointed out in Chap. 8. For future applications, we state Strichartz estimates for both (10.2) and the linear part of (11.4).

Lemma 11.1. *Let $d \geqslant 1$, and (q_1, r_1), (q_2, r_2) be admissible pairs (in the sense of Definition 7.4).*
(1) Let $V = V(x)$ be a time-independent potential satisfying Assumption 1.7, and $I \ni 0$ a finite time interval. There exist $C_{q_1}(I)$ and $C_{q_1,q_2}(I)$ independent of $\varepsilon \in]0,1]$ such that if ψ^ε solves

$$i\varepsilon\partial_t\psi^\varepsilon + \frac{\varepsilon^2}{2}\Delta\psi^\varepsilon = V(x)\psi^\varepsilon + F \quad ; \quad \psi^\varepsilon_{|t=0} = \psi^\varepsilon_0,$$

then

$$\varepsilon^{1/q_1}\|\psi^\varepsilon\|_{L^{q_1}(I;L^{r_1})} \leqslant C_{q_1}(I)\|\psi^\varepsilon_0\|_{L^2} + C_{q_1,q_2}(I)\varepsilon^{-1-1/q_2}\|F\|_{L^{q_2'}(I;L^{r_2'})}.$$

(2) Let $M \in L^\infty(\mathbb{R};\mathbb{R}^{d\times d})$ be a time-dependent symmetric matrix, bounded in time, and $s \in \mathbb{R}$ an initial time. There exist C_{q_1} and C_{q_1,q_2} independent of $s \in \mathbb{R}$ such that if u solves

$$i\partial_t u + \frac{1}{2}\Delta u = \frac{1}{2}\langle y, M(t)y\rangle u + F \quad ; \quad u_{|t=s} = u_s,$$

then for all $\eta \in [0,1]$,

$$\|u\|_{L^{q_1}([s,s+\eta];L^{r_1})} \leqslant C_{q_1}\|u_s\|_{L^2} + C_{q_1,q_2}\|F\|_{L^{q_2'}([s,s+\eta];L^{r_2'})}.$$

The lemma is a consequence of [Fujiwara (1979)] (providing local in time dispersive properties) and e.g. [Keel and Tao (1998)]. We emphasize that the estimates are necessarily local in time under the above assumptions, since typically in the case of the harmonic oscillator, eigenvalues are present: in $\mathbb{R}^d$,

$$\left(-\frac{1}{2}\Delta + \frac{|x|^2}{2}\right)g = \frac{d}{2}g, \quad \text{for } g(x) = e^{-|x|^2/2},$$

and the solution to

$$i\partial_t u + \frac{1}{2}\Delta u = \frac{|x|^2}{2}u \quad ; \quad u_{|t=0} = g,$$

is given by $u(t,x) = e^{-idt/2}g(x)$. We compute $\|u\|_{L^p(I;L^q)} = |I|^{1/p}\|g\|_{L^q}$.

For the second case, the boundedness of M as a function of time is also necessary. Suppose for instance $d = 1$ (M is a scalar function),

$$M(t) = k^2 \text{ if } 4k+1 = t_k \leqslant t \leqslant 4k+2.$$

Since we have

$$\left(-\frac{1}{2}\partial_x^2 + \frac{k^2}{2}x^2\right)e^{-kx^2/2} = \frac{k}{2}e^{-kx^2/2},$$

the function $u(t,x) = e^{-ik(t-t_k)/2-kx^2/2}$ solves the equation of (2) with $s = t_k$. If the conclusion was true for all $s \in \mathbb{R}$, we would have:

$$\|u\|_{L^p([4k+1,4k+1+\eta];L^q)} = \eta^{1/p}\left(\frac{2\pi}{kq}\right)^{1/(2q)} \leqslant C\|u(t_k)\|_{L^2} = C\left(\frac{\pi}{k}\right)^{1/4},$$

where C does not depend on k. For all $q > 2$, letting k go to infinity leads to a contradiction.

11.2.3 *Growth of Sobolev norms and momenta*

First, we have to make sure that the solution to (11.3) is defined globally in time: thanks to Strichartz inequalities, the solution is global for L^2-subcritical nonlinearities, $\sigma < 2/d$, regardless of the sign of λ, by the argument from [Tsutsumi (1987)]. For $2/d \leqslant \sigma < 2/(d-2)_+$, the assumption $\lambda \geqslant 0$ is necessary to have a global solution, since finite time blow-up is possible. It is sufficient to have a global solution $u \in C(\mathbb{R};\Sigma)$ as soon as $a \in \Sigma$: local existence is again proved as in Proposition 1.26. Global existence stems for instance from the control of the pseudo-energy considered in [Carles and Drumond Silva (2015)],

$$\mathcal{E}(t) = \frac{1}{2}\|\nabla u(t)\|_{L^2}^2 + \frac{\lambda}{\sigma+1}\|u(t)\|_{L^{2\sigma+2}}^{2\sigma+2} + \frac{1}{2}\int_{\mathbb{R}^d}|y|^2|u(t,y)|^2dy.$$

We compute

$$\frac{d}{dt}\mathcal{E}(t) = \text{Im} \int_{\mathbb{R}^d} \bar{u}(t,y)\,(y - \nabla V(t,y)) \cdot \nabla u(t,y)dy,$$

where $V(t,y) = \frac{1}{2}\langle y, M(t)y \rangle$ (V can be more general, typically at most quadratic in space, *uniformly in time*). Since $|\nabla V(t,y)| \leqslant C\langle y \rangle$ for C independent of time,

$$\frac{d}{dt}\mathcal{E}(t) \lesssim \||y|u(t)\|_{L^2}\|\nabla u(t)\|_{L^2} \lesssim \mathcal{E}(t),$$

where we have used the assumption $\lambda \geqslant 0$, and global existence follows.

We have seen in the linear case that Ehrenfest time stems from an exponential control of the growth of the third momentum of the approximate envelope, see Sec. 10.4. It will be necessary to adapt this property in the case of the nonlinear counterpart, typically (11.3), if one wants to address the large time validity of the coherent state approximation.

Definition 11.2. Let $u \in C(\mathbb{R}; \Sigma)$ be a solution to (11.3), and $k \in \mathbb{N}$. We say that $(Exp)_k$ is satisfied if there exists $C = C(k)$ such that

$$\forall \alpha, \beta \in \mathbb{N}^d, \ |\alpha| + |\beta| \leqslant k, \quad \left\|x^\alpha \partial_x^\beta u(t)\right\|_{L^2(\mathbb{R}^d)} \lesssim e^{Ct}.$$

In the linear case, we have used $(Exp)_3$. In the nonlinear case, we will use $(Exp)_3$ or $(Exp)_4$. To prove these properties, it is natural to differentiate the equation Eq. (11.3) k times, $k = 3$ or 4. We shall therefore assume $\sigma \in \mathbb{N}$. On the other hand, the nonlinearity must be energy-subcritical to grant the existence of a global solution in Σ, $\sigma < 2/(d-2)_+$. For σ integer, this implies either $d \leqslant 2$, or $d = 3$ and $\sigma = 1$, hence the assumptions in Proposition 11.3 below.

The property $(Exp)_k$ is expected to hold in a rather general setting, but its proof in this nonlinear setting is not quite straightforward. We recall some results from [Carles (2011); Carles and Drumond Silva (2015); Hari (2013)]:

Proposition 11.3. *Let $d \leqslant 3$, $\sigma \in \mathbb{N}$ with $\sigma = 1$ if $d = 3$, $a \in \mathcal{S}(\mathbb{R}^d)$ and $k \in \mathbb{N}$. Denote $M(t) = \nabla^2 V(q(t))$. Then $(Exp)_k$ is satisfied (at least) in the following cases:*

- *$\sigma = d = 1$ and $\lambda \in \mathbb{R}$ (cubic one-dimensional case).*
- *$\sigma \geqslant 2/d$, $\lambda > 0$ and $M(t)$ is diagonal with eigenvalues $\omega_j(t) \leqslant 0$.*
- *$\sigma \geqslant 2/d$, $\lambda > 0$, and $M(t)$ is diagonal with*

$$|M(t)| \leqslant \frac{C}{\langle t \rangle^\gamma} \quad \textit{for some } \gamma > 2.$$

- $\sigma \geqslant 2/d$, $\lambda > 0$ and $M(t)$ *is compactly supported.*
- $\lambda > 0$, *and there exists* $\gamma > 2$ *such that*

$$|M(t)| + \langle t \rangle \left| \frac{d}{dt} M(t) \right| \leqslant \frac{C}{\langle t \rangle^\gamma}.$$

As pointed out before, the second and third cases are rather unlikely to occur, unless $d = 1$, since no time-dependent diagonalization is needed then. The fourth case occurs typically when V is compactly supported and if the classical trajectories are not trapped. In such a situation, $(Exp)_k$ can be improved in the same spirit as the case $V = 0$ in Sec. 10.5: the Sobolev norms of u are bounded, and the L^2-norms of the momenta grow algebraically in time. The last case corresponds to Proposition 1.9 and Lemma 4.3 from [Hari (2013)], where actually only the cubic nonlinearity case ($\sigma = 1$), in dimensions $d = 2$ or 3, is considered, but the proof remains valid under the above assumptions.

Chapter 12

Power-like Nonlinearity

In this chapter, we mimic the approach of the WKB analysis presented in the first part of this book, in the context of (11.1): we show that in the subcritical case $\alpha > \alpha_c$, the linear approximation remains valid, while for $\alpha = \alpha_c$, nonlinear effects are present at leading order, through the envelope equation (11.3). We conclude this chapter with a nonlinear superposition principle: in the critical case $\alpha = \alpha_c$, if the initial datum is the sum of two coherent states with different centers in phase space, then the exact solution is well approximated by the sum of the approximate solutions associated with each initial coherent state. This is in sharp contrast with the weakly nonlinear multiphase analysis from Sec. 2.6 where, even if no new mode is created through nonlinear interaction, the modes interact nonlinearly one with another. These results were presented first in [Carles and Fermanian Kammerer (2011b)].

Like in the analysis presented in Chap. 2, two approaches are possible to prove an error estimates: work directly on the solution to the initial equation — (11.1) in this chapter — or change the unknown function, like we did in Sec. 2.2 for the WKB case, and Sec. 10.4 for the propagation of linear coherent states. However, the second approach is interesting only in the case of a single WKB state or coherent state, and with the nonlinear superposition principle in mind, we shall retain the first option from the beginning.

12.1 Subcritical case

In this section, we consider (11.1) with $\alpha > \alpha_c = 1 + d\sigma/2$, and initial datum (10.1). As suggested in Sec. 11.1, the approximate solution of the linear case is expected to be a good approximation of the solution to (11.1). We therefore resume the approximate solution (10.3) (changing the notation for future references),

$$\varphi_{\text{lin}}^{\varepsilon}(t,x) = \frac{1}{\varepsilon^{d/4}} u_{\text{lin}}\left(t, \frac{x-q(t)}{\sqrt{\varepsilon}}\right) e^{i(S(t)+p(t)\cdot(x-q(t)))/\varepsilon},$$

where (q,p) is given by the Hamiltonian flow (10.4),

$$\dot{q} = p, \quad \dot{p} = -\nabla V(q); \quad q(0) = q_0, \quad p(0) = p_0,$$

the classical action is given by (10.6),

$$S(t) = \int_0^t \left(\frac{|p(s)|^2}{2} - V(q(s))\right) ds,$$

and the envelope u_{lin} solves (10.7),

$$i\partial_t u_{\text{lin}} + \frac{1}{2}\Delta u_{\text{lin}} = \frac{1}{2}\left\langle y, \nabla^2 V(q(t)) y\right\rangle u_{\text{lin}} \quad ; \quad u_{\text{lin}|t=0} = a.$$

The approximate solution φ_{lin} solves

$$i\varepsilon\partial_t\varphi_{\text{lin}} + \frac{\varepsilon^2}{2}\Delta\varphi_{\text{lin}} = T_{q(t)}^2(x)\varphi_{\text{lin}},$$

where T_y^k stands for the Taylor polynomial of V about y, of degree k,

$$T_{q(t)}^2(x) = V(q(t)) + (x - q(t)) \cdot \nabla V(q(t))$$

$$+ \frac{1}{2}\left\langle x - q(t), \nabla^2 V(q(t))(x - q(t))\right\rangle.$$

Denote by $w^{\varepsilon} = \psi^{\varepsilon} - \varphi_{\text{lin}}^{\varepsilon}$ the error. It solves

$$i\varepsilon\partial_t w^{\varepsilon} + \frac{\varepsilon^2}{2}\Delta w^{\varepsilon} = V(x)w^{\varepsilon} + \left(V(x) - T_{q(t)}^2(x)\right)\varphi_{\text{lin}}^{\varepsilon} + \lambda\varepsilon^{\alpha}|\psi^{\varepsilon}|^{2\sigma}\psi^{\varepsilon},$$

and satisfies $w_{|t=0}^{\varepsilon} = 0$. Denote by

$$L^{\varepsilon}(t,x) = \left(V(x) - T_{q(t)}^2(x)\right)\varphi_{\text{lin}}^{\varepsilon}(t,x)$$

the linear source term, and by

$$N^{\varepsilon} = \lambda\varepsilon^{\alpha}|\psi^{\varepsilon}|^{2\sigma}\psi^{\varepsilon} = \lambda\varepsilon^{\alpha}|\varphi_{\text{lin}}^{\varepsilon} + w^{\varepsilon}|^{2\sigma}(\varphi_{\text{lin}}^{\varepsilon} + w^{\varepsilon})$$

the nonlinear source term (it is a source term in the subcritical case $\alpha > \alpha_c$, and will be a nonlinear coupling term in the critical case $\alpha = \alpha_c$). As we have seen in the linear case, we have the pointwise estimate

$$|L^{\varepsilon}(t,x)| \lesssim \varepsilon^{3/2} \times \frac{1}{\varepsilon^{d/4}}\left\langle y\right\rangle^3 |u_{\text{lin}}(t,y)|\Big|_{y=\frac{x-q(t)}{\sqrt{\varepsilon}}}.$$

If the nonlinearity is not too large (L^2-(sub)critical, $\sigma \leqslant 2/d$), we can prove an error estimate at the L^2 level, that is, without involving Sobolev norms, thanks to Strichartz estimates. This result is not present in [Carles and Fermanian Kammerer (2011b)], and the technique of proof can be seen as a preparation for the critical case.

Proposition 12.1. *Let $d \geqslant 1$, $\sigma \leqslant 2/d$, and $a \in \mathcal{S}(\mathbb{R}^d)$. Suppose that $\alpha > \alpha_c$. There exist $C, C_1 > 0$ independent of ε, and $\varepsilon_0 > 0$ such that for all $\varepsilon \in]0, \varepsilon_0]$,*

$$\|\psi^\varepsilon(t) - \varphi_{\text{lin}}^\varepsilon(t)\|_{L^2} \lesssim \varepsilon^\gamma e^{C_1 t}, \quad 0 \leqslant t \leqslant C \log \frac{1}{\varepsilon}, \quad \gamma := \min\left(\frac{1}{2}, \alpha - \alpha_c\right).$$

In particular, there exists $c > 0$ independent of ε such that

$$\sup_{0 \leqslant t \leqslant c \log \frac{1}{\varepsilon}} \|\psi^\varepsilon(t) - \varphi_{\text{lin}}^\varepsilon(t)\|_{L^2} \xrightarrow[\varepsilon \to 0]{} 0.$$

Proof. We apply Strichartz estimates for w^ε, as stated in the first point of Lemma 11.1, with $I = [0, t]$,

$$\varepsilon^{1/q_1} \|w^\varepsilon\|_{L^{q_1}(I; L^{r_1})} \lesssim \varepsilon^{-1-1/q_2} \|L^\varepsilon\|_{L^{q_2'}\left(I; L^{r_2'}\right)} + \varepsilon^{-1-1/q_3} \|N^\varepsilon\|_{L^{q_3'}\left(I; L^{r_3'}\right)},$$

where (q_j, r_j) are admissible pairs, $j = 1, 2, 3$, and can be chosen independent one from another. The easiest choice is for the linear source term, we pick $(q_2, r_2) = (\infty, 2)$, like in the energy estimate, and find

$$\varepsilon^{-1} \|L^\varepsilon\|_{L^1(0,t; L^2)} \lesssim \sqrt{\varepsilon} \|\langle y\rangle^3 u_{\text{lin}}(t)\|_{L^1(0,t; L^2)} \lesssim \sqrt{\varepsilon} \int_0^t e^{Cs} ds \lesssim \sqrt{\varepsilon} e^{Ct},$$

where we have used Lemma 10.4. Like in the proof of Proposition 7.31, we then choose $r_1 = r_3 = 2\sigma + 2 =: r$, so we have, for $j = 1$ and 3,

$$\frac{1}{r'} = \frac{1}{r} + \frac{2\sigma}{r} \quad ; \quad \frac{1}{q'} = \frac{1}{q} + \frac{2\sigma}{k},$$

with $1 < k \leqslant q = (4\sigma + 4)/(d\sigma)$, since $\sigma \leqslant 2/d$. Therefore,

$$\varepsilon^{1/q} \|w^\varepsilon\|_{L^q(I; L^r)} \lesssim \sqrt{\varepsilon} e^{Ct} + \varepsilon^{\alpha-1-1/q} \|\varphi_{\text{lin}}^\varepsilon + w^\varepsilon\|_{L^k(I; L^r)}^{2\sigma} \|\varphi_{\text{lin}}^\varepsilon + w^\varepsilon\|_{L^q(I; L^r)}$$

$$\lesssim \sqrt{\varepsilon} e^{Ct} + \varepsilon^{\alpha-1-1/q} t^{2\sigma(1/k-1/q)} \|\varphi_{\text{lin}}^\varepsilon + w^\varepsilon\|_{L^q(I; L^r)}^{2\sigma+1},$$

where we have used Hölder inequality in time for the second inequality. Estimate the last term, in the case $w^\varepsilon = 0$:

$$\|\varphi_{\text{lin}}^\varepsilon(t)\|_{L^r} = \frac{1}{\varepsilon^{d/4}} \left\| u_{\text{lin}}\left(t, \frac{\cdot - q(t)}{\sqrt{\varepsilon}}\right) \right\|_{L^r} = \varepsilon^{d/2(1/r-1/2)} \|u_{\text{lin}}(t)\|_{L^r}$$

$$= \varepsilon^{-1/q} \|u_{\text{lin}}(t)\|_{L^r},$$

where we have used the fact that (q, r) is admissible. We infer

$$\varepsilon^{\alpha-1-1/q} \left\| \varphi_{\text{lin}}^{\varepsilon} \right\|_{L^{q}(I;L^{r})}^{2\sigma+1} = \varepsilon^{\alpha-\alpha_{c}} \varepsilon^{d\sigma/2-(2\sigma+2)/q} \left\| u_{\text{lin}} \right\|_{L^{q}(I;L^{r})}^{2\sigma+1}$$

$$= \varepsilon^{\alpha-\alpha_{c}} \left\| u_{\text{lin}} \right\|_{L^{q}(I;L^{r})}^{2\sigma+1} \lesssim \varepsilon^{\alpha-\alpha_{c}} e^{C_{1}t},$$

where we have used successively, the expression of α_{c}, q, and the exponential control of u_{lin} in $L^{2\sigma+2}$, following from Lemma 10.4 and Sobolev embedding. Therefore, when $w^{\varepsilon} = 0$, the source term is controlled by $C\varepsilon^{\gamma}e^{C_{0}t}$ for some constants $C, C_{0} > 0$ (the algebraic power of t can of course be controlled by barely increasing C_{0}).

We can then use a bootstrap argument as in the proof of Proposition 7.3: so long as we have

$$\left\| w^{\varepsilon} \right\|_{L^{q}(0,t;L^{r})} \lesssim \varepsilon^{-1/q} e^{C_{1}t}, \tag{12.1}$$

then w^{ε} satisfies the same estimate as those we have used to estimate $\left\| \varphi_{\text{lin}}^{\varepsilon} \right\|_{L^{q}(I;L^{r})}^{2\sigma+1}$, and we infer

$$\varepsilon^{1/q} \left\| w^{\varepsilon} \right\|_{L^{q}(I;L^{r})} \lesssim \varepsilon^{\gamma} e^{C_{0}t}.$$

Therefore, (12.1) holds for $0 \leqslant t \leqslant C_{3} \log 1/\varepsilon$, for a suitable $C_{3} > 0$ independent of ε. Strichartz estimate with now $(q_{1}, r_{1}) = (\infty, 2)$ completes the proof.   □

In [Carles and Fermanian Kammerer (2011b)], an analogous error estimate is proved in a different setting, using different techniques: there, $d = 1$, $\sigma \in \mathbb{N}$ is arbitrary, and the analogue of Proposition 12.1 relies on energy estimates at the H^{1} level, using the embedding $H^{1} \subset L^{\infty}$, valid for $d = 1$. We do not reproduce the argument here, and move directly to the critical case, where we use estimates which make it possible to generalize Proposition 12.1 to the case $d \leqslant 3$, with $\sigma = 1$ if $d = 3$ (an energy subcritical nonlinearity is needed to apply Strichartz inequalities like we do below).

12.2   Critical case

In the critical case $\alpha = \alpha_{c} = 1 + d\sigma/2$, the definition of the approximate solution is altered:

$$\varphi^{\varepsilon}(t, x) = \frac{1}{\varepsilon^{d/4}} u \left(t, \frac{x - q(t)}{\sqrt{\varepsilon}} \right) e^{i(S(t)+p(t)\cdot(x-q(t)))/\varepsilon},$$

where, again, (q, p) is given by the Hamiltonian flow (10.4), the classical action is given by (10.6), but now the envelope u solves (11.3),

$$i\partial_t u + \frac{1}{2}\Delta u = \frac{1}{2}\langle y, \nabla^2 V(q(t)) y\rangle u + \lambda|u|^{2\sigma}u \quad ; \quad u_{|t=0} = a.$$

In view of Proposition 11.3, we first consider the cubic one-dimensional case, where the results will be the most complete, also in terms of the time interval on which we are able to prove an error estimate: we obtain a time interval of the same order of magnitude as in the linear case $(\log 1/\varepsilon)$, while in other cases (which are L^2-(super)critical), even assuming that $(Exp)_k$ is valid, the time interval given by the proof is shorter, of the order $\log\log 1/\varepsilon$. It turns out that the proof of the cubic one-dimensional case has a consequence regarding other cases which are not dealt with in [Carles and Fermanian Kammerer (2011b)]: quintic one-dimensional and cubic two-dimensional, under an extra smallness assumption on the L^2-norm of the initial envelope a.

Lemma 12.2. *Let $\lambda \in \mathbb{R}$ and $a \in L^2(\mathbb{R}^d)$.*
(1) Suppose $d = \sigma = 1$, and let $u \in C(\mathbb{R}; L^2)$ be the solution to (11.3). There exists $C > 0$ such that

$$\|u\|_{L^8(t,t+1;L^4(\mathbb{R}))} \leqslant C\|a\|_{L^2(\mathbb{R})}, \quad \forall t \in \mathbb{R}.$$

(2) Suppose $d = 1$ or 2, and $\sigma = 2/d$. There exists $\delta > 0$ such that if $\|a\|_{L^2} \leqslant \delta$, then (11.3) has a unique global solution $u \in C(\mathbb{R}; L^2) \cap L^{2+4/d}_{\text{loc}}(\mathbb{R}; L^{2+4/d}(\mathbb{R}^d))$. In addition, the (analogue of the) above conclusion remains true: There exists $C > 0$ such that

$$\|u\|_{L^{2+4/d}([t,t+1]\times\mathbb{R}^d)} \leqslant C\|a\|_{L^2(\mathbb{R}^d)}, \quad \forall t \in \mathbb{R}.$$

Proof. For $d = \sigma = 1$, we already know that (11.3) has a unique global solution $u \in C(\mathbb{R}; L^2) \cap L^8_{\text{loc}}(\mathbb{R}; L^4)$, from the proof of [Tsutsumi (1987)]. We use Strichartz estimates of the second point of Lemma 11.1, with the same Lebesgue indices as in the proof of Proposition 12.1: for $t \in \mathbb{R}$, $\tau \in [0,1]$, $I = [t, t+\tau]$,

$$\|u\|_{L^8(I;L^4)} \leqslant C\|u(t)\|_{L^2} + C \left\||u|^2u\right\|_{L^{8/7}(I;L^{4/3})}$$
$$\leqslant C\|a\|_{L^2} + C\|u\|^2_{L^{8/3}(I;L^4)}\|u\|_{L^8(I;L^4)}$$
$$\leqslant C\|a\|_{L^2} + C\sqrt{\tau}\|u\|^3_{L^8(I;L^4)},$$

where C is independent of $t \in \mathbb{R}$ and $\tau \in [0, 1]$, we have used the conservation of mass, and used Hölder inequality. Choosing $0 < \tau_0 \ll 1$, Lemma 7.22 (with $M_\tau(t) = \|u\|_{L^8(t,t+\tau;L^4)}$ function of t, with τ as a parameter) implies

$$\|u\|_{L^8(t,t+\tau_0;L^4)} \leqslant \frac{3}{2}C\|a\|_{L^2}, \quad \forall t \in \mathbb{R}.$$

We get the first point by writing

$$\|u\|_{L^8(t,t+1;L^4)} \leqslant \sum_{k=1}^{[1/\tau_0]} \|u\|_{L^8(t+k\tau_0,t+(k+1)\tau_0;L^4)}.$$

The second case of the lemma corresponds precisely with L^2-critical nonlinearity ($\sigma = 2/d$), and a smooth nonlinearity ($\sigma \in \mathbb{N}$). The latter is not necessary for this lemma, but is natural in view of $(Exp)_3$. Global existence for small L^2 data follows from the same argument as in [Cazenave and Weissler (1989)], relying essentially on the estimates below. In the case $\sigma = 2/d$, resuming the notation from the proof of Proposition 12.1, $k = q = r = 2 + 4/d$, and we have, directly with $I = [t, t+1]$,

$$\|u\|_{L^r(I;L^r)} \leqslant C\|u(t)\|_{L^2} + C\left\||u|^{4/d}u\right\|_{L^{r'}(I;L^{r'})}$$
$$\leqslant C\|a\|_{L^2} + C\|u\|_{L^r(I;L^r)}^{1+4/d}.$$

We conclude like in the first case, by invoking Lemma 7.22, whose assumptions are now satisfied provided that $\|a\|_{L^2}$ is sufficiently small. Note however that in general, we cannot replace I with $\mathbb{R}$, because here, Strichartz estimates are only (uniformly) local in time. $\quad\square$

We have seen that $(Exp)_k$ is satisfied in the first case (Proposition 11.3). The second case implies $(Exp)_k$ in the small data L^2-critical case:

Corollary 12.3. *Suppose $d = 1$ or 2, $\sigma = 2/d$. Let $k \in \mathbb{N}$ and $a \in \mathcal{S}(\mathbb{R}^d)$. There exists $\delta = \delta(k) > 0$ such that if $\|a\|_{L^2} \leqslant \delta$, then $(Exp)_k$ is satisfied.*

Proof. The proof is essentially given in Lemma 6.1 from [Carles (2011)]. We proceed by induction: $(Exp)_0$ is obviously satisfied, by the conservation of mass. Suppose $k \geqslant 1$, and that $(Exp)_{k-1}$ is satisfied: we want to prove $(Exp)_k$. To avoid a lengthy presentation, we denote by u_ℓ the family of combinations of α momenta and β space derivatives of u, with $|\alpha| + |\beta| = \ell$ ($u_0 = u$). We have, a bit loosely,

$$i\partial_t u_k + \frac{1}{2}\Delta u_k = \frac{1}{2}\langle y, M(t)y\rangle u_k + \mathsf{V}(u, u_k) + F + \sum_{\ell=0}^{k} L_\ell(w_\ell),$$

where $M(t)$ is a symmetric, bounded matrix, V is homogeneous of degree 2σ with respect to its first argument, $\mathbb{R}$-linear with respect to its second argument, F satisfies the pointwise estimate

$$|F| \lesssim \sum_{0 \leqslant \ell_j \leqslant k-1} |u_{\ell_1}| \ldots |u_{\ell_{2\sigma+1}}|,$$

where the sums carries over combinations such that in addition $\sum \ell_j = k$ ($F = 0$ in the case $k = 1$), and the L_ℓ's are linear in their argument (these terms are exactly those present as source terms in the proof of Lemma 10.4, stemming from non-trivial commutators).

Using Strichartz estimates with $q_j = r_j = 2 + 4/d = r$ and/or $(\infty, 2)$, and $I = [t, t + \tau]$ ($\tau \in [0,1]$),

$$\|w_k\|_{L^r(I;L^r) \cap L^\infty(I;L^2)} \lesssim \|w_k(t)\|_{L^2} + \|u\|^{2\sigma}_{L^r(I;L^r)} \|w_k\|_{L^r(I;L^r)}$$
$$+ \sum_{0 \leqslant \ell_j \leqslant k-1} \|w_{\ell_1}\|_{L^r(I;L^r)} \cdots \|w_{\ell_{2\sigma}}\|_{L^r(I;L^r)} \|w_{\ell_{2\sigma+1}}\|_{L^r(I;L^r)}$$
$$+ \sum_{\ell=0}^{k} \|L_\ell(w_\ell)\|_{L^1(I;L^2)},$$

where all the constants are uniform in $t \in \mathbb{R}$ and $\tau \in [0,1]$. In view of Lemma 12.2, the second term of the right-hand side is absorbed by the left-hand side, provided that $\|a\|_{L^2}$ is sufficiently small. The first sum is treated thanks to $(Exp)_{k-1}$: for $1 \leqslant j \leqslant 2\sigma + 1$,

$$\|w_{\ell_j}\|_{L^r(I;L^r)} \lesssim e^{C(t+1)},$$

and we infer

$$\|w_k\|_{L^r(I;L^r) \cap L^\infty(I;L^2)} \lesssim \|w_k(t)\|_{L^2} + e^{C_k t} + \sum_{\ell=0}^{k} \|w_\ell\|_{L^1(I;L^2)}$$
$$\lesssim \|w_k(t)\|_{L^2} + e^{\tilde{C}_k t} + \|w_k\|_{L^1(I;L^2)},$$

where we have used $(Exp)_{k-1}$. $(Exp)_k$ then follows from Gronwall lemma (in τ). $\qquad \square$

We can now slightly generalize Theorem 1.13 from [Carles and Fermanian Kammerer (2011b)]:

Theorem 12.4. *Let $\lambda \in \mathbb{R}$ and $a \in \mathcal{S}(\mathbb{R}^d)$.*
(1) *Suppose $d = \sigma = 1$. There exist $C, C_1 > 0$ independent of ε, and $\varepsilon_0 > 0$ such that for all $\varepsilon \in]0, \varepsilon_0]$,*

$$\|\psi^\varepsilon(t) - \varphi^\varepsilon(t)\|_{L^2} \lesssim \sqrt{\varepsilon} \, e^{C_1 t}, \quad 0 \leqslant t \leqslant C \log \frac{1}{\varepsilon}. \tag{12.2}$$

(2) *Suppose $d = 1$ or 2, $\sigma = 2/d$. There exists $\delta > 0$ such that if $\|a\|_{L^2} \leqslant \delta$, then there exist $C, C_1 > 0$ independent of ε, and $\varepsilon_0 > 0$ such that for all $\varepsilon \in]0, \varepsilon_0]$, (12.2) holds.*

In both situations, there exists $c > 0$ independent of ε such that

$$\sup_{0 \leqslant t \leqslant c \log \frac{1}{\varepsilon}} \|\psi^\varepsilon(t) - \varphi^\varepsilon(t)\|_{L^2} \xrightarrow[\varepsilon \to 0]{} 0.$$

Proof. The error $w^\varepsilon = \psi^\varepsilon - \varphi^\varepsilon$ solves

$$i\varepsilon\partial_t w^\varepsilon + \frac{\varepsilon^2}{2}\Delta w^\varepsilon = V(x)w^\varepsilon + \lambda\varepsilon^{\alpha_c}\left(|\psi^\varepsilon|^{2\sigma}\psi^\varepsilon - |\varphi^\varepsilon|^{2\sigma}\varphi^\varepsilon\right) + L^\varepsilon,$$

where L^ε is given like in the subcritical case, by

$$L^\varepsilon(t,x) = \left(V(x) - T_{q(t)}^2(x)\right)\varphi^\varepsilon(t,x).$$

We resume the same estimates as in the proof of Proposition 12.1: with $r = 2\sigma + 2$, (q,r) admissible, $t \geqslant 0$, $\tau \in [0,1]$ and $I = [t, t+\tau]$,

$$\varepsilon^{1/q}\|w^\varepsilon\|_{L^q(I;L^r)} \lesssim \|w^\varepsilon(t)\|_{L^2} + \varepsilon^{-1}\|L^\varepsilon\|_{L^1(I;L^2)}$$
$$+ \varepsilon^{\alpha_c - 1 - 1/q}\left\||\psi^\varepsilon|^{2\sigma}\psi^\varepsilon - |\varphi^\varepsilon|^{2\sigma}\varphi^\varepsilon\right\|_{L^{q'}(I;L^{r'})}.$$

The source term is handled like in the subcritical case, since under the assumptions of Theorem 12.4, $(Exp)_3$ is satisfied, and thus

$$\exists C > 0, \quad \varepsilon^{-1}\|L^\varepsilon\|_{L^1(I;L^2)} \lesssim \sqrt{\varepsilon}\,e^{Ct}, \quad \forall t \geqslant 0.$$

Unlike in the subcritical case, the other term on the right-hand side is not a source term, but a coupling term, and we use the pointwise estimate

$$\left||\psi^\varepsilon|^{2\sigma}\psi^\varepsilon - |\varphi^\varepsilon|^{2\sigma}\varphi^\varepsilon\right| \lesssim \left(|w^\varepsilon|^{2\sigma} + |\varphi^\varepsilon|^{2\sigma}\right)|w^\varepsilon|,$$

to infer, thanks to Hölder inequality,

$$\varepsilon^{\alpha_c - 1 - 1/q}\left\||\psi^\varepsilon|^{2\sigma}\psi^\varepsilon - |\varphi^\varepsilon|^{2\sigma}\varphi^\varepsilon\right\|_{L^{q'}(I;L^{r'})}$$
$$\lesssim \varepsilon^{d\sigma/2 - 1/q}\left(\|w^\varepsilon\|_{L^k(I;L^r)}^{2\sigma} + \|\varphi^\varepsilon\|_{L^k(I;L^r)}^{2\sigma}\right)\|w^\varepsilon\|_{L^q(I;L^r)}.$$

In view of Lemma 12.2,

$$\|\varphi^\varepsilon\|_{L^q(I;L^r)} = \varepsilon^{-1/q}\|u\|_{L^q(I;L^r)} \leqslant C\varepsilon^{-1/q}\|a\|_{L^2},$$

and therefore

$$\varepsilon^{d\sigma/2 - 1/q}\|\varphi^\varepsilon\|_{L^k(I;L^r)}^{2\sigma}\|w^\varepsilon\|_{L^q(I;L^r)}$$
$$\lesssim \underbrace{\varepsilon^{d\sigma/2 - (2\sigma+2)/q}}_{=1}\tau^{2\sigma(1/k - 1/q)}\|a\|_{L^2}^{2\sigma} \times \varepsilon^{1/q}\|w^\varepsilon\|_{L^q(I;L^r)}.$$

This yields

$$\varepsilon^{1/q}\|w^\varepsilon\|_{L^q(I;L^r)} \lesssim \|w^\varepsilon(t)\|_{L^2} + \sqrt{\varepsilon}\,e^{Ct}$$
$$+ \tau^{2\sigma(1/k - 1/q)}\left(\|a\|_{L^2}^{2\sigma} + \left(\varepsilon^{1/q}\|w^\varepsilon\|_{L^q(I;L^r)}\right)^{2\sigma}\right) \times \varepsilon^{1/q}\|w^\varepsilon\|_{L^q(I;L^r)}.$$

So long as w^ε satisfies the "same" estimate as φ^ε (up to a multiplicative constant, which we discard since w^ε is expected to be small compared to φ^ε),

$$\|w^\varepsilon\|_{L^q(I;L^r)} \leqslant \varepsilon^{-1/q}\|a\|_{L^2}, \tag{12.3}$$

we have

$$\varepsilon^{1/q}\|w^\varepsilon\|_{L^q(I;L^r)} \lesssim \|w^\varepsilon(t)\|_{L^2} + \sqrt{\varepsilon}\, e^{Ct}$$
$$+ \tau^{2\sigma(1/k-1/q)}\|a\|_{L^2}^{2\sigma} \times \varepsilon^{1/q}\|w^\varepsilon\|_{L^q(I;L^r)}.$$

We now distinguish between the two cases of Theorem 12.4:

- If $d = \sigma = 1$, then $q = 8$ and $k = 8/3$, so the power of τ is positive. Picking $0 < \tau \ll 1$, we infer

$$\varepsilon^{1/q}\|w^\varepsilon\|_{L^q(I;L^r)} \lesssim \|w^\varepsilon(t)\|_{L^2} + \sqrt{\varepsilon}\, e^{Ct}. \tag{12.4}$$

- If $d = 1$ or 2, and $\sigma = 2/d$, then $q = k = 2 + 4/d$. Picking $\|a\|_{L^2} \ll 1$, we have (12.4).

In both cases, we use Strichartz estimates again,

$$\|w^\varepsilon\|_{L^\infty(I;L^2)} \lesssim \|w^\varepsilon(t)\|_{L^2} + \sqrt{\varepsilon}\, e^{Ct} + \tau^{2\sigma(1/k-1/q)}\|a\|_{L^2}^{2\sigma} \times \varepsilon^{1/q}\|w^\varepsilon\|_{L^q(I;L^r)}$$
$$\lesssim \|w^\varepsilon(t)\|_{L^2} + \sqrt{\varepsilon}\, e^{Ct}.$$

Recalling that $I = [t, t + \tau]$, we infer, so long as (12.3) holds, $\|w^\varepsilon\|_{L^\infty(0,t;L^2)} \lesssim \sqrt{\varepsilon}\, e^{C_1 t}$, possibly for a larger constant C_1. Therefore, (12.4) holds up to $t \leqslant C \log 1/\varepsilon$ for some $C > 0$, hence the theorem. $\quad\square$

When the nonlinearity is L^2-supercritical ($\sigma > 2/d$), the known results are weaker. First, the exponential control $(Exp)_3$ is available only in specific situations, recalled in Proposition 11.3. The other reason may be only technical, and implies that the error estimate involves an extra exponential, so the error is known to be small only up to a time of order $\log\log 1/\varepsilon$ (in particular, all the time intervals independent of ε are allowed). To be more specific about this difference, resume the proof of Theorem 12.4: for $\sigma > 2/d$, we compute $k > q$, and $\|w^\varepsilon\|_{L^k(I;L^r)}$ cannot be controlled by Strichartz norms involving w^ε. To overcome this technical obstacle, we consider Sobolev norms of w^ε, which is consistent with the fact that the nonlinearity is L^2-supercritical (see Chap. 4.5).

When only one coherent state is present, we estimate ψ^ε and the error $\psi^\varepsilon - \varphi^\varepsilon$ by using vector-fields which correspond to the natural norm for the envelope u, which is the Σ-norm (this is a natural space indeed, since a quadratic potential is present in (11.3), see Sec. 2.5). Correspondingly, we consider the operators

$$A^\varepsilon(t) = \sqrt{\varepsilon}\nabla - i\frac{p(t)}{\sqrt{\varepsilon}} \quad ; \quad B^\varepsilon(t) = \frac{x - q(t)}{\sqrt{\varepsilon}}.$$

For $f \in \Sigma$, we set

$$\|f\|_{\mathcal{H}} = \|f\|_{L^2(\mathbb{R}^d)} + \|A^\varepsilon(t)f\|_{L^2(\mathbb{R}^d)} + \|B^\varepsilon(t)f\|_{L^2(\mathbb{R}^d)},$$

where we do not emphasize the fact that this norm depends on ε and t. We obviously have $\|B^\varepsilon(t)\varphi\|_{L^2} = \|yu(t)\|_{L^2}$. In view of the formula

$$A^\varepsilon(t) = \sqrt{\varepsilon}\nabla - i\frac{p(t)}{\sqrt{\varepsilon}} = \sqrt{\varepsilon}e^{i\Theta/\varepsilon}\nabla\left(e^{-i\Theta/\varepsilon}\cdot\right),$$

$$\Theta = S(t) + p(t) \cdot (x - q(t)), \tag{12.5}$$

we also have $\|A^\varepsilon(t)\varphi\|_{L^2} = \|\nabla u(t)\|_{L^2}$, so the $\mathcal{H}$-norm of coherent states centered at (q, p) correspond to the Σ-norm of their envelope.

Theorem 12.5. *Let $d \leqslant 3$, $\sigma \in \mathbb{N}$, with $\sigma = 1$ if $d = 3$, and $a \in S(\mathbb{R}^d)$. If $(Exp)_4$ is satisfied, then there exist $C, C_2 > 0$ independent of ε, and $\varepsilon_0 > 0$ such that for all $\varepsilon \in]0, \varepsilon_0]$,*

$$\|\psi^\varepsilon(t) - \varphi^\varepsilon(t)\|_{\mathcal{H}} \lesssim \sqrt{\varepsilon}\exp\left(\exp(C_2 t)\right), \quad 0 \leqslant t \leqslant C\log\log\frac{1}{\varepsilon}.$$

In particular, there exists $c > 0$ independent of ε such that

$$\sup_{0 \leqslant t \leqslant c\log\log\frac{1}{\varepsilon}} \|\psi^\varepsilon(t) - \varphi^\varepsilon(t)\|_{\mathcal{H}} \xrightarrow[\varepsilon \to 0]{} 0.$$

Proof. Resuming the same numerology as in the proof of Theorem 12.4, we consider (q, r) the admissible pair such that $r = 2\sigma + 2$. Strichartz estimates yield, for $I = [t, t + \tau]$, $t \geqslant 0$ and $\tau \in [0, 1]$,

$$\varepsilon^{1/q}\|w^\varepsilon\|_{L^q(I;L^r)} \lesssim \|w^\varepsilon(t)\|_{L^2} + \sqrt{\varepsilon}e^{Ct}$$

$$+ \varepsilon^{d\sigma/2-1/q}\left(\|w^\varepsilon\|_{L^k(I;L^r)}^{2\sigma} + \|\varphi^\varepsilon\|_{L^k(I;L^r)}^{2\sigma}\right)\|w^\varepsilon\|_{L^q(I;L^r)},$$

where we have used $(Exp)_3$ to control the source term L^ε, and k is given by (see also Proposition 7.31)

$$k = \frac{4\sigma(\sigma + 1)}{2 - \sigma(d - 2)} > q = \frac{4\sigma + 4}{d\sigma},$$

where the relation $k > q$ is equivalent to $\sigma > 2/d$, when $\sigma < 2/(d - 2)_+$. To estimate $\|w^\varepsilon\|_{L^k(I;L^r)}$, we recall (12.5), and use Gagliardo–Nirenberg inequality,

$$\|f\|_{L^r}^r = \|f\|_{L^{2\sigma+2}}^{2\sigma+2} \leqslant \frac{C}{\varepsilon^{d\sigma/2}} \|f\|_{L^2}^{2-(d-2)\sigma} \|A^\varepsilon(t)f\|_{L^2}^{d\sigma}.$$

The property $(Exp)_2$ implies

$$\|B^\varepsilon(t)\varphi\|_{L^2} + \|A^\varepsilon(t)\varphi\|_{L^2} = \|yu(t)\|_{L^2} + \|\nabla u(t)\|_{L^2} \lesssim e^{C_0 t}.$$

Recalling that $\|\varphi^\varepsilon(t)\|_{L^2} = \|a\|_{L^2}$ for all t, we replace the bootstrap argument (12.3) with

$$\|w^\varepsilon(t)\|_{L^{2\sigma+2}}^{2\sigma+2} \leqslant \varepsilon^{-d\sigma/2} e^{d\sigma C_0 t}. \tag{12.6}$$

So long as (12.6) holds, we infer

$$\varepsilon^{1/q}\|w^\varepsilon\|_{L^q(I;L^r)} \lesssim \|w^\varepsilon(t)\|_{L^2} + \sqrt{\varepsilon} e^{Ct}$$
$$+ \left(\int_t^{t+\tau} e^{c_0 s} ds\right)^{2\sigma/k} \varepsilon^{1/q}\|w^\varepsilon\|_{L^q(I;L^r)},$$

for some $c_0 > 0$. Therefore,

$$\varepsilon^{1/q}\|w^\varepsilon\|_{L^q(I;L^r)} \lesssim \|w^\varepsilon(t)\|_{L^2} + \sqrt{\varepsilon} e^{Ct} + \tau^{2\sigma/k} e^{c_1 t} \varepsilon^{1/q}\|w^\varepsilon\|_{L^q(I;L^r)}.$$

The last term is absorbed by the left-hand side provided that

$$\tau^{2\sigma/k} e^{c_1 t} \leqslant \eta,$$

for some uniform $0 < \eta \ll 1$, and we have

$$\varepsilon^{1/q}\|w^\varepsilon\|_{L^q(I;L^r)} \lesssim \|w^\varepsilon(t)\|_{L^2} + \sqrt{\varepsilon} e^{Ct}.$$

Using Strichartz estimates again, we infer

$$\|w^\varepsilon\|_{L^\infty(0,t;L^2)} \lesssim \sqrt{\varepsilon} e^{Ct} + \int_0^t e^{Cs}\|w^\varepsilon\|_{L^\infty(0,s;L^2)} ds,$$

and Gronwall lemma yields the estimate of the theorem (in L^2 at this stage), involving a double exponential.

To validate the bootstrap, we have to estimate $A^\varepsilon w^\varepsilon$: applying A^ε to the equation satisfied by w^ε, we face the commutator

$$\left[A^\varepsilon, i\varepsilon\partial_t + \frac{\varepsilon^2}{2}\Delta - V(x)\right] = \sqrt{\varepsilon}\left(\dot p(t) - \nabla V(x)\right) = \sqrt{\varepsilon}\left(\nabla V\left(q(t)\right) - \nabla V(x)\right).$$

In view of Taylor's formula, and since V is at most quadratic,

$$\left\|\left[A^\varepsilon, i\varepsilon\partial_t + \frac{\varepsilon^2}{2}\Delta - V(x)\right] w^\varepsilon\right\|_{L^p} \lesssim \varepsilon\|B^\varepsilon(t)w^\varepsilon\|_{L^p}, \quad \forall p \geqslant 1.$$

We then compute the commutator

$$\left[B^\varepsilon, i\varepsilon\partial_t + \frac{\varepsilon^2}{2}\Delta - V(x)\right] = i\sqrt{\varepsilon}\dot q(t) - \varepsilon^{3/2}\nabla = -\varepsilon A^\varepsilon(t),$$

and we get a closed system of estimates. The source term for $A^\varepsilon w^\varepsilon$ involves $A^\varepsilon L^\varepsilon$, the source term for $B^\varepsilon w^\varepsilon$ involves $B^\varepsilon L^\varepsilon$, each of these source terms is controlled like L^ε, in view of $(Exp)_4$ (instead of $(Exp)_3$ in the case of w^ε).

We estimate $A^\varepsilon w^\varepsilon$ and $B^\varepsilon w^\varepsilon$ with the same type of estimates as for w^ε, to obtain, so long as (12.6) holds,

$$\|w^\varepsilon(t)\|_{\mathcal{H}} \lesssim \sqrt{\varepsilon} \exp\left(\exp(Ct)\right).$$

The Gagliardo–Nirenberg inequality yields

$$\|w^\varepsilon(t)\|_{L^{2\sigma+2}}^{2\sigma+2} \leqslant \frac{C}{\varepsilon^{d\sigma/2}} \|w^\varepsilon(t)\|_{L^2}^{2-(d-2)\sigma} \|A^\varepsilon(t)w^\varepsilon\|_{L^2}^{d\sigma} \lesssim \varepsilon^{-d\sigma/2} \varepsilon^{\sigma+1} e^{e^{\kappa t}},$$

for some $\kappa > 0$. We conclude that (12.6) holds provided that $\varepsilon^{\sigma+1} e^{e^{\kappa t}} \ll 1$, and the result follows. $\qquad\square$

Remark 12.6. The proof of Theorem 12.4 can easily be adapted following the above lines, in order to have an error estimate involving the $\mathcal{H}$-norm instead of the L^2-norm, since $(Exp)_4$ is satisfied in this context.

Remark 12.7. Nonlinear effects are obviously present at leading order, since the envelope u solves a nonlinear equation, (11.3), and since the approximate solution φ^ε is close to the exact solution ψ^ε in $L^2(\mathbb{R}^d)$. However, as far as Wigner measures are concerned, nonlinear effects are invisible: in view of Remark 10.2, the Wigner measure of $\psi^\varepsilon(t)$ is given by

$$\mu(t, dx, d\xi) = \|a\|_{L^2(\mathbb{R}^d)}^2 \delta_{x=q(t)} \otimes \delta_{\xi=p(t)},$$

as in the linear propagation of coherent states (Sec. 10.2), since (q, p) solves the same Hamiltonian system (10.4), independent of the nonlinearity.

12.3   A nonlinear superposition principle in the critical case

As announced in the preamble of this chapter, we now consider (11.1) in the critical case $\alpha = \alpha_c$, with initial data which are sums of two different coherent states,

$$\psi^\varepsilon(0, x) = \varepsilon^{-d/4} \sum_{j=1,2} a_j \left(\frac{x - q_{0j}}{\sqrt{\varepsilon}}\right) e^{i(x - q_{0j}) \cdot p_{0j}/\varepsilon}, \qquad (12.7)$$

with $a_1, a_2 \in \mathcal{S}(\mathbb{R}^d)$, $(q_{01}, p_{01}), (q_{02}, p_{02}) \in \mathbb{R}^{2d}$, and $(q_{01}, p_{01}) \neq (q_{02}, p_{02})$. Because the envelope equation is nonlinear in the case of a single initial coherent state, (11.3), nonlinear interaction might generate terms which are not present in the case of a single initial coherent state studied in the previous section. This is not so, and the heuristics is the following: since coherent states are narrowly localized around classical trajectories, two of them do not interact much when classical trajectories meet transversally. But since

$(q_{01}, p_{01}) \neq (q_{02}, p_{02})$, classical trajectories may meet only transversally (in phase space), and so coherent states do not interact at leading order. The goal is then to quantify this loose statement, in particular when large time is considered. Denote by φ_j^ε, $j = 1, 2$, the approximate coherent state associated to each initial coherent state, with (q_j, p_j) the classical trajectory associated with the initial condition (q_{0j}, p_{0j}).

Denote $w^\varepsilon = \psi^\varepsilon - \varphi_1^\varepsilon - \varphi_2^\varepsilon$ the new error function. It solves

$$i\varepsilon \partial_t w^\varepsilon + \frac{\varepsilon^2}{2} \Delta w^\varepsilon = V(x) w^\varepsilon + L^\varepsilon + \lambda \mathcal{N}^\varepsilon \quad ; \quad w_{|t=0}^\varepsilon = 0,$$

where we have now

$$L^\varepsilon(t, x) = \left(V(x) - T_{q_1(t)}^2(x) \right) \varphi_1^\varepsilon(t, x) + \left(V(x) - T_{q_2(t)}^2(x) \right) \varphi_2^\varepsilon(t, x),$$

and

$$\mathcal{N}^\varepsilon = \varepsilon^{\alpha_c} \left(|w^\varepsilon + \varphi_1^\varepsilon + \varphi_2^\varepsilon|^{2\sigma} (w^\varepsilon + \varphi_1^\varepsilon + \varphi_2^\varepsilon) - |\varphi_1^\varepsilon|^{2\sigma} \varphi_1^\varepsilon - |\varphi_2^\varepsilon|^{2\sigma} \varphi_2^\varepsilon \right).$$

Decompose $\mathcal{N}^\varepsilon$ as the sum of a semilinear term and an interaction source term: $\mathcal{N}^\varepsilon = \mathcal{N}_S^\varepsilon + \mathcal{N}_I^\varepsilon$, where

$$\mathcal{N}_S^\varepsilon = \varepsilon^{\alpha_c} \left(|w^\varepsilon + \varphi_1^\varepsilon + \varphi_2^\varepsilon|^{2\sigma} (w^\varepsilon + \varphi_1^\varepsilon + \varphi_2^\varepsilon) - |\varphi_1^\varepsilon + \varphi_2^\varepsilon|^{2\sigma} (\varphi_1^\varepsilon + \varphi_2^\varepsilon) \right),$$

$$\mathcal{N}_I^\varepsilon = \varepsilon^{\alpha_c} \left(|\varphi_1^\varepsilon + \varphi_2^\varepsilon|^{2\sigma} (\varphi_1^\varepsilon + \varphi_2^\varepsilon) - |\varphi_1^\varepsilon|^{2\sigma} \varphi_1^\varepsilon - |\varphi_2^\varepsilon|^{2\sigma} \varphi_2^\varepsilon \right).$$

We see that the term $\mathcal{N}_S^\varepsilon$ is the exact analogue of the nonlinear term in the case of a single coherent state, where we have simply replaced φ^ε with $\varphi_1^\varepsilon + \varphi_2^\varepsilon$. We can thus repeat the previous arguments, up to the control of the new source term $\mathcal{N}_I^\varepsilon$ (the linear source term L^ε is treated as before). More precisely, we have to estimate

$$\frac{1}{\varepsilon} \|\mathcal{N}_I^\varepsilon\|_{L^1([0,t]; L^2(\mathbb{R}^d))}.$$

The first remark consists in noticing that since σ is an integer, $\mathcal{N}_I^\varepsilon$ can be estimated (pointwise) by a sum of terms of the form

$$\varepsilon^{\alpha_c} |\varphi_1^\varepsilon|^{\ell_1} \times |\varphi_2^\varepsilon|^{\ell_2}, \quad \ell_1, \ell_2 \geqslant 1, \quad \ell_1 + \ell_2 = 2\sigma + 1.$$

To be more precise, we have the control, for fixed time,

$$\frac{1}{\varepsilon} \|\mathcal{N}_I^\varepsilon(t)\|_{L^2(\mathbb{R}^d)} \lesssim \varepsilon^{d\sigma/2} \sum_{\ell_1, \ell_2 \geqslant 1, \ \ell_1 + \ell_2 = 2\sigma + 1} \left\| |\varphi_1^\varepsilon|^{\ell_1} \times |\varphi_2^\varepsilon|^{\ell_2} \right\|_{L^2(\mathbb{R}^d)}.$$

We will see below why the right-hand side must be expected to be small, when integrated with respect to time. We need to estimate

$$\varepsilon^{d\sigma/2} \left\| (\varphi_1^\varepsilon)^{\ell_1} (\varphi_2^\varepsilon)^{\ell_2} \right\|_{L^2(\mathbb{R}^d)} = \left\| u_1^{\ell_1} \left(t, y - \frac{q_1(t) - q_2(t)}{\sqrt{\varepsilon}} \right) u_2^{\ell_2}(t, y) \right\|_{L^2(\mathbb{R}_y^d)},$$

with $\ell_1, \ell_2 \geqslant 1$, $\ell_1 + \ell_2 = 2\sigma + 1$. We first quantify the fact that away from the intersection of classical trajectories, nonlinear interactions are weak.

Lemma 12.8. *Suppose $d \leqslant 3$, and $\sigma \in \mathbb{N}$. Let $T > 0$, $0 < \gamma < 1/2$, and*

$$I^\varepsilon(T) = \{t \in [0,T], \ |q_1(t) - q_2(t)| \leqslant \varepsilon^\gamma\}. \qquad (12.8)$$

Denote

$$\|f\|_{\Sigma_\varepsilon} = \|f\|_{L^2(\mathbb{R}^d)} + \|\varepsilon \nabla f\|_{L^2(\mathbb{R}^d)} + \|xf\|_{L^2(\mathbb{R}^d)}.$$

Then, for all $k > d/2$,

$$\frac{1}{\varepsilon} \int_0^T \|\mathcal{N}_I^\varepsilon(t)\|_{L^2(\mathbb{R}^d)} dt \lesssim (M_{k+1}(T))^{2\sigma+1} \left(T\varepsilon^{k(1/2-\gamma)} + |I^\varepsilon(T)| \right),$$

$$\frac{1}{\varepsilon} \int_0^T \|\mathcal{N}_I^\varepsilon(t)\|_{\Sigma_\varepsilon} dt \lesssim (M_{k+2}(T))^{2\sigma+1} \left(T\varepsilon^{k(1/2-\gamma)} + |I^\varepsilon(T)| \right) e^{CT},$$

where
$$M_k(T) = \sup \left\{ \| \langle y \rangle^\alpha \partial_y^\beta u_j \|_{L^\infty([0,T];L^2(\mathbb{R}^d))}; \quad j \in \{1,2\}, \quad |\alpha| + |\beta| \leqslant k \right\}.$$

Proof. We observe that in view of Peetre inequality (see e.g. [Alinhac and Gérard (2007)]), for $\eta \in \mathbb{R}^d$,

$$\sup_{y \in \mathbb{R}^d} \left(\langle y \rangle^{-1} \langle y - \eta \rangle^{-1} \right) \lesssim \frac{1}{\langle \eta \rangle} \lesssim \frac{1}{|\eta|}.$$

With $\eta^\varepsilon(t) = \frac{q_1(t) - q_2(t)}{\sqrt{\varepsilon}}$, we infer (forgetting the sum over ℓ_1, ℓ_2),

$$\frac{1}{\varepsilon} \int_{[0,T] \backslash I^\varepsilon(T)} \|\mathcal{N}_I^\varepsilon(t)\|_{L^2} dt$$

$$\lesssim \int_{[0,T] \backslash I^\varepsilon(T)} \left\| \langle y - \eta^\varepsilon(t) \rangle^{-k} \langle y \rangle^{-k} \langle y - \eta^\varepsilon(t) \rangle^k u_1^{\ell_1}(t, y - \eta^\varepsilon(t)) \langle y \rangle^k u_2^{\ell_2}(t, y) \right\|_{L^2}$$

$$\lesssim \left\| \langle y \rangle^k u_1^{\ell_1} \right\|_{L^\infty([0,T];L^4)} \left\| \langle y \rangle^k u_2^{\ell_2} \right\|_{L^\infty([0,T];L^4)} \int_{[0,T] \backslash I^\varepsilon(T)} \frac{dt}{|\eta^\varepsilon(t)|^k}.$$

We have, for $j \in \{1,2\}$,

$$\left\| \langle y \rangle^k u_j^{\ell_j} \right\|_{L^\infty([0,T];L^4)} \leqslant \left\| \langle y \rangle^k u_j \right\|_{L^\infty([0,T];L^4)} \|u_j\|_{L^\infty([0,T]\times\mathbb{R}^d)}^{\ell_j - 1}$$

$$\lesssim \left\| \langle y \rangle^k u_j \right\|_{L^\infty([0,T];H^1)} \|u_j\|_{L^\infty([0,T];H^k)}^{\ell_1 - 1}$$

$$\lesssim M_{k+1}(T)^{\ell_j},$$

where we have used $H^1(\mathbb{R}^d) \subset L^4(\mathbb{R}^d)$ since $d \leqslant 3$. On the other hand,

$$\int_{[0,T] \backslash I^\varepsilon(T)} \frac{dt}{|\eta^\varepsilon(t)|^k} \lesssim \int_{[0,T] \backslash I^\varepsilon(T)} \frac{\varepsilon^{k/2}}{|q_1(t) - q_2(t)|^k} dt \lesssim \varepsilon^{k(1/2-\gamma)} T.$$

On $I^\varepsilon(T)$, we simply estimate

$$\frac{1}{\varepsilon} \int_{I^\varepsilon(T)} \|\mathcal{N}_I^\varepsilon(t)\|_{L^2} dt \lesssim \|u_1\|_{L^\infty([0,T]\times\mathbb{R}^d)}^{\ell_1} \|u_2\|_{L^\infty([0,T]\times\mathbb{R}^d)}^{\ell_2-1} \|u_2\|_{L^1(I^\varepsilon(T);L^2)}$$

$$\lesssim M_k(T)^{2\sigma} |I^\varepsilon(T)| \times \|u_2\|_{L^\infty([0,T];L^2)}$$

$$\lesssim M_k(T)^{2\sigma+1} |I^\varepsilon(T)|.$$

The L^2 estimate follows, without exponentially growing factor. This factor appears when dealing with the Σ_ε-norm. Typically,

$$\|\varepsilon\nabla\varphi_j^\varepsilon(t)\|_{L^2} \lesssim \sqrt{\varepsilon}\|\nabla u_j(t)\|_{L^2} + |p_j(t)|\|u_j(t)\|_{L^2},$$

$$\|x\varphi_j^\varepsilon(t)\|_{L^2} \lesssim \sqrt{\varepsilon}\|y u_j(t)\|_{L^2} + |q_j(t)|\|u_j(t)\|_{L^2}.$$

The result then follows from the above computations, and Lemma 10.3. $\square$

At this stage, the main difficulty is to estimate the length of $I^\varepsilon(t)$. We do this in two cases: bounded t, and large time when $d = 1$. The following proposition quantifies the sparsity of intersections of classical trajectories.

Proposition 12.9. • *For $T > 0$ independent of ε, we have*

$$|I^\varepsilon(T)| = \mathcal{O}(\varepsilon^\gamma),$$

where $I^\varepsilon(T)$ is defined by (12.8).

• *In the case $d = 1$, introduce the classical energy*

$$E_j = \frac{p_j^2}{2} + V(q_j).$$

For $E_1 \neq E_2$, there exist $C, C_0 > 0$ independent of ε such that

$$|I_\varepsilon(t)| \lesssim \varepsilon^\gamma e^{C_0 t} |E_1 - E_2|^{-2}, \quad 0 \leqslant t \leqslant C \log\frac{1}{\varepsilon}.$$

Proof. The key remark is that since $(q_{01}, p_{01}) \neq (q_{02}, p_{02})$, the trajectories $q_1(t)$ and $q_2(t)$ may cross only in isolated points: by uniqueness, if $q_1(t) = q_2(t)$, then $\dot{q}_1(t) \neq \dot{q}_2(t)$. Therefore, there is only a finite numbers of such points in the interval $[0, T]$:

$$(q_1(\cdot) - q_2(\cdot))^{-1}(0) \cap [0,T] = \{t_j\}_{1\leqslant j\leqslant J}, \quad \text{where } J = J(T).$$

If we had $J = \infty$, then by compactness of $[0,T]$, a subsequence of $(t_j)_j$ would converge to some $\tau \in [0,T]$, with $q_1(\tau) = q_2(\tau)$. By uniqueness for the Hamiltonian flow, $\dot{q}_1(\tau) \neq \dot{q}_2(\tau)$: τ cannot be the limit of times where $q_1(t_j) = q_2(t_j)$, since Taylor expansion for $\delta q = q_1 - q_2$ would yield

$$0 = \delta q(t_j) = \underbrace{\delta q(\tau)}_{=0} + (\tau - t_j)\underbrace{(\dot{q}_1(\tau) - \dot{q}_2(\tau))}_{\neq 0} + \mathcal{O}((t_j - \tau)^2).$$

By uniqueness for the Hamiltonian flow, continuity and compactness, there exists $\delta > 0$ such that

$$\inf\{|\dot{q}_1(t) - \dot{q}_2(t)| \;;\; t \in \mathcal{I}(\delta, T)\} = m > 0, \quad \text{where } \mathcal{I}(\delta, T) = \bigcup_{j=1}^{J} [t_j - \delta, t_j + \delta],$$

and there exists $\varepsilon(\delta, T) > 0$ such that for $\varepsilon \in]0, \varepsilon(\delta, T)]$, $I^\varepsilon(T) \subset \mathcal{I}(\delta, T)$.

Let $t \in I^\varepsilon(T) \cap [t_j - \delta, t_j + \delta]$. Taylor's formula yields

$$q_1(t) - q_2(t) = q_1(t_j) - q_2(t_j) + (t - t_j)\left(\dot{q}_1(\tau) - \dot{q}_2(\tau)\right), \quad \tau \in [t_j - \delta, t_j + \delta].$$

We infer

$$\varepsilon^\gamma \geqslant |q_1(t) - q_2(t)| \geqslant |t - t_j| m,$$

and the first part of the proposition follows.

Suppose now $d = 1$. We consider $J^\varepsilon(t)$ an interval of maximal length included in $I^\varepsilon(t)$ and $N^\varepsilon(t)$ the number of such intervals. The result comes from the estimate

$$|I^\varepsilon(t)| \leqslant N^\varepsilon(t) \times \max |J^\varepsilon(t)|,$$

with

$$|J^\varepsilon(t)| \lesssim \varepsilon^\gamma e^{Ct} |E_1 - E_2|^{-1}, \tag{12.9}$$

$$N^\varepsilon(t) \lesssim t e^{2Ct} |E_1 - E_2|^{-1} \lesssim e^{3Ct} |E_1 - E_2|^{-1}. \tag{12.10}$$

We first prove (12.9). Let $t_1, t_2 \in J^\varepsilon(t)$. There exists $t^* \in [t_1, t_2]$ such that

$$|(q_1(t_1) - q_2(t_1)) - (q_1(t_2) - q_2(t_2))| = |t_2 - t_1| \, |p_1(t^*) - p_2(t^*)|,$$

whence

$$|t_1 - t_2| \leqslant |p_1(t^*) - p_2(t^*)|^{-1} \times 2\varepsilon^\gamma.$$

On the other hand,

$$|p_1(t^*) - p_2(t^*)| \geqslant ||p_1(t^*)| - |p_2(t^*)|| \geqslant \frac{\left||p_1(t^*)|^2 - |p_2(t^*)|^2\right|}{|p_1(t^*)| + |p_2(t^*)|}.$$

Using

$$|p_1(t^*)| + |p_2(t^*)| \lesssim e^{Ct},$$

$$|p_1(t^*)|^2 - |p_2(t^*)|^2 = 2\left(E_1 - E_2 - V(q_1(t^*)) + V(q_2(t^*))\right),$$

$$|V(q_1(t^*)) - V(q_2(t^*))| \leqslant \varepsilon^\gamma e^{Ct},$$

we get

$$\left||p_1(t^*)|^2 - |p_2(t^*)|^2\right| \gtrsim |E_1 - E_2| - \varepsilon^\gamma e^{Ct},$$

whence

$$|t_1 - t_2| \lesssim \varepsilon^\gamma e^{Ct} |E_1 - E_2|^{-1},$$

provided $\varepsilon^\gamma e^{Ct} \ll 1$.

We now prove (12.10). We use that as t is large, $N^\varepsilon(t)$ is comparable to the number $\tilde{J}^\varepsilon$ of distinct intervals of maximal size where $|q_1(t) - q_2(t)| \geqslant \varepsilon^\gamma$. We consider $J'_\varepsilon = [t_1, t_2]$ such an interval. We have

$$|q_1(t_1) - q_2(t_1)| = |q_1(t_2) - q_2(t_2)| = \varepsilon^\gamma, \text{ and } \forall t \in [t_1, t_2], \quad |q_1(t) - q_2(t)| \geqslant \varepsilon^\gamma.$$

Therefore, for $t \in [t_1, t_2]$, the quantity $q_1(t) - q_2(t)$ has a constant sign: we suppose that $q_1(t) - q_2(t)$ is positive. We then have

$$p_1(t_1) - p_2(t_1) > 0 \text{ and } p_1(t_2) - p_2(t_2) < 0.$$

Using the exponential control of $V'(q_j(t))$ for $j \in \{1, 2\}$, we obtain

$$(p_1(t_1) - p_2(t_1)) - (p_1(t_2) - p_2(t_2)) \lesssim e^{Ct} |t_1 - t_2|.$$

We write

$$p_1(t_1) - p_2(t_1) = |p_1(t_1) - p_2(t_1)| \geqslant \frac{\left| |p_1(t_1)|^2 - |p_2(t_1)|^2 \right|}{|p_1(t_1)| + |p_2(t_1)|}$$
$$\gtrsim e^{-Ct} \left| |p_1(t_1)|^2 - |p_2(t_1)|^2 \right|,$$

$$-p_1(t_2) + p_2(t_2) = |p_1(t_2) - p_2(t_2)| \geqslant \frac{\left| |p_1(t_2)|^2 - |p_2(t_2)|^2 \right|}{|p_1(t_2)| + |p_2(t_2)|}$$
$$\gtrsim e^{-Ct} \left| |p_1(t_2)|^2 - |p_2(t_2)|^2 \right|.$$

Besides, in view of

$$\frac{1}{2} \left(|p_1(t_1)|^2 - |p_2(t_2)|^2 \right) = E_1 - E_2 - V(q_1(t_1)) + V(q_2(t_1))$$
$$= E_1 - E_2 - V'(q^*)(q_1(t_1) - q_2(t_1))$$

with $q^* \in [q_2(t_1), q_1(t_1)]$, we have

$$|V'(q^*)(q_1(t_1) - q_2(t_1))| \lesssim e^{Ct}(q_1(t_1) - q_2(t_1)) \lesssim \varepsilon^\gamma e^{Ct}.$$

Therefore, if $\varepsilon^\gamma e^{Ct} \ll 1$, we have $E_1 - E_2 > 0$ and

$$\frac{1}{2} \left| |p_1(t_1)|^2 - |p_2(t_1)|^2 \right| \geqslant \frac{1}{2}(E_1 - E_2).$$

The same holds for t_2, which yields

$$(p_1(t_1) - p_2(t_1)) - (p_1(t_2) - p_2(t_2)) \gtrsim e^{-Ct}(E_1 - E_2),$$

whence the existence of a constant $c > 0$ such that

$$|t_1 - t_2| \geqslant c e^{-2CT}(E_1 - E_2) \text{ and } |J'_\varepsilon| \geqslant c e^{-2CT}(E_1 - E_2).$$

The number $\tilde{N}^\varepsilon(t)$ of intervals of the type J'_ε satisfies

$$\tilde{N}^\varepsilon(t) \times c e^{-2Ct}(E_1 - E_2) \leqslant t,$$

whence the second point of the claim. $\qquad\qquad\qquad\qquad\qquad\qquad\square$

We present the one-dimensional cubic case, for which the known result is the strongest, and simply indicate the results known in other cases.

Theorem 12.10. *Assume that $d = \sigma = 1$, and let $a_1, a_2 \in \mathcal{S}(\mathbb{R})$. Suppose that $E_1 \neq E_2$, where*

$$E_j = \frac{p_j^2}{2} + V(q_j).$$

For any $k \in \mathbb{N}$, there exist $C, C_4 > 0$ independent of ε, and $\varepsilon_0 > 0$ such that for all $\varepsilon \in]0, \varepsilon_0]$,

$$\|\psi^\varepsilon(t) - \varphi_1(t)^\varepsilon - \varphi_2^\varepsilon(t)\|_{L^2} \lesssim \varepsilon^\gamma e^{C_4 t}, \quad 0 \leqslant t \leqslant C \log \frac{1}{\varepsilon}, \quad \text{with } \gamma = \frac{k-2}{2k-2}.$$

In particular, there exists $c > 0$ independent of ε such that

$$\sup_{0 \leqslant t \leqslant c \log \frac{1}{\varepsilon}} \|\psi^\varepsilon(t) - \varphi_1^\varepsilon(t) - \varphi_2^\varepsilon(t)\|_{L^2} \xrightarrow[\varepsilon \to 0]{} 0.$$

Remark 12.11. In the above statement, k corresponds to the fact that $(Exp)_k$ is satisfied, so we can almost take $\gamma = 1/2$.

Proof. In view of the second point of Proposition 12.9, the proof becomes very similar to the proof of the first point of Theorem 12.4. As we have pointed out at the beginning of this section, the interaction term $\mathcal{N}_I^\varepsilon$ plays the role of a nonlinear source term, and is indeed estimated almost like the linear source term L^ε, up to the fact that $\sqrt{\varepsilon}$ is replaced with $\varepsilon^{(k-2)(1/2-\gamma)} + \varepsilon^\gamma$, from Lemma 12.8 and Proposition 12.9. This new error is optimized when both powers are equal,

$$(k-2)\left(\frac{1}{2} - \gamma\right) = \gamma \Longleftrightarrow \gamma = \frac{k-2}{2k-2},$$

hence the power appearing in Theorem 12.10. With this information, Theorem 12.10 is proved exactly like the first point of Theorem 12.4. $\quad\square$

Remark 12.12. A similar result is available in the quintic one-dimensional case, $d = 1$, $\sigma = 2$, provided that $\|a_1\|_{L^2}$ and $\|a_2\|_{L^2}$ are sufficiently small, for the same reasons as above. However, in the cubic two-dimensional case, we do not have a large time estimate for $I^\varepsilon(t)$ defined by (12.8).

In the L^2-supercritical case, we need estimates at the H^1 (or rather, Σ) level, like in the case of a single coherent state. However, like we faced in Chap. 2, we can no longer change frames according to the center, in phase space, of the coherent state, as soon as two of them are involved. This is why we consider the norm Σ_ε, introduced in Lemma 12.8.

Theorem 12.13. *Let* $d \leqslant 3$, $\sigma \in \mathbb{N}$ ($\sigma = 1$ *if* $d = 3$)*, and* $a_1, a_2 \in \mathcal{S}(\mathbb{R}^d)$.
(1) *For all* $T > 0$ *independent of* ε*, we have, for all* $\gamma < 1/2$:

$$\sup_{0 \leqslant t \leqslant T} \|\psi^\varepsilon(t) - \varphi_1^\varepsilon(t) - \varphi_2^\varepsilon(t)\|_{\Sigma_\varepsilon} = \mathcal{O}(\varepsilon^\gamma).$$

(2) *Assume that* $d = 1$, $\sigma \in \mathbb{N}$*, and let* $a_1, a_2 \in \mathcal{S}(\mathbb{R})$*. Suppose that* $E_1 \neq E_2$*, and that* $(Exp)_k$ *is satisfied for some* $k \geqslant 4$ *(for* u_1 *and* u_2*). There exist* $C, C_3 > 0$ *independent of* ε*, and* $\varepsilon_0 > 0$ *such that for all* $\varepsilon \in]0, \varepsilon_0]$,

$$\|\psi^\varepsilon(t) - \varphi_1(t)^\varepsilon - \varphi_2^\varepsilon(t)\|_{\Sigma_\varepsilon} \lesssim \varepsilon^\gamma \exp(\exp(C_3 t)), \quad 0 \leqslant t \leqslant C \log\log \frac{1}{\varepsilon},$$

with $\gamma = \frac{k-2}{2k-2}$.

We skip the proof of this result, which goes along the same lines as the proof of Theorem 12.5, up to two differences.

First, the new source term $\mathcal{N}_f^\varepsilon$ which, as discussed in the proof of Theorem 12.10, is handled thanks to Lemma 12.8 and Proposition 12.9 (hence an error estimate for larger time when $d = 1$).

Second, the operators A^ε and B^ε are replaced with $\varepsilon \nabla$ and x: when examining the commutators with the linear part of the equation, we again obtain a closed system of estimates. The most delicate aspect is that now Gagliardo–Nirenberg inequality yields

$$\|f\|_{L^{2\sigma+2}}^{2\sigma+2} \leqslant \frac{C}{\varepsilon^{d\sigma}} \|f\|_{L^2}^{2-(d-2)\sigma} \|\varepsilon \nabla f\|_{L^2}^{d\sigma},$$

which corresponds to a more singular power of ε compared to the analogue involving A^ε (the power is doubled). This aspect is overcome by keeping the same bootstrap assumption (12.6), by noticing that the produced error estimate is

$$\|w^\varepsilon(t)\|_{L^{2\sigma+2}}^{2\sigma+2} \lesssim \varepsilon^{-d\sigma} \|w^\varepsilon\|_{L^\infty(0,t;\Sigma_\varepsilon)}^{2\sigma+2} \lesssim \varepsilon^{(2\sigma+2)\gamma - d\sigma} e^{Ct}.$$

In the case of bounded time intervals, we can replace the term e^{Ct} by 1. For $\gamma < 1/2$ sufficiently close to $1/2$, we have $(2\sigma + 2)\gamma - d\sigma > -d\sigma/2$, since $\sigma < 2/(d-2)_+$ (energy subcritical nonlinearity).

In the case of unbounded time intervals, $d = 1$, we have indeed, since $k \geqslant 4$,

$$\gamma = \frac{k-2}{2k-2} \geqslant \frac{1}{3} > \frac{\sigma}{4\sigma + 4},$$

so the bootstrap is valid up to times of order $\log\log 1/\varepsilon$ like in the case of a single coherent state.

12.4   Comments and further results

As pointed out in Sec. 10.4, no formation of caustic is involved in the study of the propagation of coherent states, at least up to Ehrenfest time. However, a somehow similar mathematical difficulty occurs when, instead of a scalar Schrödinger equation, systems of Schrödinger equations are considered, coupled through the matricial potential V. This is a very important situation in the context of quantum chemistry. The singularity due to a caustic is replaced by the singularity due to the crossing of the eigenvalues of V. In the linear case, we refer to e.g. [Fermanian Kammerer and Lasser (2003); Fermanian Kammerer and Rousse (2008); Fermanian Kammerer and Lasser (2017); Hagedorn (1994); Hagedorn and Joye (1999); Rousse (2004)]. In the nonlinear case, very few results are available; see [Carles and Fermanian Kammerer (2011a); Hari (2013, 2016)].

In [Athanassoulis (2018)], it is proved, in the case $V = 0$, that Wigner measures ignore nonlinear effects in Eq. (11.1), for a larger range of α, and a larger class of initial data. We describe the results of Corollary 2.4 from [Athanassoulis (2018)], since they naturally generalize some of the results presented here, in particular Remark 12.7. Define the Wiener–Sobolev space A^1 by

$$\|\psi\|_{A^1(\mathbb{R}^d)} = \int_{\mathbb{R}^d} (1 + |\xi|) \, |\hat{\psi}(\xi)| d\xi,$$

and its dual A^{-1} by

$$\|\phi\|_{A^{-1}} = \sup_{\|\psi\|_{A^1}=1} |\langle \phi, \psi \rangle|.$$

Consider Eq. (11.1) with $V = 0$ and $\lambda > 0$, with initial datum (10.1). Since $V = 0$, we have

$$q(t) = q_0 + tp_0, \quad p(t) \equiv p_0.$$

The estimate proved in [Athanassoulis (2018)] reads

$$\left\| w^\varepsilon(t, x, \xi) - \|a\|^2_{L^2(\mathbb{R}^d)} \delta_{x=q(t)} \otimes \delta_{\xi=p(t)} \right\|_{A^{-1}} \lesssim (1 + t) \left(\sqrt{\varepsilon} + \varepsilon^{\alpha/2 - d\sigma/4} \right),$$

where w^ε denotes the Wigner transform of ψ^ε, whose definition was recalled in Sec. 3.4. We underline two aspects regarding this result. First, this yields

$$\mu(t, dx, d\xi) = \|a\|^2_{L^2(\mathbb{R}^d)} \delta_{x=q(t)} \otimes \delta_{\xi=p(t)},$$

provided that $\alpha > d\sigma/2 = \alpha_c - 1$. In particular, this covers the range of supercritical values, $\alpha_c - 1 < \alpha < \alpha_c$. Second, while the convergences presented in this part are justified for times of order, at best, $t \lesssim \log 1/\varepsilon$ (and sometimes only $t \lesssim \log\log 1/\varepsilon$), the above error estimate yields a small error for $t \lesssim \varepsilon^{-\min(1/2,1/(\alpha-\alpha_c+1))}$, that is, for a larger time scale, an aspect reminiscent of some results from [Bouzouina and Robert (2002)] in the linear case with a non-trivial potential. In the case of [Athanassoulis (2018)], this limitation in time is due to nonlinear effects only, since $V = 0$. The results of [Athanassoulis (2018)] also cover initial data which are not of the form (10.1), typically

$$\psi^\varepsilon(0,x) = \frac{1}{\varepsilon^{\beta d/2}} a\left(\frac{x - q_0}{\varepsilon^\beta}\right) e^{ip_0 \cdot (x-q_0)/\varepsilon}, \quad 0 < \beta < 1,$$

which generalize the case $\beta = 1/2$ of (10.1) (discussed in Sec. 10.1), but also more abstract initial data. The proof of this result relies on a change of unknown function, adapted to the classical trajectories (without the scaling aspects of our change $\psi^\varepsilon \mapsto u^\varepsilon$), and on the important remark that the functional setting based on A^{-1} makes it possible to have quantitative estimates in a nonlinear context.

Chapter 13

Hartree-Type Nonlinearity

We now consider the case of a Hartree type nonlinearity, (11.2), with a smooth kernel K, bounded as well as all its derivatives (the number of derivatives involved is stated precisely for each result below). Typically, we find in [Berloff (1999); Berloff and Roberts (1999)], in the context of Bose–Einstein condensation, and for $x \in \mathbb{R}^3$,

$$K(x) = \left(a_1 + a_2 |x|^2 + a_3 |x|^4\right) e^{-A^2 |x|^2} + a_4 e^{-B^2 |x|^2}, \quad a_j, A, B \in \mathbb{R}.$$

Throughout this chapter, no restriction is made on the dimension $d \geqslant 1$, and we recall that $V = V(x)$ satisfies Assumption 1.7. First, the existence of a unique, global solution to (11.2) is easily obtained:

Lemma 13.1. *Let V satisfy Assumption 1.7, $K \in L^\infty(\mathbb{R}^d)$ and $\psi_0 \in L^2(\mathbb{R}^d)$. There exists a unique solution $\psi \in C\left(\mathbb{R}; L^2(\mathbb{R}^d)\right)$ to*

$$i\partial_t \psi + \frac{1}{2}\Delta\psi = V(x)\psi + \left(K * |\psi|^2\right)\psi \quad ; \quad \psi_{|t=0} = \psi_0.$$

In addition, it satisfies $\|\psi(t)\|_{L^2(\mathbb{R}^d)} = \|\psi_0\|_{L^2(\mathbb{R}^d)}$ for all time $t \geqslant 0$. If $\psi_0 \in \Sigma^k$ for some integer $k \geqslant 0$, then $\psi \in C\left(\mathbb{R}; \Sigma^k\right)$, where we recall

$$\Sigma^k = \left\{\psi \in L^2(\mathbb{R}^d), \ \|\psi\|_{\Sigma^k} := \|\psi\|_{H^k} + \left\||x|^k \psi\right\|_{L^2} < \infty\right\}.$$

Except for the last assertion, the proof of this lemma can be found in [Cao and Carles (2011)]: it relies on a fixed point argument on Duhamel's formula, using Strichartz estimates stated in the first point of Lemma 11.1, and Young inequality

$$\|K * |\psi|^2\|_{L^\infty(\mathbb{R}^d)} \leqslant \|K\|_{L^\infty(\mathbb{R}^d)} \||\psi|^2\|_{L^1(\mathbb{R}^d)} = \|K\|_{L^\infty(\mathbb{R}^d)} \|\psi\|_{L^2(\mathbb{R}^d)}^2.$$

The last assertion of the lemma follows easily, by noting

$$\partial^\alpha \left(K * |\psi|^2\right) = K * \partial^\alpha |\psi|^2.$$

303

We have derived in Sec. 11.1 that the critical case corresponds to $\alpha = 1$ in

$$i\varepsilon\partial_t\psi^\varepsilon + \frac{\varepsilon^2}{2}\Delta\psi^\varepsilon = V(x)\psi^\varepsilon + \varepsilon^\alpha \left(K * |\psi^\varepsilon|^2\right)\psi^\varepsilon,$$

where we recall that the initial datum is given by (10.1). The case $\alpha > 1$ leads to a linear approximation at leading order, like in Sec. 12.1: we leave out this easy case here (it is treated in [Cao and Carles (2011)]). We first describe the critical case $\alpha = 1$, in the case of a single initial coherent state, which turns out to be much simpler than its counterpart for a power-like nonlinearity. We then focus on the case $\alpha = 0$, which is supercritical, and where it is possible to describe the propagation of one coherent state, with strong nonlinear effects. We then describe the case of two initial coherent states, in both cases $\alpha = 1$ (where nonlinear interaction occurs at leading order, contrary to the superposition result proved in the critical case for a power nonlinearity), and $\alpha = 0$ (where the interaction affects not only the envelopes, but also the centers in phase space of the coherent states). The results presented in this chapter mostly resume the content of [Cao and Carles (2011)] and [Carles (2012)], with a different organization. We note that the intermediary (supercritical) case $\alpha = 1/2$ is considered in these papers, a case that we do not present here.

13.1   Critical case

As we have seen in Sec. 11.1, the leading order approximation for ψ^ε in the case $\alpha = 1$ is given by

$$\varphi^\varepsilon(t, x) = \frac{1}{\varepsilon^{d/4}}u\left(t, \frac{x - q(t)}{\sqrt{\varepsilon}}\right)e^{i(S(t)+p(t)\cdot(x-q(t)))/\varepsilon},$$

where (q, p) is given by the Hamiltonian flow (10.4),

$$\dot{q} = p, \quad \dot{p} = -\nabla V(q); \quad q(0) = q_0, \quad p(0) = p_0,$$

the classical action is given by (10.6),

$$S(t) = \int_0^t \left(\frac{|p(s)|^2}{2} - V((q(s)))\right)ds,$$

and the envelope u solves

$$i\partial_t u + \frac{1}{2}\Delta u = \frac{1}{2}\left\langle y, \nabla^2 V(q(t))\, y\right\rangle u + K(0)\|u(t)\|_{L^2}^2 \quad ; \quad u_{|t=0} = a.$$

Since the L^2-norm of u is preserved by the flow, $\|u(t)\|_{L^2} = \|a\|_{L^2}$ for all $t \in \mathbb{R}$, we have simply

$$u(t, y) = u_{\text{lin}}(t, y)e^{-itK(0)\|a\|_{L^2}^2},$$

where the envelope u_{lin} solves the linear equation

$$i\partial_t u_{\text{lin}} + \frac{1}{2}\Delta u_{\text{lin}} = \frac{1}{2}\left\langle y, \nabla^2 V\left(q(t)\right) y\right\rangle u_{\text{lin}} \quad ; \quad u_{\text{lin}|t=0} = a.$$

Nonlinear effects consist of a purely time dependent phase shift, independent of ε, and are therefore rather weak, $\varphi^\varepsilon(t,x) = \varphi^\varepsilon_{\text{lin}}(t,x)e^{-itK(0)\|a\|^2_{L^2}}$.

Proposition 13.2. *Let $K \in W^{1,\infty}(\mathbb{R}^d;\mathbb{R})$, and $a \in \Sigma^3$. Let ψ^ε solve*

$$i\varepsilon\partial_t\psi^\varepsilon + \frac{\varepsilon^2}{2}\Delta\psi^\varepsilon = V(x)\psi^\varepsilon + \varepsilon\left(K * |\psi^\varepsilon|^2\right)\psi^\varepsilon,$$

with initial datum given by Eq. (10.1). There exist $C, C_1 > 0$ independent of ε, such that for all $\varepsilon \in]0,1]$,

$$\|\psi^\varepsilon(t) - \varphi^\varepsilon(t)\|_{L^2} \lesssim \sqrt{\varepsilon}\, e^{C_1 t}, \quad \forall t \geqslant 0.$$

In particular, there exists $c > 0$ independent of ε such that

$$\sup_{0 \leqslant t \leqslant c\log\frac{1}{\varepsilon}} \|\psi^\varepsilon(t) - \varphi^\varepsilon(t)\|_{L^2} \xrightarrow[\varepsilon\to 0]{} 0.$$

Proof. We see that φ^ε solves

$$i\varepsilon\partial_t\varphi^\varepsilon + \frac{\varepsilon^2}{2}\Delta\varphi^\varepsilon = V(x)\varphi^\varepsilon + \varepsilon\left(K * |\varphi^\varepsilon|^2\right)\varphi^\varepsilon - L^\varepsilon - N^\varepsilon,$$

where the linear source error is given by

$$L^\varepsilon(t,x) = \left(V(x) - T^2_{q(t)}(x)\right)\varphi^\varepsilon(t,x),$$

and the nonlinear source term is given by

$$N^\varepsilon(t,x) = \varepsilon^{-d/4}e^{i\phi(t,x)/\varepsilon}\times$$

$$\times \varepsilon\left(\int_{\mathbb{R}^d}\left(K(z\sqrt{\varepsilon}) - K(0)\right)|u(t,y-z)|^2 dz\right)u(t,y)\Big|_{y=(x-q(t))/\sqrt{\varepsilon}},$$

where the phase ϕ is like in Sec. 10.2. Like in the previous cases, we have the pointwise estimate

$$|L^\varepsilon(t,x)| \lesssim \varepsilon^{3/2}\times\frac{1}{\varepsilon^{d/4}}\langle y\rangle^3 |u(t,y)|\Big|_{y=(x-q(t))/\sqrt{\varepsilon}}$$

$$\lesssim \varepsilon^{3/2}\times\frac{1}{\varepsilon^{d/4}}\langle y\rangle^3 |u_{\text{lin}}(t,y)|\Big|_{y=(x-q(t))/\sqrt{\varepsilon}},$$

since u and u_{lin} have the same modulus. For N^ε, we first estimate

$$\left|\int_{\mathbb{R}^d}\left(K(z\sqrt{\varepsilon}) - K(0)\right)|u(t,y-z)|^2 dz\right| \lesssim \sqrt{\varepsilon}\int_{\mathbb{R}^d}|z||u(t,y-z)|^2 dz$$

$$\lesssim \sqrt{\varepsilon}\left(\int_{\mathbb{R}^d}|z||u(t,z)|^2 dz + |y|\|a\|^2_{L^2(\mathbb{R}^d)}\right),$$

since ∇K is bounded, to infer, for all $t \geqslant 0$,

$$\|N^\varepsilon(t)\|_{L^2(\mathbb{R}^d)} \lesssim \varepsilon^{3/2}\|u(t)\|_\Sigma^3 = \varepsilon^{3/2}\|u_{\mathrm{lin}}(t)\|_\Sigma^3 \lesssim \varepsilon^{3/2}e^{Ct},$$

where we have used Lemma 10.4. Setting $w^\varepsilon = \psi^\varepsilon - \varphi^\varepsilon$, we have $w^\varepsilon_{|t=0} = 0$ and

$$i\varepsilon\partial_t w^\varepsilon + \frac{\varepsilon^2}{2}\Delta w^\varepsilon = V(x)w^\varepsilon + \varepsilon\left(K * |\psi^\varepsilon|^2\right)\psi^\varepsilon - \varepsilon\left(K * |\varphi^\varepsilon|^2\right)\varphi^\varepsilon + L^\varepsilon + N^\varepsilon.$$

The energy estimate yields

$$\|w^\varepsilon(t)\|_{L^2} \lesssim \int_0^t \left\|\left(K * |\psi^\varepsilon|^2\right)\psi^\varepsilon - \left(K * |\varphi^\varepsilon|^2\right)\varphi^\varepsilon\right\|_{L^2} + \sqrt{\varepsilon}\int_0^t e^{Cs}ds,$$

where we have used the estimates for L^ε and N^ε in L^2. Writing

$$\left(K * |\psi^\varepsilon|^2\right)\psi^\varepsilon - \left(K * |\varphi^\varepsilon|^2\right)\varphi^\varepsilon = \left(K * |\psi^\varepsilon|^2\right)w^\varepsilon$$
$$+ \left(K * \left(|\psi^\varepsilon|^2 - |\varphi^\varepsilon|^2\right)\right)\varphi^\varepsilon,$$

and using Young and Cauchy-Schwarz inequalities, we have, in view of the conservation of mass,

$$\left\|\left(K * |\psi^\varepsilon|^2\right)\psi^\varepsilon - \left(K * |\varphi^\varepsilon|^2\right)\varphi^\varepsilon\right\|_{L^2} \lesssim \|a\|_{L^2}^2\|w^\varepsilon\|_{L^2}.$$

The proposition then follows from Gronwall lemma. $\square$

13.2   Supercritical case

We now consider the case $\alpha = 0$ in (11.2). The intermediary supercritical case $\alpha = 1/2$ is also of interest, and we refer to [Cao and Carles (2011)] and [Carles (2012)] for a presentation. Let $\psi^\varepsilon \in C(\mathbb{R}; L^2(\mathbb{R}^d))$ be the solution to

$$i\varepsilon\partial_t\psi^\varepsilon + \frac{\varepsilon^2}{2}\Delta\psi^\varepsilon = V(x)\psi^\varepsilon + \left(K * |\psi^\varepsilon|^2\right)\psi^\varepsilon, \qquad (13.1)$$

with initial datum given by (10.1). We still seek an approximation of ψ^ε under the form of a coherent state,

$$\varphi^\varepsilon(t,x) = \frac{1}{\varepsilon^{d/4}}u\left(t, \frac{x - q(t)}{\sqrt{\varepsilon}}\right)e^{i(S(t)+p(t)\cdot(x-q(t)))/\varepsilon},$$

and we resume the approach of Sec. 10.2: we do not initially require anything on q, p or S. The computations of Sec. 10.2 are complemented by the formula presented in Sec. 11.1, which we now resume: we have

$$b^\varepsilon_{\mathrm{nl}} = \left(\int_{\mathbb{R}^d} K(z\sqrt{\varepsilon})\left|u\left(t, y - z\right)\right|^2 dz\right)u(t,y),$$

and

$$\int_{\mathbb{R}^d} K(z\sqrt{\varepsilon}) \, |u\,(t, y - z)|^2 \, dz = K(0)\|u(t)\|_{L^2(\mathbb{R}^d)}^2$$

$$+ \sqrt{\varepsilon} \int_{\mathbb{R}^d} \nabla K(0) \cdot z \, |u\,(t, y - z)|^2 \, dz$$

$$+ \frac{\varepsilon}{2} \int_{\mathbb{R}^d} \langle z, \nabla^2 K(0)z \rangle \, |u\,(t, y - z)|^2 \, dz$$

$$+ \varepsilon r_{\mathrm{nl}}^\varepsilon(t, y),$$

where $r_{\mathrm{nl}}^\varepsilon$ is formally of order $\sqrt{\varepsilon}$. We have

$$\int_{\mathbb{R}^d} z \, |u\,(t, y - z)|^2 \, dz = \int_{\mathbb{R}^d} (y - z) \, |u\,(t, z)|^2 \, dz = y\|u(t)\|_{L^2(\mathbb{R}^d)}^2 - G(t),$$

where $G(t)$ denotes the center of mass of u,

$$G(t) = \int_{\mathbb{R}^d} z \, |u\,(t, z)|^2 \, dz,$$

and

$$\int_{\mathbb{R}^d} \langle z, \nabla^2 K(0)z \rangle \, |u\,(t, y - z)|^2 \, dz = \int_{\mathbb{R}^d} \langle y - z, \nabla^2 K(0)(y - z) \rangle \, |u\,(t, z)|^2 \, dz$$

$$= \langle y, \nabla^2 K(0)y \rangle \, \|u(t)\|_{L^2(\mathbb{R}^d)}^2$$

$$- 2y \cdot \nabla^2 K(0)G(t) + \int_{\mathbb{R}^d} \langle z, \nabla^2 K(0)z \rangle \, |u\,(t, z)|^2 \, dz.$$

We can then write:

$$i\varepsilon \partial_t \varphi^\varepsilon + \frac{\varepsilon^2}{2} \Delta \varphi^\varepsilon - V(x)\varphi^\varepsilon - \left(K * |\varphi^\varepsilon|^2 \right) \varphi^\varepsilon$$

$$= \frac{1}{\varepsilon^{d/4}} \left(b_0 + \sqrt{\varepsilon} b_1 + \varepsilon b_2 + \mathcal{O}\left(\varepsilon^{3/2} \right) \right) e^{i\phi/\varepsilon},$$

where

$$\phi = S(t) + p(t) \cdot (x - q(t)),$$

$$b_0 = -u \left(\dot{S}(t) - p(t) \cdot \dot{q}(t) + \frac{|p(t)|^2}{2} + V(q(t)) + K(0)\|u(t)\|_{L^2}^2 \right),$$

$$b_1 = -i\,(\dot{q}(t) - p(t)) \cdot \nabla u - y \cdot (\dot{p}(t) + \nabla V(q(t)))\, u$$

$$- y \cdot \nabla K(0)\|u(t)\|_{L^2}^2 u + \nabla K(0) \cdot G(t)u,$$

$$b_2 = i\partial_t u + \frac{1}{2}\Delta u - \frac{1}{2} \langle y, \nabla^2 V(q(t))\, y \rangle\, u - \frac{1}{2} \langle y, \nabla^2 K(0)y \rangle \, \|u(t)\|_{L^2}^2 u$$

$$+ \left(y \cdot \nabla^2 K(0)G(t) \right) u - \frac{1}{2} \left(\int_{\mathbb{R}^d} \langle z, \nabla^2 K(0)z \rangle \, |u\,(t, z)|^2 \, dz \right) u.$$

Each of the terms b_j now counts at least one term due to the presence of the nonlinearity. The term b_1 is the most delicate one: in the linear case, we had canceled the factors of ∇u and yu individually: this makes sense when the two are not related. If we want to proceed similarly here, the term $\nabla K(0) \cdot G(t)u$ falls into a no man's land. To overcome this issue, we force a different hierarchy, by moving the term $\nabla K(0) \cdot G(t)u$ to b_0, thus gaining an extra factor $\sqrt{\varepsilon}$. In other words, we cancel $b_0 + \sqrt{\varepsilon}b_1 + \varepsilon b_2$ by requiring:

$$\dot{S}(t) = p(t) \cdot \dot{q}(t) - \frac{|p(t)|^2}{2} - V((q(t)) - K(0)\|u(t)\|_{L^2}^2 + \sqrt{\varepsilon}\nabla K(0) \cdot G(t),$$

$$\dot{q}(t) = p(t), \quad \dot{p}(t) = -\nabla V(q(t)) - \nabla K(0)\|u(t)\|_{L^2}^2,$$

$$i\partial_t u + \frac{1}{2}\Delta u = \frac{1}{2}\left\langle y, \left(\nabla^2 V(q(t)) + \nabla^2 K(0)\|u(t)\|_{L^2}^2\right) y \right\rangle u$$
$$- \left(y \cdot \nabla^2 K(0)G(t)\right) u$$
$$+ \frac{1}{2}\left(\int_{\mathbb{R}^d} \langle z, \nabla^2 K(0)z \rangle |u(t,z)|^2 \, dz\right) u.$$

We remark that the coefficient of u on the right-hand side of the last (Schrödinger) equation is real-valued, so the L^2-norm of u is independent of time,

$$\|u(t)\|_{L^2} = \|a\|_{L^2},$$

as long as u is well defined (and sufficiently regular). This is the least we can expect since φ^ε is aimed at being a good approximation of ψ^ε in L^2, and $\|\varphi^\varepsilon(t)\|_{L^2} = \|u(t)\|_{L^2}$ (by definition), $\|\psi^\varepsilon(t)\|_{L^2} = \|a\|_{L^2}$ (from Lemma 13.1). We therefore consider a (possibly) modified Hamiltonian flow,

$$\begin{cases} \dot{q}(t) = p(t) \quad ; \quad q(0) = q_0, \\ \dot{p}(t) = -\nabla V(q(t)) - \nabla K(0)\|a\|_{L^2}^2 \quad ; \quad p(0) = p_0, \end{cases} \tag{13.2}$$

a modified (ε-dependent) action,

$$S^\varepsilon(t) = \int_0^t \left(\frac{1}{2}|p(s)|^2 - V(q(s)) - K(0)\|a\|_{L^2}^2 \right.$$
$$\left. + \sqrt{\varepsilon}\nabla K(0) \cdot G(s)\right)ds, \tag{13.3}$$

and an envelope given by

$$i\partial_t u + \frac{1}{2}\Delta u = \frac{1}{2}\langle y, M(t)y \rangle u - \left(y \cdot \nabla^2 K(0)G(t)\right) u$$
$$+ \frac{1}{2}\left(\int_{\mathbb{R}^d} \langle z, \nabla^2 K(0)z \rangle |u(t,z)|^2 \, dz\right) u \quad ; \quad u_{|t=0} = a, \tag{13.4}$$

where $M(t) := \nabla^2 V(q(t)) + \nabla^2 K(0)\|a\|_{L^2}^2$. We note that the action is a nonlinear function of the envelope u. Regarding the evolution of the classical trajectories, Eq. (13.2) corresponds to replacing V in Eq. (10.4) by

$$V(x) + x \cdot \nabla K(0)\|a\|_{L^2}^2,$$

which, in terms of classical mechanics, consists in adding a constant gravitational field. We make sure that the action is well defined by investigating the envelope equation below. We start with two remarks on the special structure of Eq. (13.4).

We first remark that the (nonlinear, nonlocal) potential

$$\frac{1}{2}\int_{\mathbb{R}^d} \langle z, \nabla^2 K(0)z\rangle \, |u(t,z)|^2 \, dz$$

is a function of time only. Therefore, it can formally be absorbed by a gauge transform: if u solves (13.4), then

$$v(t,y) := u(t,y)\exp\left(-\frac{i}{2}\int_0^t \int_{\mathbb{R}^d} \langle z, \nabla^2 K(0)z\rangle \, |u(t,z)|^2 \, dz\right)$$

solves

$$i\partial_t v + \frac{1}{2}\Delta v = \frac{1}{2}\langle y, M(t)y\rangle v - \left(y \cdot \nabla^2 K(0)G(t)\right) v \quad ; \quad v_{|t=0} = a, \quad (13.5)$$

where, since $|v| = |u|$,

$$G(t) = \int_{\mathbb{R}^d} z\, |u(t,z)|^2 \, dz = \int_{\mathbb{R}^d} z\, |v(t,z)|^2 \, dz.$$

Taking advantage of the special structures of the matrix M and the nonlinear term, direct (formal) computations show that is v solves Eq. (13.5), then $G(t) = \int z|v(t,z)|^2 dz$ satisfies

$$\dot{G}(t) = \text{Im}\int \bar{v}\nabla v =: J(t),$$

$$\dot{J}(t) = -\int \left(M(t)y - \nabla^2 K(0)G(t)\right)|v(t,y)|^2 dy$$

$$= -M(t)G(t) + \nabla^2 K(0)G(t)\|v\|_{L^2}^2 = -\nabla^2 V(q(t))\,G(t).$$

We have in particular:

$$\ddot{G} + \nabla^2 V(q(t))\,G = 0 \quad ; \quad G(0) = \int_{\mathbb{R}^d} z|a(z)|^2 dz, \quad \dot{G}(0) = \text{Im}\int_{\mathbb{R}^d} \bar{a}\nabla a.$$

So if the initial datum a is centered at the origin in phase space,

$$\int_{\mathbb{R}^d} z|a(z)|^2 dz = \text{Im}\int_{\mathbb{R}^d} \bar{a}\nabla a = 0,$$

then $G \equiv 0$, and Eq. (13.5) becomes linear, so we can invoke Lemma 10.4. This corresponds to the setting adopted in [Athanassoulis *et al.* (2011)], where the coherent state approximation in the supercritical case (13.1) was first justified, with the extra assumption that K is radially symmetric, so that $\nabla K(0) = 0$, a case which implies that q, p and S do not undergo nonlinear effects. Note that up to changing a to b with

$$b(y) = a(y - y_0)e^{iy \cdot \eta_0}$$

for y_0 and η_0 which can be computed explicitly, that is up to a translation in phase space, the above two cancelations hold. However, Eq. (13.5) is not invariant under this change of initial datum, and the relation between the solutions of Eq. (13.5) with initial data a and b, respectively, is not obvious. We therefore work directly on Eq. (13.5).

Lemma 13.3. *Let $k \geqslant 1$ and $a \in \Sigma^k$. Let $M \in L^\infty(\mathbb{R}; \mathbb{R}^{d \times d})$ be a bounded time-dependent, real-valued, symmetric matrix. Then (13.4) has a unique solution $u \in C(\mathbb{R}; \Sigma^k)$, where*

$$G(t) = \int_{\mathbb{R}^d} z \, |u(t, z)|^2 \, dz.$$

In addition, the mass is conserved,

$$\|u(t)\|_{L^2(\mathbb{R}^d)} = \|a\|_{L^2(\mathbb{R}^d)}, \quad \forall t \in \mathbb{R},$$

and higher order norms grow at most exponentially in time,

$$\|u(t)\|_{\Sigma^k} \lesssim e^{C_k |t|}, \quad \forall t \in \mathbb{R}.$$

Proof. In view of the above discussion, we consider Eq. (13.5), with

$$G(t) = \int_{\mathbb{R}^d} z \, |u(t, z)|^2 \, dz = \int_{\mathbb{R}^d} z \, |v(t, z)|^2 \, dz.$$

Conversely, if $v \in C(\mathbb{R}; \Sigma^k)$ solves Eq. (13.5) with $k \geqslant 1$, then u, defined by

$$u(t, y) := v(t, y) \exp\left(\frac{i}{2} \int_0^t \int_{\mathbb{R}^d} \langle z, \nabla^2 K(0)z \rangle \, |v(t, z)|^2 \, dz \right)$$

solves Eq. (13.4), and $u \in C(\mathbb{R}; \Sigma^k)$.

This equation is nonlinear, and does not seem to be directly solvable by a fixed point argument, since G involves a momentum of v. On the other hand, the iterative scheme defined by $v^{(0)}(t, y) = a(y)$ and, for $n \geqslant 1$,

$$i\partial_t v^{(n)} + \frac{1}{2}\Delta v^{(n)} = W_{n-1}(t, y)v^{(n)} \quad ; \quad v^{(n)}_{|t=0} = a, \qquad (13.6)$$

where

$$W_{n-1}(t,y) := \frac{1}{2} \langle y, M(t)y \rangle - y \cdot \nabla^2 K(0) G^{(n-1)}(t),$$

$$G^{(n-1)}(t) := \int_{\mathbb{R}^d} z \left| v^{(n-1)}(t,z) \right|^2 dz,$$

involves a linear equation at each step. More precisely, if $v^{(n-1)} \in C(\mathbb{R}; \Sigma^k)$, then W_{n-1} satisfies Assumption 1.7, up to the harmless fact that we now consider a limited smoothness in time. Therefore,

$$v^{(n)} \in C(\mathbb{R}; L^2(\mathbb{R}^d)), \quad \|v^{(n)}(t)\|_{L^2} = \|a\|_{L^2}, \quad \forall t \in \mathbb{R}, \ \forall n \in \mathbb{N}.$$

Applying the operators y and ∇ to Eq. (13.6) yields a closed system of estimates, from which we infer that $v^{(n)} \in C(\mathbb{R}; \Sigma)$, hence $G^{(n)} \in L^\infty_{\text{loc}}(\mathbb{R})$. Therefore, the scheme is well-defined. Higher order regularity can be proven similarly: for $k \geqslant 1$, by applying k times the operators y and ∇ to Eq. (13.6), we check that $v^{(n)} \in C(\mathbb{R}; \Sigma^k)$. As a matter of fact, due to the particular structure of the equation, the only informations needed to prove this property are $a \in \Sigma^k$ and $v^{(n-1)} \in C(\mathbb{R}; \Sigma)$.

To prove the convergence of this scheme we need more precise (uniform in n) estimates. We check more generally that if v solves

$$i\partial_t v + \frac{1}{2}\Delta v = \frac{1}{2} \langle y, M(t)y \rangle v + y \cdot F(t)v,$$

then $G(t) = \int z|v(t,z)|^2 dz$ satisfies formally

$$\dot{G}(t) = \text{Im} \int \bar{v} \nabla v =: J(t),$$

$$\dot{J}(t) = -\int (M(t)y + F(t)) |v(t,y)|^2 dy = -M(t)G(t) + F(t)\|v\|_{L^2}^2.$$

We have in particular:

$$\ddot{G}(t) + M(t)G(t) = -\|v\|_{L^2}^2 F(t).$$

In our case, this yields:

$$\ddot{G}^{(n)} + M(t)G^{(n)} = \|a\|_{L^2}^2 \nabla^2 K(0) G^{(n-1)}, \tag{13.7}$$

Let

$$f_n(t) = \left| \dot{G}^{(n)}(t) \right|^2 + \left| G^{(n)}(t) \right|^2.$$

We have

$$\dot{f}_n(t) \leqslant 2 \left(\left| \dot{G}^{(n)}(t) \right| \left| \ddot{G}^{(n)}(t) \right| + \left| \dot{G}^{(n)}(t) \right| \left| G^{(n)}(t) \right| \right)$$

$$\leqslant C f_n(t) + C \left| G^{(n-1)}(t) \right|^2,$$

for some C independent of t and n, since $\nabla^2 V, \nabla^2 K \in L^\infty$, and where we have used (13.7) and Young inequality. By Gronwall lemma, we infer

$$f_n(t) \leqslant f_n(0)e^{Ct} + C \int_0^t e^{C(t-s)} f_{n-1}(s)ds, \quad \forall t \geqslant 0.$$

From now on, we consider only $t \geqslant 0$, since the case $t \leqslant 0$ is similar. With our definition of the scheme, $f_n(0)$ does not depend on n:

$$f_n(0) = \left| \mathrm{Im} \int \bar{a} \nabla a \right|^2 + \left| \int z|a(z)|^2 dz \right|^2 =: C_0.$$

Therefore,

$$f_n(t) \leqslant C_0 e^{Ct} + C \int_0^t e^{C(t-s)} f_{n-1}(s)ds,$$

and by induction, we infer

$$f_n(t) \leqslant 2C_0 e^{3Ct}, \quad \forall t \geqslant 0.$$

By using energy estimates (applying successively the operators y and ∇ to the equation), we infer that there exists C_1 independent of $t \geqslant 0$ and n such that

$$\left\| v^{(n)}(t) \right\|_{\Sigma^k} \leqslant C_1 e^{C_1 t}.$$

The convergence of the scheme then follows: we check that v_n converges in $C([0,T]; \Sigma)$ if $T > 0$ is sufficiently small. Denoting by

$$H(t) = -\frac{1}{2}\Delta + \frac{1}{2} \langle y, M(t)y \rangle,$$

we have

$$
\begin{aligned}
i\partial_t \left(v^{(n)} - v^{(n-1)} \right) &= H \left(v^{(n)} - v^{(n-1)} \right) \\
&+ \left(y \cdot \nabla^2 K(0) G^{(n-1)}(t) \right) \left(v^{(n)} - v^{(n-1)} \right) \\
&+ \left(y \cdot \nabla^2 K(0) \left(G^{(n-1)}(t) - G^{(n-2)}(t) \right) \right) v^{(n-1)}.
\end{aligned}
$$

Energy estimates and the above uniform bound yield

$$
\begin{aligned}
\left\| v^{(n)}(t) - v^{(n-1)}(t) \right\|_{L^2} &\leqslant C \int_0^t \left| G^{(n-1)}(s) - G^{(n-2)}(s) \right| ds \\
&\leqslant C \int_0^t e^{C_1 s} \left\| v^{(n-1)}(s) - v^{(n-2)}(s) \right\|_\Sigma ds.
\end{aligned}
$$

By applying the operators y and ∇ to Eq. (13.6), we obtain similarly:

$$\left\| v^{(n)}(t) - v^{(n-1)}(t) \right\|_\Sigma \leqslant C \int_0^t e^{C_1 s} \left\| v^{(n-1)}(s) - v^{(n-2)}(s) \right\|_\Sigma ds,$$

where we have used Gronwall lemma since y and ∇ do not commute with the equation. Therefore, we can find $T > 0$ such that the sequence $v^{(n)}$ converges in $C([0,T]; \Sigma)$, to $v \in C([0,T]; \Sigma^k)$. The uniform bounds for the sequence $v^{(n)}$ imply that v is global in time: $v \in C(\mathbb{R}; \Sigma^k)$, with Σ^k-norms growing at most exponentially in time. □

We can then prove the expected approximation result:

Theorem 13.4. *Let* $K \in W^{3,\infty}(\mathbb{R}^d; \mathbb{R})$, *and* $a \in \Sigma^3$. *Let* ψ^ε *solve Eq. (13.1) with initial datum given by Eq. (10.1).*
1. *Suppose* $\nabla K(0) = 0$. *There exist* $C, C_1 > 0$ *such that for all* $\varepsilon \in]0,1]$,

$$\|\psi^\varepsilon(t) - \varphi^\varepsilon(t)\|_{L^2} \leqslant C\sqrt{\varepsilon}\, e^{e^{C_1 t}}, \quad \forall t \geqslant 0.$$

2. *If* $\nabla K(0) \neq 0$, *suppose in addition* $K \in W^{6,\infty}(\mathbb{R}^d; \mathbb{R})$ *and* $a \in \Sigma^6$. *There exist* $C, C_1 > 0$ *and* $\theta \in C^2(\mathbb{R}; \mathbb{R})$ *independent of* ε, *such that for all* $\varepsilon \in]0,1]$,

$$\|\psi^\varepsilon(t) - \varphi^\varepsilon(t)e^{i\theta(t)}\|_{L^2} \leqslant C\sqrt{\varepsilon}\, e^{e^{C_1 t}}, \quad \forall t \geqslant 0.$$

We note that unlike in the critical size (Proposition 13.2), the approximation is valid up to a time of order $\log\log 1/\varepsilon$ (like in Theorem 12.5), instead of the regular order of magnitude $\log 1/\varepsilon$ corresponding to Ehrenfest time.

Proof. First, we emphasize that we cannot reproduce the strategy of the proof of Proposition 13.2, since the application of Gronwall lemma would produce a factor $e^{Ct/\varepsilon}$, an issue that we encountered already in the WKB supercritical case, see Sec. 1.2. Instead of working on the coherent states $\psi^\varepsilon, \varphi^\varepsilon$ directly, we consider their envelopes, that is, u in the case of φ^ε. Recall that to define the approximate solution φ^ε, we had to modify the notion of classical trajectory and the notion of classical action: we thus rescale ψ^ε in the same fashion, and write

$$\psi^\varepsilon(t,x) = \frac{1}{\varepsilon^{d/4}}u^\varepsilon\left(t, \frac{x - q(t)}{\sqrt{\varepsilon}}\right)e^{i(S^\varepsilon(t)+p(t)\cdot(x-q(t)))/\varepsilon},$$

where (q,p) solves (13.2) and S^ε is given by Eq. (13.3), therefore involving u through G. The map $\psi^\varepsilon \mapsto u^\varepsilon$ is obviously bijective, and

$$\|\psi^\varepsilon(t) - \varphi^\varepsilon(t)\|_{L^2(\mathbb{R}^d)} = \|u^\varepsilon(t) - u(t)\|_{L^2(\mathbb{R}^d)}.$$

In addition, Lemma 13.1 shows that $u^\varepsilon \in C(\mathbb{R}; \Sigma^3)$. In terms of u^ε, Eq. (13.1) is equivalent to

$$i\partial_t u^\varepsilon + \frac{1}{2}\Delta u^\varepsilon = V^\varepsilon(t,y)u^\varepsilon + \left(K^\varepsilon * |u^\varepsilon|^2\right)u^\varepsilon$$
$$- \frac{1}{\sqrt{\varepsilon}}\nabla K(0) \cdot \left(\int_{\mathbb{R}^d} z\left(|u^\varepsilon(t,z)|^2 - |u(t,z)|^2\right)dz\right)u^\varepsilon, \tag{13.8}$$

where, like in Sec. 10.4,

$$V^\varepsilon(t,y) = \frac{1}{\varepsilon}\left(V\left(q(t)+y\sqrt{\varepsilon}\right) - V\left(q(t)\right) - \sqrt{\varepsilon}y\cdot\nabla V\left(q(t)\right)\right)$$

$$= \frac{1}{2}\int_0^1 \left\langle y, \nabla^2 V\left(q(t)+\theta y\sqrt{\varepsilon}\right)y\right\rangle(1-\theta)d\theta,$$

and

$$K^\varepsilon(y) = \frac{1}{\varepsilon}\left(K\left(y\sqrt{\varepsilon}\right) - K(0) - \sqrt{\varepsilon}y\cdot\nabla K(0)\right)$$

$$= \frac{1}{2}\int_0^1 \left\langle y, \nabla^2 K\left(\theta y\sqrt{\varepsilon}\right)y\right\rangle(1-\theta)d\theta.$$

The last term in (13.8) involves a purely time-dependent potential, which is well-defined since $u^\varepsilon, u \in C(\mathbb{R};\Sigma)$, and can be absorbed by a gauge transform: we do so, since, even if we improve the error estimate of Theorem 13.4 to an estimate in Σ instead of L^2, this time dependent potential will be $\mathcal{O}(1)$. Therefore, set

$$\theta^\varepsilon(t) = \frac{1}{\sqrt{\varepsilon}}\int_0^t \nabla K(0)\cdot\left(\int_{\mathbb{R}^d} z\left(|u^\varepsilon(s,z)|^2 - |u(s,z)|^2\right)dz\right)ds,$$

so that $\tilde{u}^\varepsilon := u^\varepsilon e^{-i\theta^\varepsilon}$ solves

$$i\partial_t\tilde{u}^\varepsilon + \frac{1}{2}\Delta\tilde{u}^\varepsilon = V^\varepsilon\tilde{u}^\varepsilon + \left(K^\varepsilon * |\tilde{u}^\varepsilon|^2\right)\tilde{u}^\varepsilon \quad ; \quad \tilde{u}^\varepsilon_{|t=0} = a. \tag{13.9}$$

At this stage, we note that V^ε and K^ε satisfy

$$|V^\varepsilon(t,y)| + |K^\varepsilon(y)| \leqslant C|y|^2,$$

for some constant C independent of ε, t and y. Repeating the argument of the proof of Lemma 13.3, we have

$$\|\tilde{u}^\varepsilon(t)\|_{\Sigma^3} \lesssim e^{Ct}, \quad \forall t \geqslant 0,$$

for some $C > 0$ independent of $\varepsilon \in]0,1]$ (recall that we already knew $\tilde{u}^\varepsilon \in C(\mathbb{R};\Sigma^3)$ from Lemma 13.1, so we only need to prove uniform estimates).

We note that we can rewrite the equation for u, Eq. (13.4), as:

$$i\partial_t u + \frac{1}{2}\Delta u = V^0 u + \left(K^0 * |u|^2\right)u$$

and thus

$$i\partial_t u + \frac{1}{2}\Delta u = V^\varepsilon u + \left(K^\varepsilon * |u|^2\right)u - L^\varepsilon - N^\varepsilon,$$

where

$$L^\varepsilon = \left(V^\varepsilon - V^0\right)u, \quad N^\varepsilon = \left(\left(K^\varepsilon - K^0\right)*|u|^2\right)u,$$

so the error $w^\varepsilon := \tilde{u}^\varepsilon - u$ solves

$$i\partial_t w^\varepsilon + \frac{1}{2}\Delta w^\varepsilon = V^\varepsilon w^\varepsilon + \left(K^\varepsilon * |\tilde{u}^\varepsilon|^2\right)\tilde{u}^\varepsilon - \left(K^\varepsilon * |u|^2\right)u + L^\varepsilon + N^\varepsilon,$$

and satisfies $w^\varepsilon_{|t=0} = 0$. Writing

$$\left(K^\varepsilon * |\tilde{u}^\varepsilon|^2\right)\tilde{u}^\varepsilon - \left(K^\varepsilon * |u|^2\right)u = \left(K^\varepsilon * |\tilde{u}^\varepsilon|^2\right)w^\varepsilon + \left(K^\varepsilon * \left(|\tilde{u}^\varepsilon|^2 - |u|^2\right)\right)u,$$

the L^2 energy estimate yields

$$\|w^\varepsilon(t)\|_{L^2} \lesssim \int_0^t \left\|\left(K^\varepsilon * \left(|\tilde{u}^\varepsilon|^2 - |u|^2\right)\right)u\right\|_{L^2}$$

$$+ \int_0^t \left(\|L^\varepsilon(s)\|_{L^2} + \|N^\varepsilon(s)\|_{L^2}\right) ds.$$

In view of Taylor's formula, Lemma 13.3 and Young inequality,

$$\|L^\varepsilon(t)\|_{L^2(\mathbb{R}^d)} + \|N^\varepsilon(t)\|_{L^2(\mathbb{R}^d)} \lesssim \sqrt{\varepsilon}\, e^{Ct}, \quad \forall t \geqslant 0.$$

Using the pointwise estimate

$$|K^\varepsilon(y-z)|\left(|\tilde{u}^\varepsilon(t,z)|^2 - |u(t,z)|^2\right) \lesssim \left(|y|^2 + |z|^2\right)|\tilde{u}^\varepsilon(t,z) - u(t,z)| \times$$
$$\times \left(|\tilde{u}^\varepsilon(t,z)| + |u(t,z)|\right),$$

we have

$$\left|\left(K^\varepsilon * \left(|\tilde{u}^\varepsilon|^2 - |u|^2\right)\right)(t,y)\right|$$

$$\lesssim |y|^2 \int_{\mathbb{R}^d} |\tilde{u}^\varepsilon(t,z) - u(t,z)| \times \left(|\tilde{u}^\varepsilon(t,z)| + |u(t,z)|\right) dz$$

$$+ \int_{\mathbb{R}^d} |\tilde{u}^\varepsilon(t,z) - u(t,z)| \times |z|^2 \left(|\tilde{u}^\varepsilon(t,z)| + |u(t,z)|\right) dz$$

$$\lesssim |y|^2 \|w^\varepsilon(t)\|_{L^2}\left(\|\tilde{u}^\varepsilon(t)\|_{L^2} + \|u(t)\|_{L^2}\right)$$

$$+ \|w^\varepsilon(t)\|_{L^2}\left(\|\tilde{u}^\varepsilon(t)\|_{\Sigma^2} + \|u(t)\|_{\Sigma^2}\right),$$

where we have used Cauchy–Schwarz inequality. We infer, in view of Lemma 13.3,

$$\left\|\left(K^\varepsilon * \left(|\tilde{u}^\varepsilon|^2 - |u|^2\right)\right)u\right\|_{L^2} \lesssim e^{Ct}\|w^\varepsilon(t)\|_{L^2}, \quad t \geqslant 0,$$

hence, by Gronwall lemma,

$$\|w^\varepsilon(t)\|_{L^2} \lesssim \sqrt{\varepsilon}e^{e^{Ct}}, \quad \forall t \geqslant 0.$$

This proves the theorem in the case $\nabla K(0) = 0$, since in that case, $\tilde{u}^\varepsilon = u^\varepsilon$.

To prove the existence of an a priori non-trivial phase shift $\theta(t)$ when $\nabla K(0) \neq 0$, we have to perform a second order expansion of $\tilde{u}^{\varepsilon}$ as $\varepsilon \to 0$. This is to be compared with the WKB supercritical case studied in Chap. 4: in order to fully describe the wave function, we had to consider the first order correction of the WKB expansion, which appears in an ε-independent modulation of the phase. We seek

$$\tilde{u}^{\varepsilon} = u + \sqrt{\varepsilon}u^{(1)} + \mathcal{O}(\varepsilon),$$

and, expanding V^{ε} and K^{ε} in powers of $\sqrt{\varepsilon}$, we find:

$$i\partial_t u^{(1)} + \frac{1}{2}\Delta u^{(1)} = V^0 u^{(1)} + \left(K^0 * |u|^2\right) u^{(1)} + 2\left(K^0 * \operatorname{Re}\left(\bar{u}u^{(1)}\right)\right) u$$

$$+ \mathcal{V}u + \left(\mathcal{K} * |u|^2\right) u \quad ; \quad u^{(1)}_{|t=0} = 0,$$

where the source terms are given by

$$\mathcal{V}(t,y) = \frac{1}{6}\nabla^3 V\left(q(t)\right) y \cdot y \cdot y, \quad \mathcal{K}(y) = \frac{1}{6}\nabla^3 K(0) y \cdot y \cdot y.$$

Suppose $K \in W^{k,\infty}$ and $a \in \Sigma^k$, $k \geqslant 3$: $u \in C(\mathbb{R}; \Sigma^k)$. The above equation is linear in $u^{(1)}$, plus a source term in $C(\mathbb{R}; \Sigma^{k-3})$. The existence of $u^{(1)}$ can be proved by considering an iterative scheme similar to the one described in the proof of Lemma 13.3, and we have, for $k \geqslant 3$,

$$\|u^{(1)}(t)\|_{\Sigma^{k-3}} \lesssim e^{Ct}, \quad \forall t \geqslant 0.$$

The approximate solution $u^{\varepsilon}_{\text{app}} = u + \sqrt{\varepsilon}u^{(1)}$ solves

$$i\partial_t u^{\varepsilon}_{\text{app}} + \frac{1}{2}\Delta u^{\varepsilon}_{\text{app}} = V^{\varepsilon}u^{\varepsilon}_{\text{app}} + \left(K^{\varepsilon} * |u^{\varepsilon}_{\text{app}}|^2\right) u^{\varepsilon}_{\text{app}} - L^{\varepsilon}_1 - N^{\varepsilon}_1,$$

where the linear source term satisfies the pointwise estimate

$$|L^{\varepsilon}_1(t,y)| \lesssim \varepsilon \langle y \rangle^4 |u(t,y)| + \varepsilon \langle y \rangle^3 |u^{(1)}(t,y)|,$$

and the nonlinear source term,

$$|N^{\varepsilon}_1| \lesssim \varepsilon \left(\langle y \rangle^4 * |u|^2\right) |u|$$

$$+ \left(\varepsilon \left|K^0 * |u^{(1)}|^2\right| + \varepsilon^{3/2} \left(\langle y \rangle^3 * |u^{(1)}|^2\right)\right) \left(|u| + \sqrt{\varepsilon}|u^{(1)}|\right).$$

In particular,

$$\|L^{\varepsilon}_1(t)\|_{L^2} + \|N^{\varepsilon}_1(t)\|_{L^2} \lesssim \varepsilon e^{Ct}, \quad \forall t \geqslant 0,$$

as soon as $k \geqslant 6$. Arguing like for the first order approximation, we have

$$\|\tilde{u}^{\varepsilon}(t) - u^{\varepsilon}_{\text{app}}(t)\|_{L^2} \lesssim \varepsilon e^{e^{Ct}}, \quad \forall t \geqslant 0. \tag{13.10}$$

Therefore, denoting $w_1^\varepsilon = \tilde{u}^\varepsilon - u_{\mathrm{app}}^\varepsilon$,

$$
\begin{aligned}
\theta^\varepsilon(t) &= \frac{1}{\sqrt{\varepsilon}} \int_0^t \nabla K(0) \cdot \left(\int_{\mathbb{R}^d} z \left(|u^\varepsilon(s,z)|^2 - |u(s,z)|^2 \right) dz \right) ds \\
&= \frac{1}{\sqrt{\varepsilon}} \int_0^t \nabla K(0) \cdot \left(\int_{\mathbb{R}^d} z \left(|\tilde{u}^\varepsilon(s,z)|^2 - |u(s,z)|^2 \right) dz \right) ds \\
&= \nabla K(0) \cdot \int_0^t \int z \left(2 \operatorname{Re} \left(\bar{u} u^{(1)} \right) + \sqrt{\varepsilon} |u^{(1)}|^2 + \frac{1}{\sqrt{\varepsilon}} |w_1^\varepsilon|^2 \right) \\
&\quad + \nabla K(0) \cdot \int_0^t \int z \left(\frac{2}{\sqrt{\varepsilon}} \operatorname{Re}(\bar{u} w_1^\varepsilon) + 2 \operatorname{Re} \left(\bar{u}^{(1)} w_1^\varepsilon \right) \right).
\end{aligned}
$$

Since $u^\varepsilon, u, u^{(1)} \in C(\mathbb{R}; \Sigma)$ uniformly in ε (with exponential growth in time), $\|w_1^\varepsilon(t)\|_\Sigma \lesssim e^{Ct}$, so Cauchy–Schwarz inequality yields, in view of (13.10),

$$
|\theta^\varepsilon(t) - \theta(t)| \lesssim \sqrt{\varepsilon} \, e^{e^{Ct}}, \quad t \geqslant 0,
$$

where

$$
\theta(t) = 2 \nabla K(0) \cdot \int_0^t \int_{\mathbb{R}^d} z \operatorname{Re} \left(\bar{u} u^{(1)} \right) (s,z) ds dz.
$$

This completes the proof of the theorem. $\qquad \square$

Remark 13.5. We see that in this supercritical case, Wigner measures may undergo nonlinear effects. More precisely, we still have

$$
\mu(t, dx, d\xi) = \|a\|_{L^2}^2 \delta_{x=q(t)} \otimes \delta_{\xi=p(t)},
$$

where (q, p) is now given by (13.2). Therefore, Wigner measures reveal nonlinear effects if and only if $\nabla K(0) \neq 0$.

13.3 The case of two initial coherent states

In this section, we resume

$$
i\varepsilon \partial_t \psi^\varepsilon + \frac{\varepsilon^2}{2} \Delta \psi^\varepsilon = V(x)\psi^\varepsilon + \varepsilon^\alpha \left(K * |\psi^\varepsilon|^2 \right) \psi^\varepsilon,
$$

and consider initial data which are sums of two coherent states with different centers in phase space,

$$
\psi^\varepsilon(0, x) = \varepsilon^{-d/4} \sum_{j=1,2} a_j \left(\frac{x - q_{j0}}{\sqrt{\varepsilon}} \right) e^{i(x - q_{j0}) \cdot p_{j0}/\varepsilon}, \tag{13.11}
$$

with $(q_{10}, p_{10}) \neq (q_{20}, p_{20})$, and of course $a_1, a_2 \not\equiv 0$. When $\alpha > 1$, the equation is linear at leading order, and so the superposition principle obviously holds. We do not analyze this case. On the other hand, for $\alpha = 1$ (critical case), we will see that, contrary to the case of power-like nonlinearity presented in Sec. 12.3, ψ^ε cannot be approximated by the sum of each approximate solution constructed in Sec. 13.1, even though nonlinear effects analyzed there were fairly weak. More precisely, the nonlinear interaction between the two coherent states induces non-trivial phase interactions (the nonlinear effects are thus qualitatively similar to what happens in the case of a single coherent state). In the supercritical case $\alpha = 0$, we will see that interactions are stronger, as can be expected: they affect the centers in phase space of the two coherent states (which then evolve according to a classical two-body motion), the actions, and the envelopes.

13.3.1   *General considerations*

We seek an approximate solution of the form

$$\psi_{\text{app}}^\varepsilon(t,x) = \varepsilon^{-d/4} \sum_{j=1,2} u_j \left(t, \frac{x - q_j(t)}{\sqrt{\varepsilon}} \right) e^{i(S_j(t) + p_j(t) \cdot (x - q_j(t)))/\varepsilon}, \quad (13.12)$$

for some profiles u_j independent of ε, and some functions $S_j(t)$ to be determined. Denote

$$\phi_j(t,x) = S_j(t) + p_j(t) \cdot (x - q_j(t)).$$

We can write

$$i\varepsilon \partial_t \psi_{\text{app}}^\varepsilon + \frac{\varepsilon^2}{2} \Delta \psi_{\text{app}}^\varepsilon - V\psi_{\text{app}}^\varepsilon - \varepsilon^\alpha \left(K * |\psi_{\text{app}}^\varepsilon|^2 \right) \psi_{\text{app}}^\varepsilon =$$

$$\varepsilon^{-d/4} \sum_{j=1,2} e^{i\phi_j(t,x)/\varepsilon} \left(b_{0j} + \sqrt{\varepsilon} b_{1j} + \varepsilon b_{2j} + \varepsilon r_j^\varepsilon \right) \left(t, \frac{x - q_j(t)}{\sqrt{\varepsilon}} \right),$$

for $b_{\ell,j}$ independent of ε. Like in the case of a single coherent state, the approximate solution $\psi_{\text{app}}^\varepsilon$ is determined by the conditions

$$b_{0j} = b_{1j} = b_{2j} = 0, \quad j = 1, 2.$$

The remaining factor r_j^ε accounts for the error between the exact solution ψ^ε and the approximate solution $\psi_{\text{app}}^\varepsilon$. To each wave packet, we associate the relevant space variable

$$y_j = \frac{x - q_j(t)}{\sqrt{\varepsilon}}, \quad j = 1, 2,$$

corresponding to the moving frame of each wave packet. The computations for the linear part of the equation are the same as before, up to the fact that we now have two space variables y_j. Recalling that the relevant space variable for u_j is y_j, we have:

$$\partial_t \phi_j = \dot{S}_j(t) + \frac{d}{dt}\left(p_j(t) \cdot (x - q_j(t))\right) = \dot{S}_j(t) + \sqrt{\varepsilon}\dot{p}_j(t) \cdot y_j - p_j(t) \cdot \dot{q}_j(t).$$

For the linear potential term, we resume the previous computation:

$$V\left(q_j(t) + y_j\sqrt{\varepsilon}\right)u_j(t, y_j) = V\left(q_j(t)\right)u_j(t, y_j) + \sqrt{\varepsilon}y_j \cdot \nabla V\left(q_j(t)\right)u_j(t, y_j)$$
$$+ \frac{\varepsilon}{2}\left\langle y_j, \nabla^2 V\left(q_j(t)\right)y_j\right\rangle u_j(t, y_j) + \varepsilon^{3/2}r_{jV}^{\varepsilon}(t, y_j),$$

with

$$\left|r_{jV}^{\varepsilon}(t, y_j)\right| \leqslant C\left\langle y_j\right\rangle^3 |u_j(t, y_j)|, \tag{13.13}$$

for some C independent of ε, t and y_j. We gather the information of the linear case $K = 0$ as

$$b_{0j}^{\text{lin}} = -u_j\left(\dot{S}_j(t) - p_j(t) \cdot \dot{q}_j(t) + \frac{|p_j(t)|^2}{2} + V\left(q_j(t)\right)\right),$$
$$b_{1j}^{\text{lin}} = -i\left(\dot{q}_j(t) - p_j(t)\right) \cdot \nabla u_j - y_j \cdot \left(\dot{p}_j(t) + \nabla V\left(q_j(t)\right)\right)u_j,$$
$$b_{2j}^{\text{lin}} = i\partial_t u_j + \frac{1}{2}\Delta u_j - \frac{1}{2}\left\langle y_j, \nabla^2 V\left(q_j(t)\right)y_j\right\rangle u_j.$$

For the nonlinear term, the computations are heavier:

$$\left(K * |\psi_{\text{app}}^{\varepsilon}|^2\right)\psi_{\text{app}}^{\varepsilon}$$
$$= \varepsilon^{-d/4}\sum_{j=1,2} e^{i\phi_j(t,x)/\varepsilon}\left(\int K(z)|\psi_{\text{app}}^{\varepsilon}(t, x - z)|^2 dz\right)u_j(t, y_j).$$

Eventually, each envelope u_j will solve a Schrödinger equation, the two equations being coupled. The precise expression of these equations depends on α, but at this stage, we can notice that for $j = 1, 2$, u_j solves an equation of the form

$$i\partial_t u_j + \frac{1}{2}\Delta u_j = \frac{1}{2}\left\langle y_j, \nabla^2 V\left(q_j(t)\right)y_j\right\rangle u_j + F_j u_j, \tag{13.14}$$

where the function F_j, accounting for nonlinear effects due to the Hartree kernel, is *real-valued*. So like before, the L^2-norm of u_j is independent of time,

$$\|u_j(t)\|_{L^2(\mathbb{R}^d)} = \|a_j\|_{L^2(\mathbb{R}^d)}, \quad \forall t \geqslant 0, \ j = 1, 2.$$

In the above sum, the variable x must be expressed in terms of y_j:

$$K * |\psi_{\text{app}}^{\varepsilon}|^2 = \int K(z)\left|\psi_{\text{app}}^{\varepsilon}\left(t, q_j(t) + \sqrt{\varepsilon}y_j - z\right)\right|^2 dz$$

$$= \varepsilon^{-d/2}\int K(z)\left|\sum_{k=1,2} e^{i\phi_k(t,x-z)/\varepsilon}u_k\left(t, y_j + \frac{q_j(t) - q_k(t)}{\sqrt{\varepsilon}} - \frac{z}{\sqrt{\varepsilon}}\right)\right|^2 dz.$$

Before changing the integration variable, we develop the squared modulus:

$$\left| \sum_{k=1,2} e^{i\phi_k(t,x-z)/\varepsilon} u_k \left(t, y_j + \frac{q_j(t) - q_k(t)}{\sqrt{\varepsilon}} - \frac{z}{\sqrt{\varepsilon}} \right) \right|^2$$

$$= \left| u_1 \left(t, y_j + \frac{q_j(t) - q_1(t)}{\sqrt{\varepsilon}} - \frac{z}{\sqrt{\varepsilon}} \right) \right|^2 + \left| u_2 \left(t, y_j + \frac{q_j(t) - q_2(t)}{\sqrt{\varepsilon}} - \frac{z}{\sqrt{\varepsilon}} \right) \right|^2$$

$$+ 2 \operatorname{Re} e^{i\delta\phi/\varepsilon} u_1 \left(t, y_j + \frac{q_j(t) - q_1(t)}{\sqrt{\varepsilon}} - \frac{z}{\sqrt{\varepsilon}} \right) \overline{u}_2 \left(t, y_j + \frac{q_j(t) - q_2(t)}{\sqrt{\varepsilon}} - \frac{z}{\sqrt{\varepsilon}} \right),$$

where $\delta\phi$ stands for $\phi_1(t, x - z) - \phi_2(t, x - z)$. To ease notations, we shall denote from now on:

$$\delta q(t) = q_1(t) - q_2(t) \quad ; \quad \delta p(t) = p_1(t) - p_2(t).$$

We can write

$$\left(K * |\psi_{\mathrm{app}}^\varepsilon|^2 \right) \psi_{\mathrm{app}}^\varepsilon = \varepsilon^{-d/4} \sum_{j=1,2} e^{i\phi_j(t,x)/\varepsilon} V_j^{\mathrm{NL}}(t, y_j) u_j(t, y_j),$$

with

$$V_1^{\mathrm{NL}} = \varepsilon^{-d/2} \int K(z) \left(\left| u_1 \left(t, y_1 - \frac{z}{\sqrt{\varepsilon}} \right) \right|^2 + \left| u_2 \left(t, y_1 + \frac{\delta q(t)}{\sqrt{\varepsilon}} - \frac{z}{\sqrt{\varepsilon}} \right) \right|^2 \right.$$

$$\left. + 2 \operatorname{Re} e^{i\delta\phi/\varepsilon} u_1 \left(t, y_1 - \frac{z}{\sqrt{\varepsilon}} \right) \overline{u}_2 \left(t, y_1 + \frac{\delta q(t)}{\sqrt{\varepsilon}} - \frac{z}{\sqrt{\varepsilon}} \right) \right) dz,$$

$$V_2^{\mathrm{NL}} = \varepsilon^{-d/2} \int K(z) \left(\left| u_1 \left(t, y_2 - \frac{\delta q(t)}{\sqrt{\varepsilon}} - \frac{z}{\sqrt{\varepsilon}} \right) \right|^2 + \left| u_2 \left(t, y_2 - \frac{z}{\sqrt{\varepsilon}} \right) \right|^2 \right.$$

$$\left. + 2 \operatorname{Re} e^{i\delta\phi/\varepsilon} u_1 \left(t, y_2 - \frac{\delta q(t)}{\sqrt{\varepsilon}} - \frac{z}{\sqrt{\varepsilon}} \right) \overline{u}_2 \left(t, y_2 - \frac{z}{\sqrt{\varepsilon}} \right) \right) dz.$$

Each nonlinear potential V_j^{NL} is the sum of three terms. The third term in each of these two expressions, involving the product $u_1 \overline{u}_2$, will be referred to as *cross term*, as opposed to *squared terms*, involving squared moduli. We prove below a general estimate on cross terms, showing that they are negligible in the limit $\varepsilon \to 0$, regardless of the value of α. Therefore, we now consider only the squared terms.

13.3.2   *The approximate solution*

Generalizing the notation of a single coherent state to

$$G_j(t) = \int_{\mathbb{R}^d} z |u_j(t, z)|^2 dz,$$

changing variables in the integrations and performing a Taylor expansion of the kernel K, we find successively:

$$\varepsilon^{-d/2} \int K(z) \left| u_1\left(t, y_1 - \frac{z}{\sqrt{\varepsilon}}\right)\right|^2 dz = \int K\left(\sqrt{\varepsilon}(y_1 - z)\right) |u_1(t,z)|^2 dz$$

$$= K(0)\|a_1\|_{L^2}^2 + \sqrt{\varepsilon}\|a_1\|_{L^2}^2 y_1 \cdot \nabla K(0) - \sqrt{\varepsilon}\nabla K(0) \cdot G_1(t)$$

$$+ \frac{\varepsilon}{2}\langle y_1, \nabla^2 K(0)y_1\rangle \|a_1\|_{L^2}^2 + \frac{\varepsilon}{2}\int \langle z, \nabla^2 K(0)z\rangle |u_1(t,z)|^2 dz$$

$$- \varepsilon\langle \nabla^2 K(0)G_1(t), y_1\rangle + \varepsilon^{3/2}\int r_{11}^\varepsilon(t, z - y_1)|u_1(t,z)|^2 dz,$$

$$\varepsilon^{-d/2} \int K(z) \left| u_2\left(t, y_1 + \frac{\delta q(t)}{\sqrt{\varepsilon}} - \frac{z}{\sqrt{\varepsilon}}\right)\right|^2 dz$$

$$= \int K\left(\delta q(t) + \sqrt{\varepsilon}(y_1 - z)\right) |u_2(t,z)|^2 dz$$

$$= K(\delta q)\|a_2\|_{L^2}^2 + \sqrt{\varepsilon}\|a_2\|_{L^2}^2 y_1 \cdot \nabla K(\delta q) - \sqrt{\varepsilon}\nabla K(\delta q) \cdot G_2(t)$$

$$+ \frac{\varepsilon}{2}\langle y_1, \nabla^2 K(\delta q)y_1\rangle \|a_2\|_{L^2}^2 + \frac{\varepsilon}{2}\int \langle z, \nabla^2 K(\delta q)z\rangle |u_2(t,z)|^2 dz$$

$$- \varepsilon\langle \nabla^2 K(\delta q)G_2(t), y_1\rangle + \varepsilon^{3/2}\int r_{12}^\varepsilon(t, z - y_1)|u_2(t,z)|^2 dz,$$

$$\varepsilon^{-d/2} \int K(z) \left| u_1\left(t, y_2 - \frac{\delta q(t)}{\sqrt{\varepsilon}} - \frac{z}{\sqrt{\varepsilon}}\right)\right|^2 dz$$

$$= K(-\delta q)\|a_1\|_{L^2}^2 + \sqrt{\varepsilon}\|a_1\|_{L^2}^2 y_2 \cdot \nabla K(-\delta q) - \sqrt{\varepsilon}\nabla K(-\delta q) \cdot G_1(t)$$

$$+ \frac{\varepsilon}{2}\langle y_2, \nabla^2 K(-\delta q)y_2\rangle \|a_1\|_{L^2}^2 + \frac{\varepsilon}{2}\int \langle z, \nabla^2 K(-\delta q)z\rangle |u_1(t,z)|^2 dz$$

$$- \varepsilon\langle \nabla^2 K(-\delta q)G_1(t), y_2\rangle + \varepsilon^{3/2}\int r_{21}^\varepsilon(t, z - y_2)|u_1(t,z)|^2 dz,$$

$$\varepsilon^{-d/2} \int K(z) \left| u_2\left(t, y_2 - \frac{z}{\sqrt{\varepsilon}}\right)\right|^2 dz$$

$$= K(0)\|a_2\|_{L^2}^2 + \sqrt{\varepsilon}\|a_2\|_{L^2}^2 y_2 \cdot \nabla K(0) - \sqrt{\varepsilon}\nabla K(0) \cdot G_2(t)$$

$$+ \frac{\varepsilon}{2}\langle y_2, \nabla^2 K(0)y_2\rangle \|a_2\|_{L^2}^2 + \frac{\varepsilon}{2}\int \langle z, \nabla^2 K(0)z\rangle |u_2(t,z)|^2 dz$$

$$- \varepsilon\langle \nabla^2 K(0)G_2(t), y_2\rangle + \varepsilon^{3/2}\int r_{22}^\varepsilon(t, z - y_2)|u_2(t,z)|^2 dz,$$

where the functions r_{jk}^ε satisfy uniform estimates of the form

$$|r_{jk}^\varepsilon(t,z)| \leqslant C(T)\langle z\rangle^3, \quad \forall z \in \mathbb{R}^d,\ t \in [0,T], \tag{13.15}$$

with $C(T)$ independent of ε, j and k, but possibly depending on T. Examine the outcome of this computation, according to the value of α.

The approximate solution in the critical case: $\alpha = 1$

When $\alpha = 1$, we have $b_{\ell j} = b_{\ell j}^{\text{lin}}$ for $\ell = 0, 1$: like in the linear case, (q_j, p_j) solves (10.4), S_j is defined as in (10.6), for $j = 1, 2$. On the other hand, the equations for the envelopes are affected by nonlinear interactions:

$$b_{21} = i\partial_t u_1 + \frac{1}{2}\Delta u_1 - \frac{1}{2}\left\langle y_1, \nabla^2 V(q_1(t)) y_1 \right\rangle u_1 - K(0)\|a_1\|_{L^2}^2 u_1$$
$$- K(\delta q(t))\|a_2\|_{L^2}^2 u_1,$$

$$b_{22} = i\partial_t u_2 + \frac{1}{2}\Delta u_2 - \frac{1}{2}\left\langle y_2, \nabla^2 V(q_2(t)) y_2 \right\rangle u_2 - K(0)\|a_2\|_{L^2}^2 u_2$$
$$- K(-\delta q(t))\|a_1\|_{L^2}^2 u_2.$$

The last term in each expression corresponds to a coupling, revealing a leading order interaction of the two wave packets. This coupling can be understood rather explicitly, since it consists of a purely time dependent potential. Solving the equations $b_{2j} = 0$, we infer, with obvious notations,

$$u_1(t, y_1) = u_1^{\text{lin}}(t, y_1) \exp\left(-itK(0)\|a_1\|_{L^2}^2 - i\|a_2\|_{L^2}^2 \int_0^t K(\delta q(s))\, ds\right),$$

$$u_2(t, y_2) = u_2^{\text{lin}}(t, y_2) \exp\left(-itK(0)\|a_2\|_{L^2}^2 - i\|a_1\|_{L^2}^2 \int_0^t K(-\delta q(s))\, ds\right).$$

The presence of these phase shifts accounts for nonlinear effects at leading order in the approximate wave packet $\psi_{\text{app}}^\varepsilon$: nonlinear effects in the case of a single wave packet, and nonlinear coupling. For the remainder terms, we have the (rough) pointwise estimate

$$|r_j^\varepsilon(t, y_j)| \leqslant C(T)\sqrt{\varepsilon}\,\langle y_j \rangle^3\, |u_j(t, y_j)| \left(1 + \sum_{k=1,2} \|u_k(t)\|_{\Sigma^2}^2\right), \quad t \in [0, T].$$

The remainder r_j^ε is the sum of the terms r_{jV}^ε and $\varepsilon^{\alpha+3/2}(r_{jk}^\varepsilon * |u_k|^2)u_j$, $k = 1, 2$, so this estimate is an easy consequence of (13.13) and (13.15). To be precise, this estimate is valid up to the cross terms that we have discarded so far, when we have developed $(K * |\psi_{\text{app}}^\varepsilon|^2)\psi_{\text{app}}^\varepsilon$.

The approximate solution in the supercritical case $\alpha = 0$

Now all the coefficients $b_{\ell j}$ are affected by the nonlinearity. Denote

$$M_1(t) = \|a_1\|_{L^2(\mathbb{R}^d)}^2 \nabla^2 K(0) + \|a_2\|_{L^2(\mathbb{R}^d)}^2 \nabla^2 K(\delta q(t)) + \nabla^2 V(q_1(t)),$$

$$M_2(t) = \|a_2\|_{L^2(\mathbb{R}^d)}^2 \nabla^2 K(0) + \|a_1\|_{L^2(\mathbb{R}^d)}^2 \nabla^2 K(-\delta q(t)) + \nabla^2 V(q_2(t)).$$

Like in the case of a single coherent state, we incorporate the last term of b_{1j} into b_{0j}, that is we modify the action as follows:

$$S_1^\varepsilon(t) = \int_0^t \left(\frac{1}{2}|p_1(s)|^2 - V(s, q_1(s)) - K(0)\|a_1\|_{L^2}^2 - K(\delta q(s))\|a_2\|_{L^2}^2 \right.$$
$$\left. + \sqrt{\varepsilon}\nabla K(0) \cdot G_1(s) + \sqrt{\varepsilon}\nabla K(\delta q(s)) \cdot G_2(s) \right) ds,$$

$$S_2^\varepsilon(t) = \int_0^t \left(\frac{1}{2}|p_2(s)|^2 - V(s, q_2(s)) - K(0)\|a_2\|_{L^2}^2 - K(-\delta q(s))\|a_1\|_{L^2}^2 \right.$$
$$\left. + \sqrt{\varepsilon}\nabla K(0) \cdot G_2(s) + \sqrt{\varepsilon}\nabla K(-\delta q(s)) \cdot G_1(s) \right) ds.$$

Note that for S_j^ε to be well defined, we have to first define u_j, for which we solve the envelope equations, given by $b_{21} = b_{22} = 0$. Canceling the terms b_{1j} yields the modified system of trajectories:

$$\begin{cases} \dot{q}_1(t) = p_1(t), \\ \dot{p}_1(t) = -\nabla V(q_1(t)) - \|a_1\|_{L^2}^2 \nabla K(0) - \|a_2\|_{L^2}^2 \nabla K(\delta q(t)), \\ \dot{q}_2(t) = p_2(t), \\ \dot{p}_2(t) = -\nabla V(q_2(t)) - \|a_2\|_{L^2}^2 \nabla K(0) - \|a_1\|_{L^2}^2 \nabla K(-\delta q(t)). \end{cases} \quad (13.16)$$

In the case where K is even, $K(-x) = K(x)$, the above system has a Hamiltonian structure, and corresponds to a classical two-body problem. Setting, for $z = (q_1, p_1, q_2, p_2)^T$,

$$H(t, z) = \alpha_1 \left(\frac{1}{2}|p_1|^2 + V(q_1) \right) + \alpha_2 \left(\frac{1}{2}|p_2|^2 + V(q_2) \right) + \alpha_1 \alpha_2 K(q_1 - q_2),$$

where $\alpha_j = \|a_j\|_{L^2}^2$, Eq. (13.16) can be written as

$$\frac{dz}{dt} = JD_z H(t, z) \quad \text{with} \quad J = \begin{pmatrix} 0 & 1/\alpha_1 & 0 & 0 \\ -1/\alpha_1 & 0 & 0 & 0 \\ 0 & 0 & 0 & 1/\alpha_2 \\ 0 & 0 & -1/\alpha_2 & 0 \end{pmatrix}.$$

Finally, the system for the envelopes is:

$$\begin{cases} i\partial_t u_1 + \dfrac{1}{2}\Delta u_1 = \dfrac{1}{2}\langle y, M_1(t)y\rangle u_1 \\[2mm] \qquad - \langle \nabla^2 K(0)G_1(t), y\rangle u_1 - \langle \nabla^2 K(\delta q)G_2(t), y\rangle u_1 \\[2mm] \qquad + \dfrac{1}{2}\left(\int \langle z, \nabla^2 K(0)z\rangle |u_1(t,z)|^2 dz\right) u_1 \\[2mm] \qquad + \dfrac{1}{2}\left(\int \langle z, \nabla^2 K(\delta q)z\rangle |u_2(t,z)|^2 dz\right) u_1, \\[3mm] i\partial_t u_2 + \dfrac{1}{2}\Delta u_2 = \dfrac{1}{2}\langle y, M_2(t)y\rangle u_2 \\[2mm] \qquad - \langle \nabla^2 K(0)G_2(t), y\rangle u_2 - \langle \nabla^2 K(-\delta q)G_1(t), y\rangle u_2 \\[2mm] \qquad + \dfrac{1}{2}\left(\int \langle z, \nabla^2 K(0)z\rangle |u_2(t,z)|^2 dz\right) u_2 \\[2mm] \qquad + \dfrac{1}{2}\left(\int \langle z, \nabla^2 K(-\delta q)z\rangle |u_1(t,z)|^2 dz\right) u_2. \end{cases} \tag{13.17}$$

13.3.3   Estimating the cross terms

In the previous section, we have discarded the cross terms, claiming that they are negligible in the limit $\varepsilon \to 0$. We now justify precisely this statement. Since the two cross terms are similar, we simply consider the first one:

$$2\varepsilon^{-d/2}\,\mathrm{Re}\int_{\mathbb{R}^d} K(z)e^{i\delta\phi/\varepsilon}u_1\left(t, y_1 - \frac{z}{\sqrt{\varepsilon}}\right)\overline{u}_2\left(t, y_1 + \frac{\delta q(t)}{\sqrt{\varepsilon}} - \frac{z}{\sqrt{\varepsilon}}\right)dz.$$

Notice that we have not yet expressed the phases ϕ_k in terms of the variable y_1, and that the expression of ϕ_k varies according to $\alpha = 1$ or $\alpha = 0$. We shall retain only a common feature though, that is,

$$\phi_k^\varepsilon(t, x) = \Theta_k^\varepsilon(t) + x \cdot p_k(t),$$

where only the purely time dependent function Θ may depend on ε (when $\alpha = 0$), and the spatial oscillations are singled out. Since $x = q_1(t) + \sqrt{\varepsilon}y_1$, we get, once the real part and the time oscillations are omitted:

$$\varepsilon^{-d/2}\int_{\mathbb{R}^d} K(z)e^{i(\sqrt{\varepsilon}y_1 - z)\cdot\delta p(t)/\varepsilon}u_1\left(t, y_1 - \frac{z}{\sqrt{\varepsilon}}\right)\overline{u}_2\left(t, y_1 + \frac{\delta q(t)}{\sqrt{\varepsilon}} - \frac{z}{\sqrt{\varepsilon}}\right)dz.$$

Changing the integration variable, and introducing more general notations, we examine:

$$I^\varepsilon(t, y) = \int_{\mathbb{R}^d} \mathcal{K}\left(\sqrt{\varepsilon}(y - z)\right)e^{iz\cdot\delta p(t)/\sqrt{\varepsilon}}u_1(t, z)\overline{u}_2\left(t, z + \frac{\delta q(t)}{\sqrt{\varepsilon}}\right)dz,$$

where (q_j, p_j) solves either (10.4) (case $\alpha = 1$) or (13.16) (case $\alpha = 0$).

Proposition 13.6. *Let $T > 0$. Suppose that $\mathcal{K} \in W^{\ell,\infty}$, $u_j \in C([0,T]; \Sigma^k)$ with $k, \ell \in \mathbb{N}$. There exists $C > 0$ independent of $\varepsilon \in]0,1]$, $\mathcal{K}$, u_1 and u_2 such that*

$$\sup_{t \in [0,T]} \|I^\varepsilon(t)\|_{L^\infty(\mathbb{R}^d)} \leqslant C\varepsilon^{\min(\ell,k)/2} \|\mathcal{K}\|_{W^{\ell,\infty}} \prod_{j=1,2} \|u_j\|_{L^\infty([0,T];\Sigma^k)}.$$

Remark 13.7. Proposition 13.6 is a refinement of Proposition 12.9, in the sense that the power of ε on the right-hand side is as large as we wish, provided that $\mathcal{K}$ is sufficiently smooth, and that the functions u_1 and u_2 are sufficiently localized in space and frequency.

A microlocal property

The proof of Proposition 13.6 is based on the following remark: the function that we integrate is localized away from the origin in *phase space*:

Lemma 13.8. *Suppose $(q_{10}, p_{10}) \neq (q_{20}, p_{20})$. In either of the cases $\alpha = 1$ or $\alpha = 0$, the following holds. For any $T > 0$, there exists $\eta > 0$ such that for all $t \in [0,T]$,*

$$|\delta q(t)| + |\delta p(t)| \geqslant 2\eta.$$

Proof. We argue by contradiction: if the result were not true, we could find a sequence $t_n \in [0,T]$ so that

$$|\delta q(t_n)| + |\delta p(t_n)| \xrightarrow[n \to \infty]{} 0.$$

By compactness of $[0,T]$ and continuity of (q_j, p_j), there would exist $t_* \in [0,T]$ such that

$$q_1(t_*) = q_2(t_*), \quad p_1(t_*) = p_2(t_*).$$

In the case $\alpha = 1$, (q_j, p_j) is given by Eq. (10.4): uniqueness for (10.4) implies $(q_{10}, p_{10}) = (q_{20}, p_{20})$, hence a contradiction.

The case $\alpha = 0$ is a bit more delicate. From (13.16), we infer:

$$\begin{cases} \dfrac{d(\delta q)}{dt} = \delta p, \\[2mm] \dfrac{d(\delta p)}{dt} = \nabla V(q_2(t)) - \nabla V(q_1(t)) + \|a_1\|_{L^2}^2 (\nabla K(-\delta q(t)) - \nabla K(0)) \\[2mm] \qquad\qquad + \|a_2\|_{L^2}^2 (\nabla K(0) - \nabla K(\delta q(t))). \end{cases}$$

Since $\nabla^2 V, \nabla^2 K \in L^\infty(\mathbb{R}^d)$, there exists C independent of t such that

$$\left| \frac{d(\delta q)}{dt} \right| + \left| \frac{d(\delta p)}{dt} \right| \leqslant C (|\delta p| + |\delta q|).$$

Gronwall lemma yields a contradiction, and the lemma is proved. $\qquad\square$

Proof of Proposition 13.6

From Lemma 13.8, for all time $t \in [0, T]$, $|\delta q(t)| \geqslant \eta$, or $|\delta p(t)| \geqslant \eta$.

First case. If $|\delta q(t)| \geqslant \eta$, we use Cauchy–Schwarz inequality to infer

$$
|I^\varepsilon(t, y)| \leqslant \|\mathcal{K}\|_{L^\infty} \int \frac{\langle z \rangle^k}{\langle z \rangle^k} |u_1(t, z)| \frac{\left\langle z + \frac{\delta q(t)}{\sqrt{\varepsilon}} \right\rangle^k}{\left\langle z + \frac{\delta q(t)}{\sqrt{\varepsilon}} \right\rangle^k} \left| u_2 \left(t, z + \frac{\delta q(t)}{\sqrt{\varepsilon}} \right) \right| dz
$$

$$
\leqslant \|\mathcal{K}\|_{L^\infty} \|u_1(t)\|_{\Sigma^k} \|u_2(t)\|_{\Sigma^k} \sup_{z \in \mathbb{R}^d} \langle z \rangle^{-k} \left\langle z + \frac{\delta q(t)}{\sqrt{\varepsilon}} \right\rangle^{-k}.
$$

In view of Peetre inequality (see e.g. [Alinhac and Gérard (2007)]),

$$
\sup_{z \in \mathbb{R}^d} \langle z \rangle^{-k} \left\langle z + \frac{\delta q(t)}{\sqrt{\varepsilon}} \right\rangle^{-k} \leqslant C_k \left(\frac{\sqrt{\varepsilon}}{|\delta q(t)|} \right)^k \leqslant \frac{C_k}{\eta^k} \varepsilon^{k/2}.
$$

Second case. If $|\delta p(t)| \geqslant \eta$, we perform repeated integrations by parts (like in the standard proof of the nonstationary phase lemma, see e.g. [Alinhac and Gérard (2007)]) relying on the relation

$$
e^{iz \cdot \delta p(t)/\sqrt{\varepsilon}} = -i \frac{\sqrt{\varepsilon}}{|\delta p(t)|^2} \sum_{\ell=1}^{d} (\delta p(t))_\ell \frac{\partial}{\partial z_\ell} \left(e^{iz \cdot \delta p(t)/\sqrt{\varepsilon}} \right).
$$

Note that since we assume $\mathcal{K} \in W^{\ell,\infty}$ and $u_j \in \Sigma^k$, we perform no more than $\min(\ell, k)$ integrations by parts, and Cauchy–Schwarz inequality yields

$$
|I^\varepsilon(t, y)| \leqslant \frac{1}{\eta^\ell} \|\mathcal{K}\|_{W^{\ell,\infty}} \|u_1(t)\|_{\Sigma^k} \|u_2(t)\|_{\Sigma^k} \varepsilon^{\min(\ell,k)/2}.
$$

13.3.4   *Error estimate in the critical case*

Having constructed the approximate solution, it is rather straightforward to prove an error estimate.

Proposition 13.9. *Let* $K \in W^{1,\infty}(\mathbb{R}^d; \mathbb{R})$, $a_1, a_2 \in \Sigma^3$. *Let* ψ^ε *solve*

$$
i\varepsilon \partial_t \psi^\varepsilon + \frac{\varepsilon^2}{2} \Delta \psi^\varepsilon = V(x) \psi^\varepsilon + \varepsilon \left(K * |\psi^\varepsilon|^2 \right) \psi^\varepsilon,
$$

with initial datum given by Eq. (13.11). For all $T > 0$, *there exists* $C = C(T) > 0$ *independent of* ε, *such that for all* $\varepsilon \in]0, 1]$,

$$
\|\psi^\varepsilon(t) - \psi^\varepsilon_{\mathrm{app}}(t)\|_{L^2} \lesssim C\sqrt{\varepsilon}, \quad \forall t \in [0, T],
$$

where $\psi_{\mathrm{app}}^{\varepsilon}$ is given by (13.12), (q_j, p_j) solves Eq. (10.4), S_j is given by Eq. (10.6), and

$$u_1(t, y_1) = u_1^{\mathrm{lin}}(t, y_1) \exp\left(-itK(0)\|a_1\|_{L^2}^2 - i\|a_2\|_{L^2}^2 \int_0^t K\left(\delta q(s)\right) ds \right),$$

$$u_2(t, y_2) = u_2^{\mathrm{lin}}(t, y_2) \exp\left(-itK(0)\|a_2\|_{L^2}^2 - i\|a_1\|_{L^2}^2 \int_0^t K\left(-\delta q(s)\right) ds \right).$$

Proof. The proof of Proposition 13.2 can be resumed with little change. The only new source terms correspond to the cross terms evoked above, of the form $\varepsilon I^\varepsilon$, where the ε factor is the one in front of the nonlinearity. In view of Proposition 13.6, $|I^\varepsilon| = \mathcal{O}(\sqrt{\varepsilon})$ on $[0, T]$. This being noticed, the proof of Proposition 13.2 can be repeated. □

In this case, the Wigner measure of ψ^ε is given by

$$\mu(t, dx, d\xi) = \sum_{j=1,2} \|a_j\|_{L^2(\mathbb{R}^d)}^2 \delta_{x=q_j(t)} \otimes \delta_{\xi=p_j(t)},$$

and thus, since the classical trajectories ignore nonlinear interactions, the Wigner measure is the same as in the linear case.

Contrary to Proposition 13.2 , we do not state a large time error estimate. This is due to our proof of Proposition 13.6, which is not quantitative for large time, since Lemma 13.8 is not.

13.3.5 *Supercritical case*

We first remark that the centers in phase space are well-defined: from Cauchy–Lipschitz Theorem, if $K \in C^k \cap W^{k,\infty}(\mathbb{R}^d; \mathbb{R})$, $k \geqslant 1$, then Eq. (13.16) has a unique solution $(q_1, p_1, q_2, p_2) \in C^k(\mathbb{R}; \mathbb{R}^{4d})$. Like in the case of a single initial coherent state, an important step is the analysis of the envelope equations, which is needed to ensure the definition of the actions.

Lemma 13.10. *Let $k \geqslant 1$, $a_1, a_2 \in \Sigma^k$. Then (13.17) has a unique solution $(u_1, u_2) \in C(\mathbb{R}; \Sigma^k)$ with initial data $u_{j|t=0} = a_j$. In addition, the following conservations hold:*

$$\|u_j(t)\|_{L^2(\mathbb{R}^d)} = \|a_j\|_{L^2(\mathbb{R}^d)}, \quad \forall t \in \mathbb{R}, \ j = 1, 2.$$

The proof of Lemma 13.3 can essentially be resumed in order to prove the above result. We do not emphasize the exponential control here, since, as in the above critical case, we will not state a large time error estimate, due to the proof of Lemma 13.8.

On the other hand, it does not seem possible to easily adapt the proof of Theorem 13.4 to the case of two (or more) coherent states, since the proof of Theorem 13.4 relies on a change of unknown function based on the moving frame of the approximate solution. In the case of two wave packets, there are two moving frames, but the actual reason which prevents us from directly adapting the proof of Theorem 13.4 is that in the case of two coherent states, the form of the equation for the exact solution is altered, due to the crossed terms analyzed above. We construct a solution to (13.1) of the form

$$\psi^\varepsilon(t,x) = \varepsilon^{-d/4} \sum_{j=1,2} u_j^\varepsilon\left(t, \frac{x-q_j(t)}{\sqrt{\varepsilon}}\right) e^{i\left(S_j^\varepsilon(t)+p_j(t)\cdot(x-q_j(t))\right)/\varepsilon}, \quad (13.18)$$

where the quantities (q_j, p_j) and S_j^ε are given by Eq. (13.2) and Eq. (13.3), respectively. We emphasize that the initial complex-valued function ψ^ε will be determined by the two complex-valued functions u_1^ε and u_2^ε: the representation (13.18) is not unique. This introduction of new unknown functions coming with extra degrees of freedom is to be compared with the approaches presented in Chap. 4 in the case of supercritical WKB analysis. Naturally, the envelopes u_1^ε and u_2^ε are determined by computations similar to those presented in Sec. 13.3.1: the determinations of (q_j, p_j) and S_j^ε being settled, u_1^ε and u_2^ε solve a system similar to (13.17), up to the fact that no "negligible" term is removed, making the corresponding formula rather heavy.

In order to shorten the formulas, we consider indices in $\mathbb{Z}/2\mathbb{Z}$: typically, q_j stands for q_1 whenever $j=1$ or 3, and we write $\delta q_j = q_j - q_{j+1}$. Following the strategy presented above, we find:

$$
\begin{cases}
i\partial_t u_j^\varepsilon + \dfrac{1}{2}\Delta u_j^\varepsilon = V_j^\varepsilon(t, y_j)u_j^\varepsilon + \left(K_{j,\mathrm{diag}}^\varepsilon * |u_j^\varepsilon|^2\right) u_j^\varepsilon \\[2mm]
\quad + \left(K_{j,\mathrm{off}}^\varepsilon * |u_{j+1}^\varepsilon|^2\right) u_j^\varepsilon \\[2mm]
\quad - \dfrac{1}{\sqrt{\varepsilon}}\nabla K(0)\cdot\left(\displaystyle\int z\left(|u_j^\varepsilon(t,z)|^2 - |u_j(t,z)|^2\right)dz\right) u_j^\varepsilon \\[2mm]
\quad - \dfrac{1}{\sqrt{\varepsilon}}\nabla K(\delta q_j)\cdot\left(\displaystyle\int z\left(|u_{j+1}^\varepsilon(t,z)|^2 - |u_{j+1}(t,z)|^2\right)dz\right) u_j^\varepsilon \\[2mm]
\quad + \dfrac{1}{\varepsilon}\left(2\,\mathrm{Re}\,W_j^\varepsilon(t,y_j)\right) u_j^\varepsilon,
\end{cases}
\quad (13.19)
$$

where

$$V_j^\varepsilon(t, y_j) = \frac{1}{\varepsilon} \left(V\left(t, q_j(t) + y_j\sqrt{\varepsilon}\right) - V\left(t, q_j(t)\right) - \sqrt{\varepsilon}y_j \cdot \nabla V\left(t, q_j(t)\right) \right)$$

$$= \int_0^1 \left\langle y_j, \nabla^2 V\left(t, q_j(t) + \theta y_j\sqrt{\varepsilon}\right) y_j \right\rangle (1 - \theta)d\theta,$$

$$K_{j,\mathrm{diag}}^\varepsilon(t, y_j) = \frac{1}{\varepsilon} \left(K\left(\sqrt{\varepsilon}y_j\right) - K(0) - \sqrt{\varepsilon}y_j \cdot \nabla K(0) \right)$$

$$= \int_0^1 \left\langle y_j, \nabla^2 K\left(\theta y_j\sqrt{\varepsilon}\right) y_j \right\rangle (1 - \theta)d\theta,$$

$$K_{j,\mathrm{off}}^\varepsilon(t, y_j) = \frac{1}{\varepsilon} \left(K\left(\delta q_j + \sqrt{\varepsilon}y_j\right) - K(\delta q_j) - \sqrt{\varepsilon}y_j \cdot \nabla K(\delta q_j) \right)$$

$$= \int_0^1 \left\langle y_j, \nabla^2 K\left(\delta q_j(t) + \theta y_j\sqrt{\varepsilon}\right) y_j \right\rangle (1 - \theta)d\theta,$$

and

$$W_j^\varepsilon(t, y_j) = e^{i\left(S_j^\varepsilon - S_{j+1}^\varepsilon - q_j \cdot p_j + q_{j+1} \cdot p_{j+1}\right)/\varepsilon} \times$$

$$\times \int K\left(\sqrt{\varepsilon}(y_j - z)\right) e^{iz \cdot (p_j - p_{j+1})/\sqrt{\varepsilon}} u_j^\varepsilon(t, z)\overline{u}_{j+1}^\varepsilon\left(t, z + \frac{\delta q_j}{\sqrt{\varepsilon}}\right) dz.$$

The main difference with the case of a single coherent state presented in Sec. 13.2 is precisely these last terms W_1^ε and W_2^ε. They are nonlinear in $(u_1^\varepsilon, u_2^\varepsilon)$, and cannot be estimated directly by invoking Proposition 13.6: since the structure of Eq. (13.19) is different from Eq. (13.17), we do not know directly that u_1^ε and u_2^ε have a sufficient regularity in y_j, uniformly in ε, to claim that the last term in Eq. (13.19) is actually not singular in the limit $\varepsilon \to 0$, and even negligible. We obtain this information by a bootstrap argument, by proving that the exact solution is close to a suitable modification of the approximate solution. Indeed, recall that in the case of a single coherent state (Theorem 13.4), a time dependent phase shift had to be incorporated, in the case $\nabla K(0) \neq 0$. In Eq. (13.19), if $\nabla K(0) = 0$, the third line vanishes, but, unless K is constant, the fourth line is non-trivial, so the presence of non-trivial time dependent phase shifts must be expected.

Theorem 13.11. *Let* $K \in W^{7,\infty}(\mathbb{R}^d; \mathbb{R})$, *and* $a \in \Sigma^7$. *Let* ψ^ε *solve Eq. (13.1) with initial datum (13.11). There exist* $\theta_1, \theta_2 \in C^3(\mathbb{R}; \mathbb{R})$ *independent of* ε, *such that for all* $T > 0$, *there exists* $C > 0$ *such that*

$$\left\| \psi^\varepsilon(t) - \sum_{j=1,2} \varphi_j^\varepsilon(t)e^{i\theta_j(t)} \right\|_{L^2} \leqslant C\sqrt{\varepsilon}, \quad \forall t \in [0, T],$$

where we have denoted

$$\varphi_j^\varepsilon(t,x) = \frac{1}{\varepsilon^{d/4}} u_j\left(t, \frac{x - q_j(t)}{\sqrt{\varepsilon}}\right) e^{i\left(S_j^\varepsilon(t) + p_j(t)\cdot(x - q_j(t))\right)/\varepsilon},$$

with (q_j, p_j) given by Eq. (13.2), S_j^ε given by Eq. (13.3), and u_j given by Eq. (13.17).
The phase shifts are such that $\theta_j(0) = \dot\theta_j(0) = \ddot\theta_j(0) = 0$, but are not trivial in general (even if $\nabla K(0) = 0$).

Remark 13.12. The above result shows that whether $\nabla K(0) = 0$ or not, the Wigner measure of ψ^ε is affected by nonlinear effects: like in the critical case, it is given by the formula

$$\mu(t, dx, d\xi) = \sum_{j=1,2} \|a_j\|_{L^2(\mathbb{R}^d)}^2 \delta_{x=q_j(t)} \otimes \delta_{\xi=p_j(t)},$$

but now the classical trajectories are given by Eq. (13.2).

Analysis of a reduced system

We outline the main steps of the proof, and refer to [Carles (2012)] for details omitted here. Like in the case of a single coherent state, we reduce the problem since the third and fourth lines of Eq. (13.19) are purely time-dependent. Denote

$$\theta_j^\varepsilon(t) = \frac{1}{\sqrt{\varepsilon}} \int_0^t \nabla K(0) \cdot \left(\int z\left(|u_j^\varepsilon(s,z)|^2 - |u_j(s,z)|^2\right) dz\right) ds$$

$$+ \frac{1}{\sqrt{\varepsilon}} \int_0^t \nabla K(\delta q_j(s)) \cdot \left(\int z\left(|u_{j+1}^\varepsilon(s,z)|^2 - |u_{j+1}(s,z)|^2\right) dz\right) ds,$$

and introduce $\tilde{u}_j^\varepsilon(t,y) = u_j^\varepsilon(t,y)e^{-i\theta_j^\varepsilon(t)}$. Then $|\tilde{u}_j^\varepsilon| = |u_j^\varepsilon|$, so u_j^ε can be replaced by $\tilde{u}_j^\varepsilon$ in the definition of θ_j^ε, and Eq. (13.19) is equivalent to:

$$i\partial_t \tilde{u}_j^\varepsilon + \frac{1}{2}\Delta \tilde{u}_j^\varepsilon = V_j^\varepsilon(t, y_j)\tilde{u}_j^\varepsilon + \left(K_{j,\text{diag}}^\varepsilon * |\tilde{u}_j^\varepsilon|^2\right) \tilde{u}_j^\varepsilon$$

$$+ \left(K_{j,\text{off}}^\varepsilon * |\tilde{u}_{j+1}^\varepsilon|^2\right) \tilde{u}_j^\varepsilon + \frac{1}{\varepsilon}\left(2\,\text{Re}\,\tilde{W}_j^\varepsilon\right)\tilde{u}_j^\varepsilon, \tag{13.20}$$

with initial data $\tilde{u}_{j|t=0}^\varepsilon = a_j$, and where we have denoted

$$\tilde{W}_j^\varepsilon = e^{i\left(\delta S_j^\varepsilon - q_j\cdot p_j + q_{j+1}\cdot p_{j+1}\right)/\varepsilon} e^{i\delta\theta_j^\varepsilon} \times$$

$$\times \int K\left(\sqrt{\varepsilon}(y_j - z)\right) e^{iz\cdot\delta p_j/\sqrt{\varepsilon}} \tilde{u}_j^\varepsilon(t,z)\overline{\tilde{u}}_{j+1}^\varepsilon\left(t, z + \frac{\delta q_j}{\sqrt{\varepsilon}}\right) dz,$$

with

$$\delta q_j = q_j - q_{j+1}, \quad \delta p_j = p_j - p_{j+1}, \quad \delta S_j^\varepsilon = S_j^\varepsilon - S_{j+1}^\varepsilon, \quad \delta\theta_j^\varepsilon = \theta_j^\varepsilon - \theta_{j+1}^\varepsilon.$$

Like in the case of Theorem 13.4, we first prove an approximation result which corresponds to Theorem 13.11, up to the phase shifts θ_j.

Theorem 13.13. *Let* $K \in W^{7,\infty}(\mathbb{R}^d; \mathbb{R})$, *and* $a_1, a_2 \in \Sigma^7$. *Let* $T > 0$. *There exists* $\varepsilon_0 > 0$ *such that for* $\varepsilon \in (0, \varepsilon_0]$, *Eq. (13.20) has a unique solution* $(\tilde{u}_1^\varepsilon, \tilde{u}_2^\varepsilon) \in C([0, T]; \Sigma^4)^2$. *Moreover, there exists* C *independent of* $\varepsilon \in]0, \varepsilon_0]$ *such that*

$$\sup_{t \in [0,T]} \|\tilde{u}_1^\varepsilon(t) - u_1(t)\|_{\Sigma^4} + \sup_{t \in [0,T]} \|\tilde{u}_2^\varepsilon(t) - u_2(t)\|_{\Sigma^4} \leqslant C\sqrt{\varepsilon}. \tag{13.21}$$

Remark 13.14. The regularity $K \in W^{7,\infty}(\mathbb{R}^d; \mathbb{R})$, $a_1, a_2 \in \Sigma^7$ could be replaced by $K \in W^{6,\infty}(\mathbb{R}^d; \mathbb{R})$, $a_1, a_2 \in \Sigma^6$ for the bootstrap argument below, as in [Carles (2012)]. But like in the proof of Theorem 13.4, we will need to examine the next term in the asymptotic expansion of $\tilde{u}_j^\varepsilon$, and this is where an extra level of regularity is needed.

Proof. Thanks to the above reductions, Theorem 13.13 can be proved quite like the corresponding counterpart of Theorem 13.4, up to the bootstrap argument, which requires to consider a higher regularity. Indeed, in view of a comparison with the exact solution (13.20), we may rewrite (13.17) as

$$i\partial_t u_j + \frac{1}{2}\Delta u_j = V_j^\varepsilon u_j + \left(K_{j,\text{diag}}^\varepsilon * |u_j|^2\right) u_j + \left(K_{j,\text{off}}^\varepsilon * |u_{j+1}|^2\right) u_j - L_j^\varepsilon - N_j^\varepsilon,$$

where the source terms are given by

$$L_j^\varepsilon = \left(V_j^\varepsilon - V_j^0\right) u_j,$$
$$N_j^\varepsilon = \left(\left(K_{j,\text{diag}}^\varepsilon - K_{j,\text{diag}}^0\right) * |u_j|^2\right) u_j + \left(\left(K_{j,\text{off}}^\varepsilon - K_{j,\text{off}}^0\right) * |u_{j+1}|^2\right) u_j.$$

In particular, we have the pointwise estimate

$$|L_j^\varepsilon(t, y_j)| \lesssim \sqrt{\varepsilon} \langle y_j \rangle^3 |u_j(t, y_j)|,$$

$$|N_j^\varepsilon(t, y_j)| \lesssim \sqrt{\varepsilon} \left(\langle y \rangle^3 + \int_{\mathbb{R}^d} \langle z \rangle^3 \left(|u_1(t, z)|^2 + |u_2(t, z)|^2\right) dz\right) |u_j(t, y_j)|,$$

on any time interval $[0, T]$, where the implicit constants depend on T, and, proceeding similarly with spatial derivatives, we have, for $k \geqslant 0$,

$$\|L_j^\varepsilon(t)\|_{\Sigma^k} + \|N_j^\varepsilon(t)\|_{\Sigma^k} \lesssim \sqrt{\varepsilon} \left(1 + \|u_1(t)\|_{\Sigma^2}^2 + \|u_2(t)\|_{\Sigma^2}^2\right) \|u_j(t)\|_{\Sigma^{k+3}}.$$

When forming the difference $\tilde{u}_j^\varepsilon - u_j$, we have another source term, due to the presence of $\tilde{W}_j^\varepsilon$. In view of Proposition 13.6, we require a Σ^3 control at least, to that show the corresponding term is small, $\mathcal{O}(\sqrt{\varepsilon})$. As a first step, proceeding as in the proof of Lemma 13.3, we can prove a local existence result, on a time interval possibly depending on ε, precisely because the presence of $\tilde{W}_j^\varepsilon$ prevents us from having directly estimates which are uniform in ε: $(\tilde{u}_1^\varepsilon, \tilde{u}_2^\varepsilon) \in C([0, \tau^\varepsilon]; \Sigma^4)$, for some $\tau^\varepsilon > 0$. As evoked above, we work directly at the Σ^4 level, in view of a higher order asymptotic expansion.

Fix $T > 0$ once and for all in the course of the proof. By Lemma 13.10, there exists $C_0 > 0$ such that

$$\sup_{t \in [0,T]} \|u_1(t)\|_{\Sigma^4} + \sup_{t \in [0,T]} \|u_2(t)\|_{\Sigma^4} \leqslant C_0.$$

Denote $w_j^\varepsilon = \tilde{u}_j^\varepsilon - u_j$. Since $w_{j|t=0}^\varepsilon = 0$ and $\tilde{u}_j^\varepsilon \in C([0, \tau^\varepsilon]; \Sigma^4)$ for some τ^ε, we can find $t^\varepsilon > 0$ such that

$$\|w_1^\varepsilon(t)\|_{\Sigma^4} + \|w_2^\varepsilon(t)\|_{\Sigma^4} \leqslant C_0 \tag{13.22}$$

for $0 \leqslant t \leqslant t^\varepsilon$. So long as (13.22) holds, we perform energy estimates, to show the estimate of Theorem 13.11, with a constant C independent of ε. It will follow that up to choosing $\varepsilon \in]0, \varepsilon_0]$ with $\varepsilon_0 > 0$ sufficiently small, (13.22) holds for $t \in [0, T]$.

As long as (13.22) holds, the triangle inequality yields

$$\|\tilde{u}_1^\varepsilon(t)\|_{\Sigma^4} + \|\tilde{u}_2^\varepsilon(t)\|_{\Sigma^4} \leqslant 2C_0,$$

and so, by Proposition 13.6,

$$\left\| \frac{1}{\varepsilon} \left(2 \operatorname{Re} \tilde{W}_j^\varepsilon \right) \tilde{u}_j^\varepsilon \right\|_{\Sigma^4} \lesssim \frac{1}{\varepsilon} \left\| \tilde{W}_j^\varepsilon \right\|_{W^{4,\infty}} \|\tilde{u}_j^\varepsilon\|_{\Sigma^4} \lesssim \sqrt{\varepsilon} \|K\|_{W^{7,\infty}}.$$

Under the assumptions of Theorem 13.13, setting $k = 4$ in the previous estimate for the source terms,

$$\|L_j^\varepsilon(t)\|_{\Sigma^4} + \|N_j^\varepsilon(t)\|_{\Sigma^4} \lesssim \sqrt{\varepsilon} \left(1 + \|u_1(t)\|_{\Sigma^2}^2 + \|u_2(t)\|_{\Sigma^2}^2 \right) \|u_j(t)\|_{\Sigma^7}$$
$$\leqslant C(T)\sqrt{\varepsilon}.$$

We can then repeat the proof of Theorem 13.4, to conclude that so long as (13.22) holds, the estimate of Theorem 13.11 are true, with a constant C independent of ε, and so the bootstrap argument is valid up to time T, completing the proof of Theorem 13.13. $\qquad\square$

Back to the initial unknown and existence of phase shifts

Once again, we fix $T > 0$ independent of ε. Theorem 13.13 implies that the functions θ_j^ε are bounded, with bounded derivatives, uniformly in ε, on any interval $[0, T]$. Like in the case of a single coherent state, we examine the first corrector in the asymptotic analysis of $\tilde{u}_j^\varepsilon$, in order to show that the θ_j^ε's actually converge. We thus seek

$$\tilde{u}_j^\varepsilon = u_j + \sqrt{\varepsilon} u_j^{(1)} + o\left(\sqrt{\varepsilon}\right).$$

Leaving out the cross terms, we find

$$i\partial_t u_j^{(1)} + \frac{1}{2}\Delta u_j^{(1)} = V_j^0 u_j^{(1)} + \left(K_{j,\text{diag}}^0 * |u_j|^2\right) u_j^{(1)} + \left(K_{j,\text{off}}^0 * |u_{j+1}|^2\right) u_j^{(1)}$$
$$+ 2\left(K_{j,\text{diag}}^0 * \text{Re}\left(\overline{u}_j u_j^{(1)}\right)\right) u_j + 2\left(K_{j,\text{off}}^0 * \text{Re}\left(\overline{u}_{j+1} u_{j+1}^{(1)}\right)\right) u_j$$
$$+ \mathcal{V}_j u_j + \left(\mathcal{K}_{j,\text{diag}} * |u_j|^2\right) u_j + \left(\mathcal{K}_{j,\text{off}} * |u_{j+1}|^2\right) u_j,$$

with Cauchy data $u_{j|t=0}^{(1)} = 0$, where we have denoted the third order Taylor expansions

$$\mathcal{V}_j(t, y) = \frac{1}{6}\nabla^3 V\left(q_j(t)\right) y \cdot y \cdot y,$$

$$\mathcal{K}_{j,\text{diag}}(y) = \frac{1}{6}\nabla^3 K(0) y \cdot y \cdot y,$$

$$\mathcal{K}_{j,\text{off}}(t, y) = \frac{1}{6}\nabla^3 K\left(q_j(t) - q_{j+1}(t)\right) y \cdot y \cdot y.$$

These equations are naturally linear in the unknown $(u_1^{(1)}, u_2^{(1)})$. In view of Lemma 13.10 and the assumptions of Theorem 13.11, the last line in the equation for $u_j^{(1)}$, which corresponds to a source term, belongs to $C(\mathbb{R}; \Sigma^4)$. This non-trivial source term makes $u_j^{(1)}$ non-zero. Like in the case of a single coherent state, there exists a unique solution $(u_1^{(1)}, u_2^{(1)}) \in C(\mathbb{R}; \Sigma^4)^2$. Denote by $v_j^\varepsilon = u_j + \sqrt{\varepsilon} u_j^{(1)}$ the second order approximate solution, and by $\tilde{w}_j^\varepsilon = \tilde{u}_j^\varepsilon - v_j^\varepsilon$ the corresponding error term. It satisfies $\tilde{w}_{j|t=0}^\varepsilon = 0$, and

$$\begin{cases} i\partial_t \tilde{w}_j^\varepsilon + \frac{1}{2}\Delta \tilde{w}_j^\varepsilon = V_j^\varepsilon \tilde{w}_j^\varepsilon + \left(K_{j,\text{diag}}^\varepsilon * |\tilde{u}_j^\varepsilon|^2\right) \tilde{u}_j^\varepsilon - \left(K_{j,\text{diag}}^\varepsilon * |v_j^\varepsilon|^2\right) v_j^\varepsilon \\ \qquad + \left(K_{j,\text{off}}^\varepsilon * |\tilde{u}_{j+1}^\varepsilon|^2\right) \tilde{u}_j^\varepsilon - \left(K_{j,\text{off}}^\varepsilon * |v_{j+1}^\varepsilon|^2\right) v_j^\varepsilon \\ \qquad + \frac{1}{\varepsilon}\left(2\,\text{Re}\,\tilde{W}_j^\varepsilon\right) \tilde{u}_j^\varepsilon + \tilde{L}_j^\varepsilon + \tilde{N}_j^\varepsilon, \end{cases}$$

where the new source term is such that

$$\sup_{t\in[0,T]} \|\tilde{L}_j^\varepsilon(t)\|_\Sigma + \sup_{t\in[0,T]} \|\tilde{N}_j^\varepsilon(t)\|_\Sigma \leqslant C\varepsilon.$$

The reason why it makes sense to leave out the cross terms is a consequence of Theorem 13.13, and explains the regularity that we have considered in this result. Indeed, Proposition 13.6 yields

$$\left\| \frac{1}{\varepsilon} \left(2\operatorname{Re} \tilde{W}_j^\varepsilon \right) \tilde{u}_j^\varepsilon \right\|_\Sigma \lesssim \frac{1}{\varepsilon} \| \tilde{W}_j^\varepsilon \|_{W^{1,\infty}} \| \tilde{u}_j^\varepsilon \|_\Sigma$$

$$\lesssim \varepsilon \| K \|_{W^{5,\infty}} \left(\| \tilde{u}_1^\varepsilon \|_{\Sigma^4} + \| \tilde{u}_2^\varepsilon \|_{\Sigma^4} \right)^3 \lesssim \varepsilon,$$

since $\tilde{u}_1^\varepsilon$ and $\tilde{u}_2^\varepsilon$ are bounded in $L^\infty([0,T]; \Sigma^4)$ uniformly in ε, from Theorem 13.13 and Lemma 13.10.

Resuming the same strategy as before, we can prove that there exists $C > 0$ such that for $\varepsilon \in]0, \varepsilon_0]$ (with $\varepsilon_0 > 0$ provided by Theorem 13.13),

$$\sum_{j=1,2} \sup_{t \in [0,T]} \left\| \tilde{u}_j^\varepsilon(t) - u_j(t) - \sqrt{\varepsilon} u_j^{(1)}(t) \right\|_\Sigma \leqslant C\varepsilon.$$

Note that unlike in the proof of Theorem 13.13, no bootstrap argument is needed at this stage, since we already have uniform estimates for $\tilde{u}_j^\varepsilon, u_j, u_j^{(1)}$ in $C([0,T]; \Sigma^4)$. We readily infer:

$$\theta_j^\varepsilon(t) = \theta_j(t) + \mathcal{O}\left(\sqrt{\varepsilon} \right) \quad \text{in } L^\infty([0,T]),$$

where θ_j is given by

$$\theta_j(t) = \int_0^t \nabla K(0) \cdot \left(2\operatorname{Re} \int z \overline{u}_j(s,z) u_j^{(1)}(s,z) dz \right) ds$$
$$+ \int_0^t \nabla K(\delta q_j(s)) \cdot \left(2\operatorname{Re} \int z \overline{u}_{j+1}(s,z) u_{j+1}^{(1)}(s,z) dz \right) ds. \tag{13.23}$$

We have obviously $\theta_j \in C^1([0,T])$, and $\theta_j(0) = \dot{\theta}_j(0) = 0$, with

$$\dot{\theta}_j(t) = \nabla K(0) \cdot \left(2\operatorname{Re} \int z \overline{u}_j(t,z) u_j^{(1)}(t,z) dz \right)$$
$$+ \nabla K(\delta q_j(t)) \cdot \left(2\operatorname{Re} \int z \overline{u}_{j+1}(t,z) u_{j+1}^{(1)}(t,z) dz \right).$$

Recall that $u_j, u_j^{(1)} \in C([0,T]; \Sigma^4)$. Examining the equations defining these functions, we also have, successively, $\partial_t u_j \in C([0,T]; \Sigma^2)$, $\partial_t^2 u_j \in C([0,T]; L^2)$, $\partial_t u_j^{(1)} \in C([0,T]; \Sigma)$, and $\partial_t^2 u_j^{(1)} \in C([0,T]; \Sigma^{-1})$, where Σ^{-1} stands for the dual space of Σ. These regularity properties yield $\theta_j \in C^3([0,T])$.

We emphasize a flaw in [Carles (2012)], where it is claimed that $\ddot\theta_j(0) \neq 0$ in general. Since $u^{(1)}_{j|t=0} = 0$, we have

$$\ddot\theta_j(0) = \nabla K(0) \cdot \left(2\,\mathrm{Re} \int z\overline{a}_j(z)\partial_t u^{(1)}_j(0,z)dz \right)$$
$$+ \nabla K(\delta q_j(0)) \cdot \left(2\,\mathrm{Re} \int z\overline{a}_{j+1}(z)\partial_t u^{(1)}_{j+1}(0,z)dz \right).$$

From the equation for the corrector,

$$i\partial_t u^{(1)}_j(0,y) = \left(\mathcal{V}_j(0,y) + \mathcal{K}_{j,\mathrm{diag}} * |a_j|^2 + \mathcal{K}_{j,\mathrm{off}} * |a_{j+1}|^2 \right) a_j(y),$$

and in particular, $\partial_t u^{(1)}_j(0) \in \mathbb{R} \times (ia_j)$, showing that $\ddot\theta_j(0) = 0$. On the other hand, we can exploit this computation to show that for $0 < |t| \ll 1$, $\ddot\theta_j(t) \neq 0$. Using the equations defining u_j and $u^{(1)}_j$, we find

$$\partial_t \,\mathrm{Re}\left(\overline{u}_{j+1}u^{(1)}_{j+1} \right) = \mathrm{Im}\, i\partial_t \left(\overline{u}_{j+1}u^{(1)}_{j+1} \right) = \frac{1}{2}\,\mathrm{Im}\left(u^{(1)}_j \Delta \overline{u}_j - \overline{u}_j \Delta u^{(1)}_j \right).$$

Using Taylor expansion $u(t) \approx u(0) + t\partial_t u(0)$, we infer, for $|t| \ll 1$,

$$\partial_t \,\mathrm{Re}\left(\overline{u}_{j+1}u^{(1)}_{j+1} \right) \approx \frac{t}{2}\,\mathrm{Im}\left(\partial_t u^{(1)}_j \Delta \overline{a}_j - \overline{a}_j \Delta \partial_t u^{(1)}_j \right),$$

hence, denoting

$$\mathcal{W}_j := \mathcal{V}_j(0,y) + \mathcal{K}_{j,\mathrm{diag}} * |a_j|^2 + \mathcal{K}_{j,\mathrm{off}} * |a_{j+1}|^2,$$

and using integrations by parts,

$$\partial_t \int_{\mathbb{R}^d} z\,\mathrm{Re}\left(\overline{u}_{j+1}u^{(1)}_{j+1} \right) \approx \frac{t}{2}\,\mathrm{Im} \int_{\mathbb{R}^d} z\left(i\mathcal{W}_j a_j \Delta \overline{a}_j - \overline{a}_j \Delta \left(i\mathcal{W}_j a_j \right) \right)$$
$$\approx \frac{t}{2}\,\mathrm{Re} \int_{\mathbb{R}^d} z\left(\mathcal{W}_j a_j \Delta \overline{a}_j - \overline{a}_j \Delta \left(\mathcal{W}_j a_j \right) \right)$$
$$\approx \frac{t}{2}\,\mathrm{Re} \int_{\mathbb{R}^d} z\mathcal{W}_j a_j \Delta \overline{a}_j - \Delta \left(z\overline{a}_j \right)\left(\mathcal{W}_j a_j \right)$$
$$\approx -t\,\mathrm{Re} \int_{\mathbb{R}^d} \mathcal{W}_j a_j \nabla \overline{a}_j$$
$$\approx -\frac{t}{2} \int_{\mathbb{R}^d} \mathcal{W}_j \nabla |a_j|^2 = \frac{t}{2} \int_{\mathbb{R}^d} |a_j|^2 \nabla \mathcal{W}_j.$$

Back to the expression of $\ddot\theta_j(t)$, this shows that (in general) $(\ddot\theta_1(t), \ddot\theta_2(t))$ is non-trivial for $0 < |t| \ll 1$, even if $\nabla K(0) = 0$.

Bibliography

Alazard, T. (2006). Low Mach number limit of the full Navier-Stokes equations, *Arch. Ration. Mech. Anal.* **180**, 1, pp. 1–73.

Alazard, T. and Carles, R. (2009a). Loss of regularity for super-critical nonlinear Schrödinger equations, *Math. Ann.* **343**, 2, pp. 397–420.

Alazard, T. and Carles, R. (2009b). Supercritical geometric optics for nonlinear Schrödinger equations, *Arch. Ration. Mech. Anal.* **194**, 1, pp. 315–347.

Alinhac, S. (1995a). *Blowup for nonlinear hyperbolic equations* (Birkhäuser Boston Inc., Boston, MA).

Alinhac, S. (1995b). Explosion géométrique pour des systèmes quasi-linéaires, *Amer. J. Math.* **117**, 4, pp. 987–1017.

Alinhac, S. (2002). A minicourse on global existence and blowup of classical solutions to multidimensional quasilinear wave equations, in *Journées "Équations aux Dérivées Partielles" (Forges-les-Eaux, 2002)* (Univ. Nantes, Nantes), pp. Exp. No. I, 33.

Alinhac, S. and Gérard, P. (2007). *Pseudo-differential operators and the Nash-Moser theorem, Graduate Studies in Mathematics*, Vol. 82 (American Mathematical Society, Providence, RI), translated from the 1991 French original by Stephen S. Wilson.

Athanassoulis, A. (2018). Semiclassical regularization of Vlasov equations and wavepackets for nonlinear Schrödinger equations, *Nonlinearity* **31**, 3, pp. 1045–1072.

Athanassoulis, A., Paul, T., Pezzotti, F. and Pulvirenti, M. (2011). Semiclassical propagation of coherent states for the Hartree equation, *Ann. Henri Poincaré* **12**, 8, pp. 1613–1634.

Bahouri, H. and Gérard, P. (1999). High frequency approximation of solutions to critical nonlinear wave equations, *Amer. J. Math.* **121**, 1, pp. 131–175.

Bambusi, D., Graffi, S. and Paul, T. (1999). Long time semiclassical approximation of quantum flows: a proof of the Ehrenfest time, *Asymptot. Anal.* **21**, 2, pp. 149–160.

Bambusi, D. and Sacchetti, A. (2007). Exponential times in the one-dimensional Gross-Pitaevskii equation with multiple well potential, *Comm. Math. Phys.* **275**, 1, pp. 1–36.

Barab, J. E. (1984). Nonexistence of asymptotically free solutions for nonlinear Schrödinger equation, *J. Math. Phys.* **25**, pp. 3270–3273.

Bégout, P. and Vargas, A. (2007). Mass concentration phenomena for the L^2-critical nonlinear Schrödinger equation, *Trans. Amer. Math. Soc.* **359**.

Ben Abdallah, N., Castella, F. and Méhats, F. (2008). Time averaging for the strongly confined nonlinear Schrödinger equation, using almost periodicity, *J. Differential Equations* **245**, 1, pp. 154–200.

Bensoussan, A., Lions, J.-L. and Papanicolaou, G. (1978). *Asymptotic analysis for periodic structures*, Vol. 5 (North-Holland Publishing Co., Amsterdam).

Berloff, N. G. (1999). Nonlocal nonlinear Schrödinger equations as models of superfluidity, *J. Low Temp. Phys.* **116**, 5–6, pp. 359–380.

Berloff, N. G. and Roberts, P. H. (1999). Motions in a Bose condensate. VI. Vortices in a nonlocal model, *J. Phys. A* **32**, 30, pp. 5611–5625.

Besse, C., Carles, R. and Méhats, F. (2013). An asymptotic preserving scheme based on a new formulation for NLS in the semiclassical limit, *Multiscale Model. Simul.* **11**, 4, pp. 1228–1260.

Bialynicki-Birula, I. and Mycielski, J. (1976). Nonlinear wave mechanics, *Ann. Physics* **100**, 1–2, pp. 62–93.

Bourgain, J. (1995). Some new estimates on oscillatory integrals, in *Essays on Fourier analysis in honor of Elias M. Stein (Princeton, NJ, 1991)*, *Princeton Math. Ser.*, Vol. 42 (Princeton Univ. Press, Princeton, NJ), pp. 83–112.

Bourgain, J. (1998). Refinements of Strichartz' inequality and applications to 2D-NLS with critical nonlinearity, *Internat. Math. Res. Notices*, 5, pp. 253–283.

Bourgain, J. (1999). *Global solutions of nonlinear Schrödinger equations*, *American Mathematical Society Colloquium Publications*, Vol. 46 (American Mathematical Society, Providence, RI).

Bouzouina, A. and Robert, D. (2002). Uniform semiclassical estimates for the propagation of quantum observables, *Duke Math. J.* **111**, 2, pp. 223–252.

Boyd, R. W. (1992). *Nonlinear Optics* (Academic Press, New York).

Brenier, Y. (2000). Convergence of the Vlasov-Poisson system to the incompressible Euler equations, *Comm. Partial Differential Equations* **25**, 3–4, pp. 737–754.

Brillouin, L. (1926). La mécanique ondulatoire de Schrödinger; une méthode générale de résolution par approximations successives, *C. r. hebd. séances Acad. sci.* **183**, pp. 24–26.

Bronski, J. C. and Jerrard, R. L. (2000). Soliton dynamics in a potential, *Math. Res. Lett.* **7**, 2–3, pp. 329–342.

Burq, N. (1997). Mesures semi-classiques et mesures de défaut, *Astérisque*, 245, pp. Exp. No. 826, 4, 167–195, séminaire Bourbaki, Vol. 1996/97.

Burq, N., Gérard, P. and Tzvetkov, N. (2004). Strichartz inequalities and the nonlinear Schrödinger equation on compact manifolds, *Amer. J. Math.* **126**, 3, pp. 569–605.

Burq, N., Gérard, P. and Tzvetkov, N. (2005). Multilinear eigenfunction estimates and global existence for the three dimensional nonlinear Schrödinger equations, *Ann. Sci. École Norm. Sup. (4)* **38**, 2, pp. 255–301.

Burq, N. and Zworski, M. (2005). Instability for the semiclassical non-linear Schrödinger equation, *Comm. Math. Phys.* **260**, 1, pp. 45–58.

Cao, P. and Carles, R. (2011). Semiclassical wave packet dynamics for Hartree equations, *Rev. Math. Phys.* **23**, 9, pp. 933–967.

Carles, R. (2000a). Focusing on a line for nonlinear Schrödinger equations in $\mathbb{R}^2$, *Asymptot. Anal.* **24**, 3–4, pp. 255–276.

Carles, R. (2000b). Geometric optics with caustic crossing for some nonlinear Schrödinger equations, *Indiana Univ. Math. J.* **49**, 2, pp. 475–551.

Carles, R. (2001a). Geometric optics and long range scattering for one-dimensional nonlinear Schrödinger equations, *Comm. Math. Phys.* **220**, 1, pp. 41–67.

Carles, R. (2001b). Remarques sur les mesures de Wigner, *C. R. Acad. Sci. Paris, t. 332, Série I* **332**, 11, pp. 981–984.

Carles, R. (2002). Critical nonlinear Schrödinger equations with and without harmonic potential, *Math. Models Methods Appl. Sci.* **12**, 10, pp. 1513–1523.

Carles, R. (2003a). Nonlinear Schrödinger equations with repulsive harmonic potential and applications, *SIAM J. Math. Anal.* **35**, 4, pp. 823–843.

Carles, R. (2003b). Semi-classical Schrödinger equations with harmonic potential and nonlinear perturbation, *Ann. Inst. H. Poincaré Anal. Non Linéaire* **20**, 3, pp. 501–542.

Carles, R. (2007a). Cascade of phase shifts for nonlinear Schrödinger equations, *J. Hyperbolic Differ. Equ.* **4**, 2, pp. 207–231.

Carles, R. (2007b). Geometric optics and instability for semi-classical Schrödinger equations, *Arch. Ration. Mech. Anal.* **183**, 3, pp. 525–553.

Carles, R. (2007c). WKB analysis for nonlinear Schrödinger equations with potential, *Comm. Math. Phys.* **269**, 1, pp. 195–221.

Carles, R. (2008). On the Cauchy problem in Sobolev spaces for nonlinear Schrödinger equations with potential, *Portugal. Math. (N. S.)* **65**, 2, pp. 191–209.

Carles, R. (2009). Rotating points for the conformal NLS scattering operator, *Dyn. Partial Differ. Equ.* **6**, 1, pp. 35–51.

Carles, R. (2011). Nonlinear Schrödinger equation with time dependent potential, *Commun. Math. Sci.* **9**, 4, pp. 937–964.

Carles, R. (2012). Interaction of coherent states for Hartree equations, *Arch. Ration. Mech. Anal.* **204**, 2, pp. 559–598.

Carles, R. and Drumond Silva, J. (2015). Large time behavior in nonlinear Schrödinger equation with time dependent potential, *Commun. Math. Sci.* **13**, 2, pp. 443–460.

Carles, R., Dumas, E. and Sparber, C. (2010). Multiphase weakly nonlinear geometric optics for Schrödinger equations, *SIAM J. Math. Anal.* **42**, 1, pp. 489–518.

Carles, R., Dumas, E. and Sparber, C. (2012). Geometric optics and instability for NLS and Davey-Stewartson models, *J. Eur. Math. Soc. (JEMS)* **14**, 6, pp. 1885–1921.

Carles, R. and Faou, E. (2012). Energy cascade for NLS on the torus, *Discrete Contin. Dyn. Syst.* **32**, 6, pp. 2063–2077.

Carles, R., Fermanian, C. and Gallagher, I. (2003). On the role of quadratic oscillations in nonlinear Schrödinger equations, *J. Funct. Anal.* **203**, 2, pp. 453–493.

Carles, R. and Fermanian Kammerer, C. (2011a). A nonlinear adiabatic theorem for coherent states, *Nonlinearity* **24**, 8, pp. 2143–2164.

Carles, R. and Fermanian Kammerer, C. (2011b). Nonlinear coherent states and Ehrenfest time for Schrödinger equations, *Commun. Math. Phys.* **301**, 2, pp. 443–472.

Carles, R. and Gallo, C. (2017). On Fourier time-splitting methods for nonlinear Schrödinger equations in the semi-classical limit II. Analytic regularity, *Numer. Math.* **136**, 1, pp. 315–342.

Carles, R. and Kappeler, T. (2017). Norm-inflation with infinite loss of regularity for periodic NLS equations in negative Sobolev spaces, *Bull. Soc. Math. France* **145**, 4, pp. 623–642.

Carles, R. and Keraani, S. (2007). On the role of quadratic oscillations in nonlinear Schrödinger equations II. The L^2-critical case, *Trans. Amer. Math. Soc.* **359**, 1, pp. 33–62.

Carles, R., Markowich, P. A. and Sparber, C. (2004). Semiclassical asymptotics for weakly nonlinear Bloch waves, *J. Stat. Phys.* **117**, 1–2, pp. 343–375.

Carles, R. and Miller, L. (2004). Semiclassical nonlinear Schrödinger equations with potential and focusing initial data, *Osaka J. Math.* **41**, 3, pp. 693–725.

Carles, R. and Nakamura, Y. (2004). Nonlinear Schrödinger equations with Stark potential, *Hokkaido Math. J.* **33**, 3, pp. 719–729.

Carles, R. and Nouri, A. (2017). Monokinetic solutions to a singular Vlasov equation from a semiclassical perspective, *Asymptot. Anal.* **102**, pp. 99–117.

Carles, R. and Rauch, J. (2002). Focusing of spherical nonlinear pulses in $\mathbb{R}^{1+3}$, *Proc. Amer. Math. Soc.* **130**, 3, pp. 791–804.

Carles, R. and Rauch, J. (2004a). Focusing of Spherical Nonlinear Pulses in $\mathbb{R}^{1+3}$ II. Nonlinear Caustic, *Rev. Mat. Iberoamericana* **20**, 3, pp. 815–864.

Carles, R. and Rauch, J. (2004b). Focusing of Spherical Nonlinear Pulses in $\mathbb{R}^{1+3}$ III. Sub and Supercritical cases, *Tohoku Math. J.* **56**, 3, pp. 393–410.

Cazenave, T. (2003). *Semilinear Schrödinger equations, Courant Lecture Notes in Mathematics*, Vol. 10 (New York University Courant Institute of Mathematical Sciences, New York).

Cazenave, T. and Haraux, A. (1998). *An introduction to semilinear evolution equations, Oxford Lecture Series in Mathematics and its Applications*, Vol. 13 (The Clarendon Press Oxford University Press, New York), translated from the 1990 French original by Yvan Martel and revised by the authors.

Cazenave, T. and Weissler, F. (1989). Some remarks on the nonlinear Schrödinger equation in the critical case, in *Lect. Notes in Math.*, Vol. 1394 (Springer-Verlag, Berlin), pp. 18–29.

Cazenave, T. and Weissler, F. (1990). The Cauchy problem for the critical nonlinear Schrödinger equation in H^s, *Nonlinear Anal. TMA* **14**, pp. 807–836.

Cazenave, T. and Weissler, F. (1992). Rapidly decaying solutions of the nonlinear Schrödinger equation, *Comm. Math. Phys.* **147**, pp. 75–100.

Chemin, J.-Y. (1990). Dynamique des gaz à masse totale finie, *Asymptotic Anal.* **3**, 3, pp. 215–220.

Chemin, J.-Y. (1998). *Perfect incompressible fluids, Oxford Lecture Series in Mathematics and its Applications*, Vol. 14 (The Clarendon Press Oxford University Press, New York), translated from the 1995 French original by I. Gallagher and D. Iftimie.

Cheverry, C. (2004). Propagation of oscillations in real vanishing viscosity limit, *Comm. Math. Phys.* **247**, 3, pp. 655–695.

Cheverry, C. (2005). Sur la propagation de quasi-singularitiés, in *Séminaire: Équations aux Dérivées Partielles. 2004–2005* (École Polytech., Palaiseau), pp. Exp. No. VIII, 20.

Cheverry, C. (2006). Cascade of phases in turbulent flows, *Bull. Soc. Math. France* **134**, 1, pp. 33–82.

Cheverry, C. and Guès, O. (2008). Counter-examples to concentration-cancellation, *Arch. Ration. Mech. Anal.* **189**, 3, pp. 363–424.

Chiron, D. and Rousset, F. (2009). Geometric optics and boundary layers for nonlinear Schrödinger equations, *Comm. Math. Phys.* **288**, 2, pp. 503–546.

Christ, M., Colliander, J. and Tao, T. (2003a). Asymptotics, frequency modulation, and low regularity ill-posedness for canonical defocusing equations, *Amer. J. Math.* **125**, 6, pp. 1235–1293.

Christ, M., Colliander, J. and Tao, T. (2003b). Ill-posedness for nonlinear Schrödinger and wave equations, archived as arXiv:math.AP/0311048.

Christ, M., Colliander, J. and Tao, T. (2003c). Instability of the periodic nonlinear Schrödinger equation, archived as arXiv:math.AP/0311227.

Cicognani, M. and Colombini, F. (2006a). Loss of derivatives in evolution Cauchy problems, *Ann. Univ. Ferrara Sez. VII Sci. Mat.* **52**, 2, pp. 271–280.

Cicognani, M. and Colombini, F. (2006b). Modulus of continuity of the coefficients and loss of derivatives in the strictly hyperbolic Cauchy problem, *J. Differential Equations* **221**, 1, pp. 143–157.

Colin, M. and Lannes, D. (2009). Short pulses approximations in dispersive media, *SIAM J. Math. Anal.* **41**, 2, pp. 708–732.

Colliander, J., Keel, M., Staffilani, G., Takaoka, H. and Tao, T. (2010). Transfer of energy to high frequencies in the cubic defocusing nonlinear Schrödinger equation, *Invent. Math.* **181**, 1, pp. 39–113.

Combescure, M. and Robert, D. (1997). Semiclassical spreading of quantum wave packets and applications near unstable fixed points of the classical flow, *Asymptot. Anal.* **14**, 4, pp. 377–404.

Combescure, M. and Robert, D. (2006). Quadratic quantum Hamiltonians revisited, *Cubo* **8**, 1, pp. 61–86.

Combescure, M. and Robert, D. (2012). *Coherent states and applications in mathematical physics*, Theoretical and Mathematical Physics (Springer, Dordrecht).

Constantin, P. and Saut, J.-C. (1988). Local smoothing properties of dispersive equations, *J. Amer. Math. Soc.* **1**, pp. 413–439.

Dalfovo, F., Giorgini, S., Pitaevskii, L. P. and Stringari, S. (1999). Theory of Bose-Einstein condensation in trapped gases, *Rev. Mod. Phys.* **71**, 3, pp. 463–512.

Delort, J.-M. (1992). *F.B.I. transformation. Second microlocalization and semilinear caustics, Lecture Notes in Mathematics*, Vol. 1522 (Springer-Verlag, Berlin).

Dereziński, J. and Gérard, C. (1997). *Scattering theory of quantum and classical N-particle systems* (Texts and Monographs in Physics, Springer Verlag, Berlin Heidelberg).

Duistermaat, J. J. (1974). Oscillatory integrals, Lagrange immersions and unfolding of singularities, *Comm. Pure Appl. Math.* **27**, pp. 207–281.

Dunford, N. and Schwartz, J. T. (1963). *Linear operators. Part II: Spectral theory. Self adjoint operators in Hilbert space*, With the assistance of William G. Bade and Robert G. Bartle (Interscience Publishers John Wiley & Sons New York-London).

Duyckaerts, T., Merle, F. and Roudenko, S. (2011). Maximizers for the Strichartz norm for small solutions of mass-critical NLS, *Ann. Sc. Norm. Super. Pisa Cl. Sci. (5)* **10**, 2, pp. 427–476.

Ehrenfest, P. (1927). Bemerkung über die angenaherte Gültigkeit der klassischen Mechanik innerhalb der Quantenmechanik, *Zeitschrift für Physik* **45**, 7-8, pp. 455–457.

Evans, L. C. (1998). *Partial differential equations, Graduate Studies in Mathematics*, Vol. 19 (American Mathematical Society, Providence, RI).

Fefferman, C. L. (1983). The uncertainty principle, *Bull. Amer. Math. Soc. (N.S.)* **9**, 2, pp. 129–206.

Fermanian Kammerer, C. and Lasser, C. (2003). Wigner measures and codimension two crossings, *J. Math. Phys.* **44**, 2, pp. 507–527.

Fermanian Kammerer, C. and Lasser, C. (2017). An Egorov theorem for avoided crossings of eigenvalue surfaces, *Comm. Math. Phys.* **353**, 3, pp. 1011–1057.

Fermanian Kammerer, C. and Rousse, V. (2008). Resolvent estimates and matrix-valued Schrödinger operator with eigenvalue crossings; application to Strichartz estimates, *Comm. Partial Differential Equations* **33**, 1-3, pp. 19–44.

Feynman, R. P. and Hibbs, A. R. (1965). *Quantum mechanics and path integrals (International Series in Pure and Applied Physics)* (Maidenhead, Berksh.: McGraw-Hill Publishing Company, Ltd., 365 p.).

Foschi, D. (2005). Inhomogeneous Strichartz estimates, *J. Hyperbolic Differ. Equ.* **2**, 1, pp. 1–24.

Fujiwara, D. (1979). A construction of the fundamental solution for the Schrödinger equation, *J. Analyse Math.* **35**, pp. 41–96.

Fujiwara, D. (1980). Remarks on the convergence of the Feynman path integrals, *Duke Math. J.* **47**, 3, pp. 559–600.

Gallagher, I. and Gérard, P. (2001). Profile decomposition for the wave equation outside a convex obstacle, *J. Math. Pures Appl. (9)* **80**, 1, pp. 1–49.

Gérard, P. (1993). Remarques sur l'analyse semi-classique de l'équation de Schrödinger non linéaire, in *Séminaire sur les Équations aux Dérivées Partielles, 1992–1993* (École Polytech., Palaiseau), pp. Exp. No. XIII, 13.

Gérard, P. (1996). Oscillations and concentration effects in semilinear dispersive wave equations, *J. Funct. Anal.* **141**, 1, pp. 60–98.

Gérard, P. (1998). Description du défaut de compacité de l'injection de Sobolev, *ESAIM Control Optim. Calc. Var.* **3**, pp. 213–233 (electronic).

Gérard, P., Markowich, P. A., Mauser, N. J. and Poupaud, F. (1997). Homogenization limits and Wigner transforms, *Comm. Pure Appl. Math.* **50**, 4, pp. 323–379.

Ginibre, J. (1995). Introduction aux équations de Schrödinger non linéaires, Cours de DEA, Paris Onze Édition.

Ginibre, J. (1997). An introduction to nonlinear Schrödinger equations, in R. Agemi, Y. Giga and T. Ozawa (eds.), *Nonlinear waves (Sapporo, 1995)*, GAKUTO International Series, Math. Sciences and Appl. (Gakkōtosho, Tokyo), pp. 85–133.

Ginibre, J. and Ozawa, T. (1993). Long range scattering for nonlinear Schrödinger and Hartree equations in space dimension $n \geq 2$, *Comm. Math. Phys.* **151**, 3, pp. 619–645.

Ginibre, J., Ozawa, T. and Velo, G. (1994). On the existence of the wave operators for a class of nonlinear Schrödinger equations, *Ann. IHP (Physique Théorique)* **60**, pp. 211–239.

Ginibre, J. and Velo, G. (1979). On a class of nonlinear Schrödinger equations. II Scattering theory, general case, *J. Funct. Anal.* **32**, pp. 33–71.

Ginibre, J. and Velo, G. (1985a). The global Cauchy problem for the nonlinear Schrödinger equation revisited, *Ann. Inst. H. Poincaré Anal. Non Linéaire* **2**, pp. 309–327.

Ginibre, J. and Velo, G. (1985b). Scattering theory in the energy space for a class of nonlinear Schrödinger equations, *J. Math. Pures Appl. (9)* **64**, 4, pp. 363–401.

Ginibre, J. and Velo, G. (1992). Smoothing properties and retarded estimates for some dispersive evolution equations, *Comm. Math. Phys.* **144**, 1, pp. 163–188.

Ginibre, J. and Velo, G. (2001). Long range scattering and modified wave operators for some Hartree type equations. III. Gevrey spaces and low dimensions, *J. Differential Equations* **175**, 2, pp. 415–501.

Grébert, B. and Thomann, L. (2012). Resonant dynamics for the quintic nonlinear Schrödinger equation, *Ann. Inst. H. Poincaré Anal. Non Linéaire* **29**, 3, pp. 455–477.

Grenier, E. (1998). Semiclassical limit of the nonlinear Schrödinger equation in small time, *Proc. Amer. Math. Soc.* **126**, 2, pp. 523–530.

Grigis, A. and Sjöstrand, J. (1994). *Microlocal analysis for differential operators*, London Mathematical Society Lecture Note Series, Vol. 196 (Cambridge University Press, Cambridge), an introduction.

Hagedorn, G. A. (1980). Semiclassical quantum mechanics. I. The $\hbar \to 0$ limit for coherent states, *Comm. Math. Phys.* **71**, 1, pp. 77–93.

Hagedorn, G. A. (1981). Semiclassical quantum mechanics. III. The large order asymptotics and more general states, *Ann. Physics* **135**, 1, pp. 58–70.

Hagedorn, G. A. (1994). Molecular propagation through electron energy level crossings, *Mem. Amer. Math. Soc.* **111**, 536, pp. vi+130.

Hagedorn, G. A. and Joye, A. (1999). Molecular propagation through small avoided crossings of electron energy levels, *Rev. Math. Phys.* **11**, 1, pp. 41–101.

Hani, Z. and Pausader, B. (2014). On scattering for the quintic defocusing nonlinear Schrödinger equation on $\mathbb{R} \times \mathbb{T}^2$, *Comm. Pure Appl. Math.* **67**, 9, pp. 1466–1542.

Hari, L. (2013). Coherent states for systems of L^2-supercritical nonlinear Schrödinger equations, *Comm. Partial Differential Equations* **38**, 3, pp. 529–573.

Hari, L. (2016). Propagation of semiclassical wave packets through avoided eigenvalue crossings in nonlinear Schrödinger equations, *J. Inst. Math. Jussieu* **15**, 2, pp. 319–365.

Hayashi, N. and Naumkin, P. (1998). Asymptotics for large time of solutions to the nonlinear Schrödinger and Hartree equations, *Amer. J. Math.* **120**, 2, pp. 369–389.

Hayashi, N. and Naumkin, P. (2006). Domain and range of the modified wave operator for Schrödinger equations with a critical nonlinearity, *Comm. Math. Phys.* **267**, 2, pp. 477–492.

Hayashi, N. and Tsutsumi, Y. (1987). Remarks on the scattering problem for nonlinear Schrödinger equations, in *Differential equations and mathematical physics (Birmingham, Ala., 1986)*, Lectures Notes in Math., Vol. 1285 (Springer, Berlin), pp. 162–168.

Helffer, B. (1984). Théorie spectrale pour des opérateurs globalement elliptiques, in *Astérisque*, Vol. 112 (Soc. Math. France).

Heller, E. J. (1975). Time dependent approach to semiclassical dynamics, *J. Chem. Phys.* **62**, 1, pp. 1544–1555.

Hepp, K. (1974). The classical limit for quantum mechanical correlation functions, *Comm. Math. Phys.* **35**, pp. 265–277.

Hörmander, L. (1994). *The analysis of linear partial differential operators* (Springer-Verlag, Berlin).

Hörmander, L. (1995). Symplectic classification of quadratic forms, and general Mehler formulas, *Math. Z.* **219**, 3, pp. 413–449.

Hunter, J. and Keller, J. (1987). Caustics of nonlinear waves, *Wave motion* **9**, pp. 429–443.

Ibrahim, S. (2004). Geometric Optics for Nonlinear Concentrating Waves in a Focusing and non Focusing two geometries, *Commun. Contemp. Math.* **6**, 1, pp. 1–23.

Jin, S., Markowich, P. and Sparber, C. (2011). Mathematical and computational methods for semiclassical Schrödinger equations, *Acta Numer.* **20**, pp. 121–209.

Joly, J.-L., Métivier, G. and Rauch, J. (1994). Coherent nonlinear waves and the Wiener algebra, *Ann. Inst. Fourier (Grenoble)* **44**, 1, pp. 167–196.

Joly, J.-L., Métivier, G. and Rauch, J. (1995a). Coherent and focusing multidimensional nonlinear geometric optics, *Ann. Sc. de l'Ecole Normale Sup.* **28**, pp. 51–113.

Joly, J.-L., Métivier, G. and Rauch, J. (1995b). Focusing at a point and absorption of nonlinear oscillations, *Trans. Amer. Math. Soc.* **347**, 10, pp. 3921–3969.

Joly, J.-L., Métivier, G. and Rauch, J. (1996a). Nonlinear oscillations beyond caustics, *Comm. Pure Appl. Math.* **49**, 5, pp. 443–527.

Joly, J.-L., Métivier, G. and Rauch, J. (1996b). Several recent results in nonlinear geometric optics, in *Partial differential equations and mathematical physics (Copenhagen, 1995; Lund, 1995)* (Birkhäuser Boston, Boston, MA), pp. 181–206.

Joly, J.-L., Métivier, G. and Rauch, J. (1997a). Caustics for dissipative semilinear oscillations, in F. Colombini and N. Lerner (eds.), *Geometrical Optics and Related Topics* (Birkäuser), pp. 245–266.

Joly, J.-L., Métivier, G. and Rauch, J. (1997b). Estimations L^p d'intégrales oscillantes, in *Séminaire Équations aux Dérivées Partielles, 1996–1997* (École Polytech., Palaiseau), pp. Exp. No. VII, 17.

Joly, J.-L., Métivier, G. and Rauch, J. (2000). Caustics for dissipative semilinear oscillations, *Mem. Amer. Math. Soc.* **144**, 685, pp. viii+72.

Kato, T. (1987). On nonlinear Schrödinger equations, *Ann. IHP (Phys. Théor.)* **46**, 1, pp. 113–129.

Kato, T. (1989). Nonlinear Schrödinger equations, in *Schrödinger operators (Sønderborg, 1988), Lecture Notes in Phys.*, Vol. 345 (Springer, Berlin), pp. 218–263.

Keel, M. and Tao, T. (1998). Endpoint Strichartz estimates, *Amer. J. Math.* **120**, 5, pp. 955–980.

Kenig, C., Ponce, G. and Vega, L. (2001). On the ill-posedness of some canonical dispersive equations, *Duke Math. J.* **106**, 3, pp. 617–633.

Keraani, S. (2001). On the defect of compactness for the Strichartz estimates of the Schrödinger equations, *J. Diff. Eq.* **175**, 2, pp. 353–392.

Keraani, S. (2002). Semiclassical limit for a class of nonlinear Schrödinger equations with potential, *Comm. Part. Diff. Eq.* **27**, 3–4, pp. 693–704.

Keraani, S. (2005). Limite semi-classique pour l'équation de Schrödinger non-linéaire avec potentiel harmonique, *C. R. Math. Acad. Sci. Paris* **340**, 11, pp. 809–814.

Keraani, S. (2006). Semiclassical limit for nonlinear Schrödinger equation with potential. II, *Asymptot. Anal.* **47**, 3–4, pp. 171–186.

Kishimoto, N. (2019). A remark on norm inflation for nonlinear Schrödinger equations, *Commun. Pure Appl. Anal.* **18**, 3, pp. 1375–1402.

Klainerman, S. (1985). Uniform decay estimates and the Lorentz invariance of the classical wave equation, *Comm. Pure Appl. Math.* **38**, 3, pp. 321–332.

Kolomeisky, E. B., Newman, T. J., Straley, J. P. and Qi, X. (2000). Low-dimensional Bose liquids: Beyond the Gross-Pitaevskii approximation, *Phys. Rev. Lett.* **85**, 6, pp. 1146–1149.

Kossioris, G. T. (1993). Formation of singularities for viscosity solutions of Hamilton-Jacobi equations in higher dimensions, *Comm. Partial Differential Equations* **18**, 7–8, pp. 1085–1108.

Kramers, H. A. (1926). Wellenmechanik und halbzahlige Quantisierung, *Zeitschrift für Physik* **39**, pp. 828–840.

Kuksin, S. B. (1995). On squeezing and flow of energy for nonlinear wave equations, *Geom. Funct. Anal.* **5**, 4, pp. 668–701.

Kwong, M. K. (1989). Uniqueness of positive solutions of $\Delta u - u + u^p = 0$ in $\mathbb{R}^n$, *Arch. Rational Mech. Anal.* **105**, 3, pp. 243–266.

Landau, L. and Lifschitz, E. (1967). *Physique théorique ("Landau-Lifchitz"). Tome III: Mécanique quantique. Théorie non relativiste* (Éditions Mir, Moscow), deuxième édition, Traduit du russe par Édouard Gloukhian.

Lasser, C. and Lubich, C. (2020). Computing quantum dynamics in the semiclassical regime, *Acta Numer.*

Lax, P. D. (1957). Asymptotic solutions of oscillatory initial value problems, *Duke Math. J.* **24**, pp. 627–646.

Lebeau, G. (1992). Contrôle de l'équation de Schrödinger, *J. Math. Pures Appl. (9)* **71**, 3, pp. 267–291.

Lebeau, G. (2001). Non linear optic and supercritical wave equation, *Bull. Soc. Roy. Sci. Liège* **70**, 4–6, pp. 267–306 (2002), hommage à Pascal Laubin.

Lebeau, G. (2005). Perte de régularité pour les équations d'ondes sur-critiques, *Bull. Soc. Math. France* **133**, pp. 145–157.

Lerner, N., Nguyen, T. and Texier, B. (2018). The onset of instability in first-order systems, *J. Eur. Math. Soc. (JEMS)* **20**, 6, pp. 1303–1373.

Lin, F. and Zhang, P. (2005). Semiclassical limit of the Gross-Pitaevskii equation in an exterior domain, *Arch. Rational Mech. Anal.* **179**, 1, pp. 79–107.

Lions, P.-L. (1996). *Mathematical topics in fluid mechanics. Vol. 1, Oxford Lecture Series in Mathematics and its Applications*, Vol. 3 (The Clarendon Press Oxford University Press, New York), incompressible models, Oxford Science Publications.

Lions, P.-L. and Paul, T. (1993). Sur les mesures de Wigner, *Rev. Mat. Iberoamericana* **9**, 3, pp. 553–618.

Lubich, C. (2008). *From quantum to classical molecular dynamics: reduced models and numerical analysis*, Zurich Lectures in Advanced Mathematics (European Mathematical Society (EMS), Zürich).

Ludwig, D. (1966). Uniform asymptotic expansions at a caustic, *Comm. Pure Appl. Math.* **19**, pp. 215–250.

Majda, A. (1984). *Compressible fluid flow and systems of conservation laws in several space variables, Applied Mathematical Sciences*, Vol. 53 (Springer-Verlag, New York).

Makino, T., Ukai, S. and Kawashima, S. (1986). Sur la solution à support compact de l'équation d'Euler compressible, *Japan J. Appl. Math.* **3**, 2, pp. 249–257.

Martinez, A. (2002). *An introduction to semiclassical and microlocal analysis*, Universitext (Springer-Verlag, New York).

Masaki, S. (2007). Semi-classical analysis for Hartree equations in some supercritical cases, *Ann. Henri Poincaré* **8**, 6, pp. 1037–1069.

Maslov, V. P. and Fedoriuk, M. V. (1981). *Semiclassical approximation in quantum mechanics, Mathematical Physics and Applied Mathematics*, Vol. 7 (D. Reidel Publishing Co., Dordrecht), translated from the Russian by J. Niederle and J. Tolar, Contemporary Mathematics, 5.

Merle, F. and Vega, L. (1998). Compactness at blow-up time for L^2 solutions of the critical nonlinear Schrödinger equation in 2D, *Internat. Math. Res. Notices*, 8, pp. 399–425.

Métivier, G. (2004a). Exemples d'instabilités pour des équations d'ondes non linéaires (d'après G. Lebeau), *Astérisque*, 294, pp. vii, 63–75.

Métivier, G. (2004b). *Small viscosity and boundary layer methods*, Modeling and Simulation in Science, Engineering and Technology (Birkhäuser Boston Inc., Boston, MA), theory, stability analysis, and applications.

Métivier, G. (2005). Remarks on the well-posedness of the nonlinear Cauchy problem, in *Geometric analysis of PDE and several complex variables, Contemp. Math.*, Vol. 368 (Amer. Math. Soc., Providence, RI), pp. 337–356.

Métivier, G. and Schochet, S. (1998). Trilinear resonant interactions of semilinear hyperbolic waves, *Duke Math. J.* **95**, 2, pp. 241–304.

Mouzaoui, L. (2013). High-frequency averaging in the semi-classical singular Hartree equation, *Asymptot. Anal.* **84**, 3–4, pp. 229–245.

Moyua, A., Vargas, A. and Vega, L. (1999). Restriction theorems and maximal operators related to oscillatory integrals in $\mathbb{R}^3$, *Duke Math. J.* **96**, 3, pp. 547–574.

Nakanishi, K. and Ozawa, T. (2002). Remarks on scattering for nonlinear Schrödinger equations, *NoDEA Nonlinear Differential Equations Appl.* **9**, 1, pp. 45–68.

Nier, F. (1996). A semi-classical picture of quantum scattering, *Ann. Sci. École Norm. Sup. (4)* **29**, 2, pp. 149–183.

Oh, T. (2017). A remark on norm inflation with general initial data for the cubic nonlinear Schrödinger equations in negative Sobolev spaces, *Funkcialaj Ekvacioj* **60**, 2, pp. 259–277.

Oh, T. and Wang, Y. (2018). On the ill-posedness of the cubic nonlinear Schrödinger equation on the circle, *An. Ştiinţ. Univ. Al. I. Cuza Iaşi. Mat. (N.S.)* **64**, 1, pp. 53–84.

Ozawa, T. (1991). Long range scattering for nonlinear Schrödinger equations in one space dimension, *Comm. Math. Phys.* **139**, pp. 479–493.

Paul, T. (1997). Semi-classical methods with emphasis on coherent states, in *Quasiclassical methods (Minneapolis, MN, 1995), IMA Vol. Math. Appl.*, Vol. 95 (Springer, New York), pp. 51–88.

Pitaevskii, L. and Stringari, S. (2003). *Bose-Einstein condensation, International Series of Monographs on Physics*, Vol. 116 (The Clarendon Press Oxford University Press, Oxford).

Rauch, J. (1991). *Partial Differential Equations, Graduate Texts in Math.*, Vol. 128 (Springer-Verlag, New York).

Rauch, J. (2012). *Hyperbolic partial differential equations and geometric optics, Graduate Studies in Mathematics*, Vol. 133 (American Mathematical Society, Providence, RI).

Reed, M. and Simon, B. (1975). *Methods of modern mathematical physics. II. Fourier analysis, self-adjointness* (Academic Press [Harcourt Brace Jovanovich Publishers], New York).

Robert, D. (1987). *Autour de l'approximation semi-classique, Progress in Mathematics*, Vol. 68 (Birkhäuser Boston Inc., Boston, MA).

Robert, D. (1998). Semi-classical approximation in quantum mechanics. A survey of old and recent mathematical results, *Helv. Phys. Acta* **71**, 1, pp. 44–116.

Robert, D. (2010). On the Herman-Kluk semiclassical approximation, *Rev. Math. Phys.* **22**, 10, pp. 1123–1145.

Rousse, V. (2004). Landau-Zener transitions for eigenvalue avoided crossings in the adiabatic and Born-Oppenheimer approximations, *Asymptot. Anal.* **37**, 3–4, pp. 293–328.

Sacchetti, A. (2005). Nonlinear double well Schrödinger equations in the semiclassical limit, *J. Stat. Phys.* **119**, 5–6, pp. 1347–1382.

Schubert, R., Vallejos, R. O. and Toscano, F. (2012). How do wave packets spread? Time evolution on Ehrenfest time scales, *J. Phys. A* **45**, 21, p. 215307.

Schwartz, J. T. (1969). *Nonlinear functional analysis* (Gordon and Breach Science Publishers, New York), notes by H. Fattorini, R. Nirenberg and H. Porta, with an additional chapter by Hermann Karcher, Notes on Mathematics and its Applications.

Serre, D. (1997). Solutions classiques globales des équations d'Euler pour un fluide parfait compressible, *Ann. Inst. Fourier* **47**, pp. 139–153.

Sjöstrand, J. (1982). Singularités analytiques microlocales, in *Astérisque*, Vol. 95 (Soc. Math. France, Paris), pp. 1–166.

Sone, Y., Aoki, K., Takata, S., Sugimoto, H. and Bobylev, A. V. (1996). Inappropriateness of the heat-conduction equation for description of a temperature field of a stationary gas in the continuum limit: examination by asymptotic analysis and numerical computation of the Boltzmann equation, *Phys. Fluids* **8**, 2, pp. 628–638.

Sparber, C., Markowich, P. A. and Mauser, N. J. (2003). Wigner functions versus WKB-methods in multivalued geometrical optics, *Asymptot. Anal.* **33**, 2, pp. 153–187.

Stein, E. M. (1993). *Harmonic analysis: real-variable methods, orthogonality, and oscillatory integrals, Princeton Mathematical Series*, Vol. 43 (Princeton University Press, Princeton, NJ), with the assistance of Timothy S. Murphy, Monographs in Harmonic Analysis, III.

Strauss, W. A. (1974). Nonlinear scattering theory, in J. Lavita and J. P. Marchand (eds.), *Scattering theory in mathematical physics* (Reidel).

Strauss, W. A. (1981). Nonlinear scattering theory at low energy, *J. Funct. Anal.* **41**, pp. 110–133.

Sulem, C. and Sulem, P.-L. (1999). *The nonlinear Schrödinger equation, Self-focusing and wave collapse* (Springer-Verlag, New York).

Swart, T. and Rousse, V. (2009). A mathematical justification for the Herman-Kluk propagator, *Comm. Math. Phys.* **286**, 2, pp. 725–750.

Szeftel, J. (2005). Propagation et réflexion des singularités pour l'équation de Schrödinger non linéaire, *Ann. Inst. Fourier (Grenoble)* **55**, 2, pp. 573–671.

Tao, T. (2006). *Nonlinear dispersive equations, CBMS Regional Conference Series in Mathematics*, Vol. 106 (Published for the Conference Board of the Mathematical Sciences, Washington, DC), local and global analysis.

Tao, T. (2009). A pseudoconformal compactification of the nonlinear Schrödinger equation and applications, *New York J. Math.* **15**, pp. 265–282.

Taylor, M. (1981). *Pseudodifferential operators, Princeton Mathematical Series*, Vol. 34 (Princeton University Press, Princeton, N.J.).

Taylor, M. (1997). *Partial differential equations. III, Applied Mathematical Sciences*, Vol. 117 (Springer-Verlag, New York), nonlinear equations.

Teufel, S. (2003). *Adiabatic perturbation theory in quantum dynamics, Lecture Notes in Mathematics*, Vol. 1821 (Springer).

Thirring, W. (1981). *A course in mathematical physics. Vol. 3* (Springer-Verlag, New York), quantum mechanics of atoms and molecules, Translated from the German by Evans M. Harrell, Lecture Notes in Physics, 141.

Thomann, L. (2008). Instabilities for supercritical Schrödinger equations in analytic manifolds, *J. Differential Equations* **245**, 1, pp. 249–280.

Tsutsumi, Y. (1987). L^2-solutions for nonlinear Schrödinger equations and nonlinear groups, *Funkcial. Ekvac.* **30**, 1, pp. 115–125.

Tsutsumi, Y. and Yajima, K. (1984). The asymptotic behavior of nonlinear Schrödinger equations, *Bull. Amer. Math. Soc. (N.S.)* **11**, 1, pp. 186–188.

Weinstein, M. I. (1985). Modulational stability of ground states of nonlinear Schrödinger equations, *SIAM J. Math. Anal.* **16**, 3, pp. 472–491.

Wentzel, G. (1926). Eine Verallgemeinerung der Quantenbedingungen für die Zwecke der Wellenmechanik, *Zeitschrift für Physik* **38**, 518–529.

Whitham, G. B. (1999). *Linear and nonlinear waves*, Pure and Applied Mathematics (New York) (John Wiley & Sons Inc., New York), reprint of the 1974 original, A Wiley-Interscience Publication.

Xin, Z. (1998). Blowup of smooth solutions of the compressible Navier-Stokes equation with compact density, *Comm. Pure Appl. Math.* **51**, pp. 229–240.

Yajima, K. (1979). The quasiclassical limit of quantum scattering theory, *Comm. Math. Phys.* **69**, 2, pp. 101–129.

Yajima, K. (1987). Existence of solutions for Schrödinger evolution equations, *Comm. Math. Phys.* **110**, pp. 415–426.

Yajima, K. (1996). Smoothness and non-smoothness of the fundamental solution of time dependent Schrödinger equations, *Comm. Math. Phys.* **181**, 3, pp. 605–629.

Zakharov, V. E. and Shabat, A. B. (1971). Exact theory of two-dimensional self-focusing and one-dimensional self-modulation of waves in nonlinear media, *Ž. Èksper. Teoret. Fiz.* **61**, 1, pp. 118–134.

Zworski, M. (2012). *Semiclassical analysis, Graduate Studies in Mathematics*, Vol. 138 (American Mathematical Society, Providence, RI), doi:10.1090/gsm/138.

Index

Printed in the United States
by Baker & Taylor Publisher Services